Beyond-CMOS

SCIENCES

Electronics Engineering, Field Directors – Francis Balestra

Components and Manufacturing Techniques,
Subject Head – Francis Balestra

Beyond-CMOS

State of the Art and Trends

Coordinated by
Alessandro Cresti

iSTE

WILEY

First published 2023 in Great Britain and the United States by ISTE Ltd and John Wiley & Sons, Inc.

ISTE Ltd
27-37 St George's Road
London SW19 4EU
UK

www.iste.co.uk

John Wiley & Sons, Inc.
111 River Street
Hoboken, NJ 07030
USA

www.wiley.com

Library of Congress Control Number: 2022947446

British Library Cataloguing-in-Publication Data
A CIP record for this book is available from the British Library
ISBN 978-1-78945-127-6

ERC code:
PE7 Systems and Communication Engineering
PE7_5 (Micro- and nano-) electronic, optoelectronic and photonic components
PE3 Condensed Matter Physics
PE3_4 Electronic properties of materials, surfaces, interfaces, nanostructures, etc.
PE3_5 Physical properties of semiconductors and insulators

Contents

Preface

Alessandro CRESTI
CNRS, Grenoble INP, IMEP-LAHC, Université Grenoble Alpes, Université Savoie Mont Blanc, Grenoble, France

Going *Beyond CMOS* means going beyond Moore's law (Moore 1965), which is basically an economic law from previous years (Englebart 1960) and described in more detail by Dennard et al. (1974) in terms of scaling of size and power at each new generation of transistors. Industry has followed this paradigm for many decades, beyond physical limits such as heating and short-channel effects. In this context, this research has enabled the development of this line using the so-called *More Moore* approach, which was recently described in the *International Technology Roadmap for Semiconductors* (ITRS, see: http://www.itrs2.net), which lasted until 2016.

These limits, on the contrary, have allowed the development of alternative research guides known as *More-than-Moore* (Arden et al. 2010) and *Beyond CMOS* (Nikonov and Young 2013). The first is a shift of the paradigm to different devices, which goes beyond transistors, and includes, for example, the development of sensors and energy harvesters. *Beyond CMOS*, on the contrary, is still mainly focused on logic devices, but includes different types of transistors complementary to metal–oxide–semiconductors. It consists of new technologies (such as tunnel FETs, spintronics, memristors or optical devices) and materials (such as III–V or 2D materials) that can be used as an alternative to silicon. Such an interesting development has also generated new scientific and economic advancements. Since 2016, the *International Roadmap for Devices and Systems* (IRDS, see: https://irds.ieee.org) has focused on these new technologies. Moreover, many important books (Balestra 2014a, 2014b; Brozek 2014; Chen et al. 2015; King Liu and

Kuhn 2015; Topaloglu and Wong 2019; Dragoman and Dragoman 2021) have been published on the subject, whose contents are of primary interest.

This book focuses on some specific arguments detailed by leading and pioneering researchers in this field. All of the chapters contain a short introduction accessible to not-too-specialized readers and a discussion about the recent progresses and trends.

The first four chapters are devoted to steep-slope devices and low-power applications, which are one of the most important domains of *Beyond CMOS* electronics. The other chapters present more unconventional emerging technologies.

More specifically, Chapter 1, written by Marco Pala (CN2, Paris, France), presents the possible use of III–V semiconductors in tunnel field-effect transistors for steep-slope applications, in particular regarding their limits and advantages by numerical simulations of these devices.

Chapter 2, written by Mathieu Luisier (ETH, Zurich, Switzerland) and co-workers, illustrates the possible use of 2D materials for field-effect transistors, not restricted to transition metal dichalcogenides. 2D materials have the advantage of being extremely thin (few atomic layers), thus allowing a good electrostatic control and flexible applications.

In Chapter 3, Kaustav Banerjee (UC Santa Barbara, USA) illustrate the working mechanism and the implementation of steep-slope devices based on the negative capacitance properties of the gate obtained using ferroelectric materials. The author has a particularly critical view on the subject, which will be helpful to present the practical applications of this technology to the reader.

In Chapter 4, Joris Lacord (CEA-Leti, Grenoble, France) presents the exotic zero subthreshold swing and zero impact ionization field-effect transistor (Z2-FET) based on the silicon-on-insulator technology. This kind of steep-slope device is particularly promising, among other things, for applications in memories.

In Chapter 5, the field of *2D spintronics* is deeply analyzed by Matthieu Jamet (Spintec, Grenoble, France), Diogo C. Vaz (CIC nanoGUNE BRTA, Donostia-San Sebastian, Spain), Juan F. Sierra, Josef Světlík and Sergio O. Valenzuela (ICN2, CSIC and BIST, Barcelona, Spain), Bruno Dlubak and Pierre Seneor (Thales, Palaiseau, France), Frédéric Bonell (Spintec, Grenoble, France), Thomas Guillet (ICN2, CSIC and BIST, Barcelona, Spain) and their collaborators, especially from the experimental point of view. In this chapter, the term “2D” refers to both 2D materials, 2D electron gases and 3D topological insulators, with applications ranging from memories to spin transport logic.

A more exotic subject is presented in Chapter 6, written by Steven A. Vitale (MIT, Lexington, Massachusetts, U.S.A.), which illustrates the possible use of the valley degree of freedom and valley excitons in electronics. This innovative technology, based on the fact that an electron can occupy different valleys in the band structure, is still premature, but has important perspectives for logic operations and beyond, up to quantum computing.

Electronics can also be implemented through molecular systems. Chapter 7, written by Dominique Vuillaume (IEMN, Lille, France), is devoted to this emerging technology, which is very promising for the ultimate miniaturization of electronics. The strength of such a technology is incredibly large and varies from the implementation of standard logic devices to plasmonics and thermal transport.

Another innovative field is *superconducting quantum electronics*, presented in Chapter 8, written by Pascal Febvre (IMEP-LaHC, Le Bourget du Lac, France) and Sasan Razmkhah (TOBB university of Economy and Technology, Ankara, Turkey). Several very promising applications of superconductivity in the field of beyond CMOS electronics, including the realization of logic devices, neuromorphic computing and superconducting qubits, are presented.

Finally, a completely alternative technology, which consists of using light instead of electrons for the logic operations, is described in Chapter 9, written by Wolfram Pernice (Universität Münster, Germany) and co-workers. Being able to integrate these operations in a single chip is a major advance towards applications, including massively parallel computing and efficient neural networks.

The aim of this book is to provide the reader with a fresh overview of the major research advances in this field, especially in *Beyond CMOS* logic devices, and a detailed physical insight into new trends that are inspirational for future devices.

September 2022

References

Arden, W., Michel Brillouët, M., Patrick Cogez, P., Mart Graef, M., Bert Huizing, B., Mahnkopf, R. (2010). "More-than-Moore". White Paper [Online]. Available at: http://www.itrs2.net/uploads/4/9/7/7/49775221/irc-itrs-mtm-v2_3.pdf.

Balestra, F. (2014a). *Beyond-CMOS Nanodevices 1*. ISTE, London, and John Wiley & Sons, New York.

Balestra, F. (2014b). *Beyond-CMOS Nanodevices 2*. ISTE, London, and John Wiley & Sons, New York.

Brozek, T. (2014). *Micro- and Nanoelectronics: Emerging Device Challenges and Solutions*. CRC Press, Boca Raton.

Chen, A., Hutchby, J., Zhirnov, V., Bourianoff, G. (2015). *Emerging Nanoelectronic Devices*. John Wiley & Sons, Chichester.

Dennard, R.H., Gaensslen, F.H., Yu, H.N., Rideout, V.L., Bassous, E., LeBlanc, A.R. (1974). Design of ion-implanted MOSFET's with very small physical dimensions. *IEEE Journal of Solid-State Circuits*, 9(5), 256–268.

Dragoman, M. and Dragoman, D. (2021). *Beyond CMOS, Atomic-Scale Electronics*. Springer International Publishing, Cham.

Englebart, D. (1960). Microelectronics and the art of similitude. *IEEE International Solid-State Circuits Conference. Digest of Technical Papers*, 76–77.

King Liu, T.J. and Kuhn, K. (2015). *CMOS and Beyond*. Cambridge Univesity Press, Cambridge.

Moore, G.E. (1965). Cramming more components onto integrated circuits. *Electronics*, 38(8), 114–117.

Nikonov, D.E. and Young, I.A. (2013). Overview of beyond-CMOS devices and a uniform methodology for their benchmarking. *Proceedings of the IEEE*, 101(12), 2498–2533.

Topaloglu, R.O. and Wong, H.-S.P. (2019). *Beyond-CMOS Technologies for Next Generation Computer Design*. Springer International Publishing, Cham.

1

Tunnel Field-Effect Transistors Based on III–V Semiconductors

Marco PALA
Université Paris-Saclay, CNRS, Centre de Nanosciences et de Nanotechnologies, Palaiseau, France

1.1. Introduction

Nowadays, in the era of the Internet of Things (IoTs), the efficiency of integrated circuits (ICs) is limited by energy consumption. Consequently, a great effort is dedicated to finding innovative transistors that are able to support ultra-low power consumption and high clock frequency. The main contributions to power dissipation in modern CMOS technology ICs are due to transistor switching and leakage, namely dynamic and static power, respectively. The former can be approximated as

$$P_{dyn} = \alpha_{0\to1} f C_L V_{DD}^2, \qquad [1.1]$$

where $\alpha_{0\to1}$ is the switching probability of the gate working at frequency f, C_L is the load capacitance and V_{DD} is the supply voltage. The latter can be estimated as

$$P_{stat} = I_{off} T_D V_{DD}, \qquad [1.2]$$

where I_{off} is the off-state current and T_D is the digital gate delay. From equations [1.1] and [1.2], it results that the power of ICs can be effectively reduced by scaling the supply voltage. Reducing V_{DD} by a factor 10 would imply a factor 100 in the

For a color version of all figures in this chapter, see www.iste.co.uk/cresti/beyondcmos.zip.

Beyond-CMOS, coordinated by Alessandro CRESTI.

dynamic power. A serious difficulty in pursuing the V_{DD} scaling is the requirement to maintain the on/off current ratio as high as 10^6, since this ratio is determined by the subthreshold swing (SS) of the transistor, which measures its ability to switch on/off for a given gate-voltage modulation.

The fundamental brick of CMOS technology, the MOSFET, is limited by the so-called "Boltzmann tyranny", the impossibility to provide an SS smaller than the thermionic limit of 60 mV/dec at room temperature. From a physical point of view, this limitation arises from the injection of electrons belonging to the tail of the Boltzmann distribution of the source contact, which can be modulated, but not eliminated by varying the channel barrier with the gate voltage. In order to get rid of this limit of MOSFETs, alternative transistor architectures have been investigated in the last decades, having the common properties to provide SSs smaller than the thermionic limit. Among the various proposals for steep-slope devices, it is worth mentioning those exploiting Mott transition (Nakano et al. 2012), impact ionization (Gopalakrishnan et al. 2005), negative capacitance (Salahuddin and Datta 2008), cold sources (Qiu et al. 2018) and band-to-band-tunneling (BTBT) (Seabaugh and Zhang 2010).

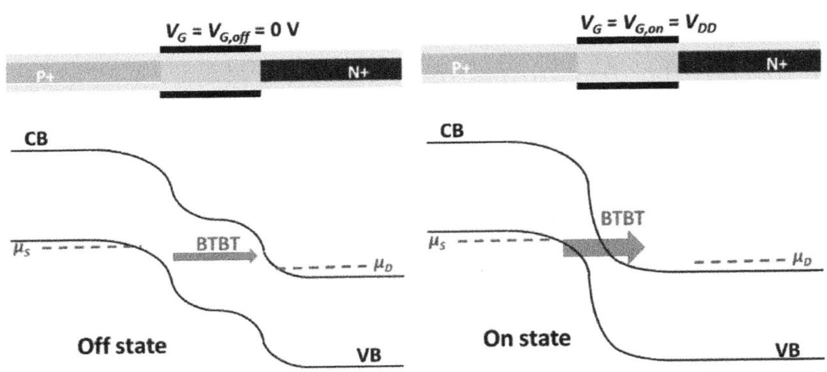

Figure 1.1. *Sketch of the TFET working principle. In the off state (V_G = 0 V), the BTBT is suppressed by imposing the CB in the channel region above the VB in the source region. In the on state, instead, the tunneling path is decreased by decreasing the CB in the channel region below the VB in the source*

Transistors based on BTBT, usually named tunnel field-effect transistors (TFETs), represent one of the most promising device designs capable of achieving a steep turn-on and therefore reduced power consumption (Ionescu and Riel 2011). The working principle of the TFET is to exploit a p–i–n diode in the reverse bias condition, with a band alignment such that the thermionic current is completely suppressed by the semiconductor gap, as illustrated in Figure 1.1, whereas the whole electron current is driven by band-to-band tunneling (BTBT) from the source valence states to the channel conduction states. This implies the energy filtering of electrons in the tail of the Fermi distribution in the off-state and allows sub-thermionic values of SS to be attained. The switching to the on-state is obtained by shortening the tunneling path the carriers have to take by means of the external gate voltage.

Because carriers have to cross a wide tunneling barrier, Si-based TFETs provide on-state currents (I_{on}) much smaller than the MOSFET, thus impeding their use in electronic circuits. In order to boost the BTBT current, the use of semiconductors with energy band-gaps smaller than silicon, such as III–V materials, has been largely investigated, as well as the use of staggered and broken-gap III–V heterojunctions (HJs). However, even if significant progress has been achieved during the last few years in obtaining TFETs with sub-thermionic behavior and large drive current, there is not, to date, any experimental demonstration of TFETs capable of outperforming the MOSFET at least $V_{DD} < 0.5$ V.

In order to clarify the main issues in realizing efficient III–V TFETs, this chapter will discuss the state of the art of the research in these devices. It is organized as follows: in section 1.2, we will describe the most relevant experimental realization of TFETs, while in section 1.3, we will briefly present the numerical methods and physical models used to simulate them; in section 1.4, we will discuss the main problems in obtaining small values of the SS and, finally, in section 1.5, we will illustrate the most promising design options to boost the I_{on} of TFETs based on III–V semiconductors.

1.2. Experiments

Experimental demonstrations of TFETs were first reported for silicon-based devices (Mayer et al. 2008; Gandhi et al. 2011; Huang et al. 2012), because of the maturity of the CMOS technology. However, these transistors suffered from low on-state currents due to the large and indirect band-gap of silicon. The performance of group-IV-based TFETs was improved by using strained Si (Knoll et al. 2013) and by integrating Ge (Krishnamohan et al. 2008), which presents a smaller and almost direct band-gap than Si. TFETs using GeSn (Yang et al. 2012) and strained SiGe (Villalon et al. 2012) materials reported further current enhancements.

These results encouraged the fabrication and exploration of the potentialities of TFETs based on III–V materials. First, experimental results were obtained with homojunction devices based on InGaAs (Mookerjea et al. 2009; Noguchi et al. 2015) and InAs (Ford et al. 2011). These devices showed higher on-state currents than group-IV TFETs, as well as minimum SS smaller than 60 mV/dec. In order to improve the drive current of III–V-based TFETs, devices exploiting heterojunctions such as InGaAs/GaAsSb (Mohata et al. 2011; Bijesh et al. 2013), InAs/GaSb (Guangle Zhou et al. 2012), InAsSb/GaSb (Dey et al. 2013) and InGaAs/InGaAs (Dewey et al. 2011) were fabricated. They provided high on-currents of the order of several tens of μA/μm, as well as poor subthreshold swings. This resulted from the difficulty of obtaining large on-currents and sub-thermionic behavior at the same time.

Achieving subthreshold swings smaller than 60 mV/dec has been pursued with the use of optimized gate stacks with a better passivation and electrostatic control. This was realized by means of TFETs based on multi-gated nanowires (NWs). Besides the SS and the drive current, another useful figure of merit for steep-slope devices is the I_{60}, namely the highest current where the SS shows a transition from sub- to super-60 mV/dec (Vandenberghe et al. 2013). In this respect, the III–V NW TFET based on InP/GaAs (Ganjipour et al. 2012) showed an encouraging result with an SS of 20 mV/dec and an I_{60} of about 10 pA/μm, while the Si/InAs heterojunction in Tomioka et al. (2012) also exhibited an SS of approximatively 20 mV/dec.

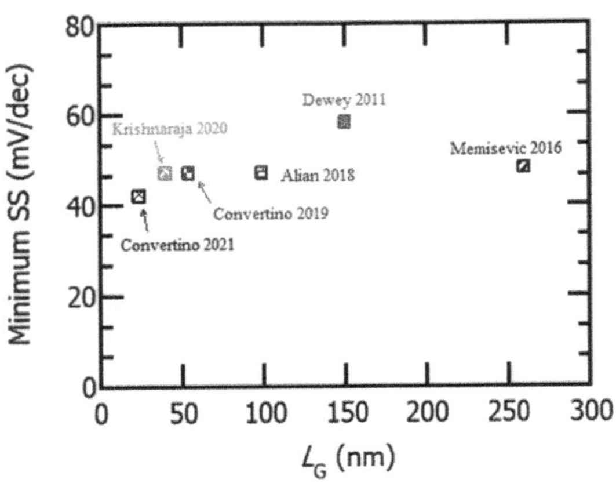

Figure 1.2. *Minimum subthreshold swing versus the gate length of III–V NW TFETs measured in recent experiments. The ability to obtain subthermionic devices at gate lengths smaller than 50 nm can be noted*

During the last few years, subthreshold swings (SS) as low as 42 mV/dec over a small current range have been demonstrated in the GaSb/InAs system by different research groups (Memisevic et al. 2016; Alian et al. 2018; Convertino et al. 2019; Krishnaraja et al. 2020; Convertino et al. 2021). Figure 1.2 summarizes these results, illustrating a comparison of the minimum SS versus the gate length reported by various experimental works on TFETs based on GaSb/InAs heterojunction nanowires. As can be seen, these transistors combine good scalability and sub-thermionic behavior. Similarly, Figure 1.3 illustrates the good scalability of the I_{60} with the transistor length.

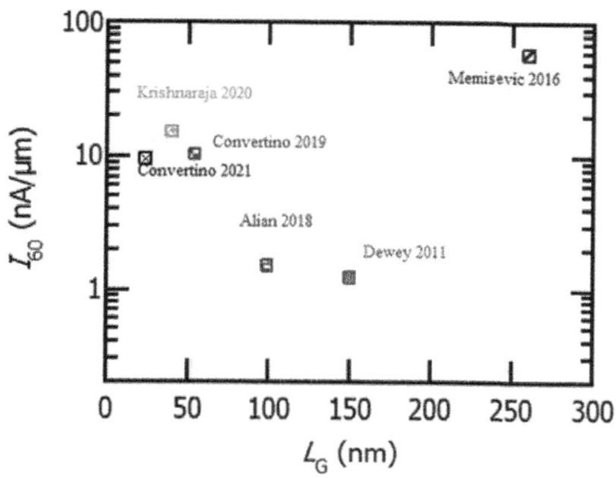

Figure 1.3. *I_{60} versus the gate length measured in recent experiments on III–V NW TFETs. I_{60} is the maximal current value for which the TFET exhibits a sub-thermionic behavior at room temperature*

At the present time, the main issue of this research field is represented by the relatively low on-state currents, which remain well below the values attainable with MOSFETs. In this respect, the potentiality of the broken-gap InAs/GaSb heterojunction as channel material for nanowire Esaki diodes with diameters smaller than 10 nm has been recently demonstrated in Shao et al. (2022), showing negative differential resistance (NDR) peak current densities as large as 1 MA/cm^2.

1.3. Simulation of III–V-based TFETs

From the first proposal of the TFET, simulations have been used to find the optimal design of these devices. As reported in experiments, the exhibition of both an SS smaller than 60 mV/dec and a high I_{on} requires a difficult trade-off that can be assessed by using accurate numerical simulations (Esseni et al. 2017). By considering that the working principle of TFETs is a quantum mechanical effect, it is quite natural to simulate them with a full quantum approach consisting of the non-equilibrium Green's-function (NEGF) formalism (Datta 2005). In the ballistic regime, the NEGF formalism provides results consistent with the Landauer–Büttiker description of mesoscopic phenomena (Datta 2005), but it is also able to address a general transport theory describing dissipative transport and non-equilibrium phenomena.

III–V heterojunction TFETs were studied in Knoch and Appenzeller (2010) by means of the NEGF method within the effective mass approximation (EMA) and the Büttiker probes to mimic inelastic scattering. The authors found that the best condition to attain the best subthreshold swing was to impose a staggered band lineup and a relatively large doping concentration. This last condition prevents the carriers injected from the source from belonging to the tail of the Fermi distribution function and hindering the filtering of high-energy electrons.

Band structure effects, such as non-parabolicity, inter-band scattering and quantization effects on the energy dispersion in the gap, are naturally accounted for in the **k.p** approximation (Shin 2009; Conzatti et al. 2011; Verreck et al. 2016; Visciarelli et al. 2016) and make it preferable to the EMA to study III–V TFETs. This approach, together with the basics of the NEGF method, will be discussed in more detail in section 1.3.1.

An atomistic description of III–V TFETs exploiting the empirical tight-binding (TB) method was used in Luisier and Klimeck (2009) to evaluate various design options for InAs TFETs, such as single-gate (SG), and dual-gate (DG), ultrathin-body (UTB) and gate-all-around (GAA) nanowire devices. The importance of having efficient control of the electrostatic potential via the gate electrode in order to achieve a sub-thermionic SS was pointed out. In particular, to maintain an SS smaller than 60 mV/dec, the GAA NWs can have diameters larger than 10 nm, while the body thickness of DG and SG UTBs must be reduced to 7 and 4 nm, respectively. The empirical TB method is able to address a full-band simulation of the semiconductors, but at the cost of increasing the required computational resources and limiting the size of the simulated devices.

Other sophisticated methods able to address a full band description of III–V TFETs are based on empirical pseudopotential (EPM) Hamiltonians (M'foukh et al. 2020) or on the density functional theory (DFT) (Lee et al. 2021), but, again, their numerical capabilities are limited and can address the simulation of devices with lateral sections of only a few nanometers.

1.3.1. *The k.p model in the NEGF formalism*

When simulating III–V-based NW TFETs, the best compromise between model accuracy and numerical efficiency is probably to adopt an NEGF approach based on an 8-band **k.p** Hamiltonian, such as the one proposed by Bahder (1990). This Hamiltonian accurately describes the semiconductor properties around the Γ valley, where both the minimum of the conduction band (CB) and the maximum of the valence band (VB) are located for all of the III–V materials of practical interest. Moreover, it is well suited to describe the effect of mechanical stress along all different crystallographic orientations. On the other hand, it is computationally less expensive that either atomistic simulations based on the empirical tight-binding (Luisier and Klimeck 2009) or ab initio methods such as density functional theory (Pala et al. 2020), therefore enabling the simulation of physical systems with realistic sizes.

The explicit form of the eight-band Hamiltonian matrix is

$$H_8 = \begin{pmatrix} C & 0 & P^+/\sqrt{6} & 0 & P^-/\sqrt{2} & -\sqrt{\frac{2}{3}}P_x & -P_x/\sqrt{3} & P^+/\sqrt{3} \\ & C & -\sqrt{\frac{2}{3}}P_x & -P^+/\sqrt{2} & 0 & -P^-/\sqrt{6} & P^-/\sqrt{6} & P_z/\sqrt{3} \\ & & -P+Q & -S^* & R & 0 & 2\sqrt{3}S & -\sqrt{2}Q \\ & & & -P-Q & 0 & R & -\sqrt{2}R & S/\sqrt{2} \\ & & & & -P-Q & S^* & S^*/\sqrt{2} & \sqrt{2}R^* \\ & & & & & -P+Q & \sqrt{2}Q & \sqrt{\frac{3}{2}}S^* \\ & & & & & & -P-\Delta_{SO} & 0 \\ & & & & & & & -P-\Delta_{SO} \end{pmatrix} \quad [1.3]$$

where

$$\begin{aligned} C &= E_g + \frac{1}{2}\frac{\hbar^2}{2m_0}\gamma_c\left(k_x^2+k_y^2+k_z^2\right) \\ P &= \frac{1}{2}\frac{\hbar^2}{2m_0}\gamma_1\left(k_x^2+k_y^2+k_z^2\right) \\ Q &= \frac{1}{2}\frac{\hbar^2}{2m_0}\gamma_2\left(k_x^2+k_y^2+k_z^2\right) \\ R &= \sqrt{3}\frac{\hbar^2}{2m_0}\left[\gamma_2\left(k_x^2-k_y^2\right)-2i\gamma_3 k_x k_y\right] \\ S &= \frac{\hbar^2}{2m_0}2\sqrt{3}\gamma_3\left(k_x-ik_y\right)k_z \\ P^{\pm} &= P_0\left(k_x \pm ik_y\right) \\ P_z &= P_0 k_z \\ P_0 &= \sqrt{E_p\hbar^2/2m_0} \\ \gamma_c &= \frac{1}{m_c}-\frac{E_p}{3}\left(\frac{2}{E_g}+\frac{1}{E_g+\Delta_{SO}}\right) \end{aligned} \quad [1.4]$$

and the modified Luttinger parameters $\gamma_{1,2,3}$ are related to the Luttinger parameters as $\gamma_1 = \gamma_1^L - \frac{E_p}{3E_g}$; $\gamma_2 = \gamma_2^L - \frac{E_p}{6E_g}$; $\gamma_3 = \gamma_3^L - \frac{E_p}{6E_g}$.

We note here that spurious solutions can be found whenever a lateral confinement is added to **k.p** Hamiltonians as a result of the possible non-ellipticity of the operator (Veprek et al. 2007) induced by the confining potential. However, such an inconvenience can be easily removed by reducing the parameter E_p, which accounts for the mixing between VB and CB, when confined systems are simulated. The **k.p** parameters for III–V materials can be taken from Vurgaftman et al. (2001).

The NEGF formalism addresses a microscopic theory of quantum transport. When considering a steady-state analysis, it is convenient to perform the Fourier transform of all the operators, and hence to describe all physical quantities of interest in terms of space and energy coordinates instead of time. Further, since transport along a specific direction (i.e. the x-axis) requires working in the real space due to the break of periodic conditions along that direction, a matrix notation with respect to the discrete spatial coordinates is adopted. Hence, the kinetic equations describing the retarded, lesser and greater Green's functions, G^r, $G^<$, $G^>$, read

$$\begin{aligned}&\left[(E+i\eta)I - H - \Sigma^r\right]G^r = I\\&\left[(E+i\eta)I - H - \Sigma^r\right]G^{<,>} = \Sigma^{<,>}G^a\end{aligned} \qquad [1.5]$$

where η is a positive infinitesimal, I is the identity matrix and H is the Hamiltonian matrix describing the device, which can be derived starting from equation [1.3] and inserting the electrostatic potential obtained by self-consistently coupling the NEGF equations with the non-linear Poisson equation (Conzatti et al. 2012).

In the presence of the electron–phonon interaction, the two equations are coupled through the definition of the retarded, lesser and greater self-energies, Σ^r, $\Sigma^<$, $\Sigma^>$, respectively. Each self-energy results from the sum of three terms $\Sigma^\alpha = \Sigma_R^\alpha + \Sigma_L^\alpha + \Sigma_{PH}^\alpha$ accounting for the contribution of the right and left leads and electron–phonon interaction, being $\alpha = r, >, <$.

From a numerical point of view, the kinetic equations for Green's function can be efficiently solved thanks to recursive algorithms (Svizhenko and Anantram 2003) that reside on the block-tridiagonal shape of the Hamiltonian discretized in the real space. The numerical burden, in terms of both CPU time and memory requirements, is ruled by the order of the submatrices $H(x_i, x_j)$, namely 8 N_y N_z, being $N_{y,z}$ the number of discretization points along the y, z axis. The order of these matrix blocks can be further reduced by adopting the so-called coupled-mode space (CMS) approach (Poli et al. 2008; Shin 2009). This consists of defining for the i-th discretization point along the x-axis a rectangular unity matrix

$$U_i = \left[\chi_A^i \chi_2^i \cdots \chi_M^i\right], \qquad [1.6]$$

where the mode χ_n^i is the n-th solution of the 2D Schrödinger problem in the y-z plane. Hence, the rank of the block matrix can be reduced by working in the hybrid space composed of the real space along the x axis and the transverse modes along the plane orthogonal to the transport. The block matrices in the CMS are defined by

$$H_{MS}(x_i, x_j) = U_i^+ H(x_i, x_j) U_j \qquad [1.7]$$

The self-energy terms describing the electron–phonon interaction can be efficiently implemented in the recursive algorithms to solve equation [1.5] by using local self-energies within the self-consistent Born approximation (SCBA). By assuming the elastic approximation for acoustic phonons, their CMS lesser-than self-energy for the n-th mode reads

$$\Sigma_{AC}^{<,n,n}\left(x_i, x_i, E\right) = \frac{D_{AC}^2 K_B T}{\rho v_s^2} \sum_m I^{m,n}\left(x_i, x_i\right) G_{MS}^{<,m,m}\left(x_i, x_i, E\right), \qquad [1.8]$$

where D_{AC} is the acoustic phonon deformation potential, ρ is the material density, v_s is the sound velocity and $I^{m,n}$ is the form factor. Similarly, the CMS lesser-than self-energy for dispersionless optical phonons for the n-th mode reads

$$\Sigma_{OP}^{<,n,n}\left(x_i, x_i, E\right) = \frac{\hbar D_{OP}^2}{2\rho\omega_j} \sum_m I^{m,n}\left(x_i, x_i\right) G_{MS}^{<,m,m}\left(x_i, x_i, E \pm \hbar\omega_j\right)\left[N_{\omega_j} + \tfrac{1}{2} \pm \tfrac{1}{2}\right], \qquad [1.9]$$

where D_{OP} is the optical phonon deformation potential, ω_j is the phonon frequency of the j-th optical branch and $N_{\omega j}$ is the corresponding Bose–Einstein distribution function.

When dealing with III–V TFETs, it turns out that the phonon scattering induces an increase of the I_{off} due to the phenomenon of phonon absorption in the source region and a reduction of the I_{on} due to the increased scattering in the channel (Conzatti et al. 2012). However, in the off state, such a phonon absorption results in a tiny current increase that hardly affects the I–V curves. This can be explained by considering that, in clean III-V TFETs, the current is dominated by direct tunneling. As we will see when discussing the impact of traps on these devices, the electron–phonon interaction will play a much more relevant role in changing the electron current in the presence of electrically active defects.

1.4. SS degradation mechanisms

Not just any kind of TFET is able to provide a sub-thermionic behavior. It turns out that several not-trivial conditions must be satisfied in order to attain values of SS smaller than 60 mV/dec at T = 300 K. In the following sub-sections, the most relevant mechanisms responsible for the SS degradation of TFETs based on III–V semiconductors will be discussed.

1.4.1. *Electrostatic integrity*

The idea beyond the TFET is to exploit the level mismatch between the VB at the source and the CB in the channel. The efficiency of switching the current off/on depends on the flatness of the bands and can be influenced by material properties, doping and gate stack. Consequently, the ability to attain small values of SS critically depends on the electrostatic control of the channel region, especially when considering short gate lengths. In order to maximize the electrostatic integrity, multi-gate low-dimensional systems such as gate-all-around (GAA) nanowires are typically considered. The scaling of InAs nanowire TFETs has been investigated in Brocard et al. (2013) by means of the method presented in section 1.3.1. The optimization of TFET parameters requires analyzing the dependence on both the lateral width W and the longitudinal gate length L_G. Nanowires with smaller W are expected to provide a better electrostatic control, as well as a lower current, while devices with smaller L_G have higher currents but worse electrostatic integrity.

Figure 1.4 shows the transfer characteristics of InAs NW TFETs for W varying from 5 to 10 nm. The gate work-function is the same for all of the devices to highlight the impact of the lateral size on the threshold voltage (V_T). As far as W increases, the I–V curves shift towards smaller V_{GS} values as a consequence of the electrostatic degradation induced by the smaller confinement. Moreover, the SS of larger devices clearly tends to increase with W. For the given gate length of 17 nm, only the smallest device with W = 5 nm presents a sub-thermionic behavior with an SS of about 40 mV/dec, but this device exhibits significantly lower on-state current with respect to thicker TFETs.

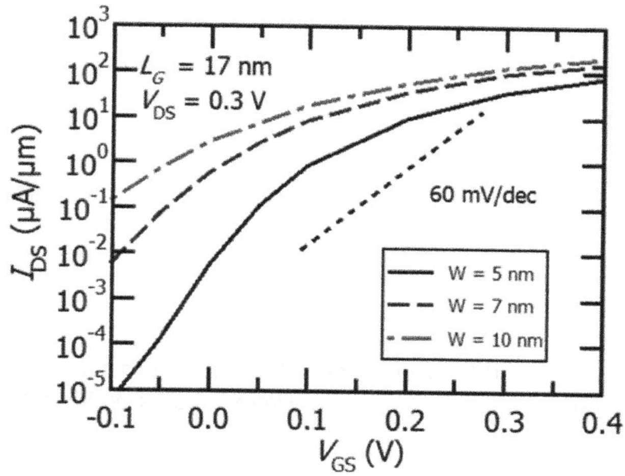

Figure 1.4. *Transfer characteristics at V_{DS} = 0.3 V for InAs rectangular nanowire TFETs for different widths W = 5, 7, 10 nm and L_G = 17 nm. The current is renormalized with respect to W. V_{DD} = 0.3 V*

In order to study the electrostatic integrity of InAs TFETs, it is also useful to analyze their channel-length dependence. To this purpose, Figure 1.5 shows the drain-induced barrier thinning (DIBT) and the SS of InAs NWs as a function of L_G. The DIBT is a figure of merit that generalizes the drain-induced barrier lowering used in MOSEFTs and estimates short-channel effects in TFETs [90]. For a given variation of the drain voltage, it reads

$$DIBT = \frac{\Delta V_T}{\Delta V_{DS}} . \qquad [1.10]$$

Figure 1.5 shows that, below L_G = 17 nm, the thickest TFET with W = 10 nm is not able to reach SS values below the 60 mV/dec limit and it is therefore not interesting for short-channel devices. Nanowires with W = 5 nm and 7 nm, however, show adequate SS and DIBT, but a further decrease in W would lead to current values that are too low. It should be remarked that the reduction of the gate length below 15 nm would deteriorate both the SS and DIBT, illustrating the difficulties in scaling III–V TFETs to gate lengths shorter than 10 nm.

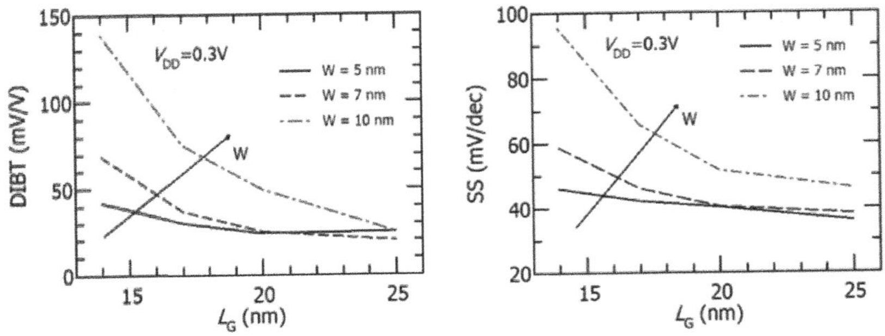

Figure 1.5. *(Left) DIBT and (right) SS as a function of L_G for InAs rectangular nanowire TFETs with different widths W. V_{DD} = 0.3 V*

The off-state physics and electrostatic integrity of TFETs have been extensively discussed in Esseni et al. (2015) and can be quantitatively assessed by using the concept of the natural length (Colinge 2004). The transistor natural length λ has different expressions depending on the device dimensionality, shape and number of gates. For instance, for the square section nanowire, it reads

$$\lambda = \sqrt{\frac{\varepsilon_{sc}}{4\varepsilon_{ox}}\left[1+\frac{\varepsilon_{ox}}{4\varepsilon_{sc}}\frac{W}{t_{ox}}\right]Wt_{ox}}\,, \qquad [1.11]$$

where $\varepsilon_{sc(ox)}$ is the semiconductor(oxide) dielectric constant and t_{ox} is the oxide thickness. Physically, it represents the penetration length of the electric field lines of the drain in the channel region.

In order to highlight the different scaling of TFETs and MOSFETs, Figure 1.6 reports the SS values as a function of the L_G normalized to the natural length λ for the InAs MOSFETs and InAs TFETs with different values of channel and oxide thickness. Here, the natural length has been computed with equation [1.11] and, for the InAs transistors with W = 5 nm and t_{ox} = 1 nm, equals λ = 2.53 nm, while for the InAs transistors with W = 4 nm and t_{ox} = 0.6 nm, equals λ = 1.82 nm. Concerning the physical lengths, most of the transistors have a gate length between 8 nm and 25 nm, while only the two sets with W = 7 nm have minimum L_G = 10 nm instead of 8 nm.

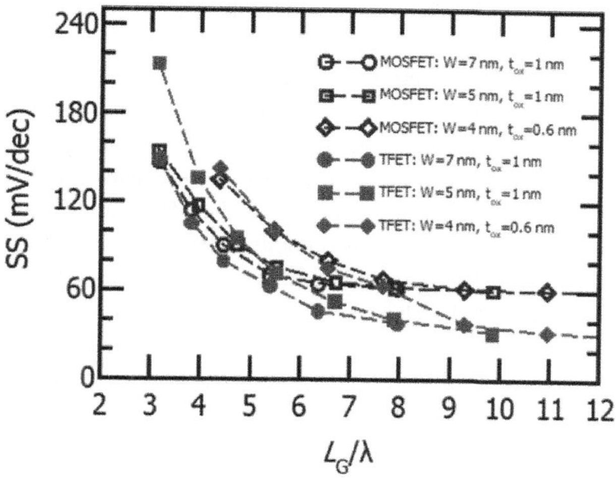

Figure 1.6. *The SS as a function of L_G over the natural length λ of InAs NW (red symbols) TFETs and (blue symbols) MOSFETs with different geometrical parameters. V_{DD} = 0.5 V*

A first important result visible in Figure 1.6 is that there are several MOSFETs and TFETs with an SS larger than a thermionic limit of 60 mV/dec, even for relatively large L_G/λ ratios. These transistors have a degraded SS value reaching up to 130 mV/dec for L_G/λ of about 4. The reason for such an SS degradation is not related to the poor electrostatic integrity, but to the non-null tunneling current, which is particularly high when considering small direct gap semiconductors such as InAs. It turns out that these devices have L_G values between 12 and 8 nm, so that the drain-to-source current in the off state is influenced by the tunneling current even for MOSFETs. A second important remark about Figure 1.6 is that the TFETs can achieve SS values below 60 mV/dec only for L_G/λ ratios larger than about five, which are relatively large values for which thermionic MOSFETs still offer almost ideal SS values of 60 mV/dec. For shorter L_G/λ values, the SS of TFETs increases at least as steeply as the SS of MOSFETs, showing that designing a well-tempered device is even more important for TFETs than for MOSFETs.

1.4.2. *Trap-assisted tunneling*

The effect of traps in nanowires can be accounted for within a full-quantum approach using the **k.p** Hamiltonian and the NEGF method presented in section 1.3.1. Traps can be included in the simulations with the model presented in Pala et al. (2012) and Pala and Esseni (2013), where they are inserted by superimposing cubic potential wells $V_t(\mathbf{r},\mathbf{r}_t)$ to the conduction band profile with $V_t(\mathbf{r}, \mathbf{r}_t) = -V_{depth}$ if $(\mathbf{r}-\mathbf{r}_t) \in C_t$ and $V_t(\mathbf{r}, \mathbf{r}_t) = 0$; otherwise, C_t being a 1 nm^3 volume cube. These wells give rise to electrically active 0D states that account for the presence of traps both for the electrostatics and the transport. Such states can either be populated or not depending on the quasi-Fermi level of the system with respect to the trap energy. They can participate in the transport thanks to the effect of the electron–phonon interaction, which can open additional tunneling paths, for instance by promoting electrons from the VB to the trap level and hence to the CB. Another advantage of this model is to consider the discrete nature of defects, which is useful when considering device variability studies. In Esseni and Pala (2013) and Pala and Esseni (2013), the traps were placed at the interface between the semiconductor and the oxide, but they could be localized anywhere in the device.

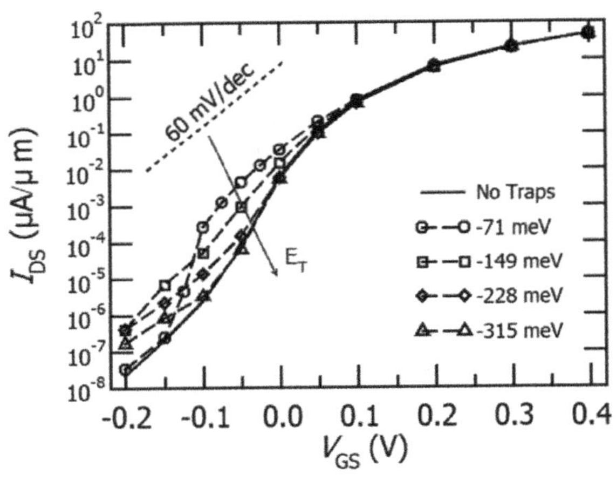

Figure 1.7. *Transfer characteristics for an InAs NW TFET at V_{DS} = 0.4 V with and without the presence of a single trap located at 4 nm inside the channel for different E_T values. It can be noted that the impact of the trap decreases for deeper trap levels*

It turns out that the TFETs are more sensitive than the MOSFET to the presence of traps. As shown in Figure 1.7, even a single trap can significantly modify the subthreshold current of an InAs NW TFET, depending on its position and energy level. In Figure 1.7, the trap is placed at 4 nm inside the channel, starting from the source/channel interface and may or may not affect the transfer characteristics according to the trap energy E_T, which is indirectly varied by changing V_{depth}. As can be seen, the trap can modulate both the I_{off} and the SS of the transistor and, moreover, shallow traps have a larger impact than deep energy traps on the I–V curves. The reason for this behavior is that shallow traps are energetically located in the vicinity of the source Fermi level, which is the energy window important for BTBT. Another interesting feature of Figure 1.7 is that there are specific trap energies, namely E_T = -71 meV and V_{GS} intervals (-0.15 V ≤ V_{GS} ≤ -0.1 V) for which the current characteristics have a steeper SS than the case with no traps. This happens when the density of states (DOS) of the trap is found in the energy window relevant for the BTBT. However, in real devices, it is almost impossible to observe this phenomenon, because many traps participate in the electron transport and the total current is given by the sum of the contributions of traps with different energy levels, which leads to the general degradation of the SS compared to the TFET without traps.

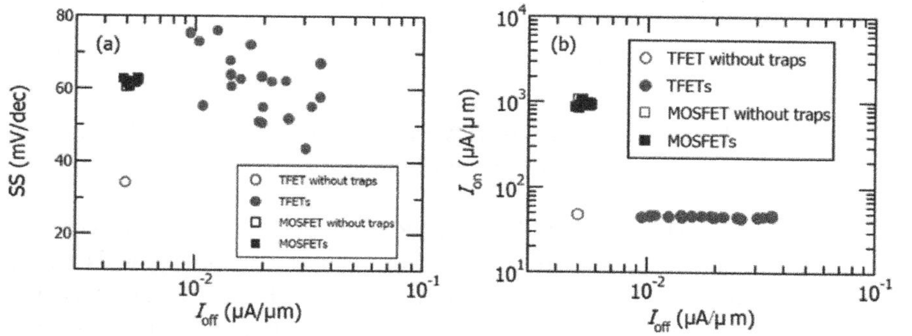

Figure 1.8. *(a) The SS (averaged for I_{DS} values between 1 nA/μm and 100 nA/μm) versus the I_{off} for 20 realizations of (red circles) InAs NW TFETs and (blue squares) InAs NW MOSFET with a spatially random distribution of traps with trap density D_T = 2×10^{12} cm^{-2}. (b) The I_{on} versus the I_{off} for the same 20 realizations of InAs TFETs and InAs MOSFET. Empty symbols refer to devices without traps. V_{DD} = 0.4 V*

As even a single trap is able to strongly modify the I–V curves of TFETs, it is also interesting to study the device variability of III–V TFETs and compare it to that of corresponding MOSFETs (Esseni and Pala 2013). Figure 1.8 presents a statistical analysis of the I_{on}, the I_{off} and the SS of InAs TFETs by comparing them to those of InAs MOSFETs. The devices under investigation are GAA NW with a lateral width of W = 5 nm and an equivalent oxide thickness of 0.7 nm. As can be seen, the traps induce a significant variability of both SS and I_{off} in TFETs, these quantities being largely deteriorated compared to the clean device. The trap-induced variability of these metrics for MOSFETs is much smaller, whereas the I_{on} variability is smaller in TFETs than in MOSFETs. This indicates that TFETs are inherently sensitive to spatial fluctuations that can alter the BTBT probability. More specifically, traps can act as stepping stones for the BTBT from the VB in the source to the CB in the channel, therefore dramatically changing the I_{off} and SS of TFETs.

1.4.3. *Surface roughness*

Surface-roughness (SR) scattering is known as one of the major physical mechanisms affecting the electrical performance and variability of NW MOS transistors (Wang et al. 2005; Poli et al. 2008; Buran et al. 2009; Cresti et al. 2011). As a consequence of the different working principles of TFETs and MOSFETs, it is very important to assess the impact of SR on the device physics of III–V TFETs. This analysis has been addressed for InAs NW TFETs in Conzatti et al. (2012) by means of the full-quantum approach described in section 1.3.1, where the effect of SR was included by generating random fluctuations of the interfaces according to a Gaussian autocovariance function

$$C(r) = \Delta_m \exp\left(-r^2 / L_m^2\right), \quad [1.12]$$

with Δ_m being the root mean square (RMS) of the fluctuation and L_m being the correlation length with values of 0.4 nm and 1 nm, respectively.

It turns out that, while in MOSFETs, the SR is especially important in decreasing the current at high gate overdrives, when the carrier wave-function is pushed towards the rough interfaces by the gate voltage, in TFETs its impact is more important in the subthreshold regime. This behavior can be explained by inspecting Figure 1.9, showing the subband profiles of both a smooth and a rough nanowire TFET, together with the corresponding current spectra. For the sake of clarity, in Figure 1.9, the SR is non-null only in the channel region. It induces significant fluctuations of the subband profiles that can modulate the tunneling path of electrons. However, these fluctuations are more effective in reducing the current spectrum of the off state, since in the on state the tunneling path is limited to the source/channel junction.

Concerning the effect of SR on the I–V curves of TFETs, it determines an important variation of the I_{on} and I_{off} metrics. To inspect this phenomenon, Figure 1.10 shows the I_{on} and the I_{off} variability of InAs NW TFETs and compares them with the corresponding quantities for strained Si (sSi) NW MOSFETs. About 50 random realizations of rough interfaces were considered for both sets of devices. It can be remarked that the TFETs have a similar I_{on} variability, but much worse I_{off} variability than MOSFETs. Moreover, the variability of I_{off} and I_{on} values for the MOSFETs is highly correlated due to the fact that the MOSFETs operate in the subthreshold regime for the small supply voltage considered here (V_{DD} = 0.3 V), while for the TFETs, there is no apparent correlation.

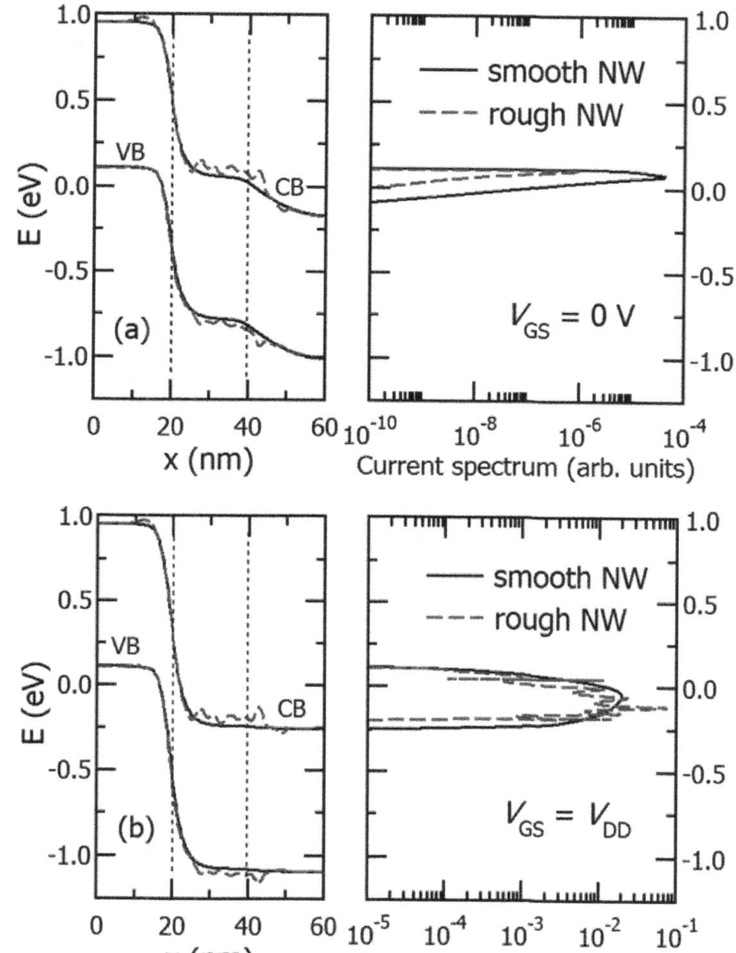

Figure 1.9. *(a) Spatial profile of the highest valence subband and of the lowest conduction subband and the corresponding current spectra for a smooth and rough InAs rectangular NW in the off-state configuration. (b) Same as in (a), but for the on-state configuration.* Δ_m *= 0.4 nm,* L_G *= 17 nm,* W *= 5 nm and* t_{ox} *= 1 nm.* V_{DD} *= 0.3 V*

Physically, the impact of SR on device variability can be explained by considering that in the off-state regime the tunneling path governing the current is much longer than in the on-state and therefore more influenced by the spatial fluctuations induced by the rough surfaces. Hence, the larger I_{off} than I_{on} variability in these devices.

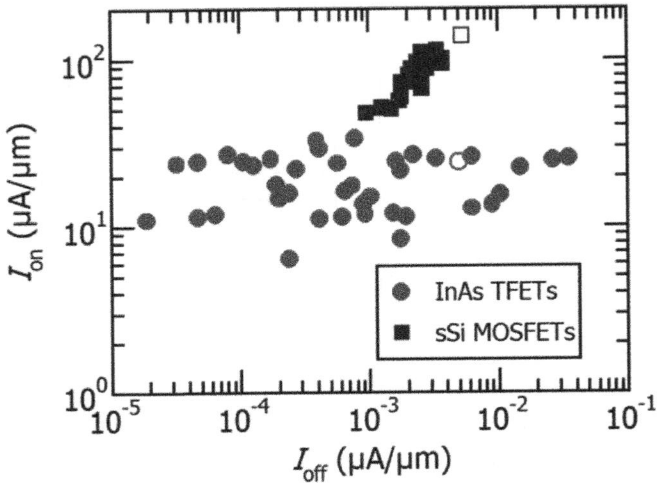

Figure 1.10. *The I_{on} versus the I_{off} for (red circles) the InAs TFETs and for (blue squares) the sSi MOSFETs. Empty symbols represent the smooth device values. Empty symbols represent the smooth devices. The devices are GAA NWs with width W = 5 nm, L_G = 20 nm, t_{ox} =1 nm. V_{DD} = 0.3 V*

1.5. Strategies to improve the on-state current

Since the first experimental realization of TFETs, it has been clear that specific design strategies were needed in order to obtain a sufficiently large on current from these devices. This was indeed the first motivation to use III–V materials instead of Si for TFETs (Knoch and Appenzeller 2010). However, even if using a small and direct band-gap material such as InAs, the I_{on} of TFETs remains much lower than the one of Si MOSFETs, even for low supply voltages (Conzatti et al. 2011). Physically, the limited on-state current of TFETs arises from the difficulties to attain transmission probabilities close to one due to the barrier at the source–channel junction.

In the following sections, different options that have been proposed to overcome this problem are reported.

1.5.1. *Strain*

Similar to strategies used in the past to improve the mobility of Si MOSFETs, one possible strategy to improve the electrical performance of III–V TFETs is to use mechanical stress/strain. The effect of strain can be naturally included in the 8-band **k.p** Hamiltonian by means of an interaction matrix (Bahder 1990). Explicitly, this reads

$$H_{8,\varepsilon}=\begin{pmatrix} a_c e & 0 & -v^* & 0 & -\sqrt{3}v & \sqrt{2}u & u & -\sqrt{2}v^* \\ & a_c e & \sqrt{2}u & -\sqrt{3}v^* & 0 & v & -\sqrt{2}v & -u \\ & & -p+q & -s^* & r & 0 & \sqrt{3/2}s & -\sqrt{2}q \\ & & & -p-q & 0 & r & -\sqrt{2}r & s/\sqrt{2} \\ & & & & -p-q & s^* & s^*/\sqrt{2} & \sqrt{2}r^* \\ & & & & & -p+q & \sqrt{2}q & \sqrt{3/2}s^* \\ & & & & & & -a_v e & 0 \\ & & & & & & & -a_v e \end{pmatrix} \quad [1.13]$$

where

$$\begin{aligned} e &= \varepsilon_{c,xx}+\varepsilon_{c,yy}+\varepsilon_{c,zz} \\ p &= a_v e \\ q &= b\left[\varepsilon_{c,zz}-\left(\varepsilon_{c,xx}+\varepsilon_{c,yy}\right)/2\right] \\ u &= \tfrac{P_0}{\sqrt{3}}\left(\varepsilon_{c,xz}k_x+\varepsilon_{c,yz}k_y+\varepsilon_{c,zz}k_z\right) \\ v &= \tfrac{P_0}{6}\left[\left(\varepsilon_{c,xx}-i\varepsilon_{c,xy}\right)k_x+\left(\varepsilon_{c,xy}-i\varepsilon_{c,yy}\right)k_y+\left(\varepsilon_{c,xz}-i\varepsilon_{c,yz}\right)k_z\right] \\ r &= \tfrac{\sqrt{3}}{2}b\left(\varepsilon_{c,xx}-\varepsilon_{c,yy}\right)-id\varepsilon_{c,xy} \\ s &= -d\left(\varepsilon_{c,xz}-i\varepsilon_{c,yz}\right) \end{aligned} \quad [1.14]$$

and $\varepsilon_{c,ij}$ is the generic component of the strain tensor. The values of deformation potentials in previous equations can be found in Vurgaftman et al. (2001) and have been calibrated with respect to experimental measurements.

The effect of using different stress configurations is illustrated in Figure 1.11, reporting the band structures of an InAs rectangular NW as a function of the real and imaginary parts of the wavevector. The latter was computed by solving the inverse energy wavevector problem. The imaginary branch connecting the top of the VB with the bottom of the CB rules the band-to-band transmission probability and hence it can be used to estimate the current in TFETs. In Figure 1.11, it is possible to observe that both uniaxial compressive and biaxial tensile stress push up the valence subband, but only the biaxial stress moves down the conduction one, which results in the largest reduction of the energy gap and of the imaginary wavevector. Hence, an important enhancement of the TFET current is expected for non-null biaxial tensile stress.

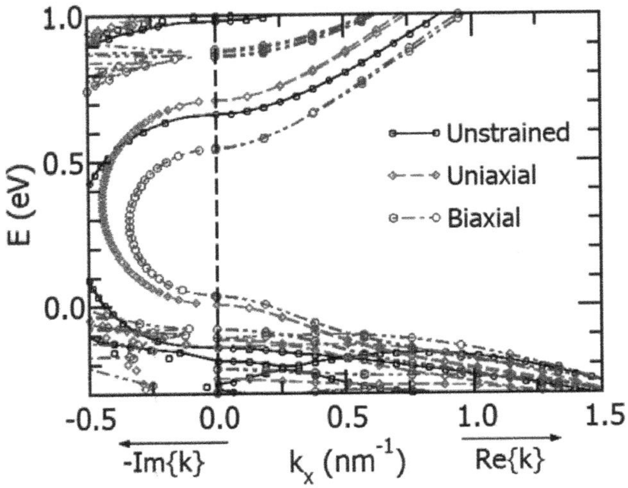

Figure 1.11. *Band structure of an InAs rectangular nanowire with W = 5 nm for the transport direction along [100] and for different stress configurations. The curves report the system without stress, the one with compressive uniaxial stress (T_{xx} = -2 GPa) and the one with tensile biaxial stress (T_{yy} = T_{zz} = 2 GPa). The energy dispersions are strongly modified by mechanical stress*

The impact of strain on the electrical performance of TFET based on InAs GAA nanowire was studied in detail in Conzatti et al. (2012). Figure 1.12 compares the transfer characteristics for different stress conditions of a device with a gate length of 20 nm. The uniaxial stress has a small impact on the transfer characteristics inducing a reduction of the V_T due to the lowering of the conduction band observed in Figure 1.11. On the contrary, the biaxial stress induces both a V_T shift and a remarkable increase of I_{DS} as a consequence of the smaller energy gap and imaginary wavevector.

The physical mechanism responsible for the enhancement of the I_{on} is illustrated in Figure 1.13. This plot compares the subband profiles and the current spectra of the TFET for the different strain configurations in the on state ($V_{GS} = V_{DS} = V_{DD} = 0.3$ V). As can be seen, the origin of the larger current density for the biaxial stress is not only a modified subband profile, but also the enhanced tunneling probability due to the smaller imaginary wavevector visible in Figure 1.11. As a drawback of the use of strain in TFETs, it is important to report a possible degradation of the SS, as visible also in Figure 1.12. As can be seen in Figure 1.13, this is due to both the gap reduction and the increase of the valence subband level in the source region, which implies that the electrons belonging to the tail of the Fermi function distribution in the source region participate in the BTBT.

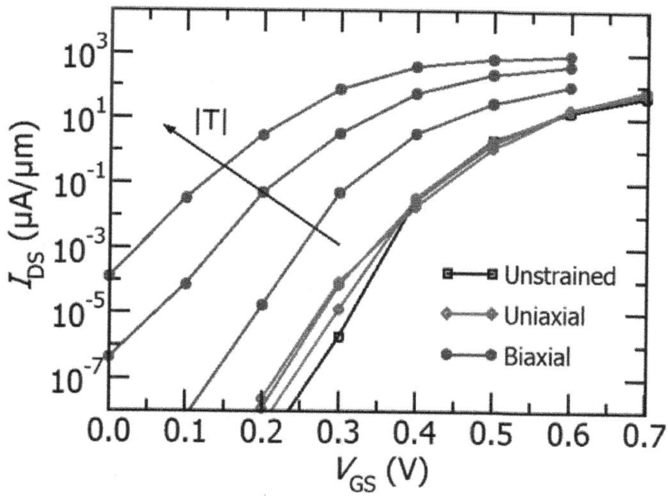

Figure 1.12. *Transfer characteristics of an InAs NW TFET for different strain conditions. The compressive uniaxial stress is T_{xx} = −1, −2, and −3 GPa, and the tensile biaxial stress is T_{yy} = T_{zz} = 1, 2, and 3 GPa. W = 5 nm, L_G = 20 nm, V_{DS} = 0.3 V*

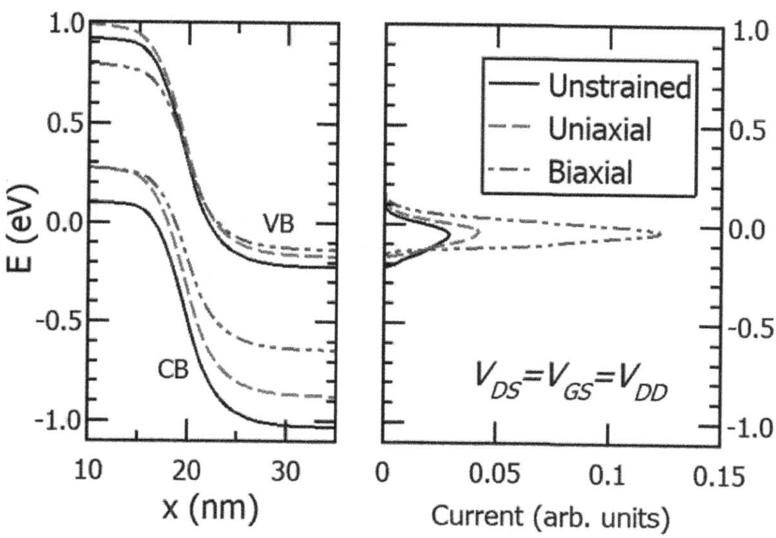

Figure 1.13. *Subband profiles of the InAs NW TFET for different stress configurations. The curves correspond to the device without stress, with compressive uniaxial stress (T_{xx} = -2 GPa) and tensile biaxial stress (T_{yy} = T_{zz} = 2 GPa)*

1.5.2. *Broken-gap hetero-structures*

In homojunction TFETs, the mismatch between the minimal conduction subband and the maximal valence subband at the source–channel interface is determined by the different dopant concentrations in the source and the channel region. The minimal tunneling path at this interface determines the tunneling probability and hence the maximal current value in the on state (see, for instance, Figure 1.13). In order to boost the I_{on} of TFETs, it is possible to design a much more abrupt mismatch by considering a hetero-structure formed by two different semiconductors in the source and the channel. This permits us to obtain almost ideal subband profiles with a very small tunneling path. Among the different possibilities, the most investigated hetero-junction (HJ) for TFETs is the one given by the GaAs/In(Sb)As couple. This choice has two main advantages: it provides a broken, or near-broken bandgap system (i.e. the bottom of the CB is lower than the top of the VB) and it is composed of two materials with a very similar lattice parameter (Vurgaftman et al. 2001). As already mentioned, the first property is important to increase the BTBT in the device, the second to prevent the presence of dislocation at the hetero-interface, which could induce technological problems due to defects and traps (see the discussion in section 1.4.2).

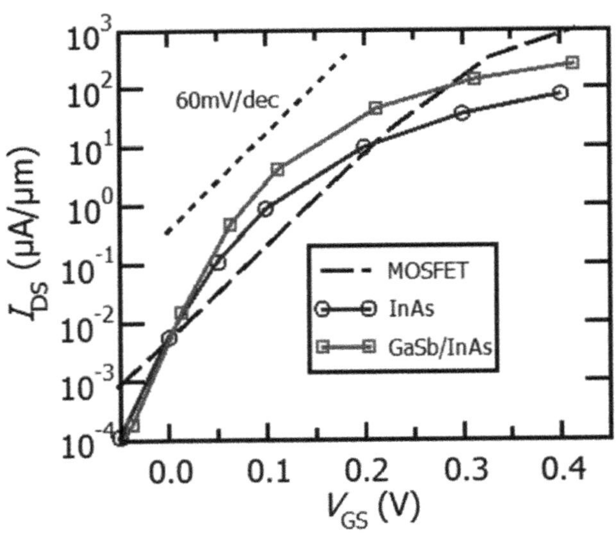

Figure 1.14. *Transfer characteristics for two rectangular nanowire TFETs and for the InAs MOSFET (L_G = 17 nm, W = 5 nm and I_{off} = 5 nA/μm). V_{DD} = 0.4 V. GaSb/InAs TFETs can improve the I_{on} of the homojunction InAs TFET without degrading the SS*

The use of the GaSb/InAs heterojunction in nanowire TFET was studied in Brocard et al. (2013) by means of self-consistent NEGF simulations and the 8-band **k.p** Hamiltonian from section 1.3.1. The HJ was described directly in the real-space Hamiltonian by using position-dependent material parameters taken from Vurgaftman et al. (2001). As shown in Figure 1.14, the broken gap GaSb/InAs HJ enhances the I_{on} value with respect to the InAs TFET, while providing an almost identical SS value. This means that the use of the broken-gap heterojunction as an on-current booster can be preferable with respect to other options such as the use of mechanical strain, which also implies an SS degradation (see Figure 1.12).

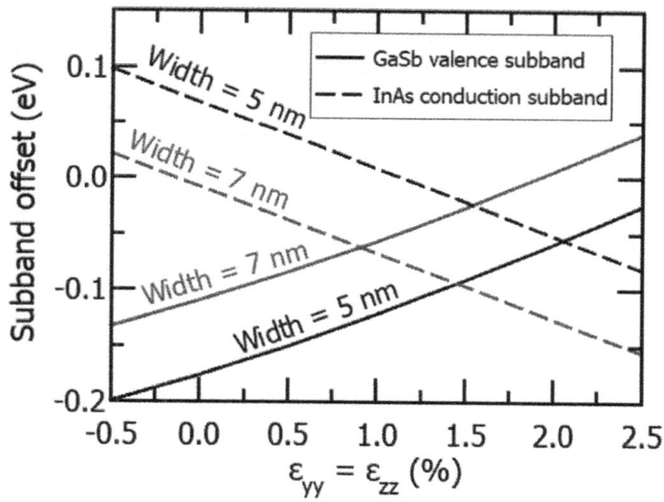

Figure 1.15. *Highest valence subband of GaSb and lowest conduction subband of InAs as a function of the biaxial strain ($\varepsilon_{yy} = \varepsilon_{zz}$) for rectangular nanowires with W = 5 nm and 7 nm. The tensile strain induces a broken-gap alignment in the GaSb/InAs HJ*

However, it should be noted that the beneficial effect of using the GaSb/InAs system in nanowires can be limited by quantum confinement effects. Indeed, this heterojunction is broken-gap in the bulk, but it can become of staggered type in nanowires. This means that the use of confined nanostructures, which, as discussed in section 1.4.1, is almost compulsory in TFETs, can reduce the current enhancement of the hetero-structure. To compensate for the impact of lateral confinement on the subband configuration, it is advisable to combine the heterojunction and mechanical strain, which, as seen in section 1.5.1, can shrink the semiconductor bandgap. This behavior is illustrated in Figure 1.15, showing the evolution of the lowest conduction and of the highest valence subbands in the flat band condition versus the strain level. The tensile biaxial strain increases the GaSb valence subband and reduces the InAs conduction one, helping the nanowire to recover a broken bandgap configuration.

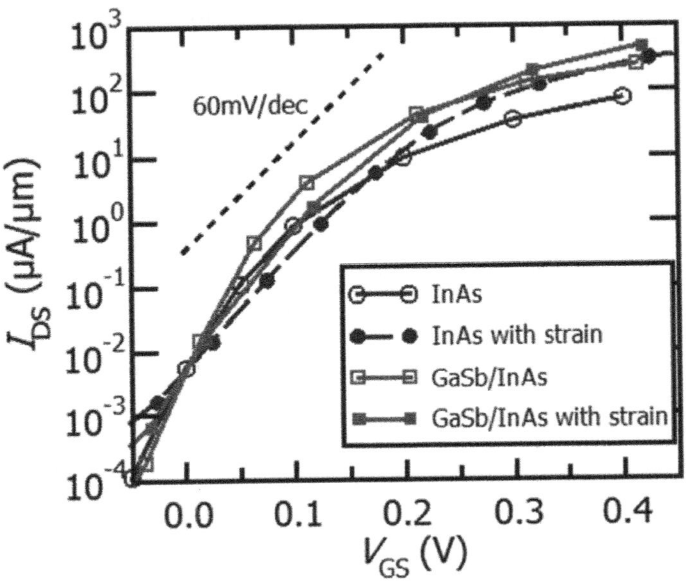

Figure 1.16. *Transfer characteristics of III–V NW TFETs with different design options (L_G = 17 nm, W = 5 nm and I_{off} = 5 nA/μm). V_{DD} = 0.4 V. The strain increases the I_{on}, but deteriorates the SS*

The impact of using strained broken-gap HJ on the I–V curves is shown in Figure 1.16. The biaxial tensile strain induces an on-current boost, in both the homojunction InAs and the heterojunction GaSb/InAs devices. The device with the GaSb/InAs HJ and strain can even provide an on/off current ratio larger than 10^6 for V_{DD} = 0.4 V. Again, the drawback of this choice is that the strain degrades the SS value. Physically, this behavior can be understood by considering that the tensile strain reduces the energy gap not only in the source region, which boosts the on-state current, but also in the channel region, which eventually results in an increase of the off-state current. As suggested in Conzatti et al. (2012), a possible solution could be to impose a local strain only in the source region.

1.5.3. *Molar fraction grading of the source material*

As can be seen in Figure 1.17, a relevant portion of the tunneling path in a GaSb/InAs heterojunction TFET is located in the depletion region of the source. Because of the source depletion, a non-negligible part of the source highest valence subband is below the source Fermi level, which is set to 0 eV. Hence, a relevant part of the transmission spectrum remains unexploited, because of the absence of charge carriers in this energy window. In order to attain a better exploitation of the transmission spectrum, it would be import to repopulate the source region with carriers with energies above the source Fermi level. This would result in a shorter depletion length and tunneling path, and consequently a larger transmission. Unfortunately, such a goal cannot be attained by increasing the dopant density in the source region, either for technological problems due to the limited dopant dilution or because this would degrade the SS (Knoch and Appenzeller 2010).

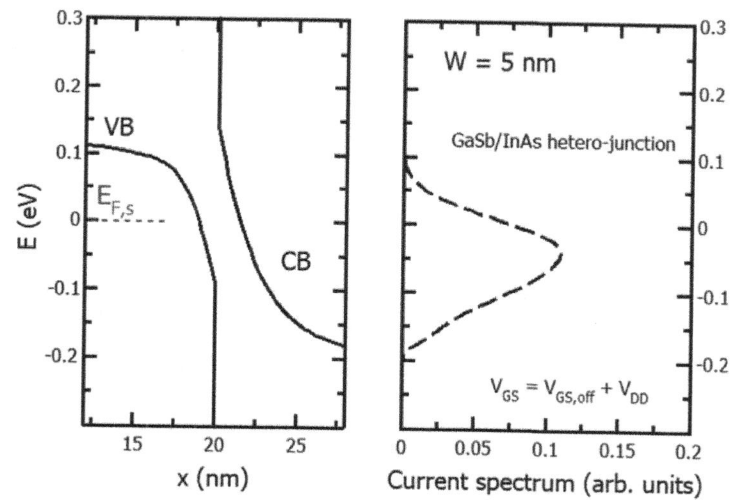

Figure 1.17. *(Left) Subband profiles and current spectrum and (right) current spectrum of a GaSb/InAs NW TFET in the on-state regime*

An alternative solution to shorten the depletion length is to use the properties of the AlGaSb ternary alloy as proposed in Brocard et al. (2014). The idea is to use a linearly graded source region of In $Al_xGa_{(1-x)}Sb$ starting with molar fraction $x = 0$ and ending with $x = 1$, in order to reduce the extension of the depletion region in the source/channel interface. Indeed, there is an energy difference of almost 400 meV between the AlSb and the GaSb, which can be used to counteract the subband bending in the source region and, moreover, the lattice parameter of the AlGaSb is practically independent of x, which is crucial to avoid undesired strain effects and the formation of dislocations.

Figure 1.18 compares the transfer characteristics of the graded source device for various thickness values of the graded region, T_{grad}, with the one of the regular GaSb/InAs TFET. A different gate work-function is assumed for each transistor in order to impose the same off-state current I_{off} = 5 nA/μm at V_{GS} = 0 V. As can be seen, the use of the molar fraction grading in the source region can provide a significantly higher (around 80% for T_{grad} = 5 and 7 nm) I_{on} in the AlGaSb TFET with respect to the GaSb/InAs TFET, with a virtually identical SS of about 30 mV/dec. Thanks to this design option, on/off current ratios larger than 10^6 can be attained with a V_{DD} as small as 0.4 V.

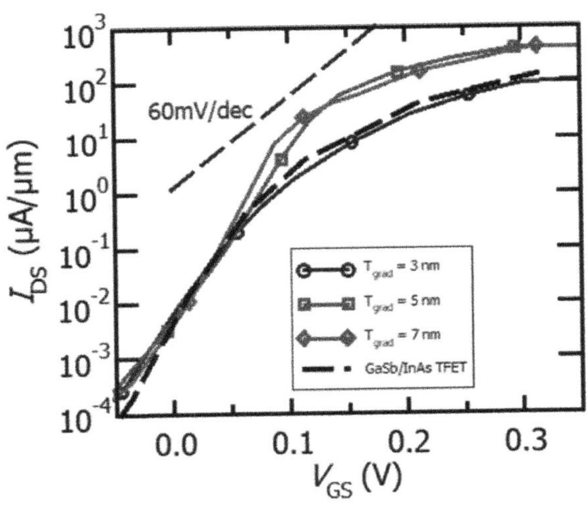

Figure 1.18. *Transfer characteristics at V_{DS} = 0.3 V for three thicknesses of the graded region in $Al_xGa_{(1-x)}Sb$/InAs TFET for a fixed I_{off} = 5 nA/μm. The GaSb/InAs TFET with no grading is shown for comparison*

A better understanding of the physical mechanism responsible for this current enhancement is illustrated in Figure 1.19 representing the local density of states (LDOS) along the transport direction of the AlGaSb/InAs NW TFET for different T_{grad} values. In Figure 1.19(a) and (b), it can be seen that the grading pushes the carriers against the source–channel junction, hence helping the tunneling in the InAs region. In Figure 1.19(c), it is also noticeable that the shortest T_{grad} of 3 nm has the undesirable effect of driving the carrier density to lower energies, thus depopulating the energy window relevant to tunneling. This explains the poor value of I_{on} for T_{grad} = 3 nm in Figure 1.18.

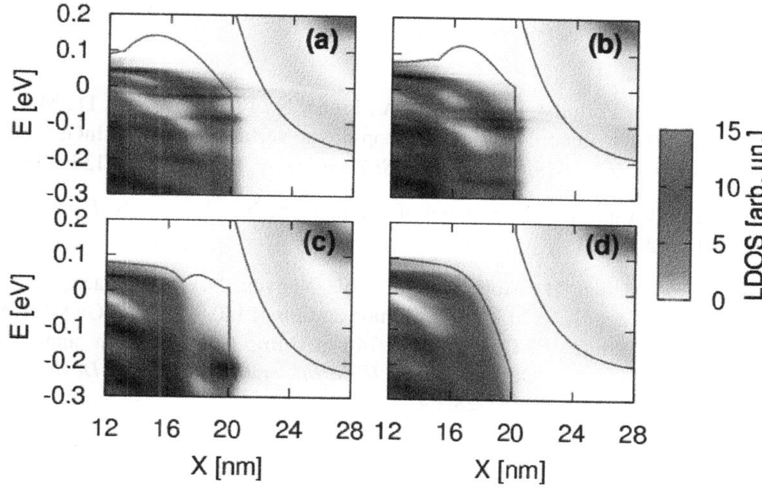

Figure 1.19. *Local density of states in the region close to the source–channel junction for $Al_xGa_{(1-x)}Sb/InAs$ TFETs*

NOTES ON FIGURE 1.19.– *(a) T_{grad} = 7 nm, (b) T_{grad} = 5 nm, (c) T_{grad} = 3 nm and (d) for the GaSb/InAs TFET. The corresponding subband profiles are superimposed as red lines. The source–channel junction is located at x = 20 nm, and the source Fermi level is set to 0 eV. $V_{GS} = V_{DS} = V_{DD} = 0.3$ V.*

Alternative design options using source engineering to boost the current of TFETs are discussed in Pala and Brocard (2015).

1.6. Conclusion

TFETs based on III–V semiconductors can be a viable option to replace MOSFETs for low-power applications, only if specific design prescriptions are fulfilled and efficient current boosters are implemented. In particular, it would be necessary to use low-dimensional multi-gate devices such as GAA NWs to assure perfect electrostatic integrity, as well as interfaces and materials with very low defect densities that can otherwise compromise the off-state physics of TFETs. Finally, even if still challenging from a technological point of view, the use of broken-gap heterojunctions in conjunction with strain or molar fraction grading can eventually lead to on-current values competitive with the MOSFET and contribute to fabricate integrated circuits working at V_{DD} smaller than 0.5 V.

1.7. References

Alian, A., Kazzi, S.E., Verhulst, A., Milenin, A., Pinna, N., Ivanov, T., Lin, D., Mocuta, D., Collaert, N. (2018). Record 47 mV/dec top-down vertical nanowire InGaAs/GaAsSb tunnel FETs. *2018 IEEE Symposium on VLSI Technology*, Honolulu, HI, 133–134.

Bahder, T.B. (1990). Eight-band k.p model of strained zinc-blende crystals. *Physical Review B*, 41(17), 11992–12001.

Bijesh, R., Liu, H., Madan, H., Mohata, D., Li, W., Nguyen, N.V., Gundlach, D., Richter, C.A., Maier, J., Wang, K. et al. (2013). Demonstration of In0.9Ga0.1As/GaAs0.18Sb0.82 near broken-gap tunnel FET with ION=740 μA/μm, GM=70μS/μm and gigahertz switching performance at VDS=0.5V. *2013 IEEE International Electron Devices Meeting (IEDM)*, Washington, DC, 28.2.1-28.2.4.

Brocard, S., Pala, M.G., Esseni, D. (2013). Design options for hetero-junction tunnel FETs with high on current and steep sub-threshold voltage slope. *2013 IEEE International Electron Devices Meeting (IEDM)*, 5.4.1-5.4.4.

Brocard, S., Pala, M.G., Esseni, D. (2014). Large on-current enhancement in hetero-junction tunnel-FETs via molar fraction grading. *IEEE Electron Device Letters*, 35(2), 184–186.

Buran, C., Pala, M.G., Bescond, M., Dubois, M., Mouis, M. (2009). Three-dimensional real-space simulation of surface roughness in silicon nanowire FETs. *IEEE Transactions on Electron Devices*, 56(10), 2186–2192.

Colinge, J.-P. (2004). Multiple-gate SOI MOSFETs. *Solid-State Electronics*, 48(6), 897–905.

Convertino, C., Zota, C.B., Baumgartner, Y., Staudinger, P., Sousa, M., Mauthe, S., Caimi, D., Czornomaz, L., Ionescu, A.M., Moselund, K.E. (2019). Sub-thermionic scalable III-V tunnel field-effect transistors integrated on Si (100). *2019 IEEE International Electron Devices Meeting (IEDM)*, San Francisco, CA, 37.1.1-37.1.4.

Convertino, C., Zota, C.B., Schmid, H., Caimi, D., Czornomaz, L., Ionescu, A.M., Moselund, K.E. (2021). A hybrid III–V tunnel FET and MOSFET technology platform integrated on silicon. *Nature Electronics*, 4(2), 162–170.

Conzatti, F., Pala, M.G., Esseni, D., Bano, E., Selmi, L. (2011). A simulation study of strain induced performance enhancements in InAs nanowire Tunnel-FETs. *2011 IEEE International Electron Devices Meeting (IEDM)*, 5.2.1-5.2.4.

Conzatti, F., Pala, M.G., Esseni, D. (2012a). Surface-roughness-induced variability in nanowire InAs tunnel FETs. *IEEE Electron Device Letters*, 33(6), 806–808.

Conzatti, F., Pala, M.G., Esseni, D., Bano, E. (2012b). Investigation of localized versus uniform strain as a performance booster in InAs Tunnel-FETs. *2012 13th International Conference on Ultimate Integration on Silicon (ULIS)*, IEEE, Grenoble, 183–186.

Conzatti, F., Pala, M.G., Esseni, D., Bano, E., Selmi, L. (2012c). Strain-induced performance improvements in InAs nanowire tunnel FETs. *IEEE Transactions on Electron Devices*, 59(8), 2085–2092.

Cresti, A., Pala, M.G., Poli, S., Mouis, M., Ghibaudo, G. (2011). A comparative study of surface-roughness-induced variability in silicon nanowire and double-gate FETs. *IEEE Transactions on Electron Devices*, 58(8), 2274–2281.

Datta, S. (2005). *Quantum Transport – Atom to Transistor*. Cambridge University Press, Cambridge.

Dewey, G., Chu-Kung, B., Boardman, J., Fastenau, J.M., Kavalieros, J., Kotlyar, R., Liu, W.K., Lubyshev, D., Metz, M., Mukherjee, N. et al. (2011). Fabrication, characterization, and physics of III-V heterojunction tunneling field effect transistors (H-TFET) for steep sub-threshold swing. *2011 International Electron Devices Meeting (IEDM)*, Washington, DC, 33.6.1-33.6.4.

Dey, A.W., Borg, B.M., Ganjipour, B., Ek, M., Dick, K.A., Lind, E., Thelander, C., Wernersson, L. (2013). High-current GaSb/InAs(Sb) nanowire tunnel field-effect transistors. *IEEE Electron Device Letters*, 34(2), 211–213.

Esseni, D. and Pala, M.G. (2013). Interface traps in InAs nanowire tunnel FETs and MOSFETs – Part II: Comparative analysis and trap-induced variability. *IEEE Transactions on Electron Devices*, 60(9), 2802–2807.

Esseni, D., Pala, M.G., Rollo, T. (2015). Essential physics of the OFF-state current in nanoscale MOSFETs and tunnel FETs. *IEEE Transactions on Electron Devices*, 62(9), 3084–3091.

Esseni, D., Pala, M., Palestri, P., Alper, C., Rollo, T. (2017). A review of selected topics in physics based modeling for tunnel field-effect transistors. *Semiconductor Science and Technology*, 32(8), 083005.

Ford, A.C., Yeung, C.W., Chuang, S., Kim, H.S., Plis, E., Krishna, S., Hu, C., Javey, A. (2011). Ultrathin body InAs tunneling field-effect transistors on Si substrates. *Applied Physics Letters*, 98(11), 113105.

Gandhi, R., Chen, Z., Singh, N., Banerjee, K., Lee, S. (2011). Vertical Si-Nanowire n-type tunneling FETs with low subthreshold swing (<< 50 mV/decade) at room temperature. *IEEE Electron Device Letters*, 32(4), 437–439.

Ganjipour, B., Wallentin, J., Borgström, M.T., Samuelson, L., Thelander, C. (2012). Tunnel field-effect transistors based on InP-GaAs heterostructure nanowires. *ACS Nano*, 6(4), 3109–3113.

Gopalakrishnan, K., Griffin, P.B., Plummer, J.D. (2005). Impact ionization MOS (I-MOS)– Part I: Device and circuit simulations. *IEEE Transactions on Electron Devices*, 52(1), 69–76.

Huang, Q., Huang, R., Zhan, Z., Qiu, Y., Jiang, W., Wu, C., Wang, Y. (2012). A novel Si tunnel FET with 36mV/dec subthreshold slope based on junction depleted-modulation through striped gate configuration. *2012 International Electron Devices Meeting (IEDM)*, San Francisco, CA, 8.5.1-8.5.4.

Ionescu, A.M. and Riel, H. (2011). Tunnel field-effect transistors as energy-efficient electronic switches. *Nature*, 479(9), 329–337.

Knoch, J. and Appenzeller, J. (2010). Modeling of high–performance p-Type III–V heterojunction tunnel FETs. *IEEE Electron Device Letters*, 31(4), 305–307.

Knoll, L., Zhao, Q.-T., Nichau, A., Trellenkamp, S., Richter, S., Schafer, A., Esseni, D., Selmi, L., Bourdelle, K.K., Mantl, S. (2013). Inverters with strained Si nanowire complementary tunnel field-effect transistors. *IEEE Electron Device Letters*, 34(6), 813–815.

Krishnamohan, T., Kim, D., Raghunathan, S., Saraswat, K. (2008). Double-gate strained-ge heterostructure tunneling FET (TFET) with record high drive currents and < 60mV/dec subthreshold slope. *2008 IEEE International Electron Devices Meeting (IEDM)*, San Francisco, CA, 1–3.

Krishnaraja, A., Svensson, J., Memisevic, E., Zhu, Z., Persson, A.R., Lind, E., Wallenberg, L.R., Wernersson, L.-E. (2020). Tuning of source material for InAs/InGaAsSb/GaSb application-specific vertical nanowire tunnel FETs. *ACS Applied Electronic Materials*, 2(9), 2882–2887.

Lee, H., Cho, Y., Jeon, S., Shin, M. (2021). First-principles-based quantum transport simulations of interfacial point defect effects on InAs nanowire tunnel FETs. *IEEE Transactions on Electron Devices*, 68(11), 5901–5907.

Luisier, M. and Klimeck, G. (2009). Atomistic full-band design study of InAs band-to-band tunneling field-effect Transistors. *IEEE Elec. Dev. Lett.*, 30(6), 602–604.

Mayer, F., Le Royer, C., Damlencourt, J.-F., Romanjek, K., Andrieu, F., Tabone, C., Previtali, B., Deleonibus, S. (2008). Impact of SOI, Si1-xGexOI and GeOI substrates on CMOS compatible Tunnel FET performance. *2008 IEEE International Electron Devices Meeting (IEDM)*, San Francisco, CA, 1–5.

Memisevic, E., Svensson, J., Hellenbrand, M., Lind, E., Wernersson, L.-E. (2016). Vertical InAs/GaAsSb/GaSb tunneling field-effect transistor on Si with S = 48 mV/decade and Ion = 10 μA/μm for Ioff = 1 nA/μm at Vds = 0.3 V. In *2016 IEEE International Electron Devices Meeting (IEDM)*, San Francisco, CA, 19.1.1-19.1.4.

M'foukh, A., Pala, M.G., Esseni, D. (2020). Full-band quantum transport of heterojunction electron devices with empirical pseudopotentials. *IEEE Transactions on Electron Devices*, 67(12), 5662–5668.

Mohata, D.K., Bijesh, R., Mujumdar, S., Eaton, C., Engel-Herbert, R., Mayer, T., Narayanan, V., Fastenau, J.M., Loubychev, D., Liu, A.K., Datta, S. (2011). Demonstration of MOSFET-like on-current performance in arsenide/antimonide tunnel FETs with staggered hetero-junctions for 300mV logic applications. *2011 International Electron Devices Meeting (IEDM)*, Washington, DC, 33.5.1-33.5.4.

Mookerjea, S., Mohata, D., Krishnan, R., Singh, J., Vallett, A., Ali, A., Mayer, T., Narayanan, V., Schlom, D., Liu, A., Datta, S. (2009). Experimental demonstration of 100nm channel length In0.53Ga0.47As-based vertical inter-band tunnel field effect transistors (TFETs) for ultra-low-power logic and SRAM applications. *2009 IEEE International Electron Devices Meeting (IEDM)*, Baltimore, MD, 1–3.

Nakano, M., Shibuya, K., Okuyama, D., Hatano, T., Ono, S., Kawasaki, M., Iwasa, Y., Tokura, Y. (2012). Collective bulk carrier delocalization driven by electrostatic surface charge accumulation. *Nature*, 487(7408), 459–462.

Noguchi, M., Kim, S., Yokoyama, M., Ichikawa, O., Osada, T., Hata, M., Takenaka, M., Takagi, S. (2015). High I_{on}/I_{off} and low subthreshold slope planar-type InGaAs tunnel field effect transistors with Zn-diffused source junctions. *Journal of Applied Physics*, 118(4), 045712.

Pala, M.G. and Brocard, S. (2015). Exploiting hetero-junctions to improve the performance of III-V nanowire tunnel-FETs. *IEEE Journal of the Electron Devices Society*, 3(3), 115–121.

Pala, M.G. and Esseni, D. (2013). Interface traps in InAs nanowire tunnel-FETs and MOSFETs – Part I: Model description and single trap analysis in tunnel-FETs. *IEEE Transactions on Electron Devices*, 60(9), 2795–2801.

Pala, M.G., Esseni, D., Conzatti, F. (2012). Impact of interface traps on the IV curves of InAs Tunnel-FETs and MOSFETs: A full quantum study. *2012 IEEE International Electron Devices Meeting (IEDM)*, 6.6.1-6.6.4.

Pala, M.G., Giannozzi, P., Esseni, D. (2020). Unit cell restricted bloch functions basis for first-principle transport models: Theory and application. *Phys. Rev. B.*, 102(4), 045410.

Poli, S., Pala, M.G., Poiroux, T., Deleonibus, S., Baccarani, G. (2008). Size dependence of surface-roughness-limited mobility in silicon-nanowire FETs. *IEEE Transactions on Electron Devices*, 55(11), 2968–2976.

Qiu, C., Liu, F., Xu, L., Deng, B., Xiao, M., Si, J., Lin, L., Zhang, Z., Wang, J., Guo, H. et al. (2018). Dirac-source field-effect transistors as energy-efficient, high-performance electronic switches. *Science*, 361(6400), 387–392.

Salahuddin, S. and Datta, S. (2008). Use of negative capacitance to provide voltage amplification for low power nanoscale devices. *Nano Letters*, 8(2), 405–410.

Seabaugh, A.C. and Zhang, Q. (2010). Low-voltage tunnel transistors for beyond CMOS logic. *Proceedings of the IEEE*, 98(12), 2095–2110.

Shao, Y., Pala, M., Esseni, D., del Alamo, J.A. (2022). Scaling of GaSb/InAs vertical nanowire esaki diodes down to Sub-10-nm diameter. *IEEE Transactions on Electron Devices*, 69(4), 2188–2195.

Shin, M. (2009). Full-quantum simulation of hole transport and band-to-band tunneling in nanowires using the k.p method. *Journal of Applied Physics*, 106(5), 054505.

Svizhenko, A. and Anantram, M.P. (2003). Role of scattering in nanotransistors. *IEEE Transactions on Electron Devices*, 50(6), 1459.

Tomioka, K., Yoshimura, M., Fukui, T. (2012). Steep-slope tunnel field-effect transistors using III-V nanowire/Si heterojunction. *2012 IEEE Symposium on VLSI Technology (VLSIT)*, Honolulu, HI, 47–48.

Vandenberghe, W.G., Verhulst, A.S., Sorée, B., Magnus, W., Groeseneken, G., Smets, Q., Heyns, M., Fischetti, M.V. (2013). Figure of merit for and identification of sub-60 mV/decade devices. *Applied Physics Letters*, 102(1), 013510.

Veprek, R.G., Steiger, S., Witzigmann, B. (2007). Ellipticity and the spurious solution problem of k.p envelope equations. *Physical Review B*, 76(16), 165320.

Verreck, D., Verhulst, A.S., Van de Put, M.L., Soree, B., Collaert, N., Mocuta, A., Thean, A., Groeseneken, G. (2016). Uniform strain in heterostructure tunnel field-effect transistors. *IEEE Electron Device Letters*, 37(3), 337–340.

Villalon, A., Le Royer, C., Casse, M., Cooper, D., Previtali, B., Tabone, C., Hartmann, J.-M., Perreau, P., Rivallin, P., Damlencourt, J.-F. et al. (2012). Strained tunnel FETs with record ION: First demonstration of ETSOI TFETs with SiGe channel and RSD. *2012 IEEE Symposium on VLSI Technology (VLSIT)*, Honolulu, HI, 49–50.

Visciarelli, M., Gnani, E., Gnudi, A., Reggiani, S., Baccarani, G. (2016). Impact of strain on tunneling current and threshold voltage in III–V nanowire TFETs. *IEEE Electron Device Letters*, 37(5), 560–563.

Vurgaftman, I., Meyer, J.R., Ram-Mohan, L.R. (2001). Band parameters for III–V compound semiconductors and their alloys. *Journal of Applied Physics*, 89(11), 5815–5875.

Wang, J., Polizzi, E., Ghosh, A., Datta, S., Lundstrom, M. (2005). Theoretical investigation of surface roughness scattering in silicon nanowire transistors. *Applied Physics Letters*, 87(4), 043101.

Yang, Y., Su, S., Guo, P., Wang, W., Gong, X., Wang, L., Low, K.L., Zhang, G., Xue, C., Cheng, B. et al. (2012). Towards direct band-to-band tunneling in P-channel tunneling field effect transistor (TFET): Technology enablement by Germanium-tin (GeSn). *2012 IEEE International Electron Devices Meeting (IEDM)*, San Francisco, CA, 16.3.1-16.3.4.

Zhou, G., Li, R., Vasen, T., Qi, M., Chae, S., Lu, Y., Zhang, Q., Zhu, H., Kuo, J.-M., Kosel, T., Wistey, M., Fay, P., Seabaugh, A., Huili, X. (2012). Novel gate-recessed vertical InAs/GaSb TFETs with record high ION of 180 μA/μm at VDS = 0.5 V. *2012 IEEE International Electron Devices Meeting (IEDM)*, San Francisco, CA, 32.6.1-32.6.4.

2

Field-Effect Transistors Based on 2D Materials: A Modeling Perspective

Mathieu LUISIER, Cedric KLINKERT, Sara FIORE, Jonathan BACKMAN, Youseung LEE, Christian STIEGER and Áron SZABÓ

Integrated Systems Laboratory, ETH Zurich, Switzerland

2.1. Introduction

2.1.1. *Future of Moore's law*

In 1965, Gordon Moore, one of Intel's co-founders, formulated in Moore (2006) the now famous law that states that the number of transistors per integrated circuit (IC) doubles every 18–24 months. This miracle has been made possible over the last 50+ years thanks to an aggressive scaling of the dimensions of silicon-based metal-oxide-semiconductor field-effect transistors (MOSFETs), as reviewed in Khan et al. (2018). Until 2003, this miniaturization followed Dennard's scaling law (Dennard et al. 1974), which consisted of reducing the spatial dimensions (width, gate length, oxide thickness) and power supply of every new MOSFET generation by 30%. In other words, these quantities were multiplied by a factor of 0.7x from one generation to the other. As a consequence, the power density of ICs stayed

---

For a color version of all figures in this chapter, see www.iste.co.uk/cresti/beyondcmos.zip.

Beyond-CMOS, coordinated by Alessandro CRESTI.

constant over the years, while their performance continued to increase, driven by the shortening of the transistor gate length. Dennard's scaling stopped, however, in 2003 at the so-called 130 nm technology node (TN), because the supply voltage of transistors could not be reduced at the same pace as their dimensions anymore. The subthreshold slope (*SS*), which indicates how rapidly the electrical current of logic switches can be increased between their OFF and ON states, explains this phenomenon. In conventional MOSFETs, it is limited to 60 mV/dec at room temperature: the gate voltage must be swept by at least 60 mV to vary the current by one order of magnitude.

The impossibility to push *SS* below this limit in MOSFETs forced the semiconductor industry to maintain relatively large supply voltages (above 1 V), thus leading to significant increases of the power and heat dissipation of electronic devices; see Pop et al. (2006). At the circuit level, the end of Dennard's law could be partly compensated by decreasing the clock frequency of ICs to reduce the power dissipation and by combining multiple cores together with shared memory to augment the computational capabilities. This was the beginning of the "multicore crisis", an on-going era with energy-efficient, parallel, but sequentially slower multicore computers compared to those at the beginning of the 2000s. At the device level, from 2003 onwards, it was observed that "simply" scaling the size of transistors was no more sufficient in enhancing their operation, in particular their switching speed. Since then, different technology boosters have therefore been introduced to ensure that the performance improvements historically brought by Moore's scaling law could continue, as summarized in Kuhn (2012):

– strain engineering as in Thompson et al. (2002): from the 90 nm TN, strain has been used in Si MOSFETs to alter the band structure of electrons and holes, with an increase of their channel mobility as a result;

– high-κ oxide layers as in Mistry et al. (2007): at the 45 nm TN, SiO_2, the native oxide of Si was replaced by high-κ dielectric layers such as HfO_2 that can provide larger gate capacitances together with lower leakage currents;

– 3D FinFETs as in Hisamoto et al. (1998): up until the 32 nm TN, transistors were 2D planar structures with a single-gate contact. In 2011, they became 3D FinFETs with a triple-gate configuration, offering a higher immunity against short-channel effects, in particular source-to-drain tunneling.

The benefits of multicore architectures, strain, high-κ dielectrics and FinFETs are best visible in the electronic products that we use on a daily basis, whether that be cell phones, tablets or laptops. Quantitatively, these benefits can be assessed by considering the Top500 list of supercomputers (Top500 2020). This list ranks all participating machines based on their performance when running the LINPACK benchmark; see Dongarra et al. (2003). What is measured is the number of floating point operations (Flop) that are processed per second (Flop/s) to solve a linear system of equations on the hardware of interest. Back in 1993 (the first Top500 list), the largest supercomputer in the world achieved 60x10^9 Flop/s, i.e. 60 GFlop/s. Today's cell phones, with a peak performance larger than 10^{12} Flop/s (1 TFlop/s), are about 15–20 times more powerful than this machine. For its part, the current largest supercomputer in the world, as of November 2020, reaches 440x10^{15} Flop/s (440 PFlop/s), according to the Top500 (Top500 2020).

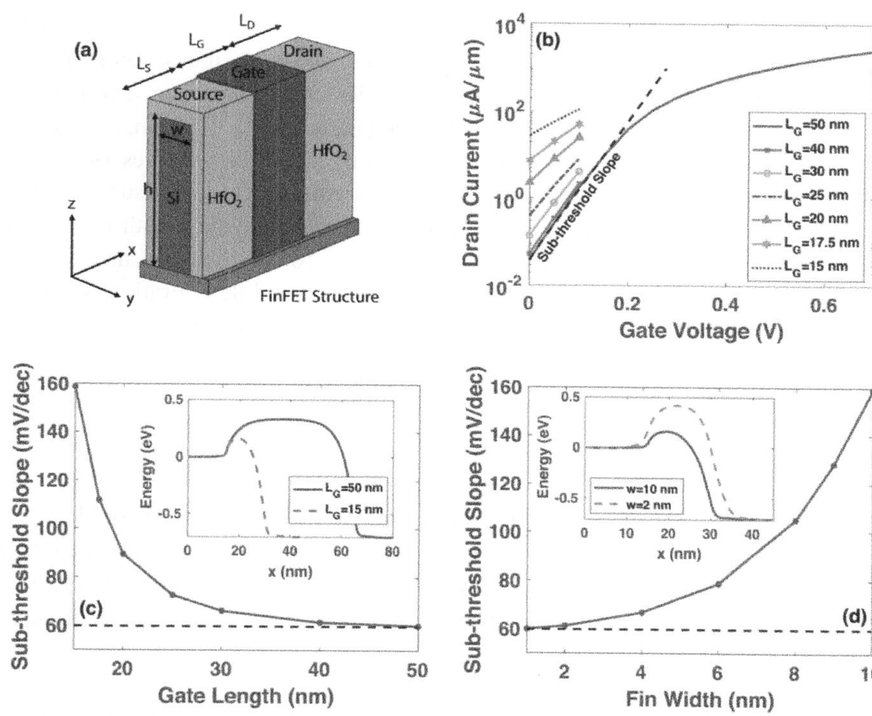

Figure 2.1. *Influence of the gate length and width scaling of 3D FinFETs on their "current vs. voltage" characteristics and sub-threshold slope*

NOTES ON FIGURE 2.1.– *(a) Schematic view of an n-type 3D FinFET with a fin height h=40 nm and width w comprised between 1 and 10 nm. The* $L_S = 15$ *nm,* L_G*, and* $L_D = 15$ *nm quantities refer to the length of the source extension, gate region and drain extension, respectively. The source and drain are doped with a donor concentration* $N_D = 10^{20}$ cm^{-3}*. The Si channel (red) is aligned with the <110> crystal axis and is surrounded by* HfO_2 *layers of thickness* $t_{ox} = 4$ *nm. (b) Transfer* *characteristics I_D-V_{GS} at V_{DS} = 0.7 V of the FinFET in (a) with w = 10 nm and L_G = 50 nm (blue line). The current was normalized by 2xh+w. The currents at low V_{GS} are also given for FinFETs with 15 ≤ L_G ≤ 40 nm. The subthreshold slope SS is indicated by the dashed black line. (c) Subthreshold slope of the FinFET in (a) as a function of L_G for w = 10 nm. Inset: electrostatic potential energy at L_G = 15 and 50 nm. (d) Same as (c), but as a function of w for L_G = 15 nm. Inset: electrostatic potential energy at w = 10 and 2 nm.*

While Si FinFETs have established themselves as the transistors of reference since 2011, they might lose this position in the future, when their gate length shrinks below 15 nm. One of the main challenges associated with the scaling of such devices is illustrated in Figure 2.1 using a simplified device structure with realistic dimensions, as schematized in sub-plot (a). At a gate length L_G = 50 nm, a fin width of 10 nm together with a fin height of 40 nm yields a subthreshold slope *SS* = 60.8 mV/dec at room temperature, close to the theoretical limit (sub-plot (b)). If the cross-sectional dimensions are kept the same, but the gate length is reduced from 50 to 15 nm, *SS* rapidly explodes, reaching 160 mv/dec for the shortest considered L_G, as can be seen in sub-plot (c). The observed deterioration is caused by a loss of electrostatic control, which can be related to a parameter called "natural channel length" λ_N and defined in Appenzeller et al. (2008) as:

$$\lambda_N = \sqrt{\frac{\epsilon_{Si}}{N\,\epsilon_{ox}} T_{Si} T_{ox}}, \qquad [2.1]$$

where ϵ_{Si}, ϵ_{ox}, N, T_{Si} and T_{ox} are the relative permittivity of Si, that of the oxide layer, the number of gate contacts, the thickness of the Si channel, and that of the oxide surrounding it, respectively. Short-channel effects can be prevented if the gate length L_G is at least 6x larger than λ_N. Hence, λ_N should be made as small as possible to conveniently scale transistors.

In FinFETs, the number of gate contacts N is equal to 3. If Si remains the channel material and HfO_2 the dielectric of choice, only T_{Si} and T_{ox} can be made thinner. Because decreasing T_{ox} would ultimately lead to high OFF-state gate leakage currents, the only viable solution to scale the L_G of FinFETs appears to be a reduction of T_{Si}, which corresponds to the fin width w. This is what has been done in Figure 2.1(d). When pushing w down to 1 nm, an *SS* value of ≈ 60 mV/dec can be achieved, which would readily allow pushing the gate length of FinFETs (well) below 15 nm while maintaining a good electrostatics integrity. However, reliably fabricating fin structures with a width of 1 nm only is extremely demanding. Surface roughness, single impurities, or interface traps are all expected to play a non-negligible role at this scale and to negatively impact the Si channel mobility;

see Uchida et al. (2007). Besides these effects, device-to-device variability might become a real issue for ultra-scaled FinFET sizes.

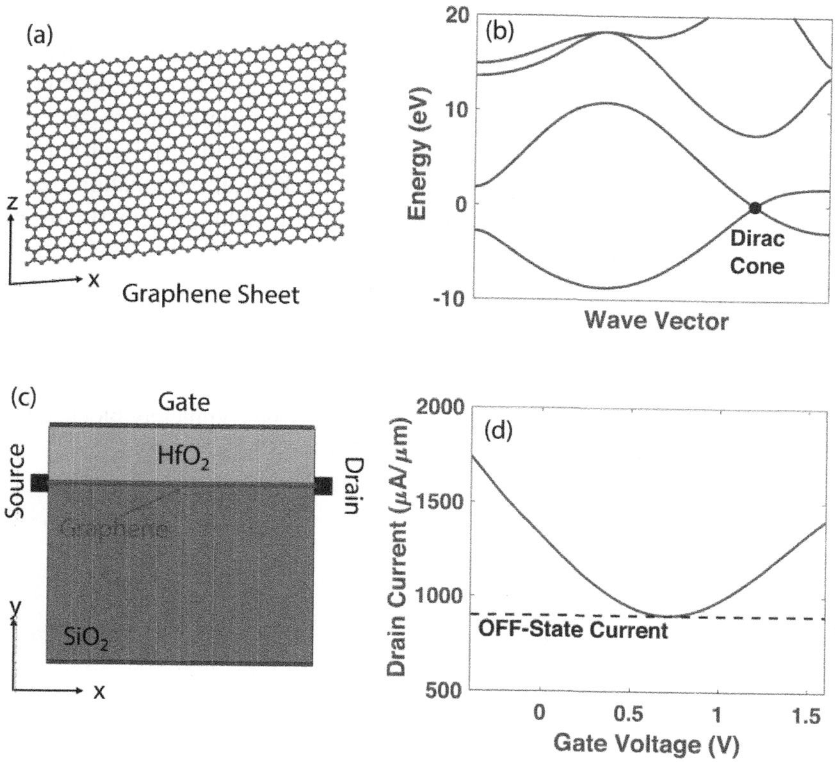

Figure 2.2. *Simulation of the bandstructure and transistor characteristics of graphene sheets*

NOTES ON FIGURE 2.2.– *(a) 2D graphene flake. The red dots represent atoms, and the black lines represent atomic bonds. (b) Band structure of graphene along the M–Γ–K–M line of the Brillouin zone. The location of the Dirac cone is indicated by a black dot. (c) Schematic view of a single-gate graphene field-effect transistor (GFET) with a gate length* $L_G = 15$ *nm. Schottky contacts are assumed for the source and drain with a barrier height of 0.25 eV. The red dots refer to carbon atoms. The channel is separated from the gate contact by a* HfO_2 *layer of thickness* $t_{ox} = 3$ *nm and is deposited on a* SiO_2 *substrate. (d) Transfer characteristics* I_D-V_{GS} *of the GFET in (c) at* $V_{DS} = 0.2$ *V. The OFF-state current cannot be brought below 900 μA/μm, a value much too high.*

2.1.2. *The potential of 2D materials*

Instead of thinning the width of semiconductors that normally have a 3D unit cell, 2D materials with a naturally flat atomic structure might be more promising to create tomorrow's transistors with a gate length of 15 nm and below (Fiori et al. 2014). Such compounds are characterized by an almost perfect electrostatic control, no surface roughness and no dangling bonds. Graphene, a carbon monolayer with a honeycomb lattice, is an excellent example of a 2D material. Its existence was confirmed experimentally in 2004 through mechanical exfoliation, as the dependence of its conductance on an external electric field; see Novoselov et al. (2004). This finding motivated the fabrication of graphene field-effect transistors (GFETs), as in Lemme et al. (2007), but the absence of a band gap in graphene does not enable us to fully switch these devices off, as indicated in Figure 2.2. A dispersionless Dirac cone where the conduction and valence bands touch each other can be noted in the band structure plot, at the *K*-point of the Brillouin zone. Because of this, the current flow from source to drain can never be completely blocked by the gate-modulated potential barrier, as in Si or III–V MOSFETs.

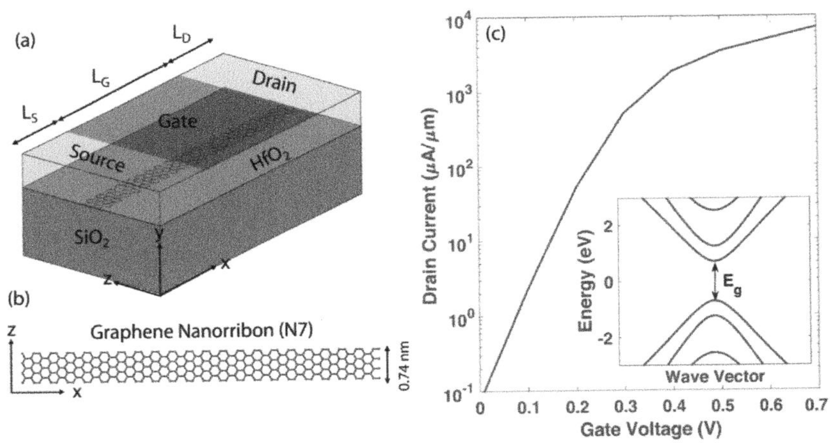

Figure 2.3. *Simulation of the bandstructure and transistor characteristics of a graphene nanoribbon*

NOTES ON FIGURE 2.3.– *(a) Schematic view of a graphene nanoribbon field-effect transistor (GNRFET). It is identical to the structure in Figure 2.2, except that the 2D graphene flake was replaced by a N7 nanoribbon of width w=0.74 nm and that the source and drain extensions of length* $L_S=L_D=25$ *nm are doped with a donor concentration* $N_D \approx 10^{13}$ cm^{-2}*. (b) Atomic structure of the N7 graphene nanoribbon in (a). (c) Transfer characteristics* I_D*-*V_{GS} *at* $V_{DS} = 0.7$ *V of the GNRFET in (a). The*

OFF-state current is fixed to I_{OFF} = 0.1 µA/µm (normalized by the nanoribbon width w=0.74 nm). The inset shows the band structure of the N7 nanoribbon, whose band gap E_g = 1.408 eV.

A band gap can be opened up in graphene if it is patterned into a quasi-1D nanoribbon, either lithographically as in Han et al. (2007) or chemically as in Cai et al. (2010). While the former approach is more straightforward to realize, it tends to produce graphene nanoribbons (GNRs) wider than 20 nm, whose transport properties suffer from detrimental line edge roughness. The width of chemically derived GNRs can be reduced to below 1 nm with an excellent control of their edges and reproducible electrical characteristics. Transistors made of such 1D nanostructures can exhibit ON/OFF current ratios greater than 10^6 as well as ON-state currents of the order of 2000 µA/µm (Wang et al. 2008). The simulation of an N7 GNR field-effect transistor (GNRFET) with a width w = 0.74 nm is shown in Figure 2.3 as an example. The investigated logic switch has a gate length L_G = 15 nm. It provides a steep subthreshold slope SS = 67.5 mV/dec, a high ON-current I_{ON} = 6.87 mA/µm at an OFF-current I_{ON} = 74 nA/µm and a supply voltage V_{DD} = 0.7 V. To deliver such a performance, the source and drain extensions of the GNRFET were doped with a donor concentration of $N_D \approx 10^{13}$ cm^{-2}, which might not be attainable experimentally. Furthermore, it is rather difficult to obtain low contact resistances in GNRFETs and the mass production of ultra-narrow structures is very tedious. For this we need materials that present themselves in the form of relatively easily manufacturable large-scale flakes, such as graphene, but also display a band gap compatible with logic applications.

Monolayers of transition metal dichalcogenides (TMD) of MX_2 composition, where M is a transition metal and X a chalcogen, fulfill these conditions with their thickness below 1 nm, band gap between 1 and 2 eV, high carrier mobility and availability as large flakes (Akinwande et al. 2019). They are therefore often seen as serious contenders to continue Moore's scaling law in the "more-than-Moore" category (Schwierz et al. 2015). In TMDs, each layer is composed of M atoms surrounded by two X atoms. The inter-atomic bonds within each layer are covalent, whereas van der Waals forces maintain adjacent layers together in few-layer structures. TMDs can adopt different crystal lattices (and symmetry groups): from the usually semiconducting 2H (hexagonal) phase to the typically metallic 1T (trigonal) or 1T' (modified trigonal) phase, going through other phases such as 3R (rhombohedral) or 2M (monoclinic). One lattice is generally more stable than the others, but phase transition can be triggered by external fields, strain or doping; see Duerloo et al. (2014).

Among all existing TMD materials and configurations, a single-layer MoS_2 was exfoliated for the first time in 1986 with scotch tape (Joensen et al. 1986), but it was only after the first experimental demonstration in 2011 of a properly working 2H single-gate monolayer MoS_2 transistor with $SS = 74$ mV/dec and an ON/OFF current ratio larger than 10^8 in Radisavljevic et al. (2011) that TMD started to receive a wide attention from the scientific community. Since then, transistors with a WSe_2 (Fang et al. 2012), WS_2 (Ovchinnikov et al. 2014), $MoTe_2$ (Fathipour et al. 2014), $MoSe_2$ (Meng et al. 2017), ReS_2 (Liu et al. 2015), $HfSe_2$ or $ZrSe_2$ (Mleczko et al. 2017) channel have also been reported, to cite a few examples. TMD monolayers were used in most cases, except for the $MoTe_2$ (down to 6L), $HfSe_2$ and $ZrSe_2$ (down to 3L) ones, which relied on few-layer structures. Among these 2D semiconductors, some are better suited to obtain *n*-type transistors, for example, MoS_2, ReS_2, $HfSe_2$ or $ZrSe_2$; others lend themselves more naturally to *p*-type devices (WSe_2 and $MoTe_2$), whereas $MoSe_2$ is rather ambipolar once crystal defects have been repaired. The type of each TMD is primarily determined by two effects: the metallic contacts attached to it, which affect the Schottky barrier height at the metal–semiconductor interfaces, and the dielectric environment around it, which can transfer electrons or holes to the channel.

The complementary logic at the core of electronic circuits requires both *n*- and *p*-type transistors. Hence, the polarity of 2D TMDs should be modifiable, which can be done through doping. For instance, a potassium-doped WSe_2 channel becomes *n*-type. It can then be combined with a *p*-type WSe_2 transistor to give rise to a fully 2D inverter as in Tosun et al. (2014). Doping TMDs and more generally 2D materials remains, however, a difficult task. Several options exist for that. A back-gate can be inserted to modulate the band edges of the source and drain extensions with respect to the contact Fermi levels (electrostatically induced doping). Alternatively, ion implantation (Suh et al. 2018), ion intercalation (Kappera et al. 2014) or charge transfer from, for example, ionic liquid (Allain and Kis 2014) can be used. Nevertheless, none of this approach currently allows reaching doping concentrations as high as in Si.

Apart from doping, TMDs face several other challenges that prevent device engineers from accessing their intrinsic performance and compare it to that of Si FinFETs. One of them is the obtention of high-quality, large-scale monolayers. Already in the 1960s, mechanical exfoliation via scotch tape was applied to isolate layered materials (Frindt and Yoffe 1963). Only small flakes can be gained with this rather straightforward technique, thus hindering the mass production of TMD transistors. Lately, impressive progresses have been made with the help of the chemical vapor deposition (CVD) (Najmaei et al. 2013) and metal-organic CVD (Kang et al. 2015), which has been shown to produce large monolayer areas with high carrier mobility. High-quality TMD flakes in the μm^2 range can also be generated with atomic layer deposition (ALD) (Balasubramanyam et al. 2020).

Other challenges related to the fabrication of high-performance, TMD-based transistors will be discussed in section 2.4. As addressing most of them will necessitate significant technological improvements, device simulation can be used in the meantime to provide invaluable insights into the physics of these logic switches, predict their performance limit under ideal conditions and support the on-going experimental activity. This chapter will first introduce a suitable modeling approach for that in section 3.2, before presenting results of recently undertaken theoretical investigations in section 2.3. The latter are not restricted to TMDs, but will also explore novel 2D semiconductors that could become the channel material of future, ultra-scaled *n*- and *p*-type transistors.

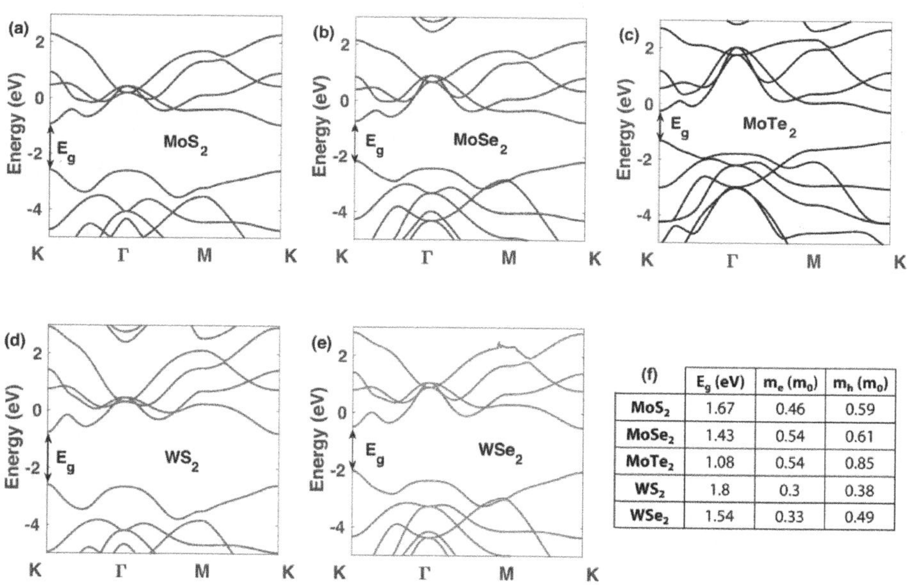

(f)	E_g (eV)	m_e (m_0)	m_h (m_0)
MoS_2	1.67	0.46	0.59
$MoSe_2$	1.43	0.54	0.61
$MoTe_2$	1.08	0.54	0.85
WS_2	1.8	0.3	0.38
WSe_2	1.54	0.33	0.49

Figure 2.4. *Bandstructure and band parameters of selected transition metal dichalcogenides*

NOTES ON FIGURE 2.4.– *(a) Band structure of MoS_2 as computed with density-functional theory (DFT). (b) Same as (a), but for $MoSe_2$. (c) Same as (a), but for $MoTe_2$. (d) Same as (a), but for WS_2. (e) Same as (a), but for WSe_2. (f) Table summarizing the electron and hole effective masses of the conduction and valence band extrema as well as the band gap of the 2D materials in (a)–(e).*

2.2. Modeling approach

2.2.1. *Requirements and state of the art*

A critical ingredient of device simulation, regardless of the chosen approach, is the band structure of the different materials that constitute the domain of interest. The band structure enters either directly or indirectly into the physical model, via the Hamiltonian matrix H or the effective mass m^* of the different components, respectively. These quantities are connected through Schrödinger's equation:

$$H(k)\psi_k(r) = E(k)\psi_k(r), \qquad [2.2]$$

$$\frac{1}{m^*} = \frac{1}{\hbar^2}\frac{d^2E(k)}{dk^2}\Big|_{dE/dk=0}, \qquad [2.3]$$

where k refers to the electron wave vector, $\psi_k(r)$ refers to the wave function of the system at position r and wave vector k, $E(k)$ refers to the corresponding k-dependent band dispersion, and ħ refers to Planck's reduced constant. The effective masses are extracted at band extrema that are characterized by the condition $dE/dk = 0$. In Figures 2.1–2.3, the Hamiltonian matrices of the investigated FinFET, GFET and GNRFET were constructed in the effective mass approximation (EMA) for Si and in the single-p_z orbital scheme of Wallace (1947) for graphene. Both models are computationally very attractive, pretty accurate for the systems mentioned above, but unfortunately not ideal for most 2D materials, as will be explained in the following paragraphs.

The band structure of selected TMDs is shown in Figure 2.4, together with their band gap and effective masses. All these quantities were calculated with density-functional theory (DFT), as proposed by Kohn and Sham (1965), an ab initio (from first-principles) method that does not require any input parameters, except for the initial atomic unit cell (AUC) of the material under consideration. This provided AUC is first relaxed so that all atoms occupy stable positions. The corresponding electron density is then self-consistently computed with Poisson's equation. Finally, all electronic bands are extracted from the obtained DFT Hamiltonian H_{DFT}. Although very accurate, DFT still relies on several approximations, among them the exchange–correlation functional, here the PBE one of Perdew et al. (1996).

What can be clearly seen in Figure 2.4 is that the band structures of TMDs exhibit complex features such as multiple valleys separated by a small energy interval, strongly non-parabolic bands and in some cases band anisotropy (the effective masses depend on the crystal orientation). To capture all these effects, a quantum mechanical simulation approach is absolutely necessary. In other words, a

Hamiltonian matrix as in equation [2.2] must be assembled to describe the electronic properties of the desired device. On their side, neither classical nor semi-classical methods can shed light on the physics of ultra-thin 2D materials where electrons and holes are confined over narrow dimensions of below 0.5 nm. Despite these shortcomings, drift-diffusion calculations can reproduce experimental data fairly well for large-scale TMD flakes, if the available material parameters are properly calibrated (Knobloch et al. 2017).

The first quantum mechanical study of a monolayer MoS_2 transistor was reported in Yoon et al. (2011): it was shown that this material could outperform Si as logic switch with a sub-15 nm gate length. To come to this conclusion, a quantum transport simulator based on the EMA and the non-equilibrium Green's function (NEGF) formalism proposed by Datta (1995) was employed. The functionality of this simulator is similar to the one that produced the data in Figure 2.1. We would like to emphasize that NEGF is one of the most popular and powerful techniques to examine charge transport in nanoscale devices. The basic NEGF equations will be reviewed in section 2.2.3. At about the same time, a simplified solver based on the straightforward top-of-the-barrier model confirmed these results and revealed that other TMDs might be competitive as well (Liu et al. 2011). However, because EMA assumes parabolic bands, the satellite conduction band valleys of TMDs could not be properly accounted for in these works. Furthermore, the atomic granularity of single-layer crystals was totally ignored, EMA being continuous.

Device simulators implementing a semi-empirical full-band model represent the next level of accuracy and a significant improvement over the EMA as they include more than a single parabolic band. Hence, the k·p as in Kormányos et al. (2015) or tight-binding (TB) method as in Zahid et al. (2013) or Ilatikhameneh et al. (2015) can be used to construct the Hamiltonian matrix of TMDs. As each atom or discretization point is described by a set of N_{orb} orbitals, the simulation time is in the order of N_{orb}^3 times longer than with EMA. Such an increase is still acceptable from a computational point of view because the number of neighbors per point, which determines the bandwidth of the Hamiltonian matrix and the size of the numerical blocks to manipulate usually remains small. On the negative side, both k·p and tight-binding models must be first parameterized, a sometimes tedious operation that is not always successful. Moreover, getting physically meaningful parameters to connect two different materials, for example, a TMD and the metallic contact attached to it, can be rather complicated. Finally, k·p in its usual forms (4×4 to 8×8) tends to be restricted to one symmetry point in the Brillouin zone and does not capture the atomic nature of TMDs.

Going up the ladder, a DFT solver relying on a localized basis set can be coupled to a quantum transport simulator based on the NEGF formalism (Leong et al. 2015). DFT+NEGF ticks all accuracy boxes (atomistic, full-band, ab initio, capable of treating interfaces and defects, etc.), but the computational burden of such an approach explodes for large structures composed of several hundreds of atoms. On the one hand, more orbitals per atom are needed. On the other hand, the inter-atomic interactions extend over much longer distances than in tight-binding or k·p. In summary, neither H_{EMA}, $H_{k\cdot p}$, H_{TB}, nor H_{DFT} Hamiltonian matrices are optimal to simulate the electrical behavior of transistors with a 2D channel material.

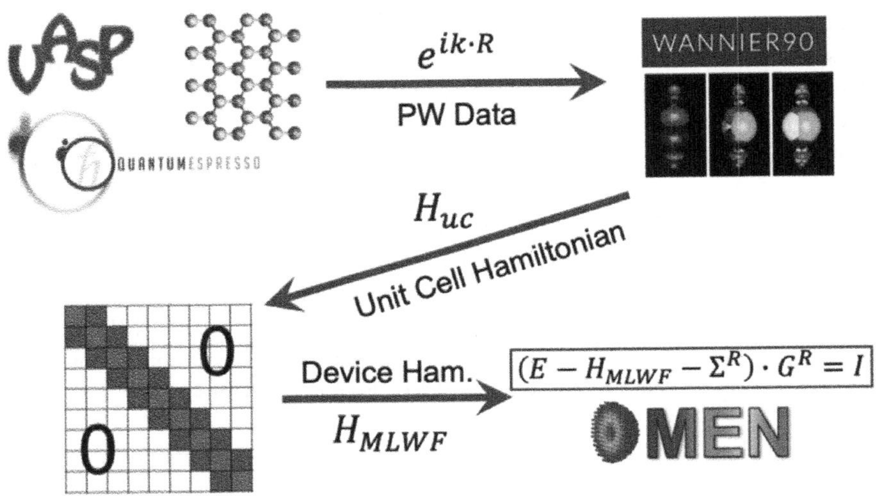

Figure 2.5. *Ab initio scheme to simulate transistors with a 2D channel material*

NOTES ON FIGURE 2.5.– *A DFT calculation of a representative 2D unit cell is first performed with a plane-wave (PW) code such as VASP (Kresse and Furthmüller 1996) or Quantum ESPRESSO (Giannozzi et al. 2009). The results are then converted to a maximally localized Wannier function basis with the wannier90 tool (Mostofi et al. 2008), which applies a unitary transformation. Next, the produced unit cell Hamiltonian,* H_{uc}*, is scaled up to the device dimensions, as in Stieger and Luisier (2020). Finally, the obtained* H_{MLWF} *is passed to a quantum transport simulator, for example, OMEN (Brück et al. 2017).*

2.2.2. *Maximally localized Wannier functions (MLWFs)*

A model that has the same accuracy as DFT and a computational complexity comparable to tight-binding or k·p would be ideal to probe 2D materials as next-generation logic switches. Maximally localized Wannier functions (MLWFs), as introduced by Marzari and Vanderbilt (1997), satisfy both conditions, as was first demonstrated in Chang et al. (2014) for various TMDs. Since this early work, MLWFs have become very popular among the 2D research community, as in Bruzzone et al. (2014), Szabó et al. (2015a), Pizzi et al. (2016), Wang et al. (2016), Lu et al. (2017), AlMutairi et al. (2018), Dong and Littlewood (2019), and Afzalian (2021), to mention a few relevant examples. The MLWF method is as accurate as DFT plane-wave (PW) calculations from which they are derived; they require a small number of basic elements per atom, and the distance over which they decay is very short. The numerical treatment of the resulting Hamiltonian matrices, H_{MLWF}, is therefore facilitated. Thanks to these unique features, MLWFs allow us to simulate large transistor structures made of thousands of atoms within reasonable computational times. They can be seen as a first step towards fully ab initio device investigations.

The principle of MLWF-based quantum transport simulations is summarized in Figure 2.5. The whole process starts with a plane-wave DFT calculation of a primitive unit cell that best represents the geometry of the targeted atomic system. Different tools such as VASP (Kresse and Furthmüller 1996) or Quantum ESPRESSO (Giannozzi et al. 2009) can be used for this purpose. The produced eigenenergies and eigenvectors are then transformed into a set of MLWFs with the wannier90 package (Mostofi et al. 2008). The required unitary transformation is exact so that the band structures obtained in the PW and MLWF basis sets are theoretically identical, the only difference being that MLWFs only return a sub-set of all PW bands, in the present case those required to evaluate transport properties. Practically, small discrepancies might emerge due to the truncation of long-ranging interactions. They have a limited impact on the results. As illustrations, the band structures of MoS_2, $MoTe_2$ and WSe_2 are plotted in Figure 2.6, as computed with DFT and after a transformation into MLWFs. Excellent agreement between both data sets can be observed.

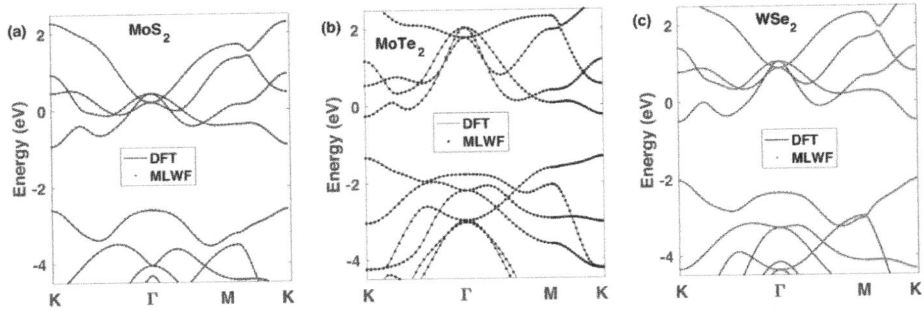

Figure 2.6. *(a) Band structure of MoS_2 as computed with DFT using the VASP tool (solid black lines) and after a transformation of the plane-wave results into a set of maximally localized Wannier functions (MLWFs, blue dots). (b) Same as (a), but for a $MoTe_2$ monolayer. (c) Same as (a) and (b), but for WSe_2*

This PW-to-MLWF conversion gives rise to small Hamiltonian blocks that describe the coupling of the chosen unit cell with itself and with its neighboring cells. Those blocks must be upscaled to form a block tri-diagonal Hamiltonian matrix corresponding to the device to be simulated. Such an upscaling scheme is described in Stieger and Luisier (2020). The obtained H_{MLWF} is perfectly suitable for quantum transport simulations, its increased bandwidth being partly compensated by the fact that less orbitals per atom are required. As a consequence, the same numerical algorithms as with tight-binding can still be employed to solve the NEGF equations. A quantum transport solver such as OMEN (Brück et al. 2017) can do this as it has been specifically designed to handle large-scale nanostructures from the first-principles (Calderara et al. 2015). Note that the procedure outlined in Figure 2.5 works for any exchange–correlation functional, for example, the generalized gradient approximation with PBE parameterization (*GGA-PBE*) of Perdew et al. (1996), hybrid functionals (*HSE06*) of Heyd et al. (2003) or the *GW* plus Bethe–Salpeter equation (GW-BSE) of Hybertsen, Louie (1986). The initial DFT run will be longer with *HSE06* or *GW*, but the time for the transport calculation is not affected by this choice.

2.2.3. *Towards ab initio quantum transport simulations*

In the previous section, the construction of an MLWF-based Hamiltonian matrix was presented. As the last step, H_{MLWF} should be passed to a quantum transport (QT) solver to perform ab initio device simulations. Various QT methods have been proved effective, among them the solution of the Wigner transport equation (Nedjalkov et al. 1996), Pauli's master equation (Fischetti 1998), the quantum transmitting boundary method (QTBM) (Lent and Kirkner 1990) and, of course, the non-equilibrium Green's function (NEGF) formalism (Datta 1995) that has been mentioned above. As all results in section 2.3 have been obtained with NEGF, the equations governing this transport approach will now be briefly introduced, with emphasis on 2D material applications.

To describe electron transport within the NEGF framework, the following nonlinear system of equations must be solved:

$$\begin{cases} \left(E - H_{MLWF}(k_z) - \Sigma^{RB}(E,k_z) - \Sigma^{RS}(E,k_z)\right) \cdot G^R(E,k_z) = I \\ G^{\gtrless}(E,k_z) = G^R(E,k_z) \cdot \left(\Sigma^{\gtrless B}(E,k_z) + \Sigma^{\gtrless S}(E,k_z)\right) \cdot G^A(E,k_z)\,. \end{cases} \qquad [2.4]$$

In equation [2.4], the G's represent the electron Green's functions. They depend on the electron energy E and momentum k_z as well as on the MLWF Hamiltonian matrix H_{MLWF}. The k_z momentum models the flake direction that is orthogonal to the transport axis and is assumed periodic. The G's can be of four different types, retarded (R), advanced (A), lesser (<) or greater (>). The Σ's refer to the corresponding self-energies where the superscript B stands for boundary and S for scattering. With Σ^B, the coupling of the simulation domain with contact electrodes is captured. This self-energy can be computed with so-called decimation techniques (Sancho et al. 1984) or more advanced schemes, for example, through eigenvalue problems (Luisier et al. 2006) or contour integrals, as in Beyn (2012). Its scattering counterpart Σ^S can include different interaction mechanisms such as electron–phonon, impurity or interface roughness scattering; see Lake et al. (1997).

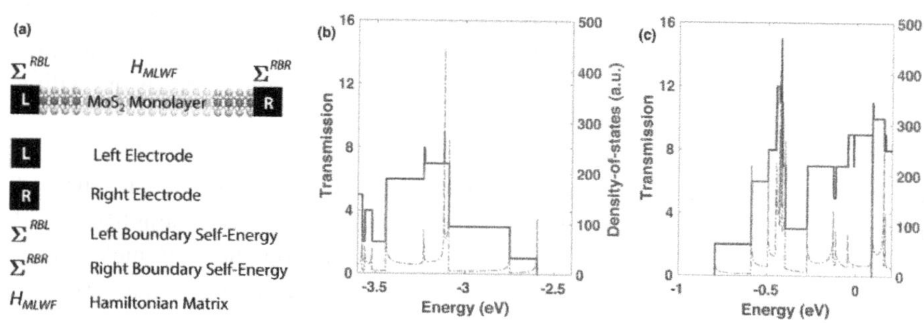

Figure 2.7. *Calculation of the transmission function through a MoS2 monolayer*

NOTES ON FIGURE 2.7.– *(a) Schematic view of a MoS_2 monolayer described by a Hamiltonian matrix H_{MLWF}. The plot shows how contacts are accounted for through boundary self-energies (Σ matrices). (b) Transmission function (blue line) and* *density-of-states (dashed red line) of the valence band of a MoS_2 monolayer under flat band conditions at $k_z = 0$. These results were obtained with the computational scheme of Figure 2.5. (c) Same as (b), but for the conduction band of a MoS_2 monolayer.*

Equation [2.4] can be efficiently solved with a recursive algorithm that constructs Green's functions from one side of the device to the other in two steps; see Svizhenko et al. (2002). In the case of ballistic transport, i.e. in the absence of interactions with other carriers, impurities, rough surfaces or crystal vibrations, only the retarded Green's function G^R needs to be calculated. With its knowledge, both the density-of-states (DOS) and transmission function (*TE*) of the considered system can be evaluated, from which the carrier density and the electronic current can be derived as in Datta (1995). The DOS and *TE* of a MoS_2 monolayer are provided in Figure 2.7 under flat band conditions, evaluated at a single momentum point, $k_z = 0$, for both the conduction and valence bands. The role of H_{MLWF} and $\Sigma^{RBL}/\Sigma^{RBR}$, the boundary self-energies, is highlighted as well. As expected under such circumstances, the transmission function displays a step-like behavior and effectively counts the number of bands available in the left and right contacts. Each time a transmission channel turns on, the DOS peaks, followed by a rapid decay that is only interrupted by the next peak.

If scattering should be accounted for, the lesser and greater Green's functions $G^{\gtrless}$ in equation [2.4] must also be computed. These quantities must be solved self-consistently with the scattering self-energies $\Sigma^{\gtrless S}$ as they depend on each other. This has been done, for example, in Szabó et al. (2015b) for single-, double- and triple-layer MoS_2 where electron–phonon scattering was treated at the ab initio level. The same approach was extended in Stieger et al. (2017) to model self-heating effects and the formation of local hot spots in various field-effect transistors with a TMD monolayer as channel material.

All results presented in this chapter have been obtained in the presence of electron–phonon scattering, but the chosen model relies on a simplified and phenomenological approach with one single phonon energy $\hbar\omega = 40$ meV and a scattering strength $D_{e\text{-}ph}$ comprised between 25 and 125 meV (Klinkert et al. 2020):

$$\Sigma^{\gtrless S}(k_z, E) = D_{e-ph}^2[n_\omega G^{\gtrless}(k_z, E + \hbar\omega) + (n_\omega + 1)G^{\gtrless}(k_z, E - \hbar\omega)]. \quad [2.5]$$

In equation [2.5], n_ω is the phonon's Bose–Einstein distribution function. A dissipative scattering mechanism is needed to avoid a negative differential resistance (NDR) behavior in the I_D-V_{DS} output characteristics of the 2D FETs, which has never been experimentally observed at room temperature. NDR originates from the band structure of 2D materials, which often exhibits several narrow bands that cannot propagate if the electrostatic potential undergoes large variations from source to drain (Gnani et al. 2008). It is an artifact of the ballistic approximation. The inclusion of electron–phonon scattering helps to get rid of this non-physical effect by connecting bands that would otherwise be independent of each other (Szabó et al. 2015b). Accounting for the "real" electron–phonon interactions would be more accurate, but gathering the required phonon band structures and coupling elements is computationally very demanding, as the self-consistent calculation of the scattering self-energies. For all these reasons, the model of equation [2.5] was adopted.

Finally, it should be noted that all simulations were performed at room temperature with the metal gate work function adjusted so that the OFF-state current is fixed to 0.1 μA/μm. Perfectly ohmic contacts are assumed (no resistance), except if mentioned otherwise.

2.3. 2D device performance analysis

2.3.1. *MoS_2 and other TMDs*

The first 2D material under the ab initio microscope is MoS_2; the TMD whose monolayer form was initially shown to provide excellent transistor characteristics is described in Radisavljevic et al. (2011). Before assessing the performance of MoS_2 transistors with respect to other 2D materials, we would like to underline the importance of the channel thickness. In Figure 2.8, a simplified device structure is schematized. Its channel is either made of a MoS_2 monolayer, bilayer or trilayer, whose thickness is approximately 0.6, 1.2 and 1.8 nm, respectively. The corresponding transfer characteristics are shown on the right side of the plot. An MLWF Hamiltonian matrix was constructed for each channel configuration. The ab initio simulations reveal that the monolayer structure has the highest ON-state current (1.15 mA/μm), followed by the bilayer (1.06 mA/μm) and finally the trilayer one (0.73 mA/μm). Normally, the opposite order would be expected as the transport effective mass m_{trans} decreases as the number of stacked layers increases. However, the gate contact loses part of its control efficiency at larger channel thicknesses, as already demonstrated in Figure 2.1.

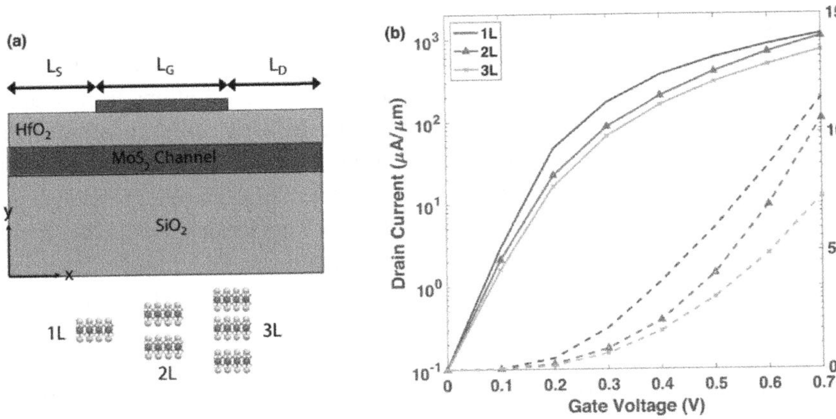

Figure 2.8. *(a) Schematic view of an n-type single-gate transistor with a single- (1L), bi- (2L) and triple- (3L) layer of MoS_2 as the channel. The gate length L_G measures 15 nm, the source (L_S) and drain (L_D) extensions 10 nm with a donor concentration $N_D = 5x10^{13}$ cm^{-2}. (b) Transfer characteristics I_D-V_{GS} at $V_{DS} = 0.7$ V of the MoS_2 transistors in (a), on a logarithmic (solid lines) and linear (dashed lines) scale*

This deterioration is best reflected in the *SS* value of each transistor, which goes from 68.6 mV/dec in the monolayer case to 75.6 mV/dec in the bilayer and 82.8 mV/dec in the trilayer. The gate contact can very well modulate the height of the potential barrier in the layer that is the closest to it, but its influence decreases as carriers are situated away from it. Hence, the benefit of smaller effective masses in few-layer structures is washed out by the poorer electrostatics of thicker channels. If we compare these results to those of Figure 2.1(d), we note a substantially larger *SS* than in the FinFET with a 1-nm wide fin, even for the monolayer FET. The presence of a triple gate in the FinFET case explains the better scalability of this device. Adding a second gate to the 2D MoS_2 FETs produces the same effect and leads to an almost perfect electrostatic control in mono-, bi- and trilayer structures down to a gate length $L_G = 10$ nm (Szabó et al. 2015b). Note that in this publication, different DFT models were used in comparison to here.

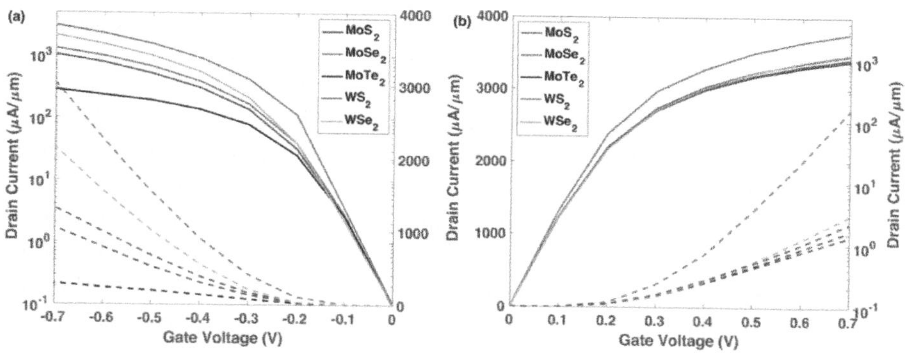

Figure 2.9. *(a) Transfer characteristics I_D-V_{GS} at V_{DS} = -0.7 V of p-type monolayer TMD field-effect transistors (FETs) with a gate length L_G = 15 nm, both on a linear (dashed lines) and logarithmic (solid lines) scale. The following TMDs are represented: MoS_2 (blue), $MoSe_2$ (red), $MoTe_2$ (black), WS_2 (magenta) and WSe_2 (cyan). (b) Same as (a), but for n-type monolayer TMD FETs at V_{DS} = 0.7 V*

MoS_2 has become the most popular TMD, as confirmed by the number of publications dedicated to it, but it is not necessarily the most promising one. Experimentally, no ON-state current larger than 700 μA/μm has ever been reported for monolayer MoS_2 (McClellan et al. 2021). This value was obtained at V_{DS} = 5 V, V_{GS} = 30 V, and in a device with a gate length L_G = 380 nm and an ON/OFF current ratio larger than 10^6. Other TMDs have therefore also been investigated. The proposed ab initio simulation approach can also be applied to them, as shown in Figure 2.6. The MLWF results perfectly reproduce the plane-wave DFT calculations for all TMDs. Taking advantage of that, the transfer characteristics I_d-V_{gs} at V_{ds} = 0.7 V of *n*- and *p*-type $MoSe_2$, $MoTe_2$, WS_2 and WSe_2 field-effect transistors were computed as well. They are displayed in Figure 2.9 and compared to those of MoS_2. All devices have a structure similar to that in Figure 2.8(a) with a gate length L_G = 15 nm and a TMD monolayer as channel material.

Firstly, it can be seen that almost all 2D FETs have a subthreshold slope of the order of 70 mV/dec, despite the relatively short L_G and the presence of a single gate contact. There is one exception, WS_2, whose *SS* is smaller (~65 mV/dec). This is not a consequence of a better electrostatic control, but of the influence of narrow energy bands. As explained above, they can lead to the presence of NDR as well as to *SS* values which are too low, even below 60 mV/dec. A physical parameter called a "pass factor" allows quantifying the importance of these bands, as explained in Klinkert et al. (2020). The electron–phonon scattering model of equation [2.5] is expected to eliminate these artefacts, but for some 2D materials, for example, WS_2, it does not fully succeed. Increasing the electron–phonon coupling can improve the situation. However, at the same time, this affects the ON-state current, preventing a fair comparison with other 2D materials.

The ON-state current values are more broadly distributed than the subthreshold slopes. Generally, it can be observed that the W-based TMDs perform better than the Mo-based ones due to lower effective masses and therefore faster carriers, as illustrated in Figure 2.4. This trend is not altered if electron–phonon scattering is included, because W atoms have a larger mass than that of Mo, their oscillation amplitude is smaller, as is their probability to interact with free carriers (Luisier et al. 2016). Another ab initio theoretical study came to a different conclusion, predicting that the mobility of WS_2 would be significantly impacted by electron–phonon scattering (Sohier et al. 2018). The small energy difference between the conduction band minimum of this material and its satellite valleys is the reason behind this discrepancy. As this energy separation strongly depends on the choice of the DFT functional, different calculations may have different outcomes.

	MoS_2	$MoSe_2$	$MoTe_2$	WS_2	WSe_2
n-I_{ON} (mA/μA)	1.15	1.04	1.11	2.71	1.25
n-SS (mV/dec)	68.6	70.1	71	65.3	69.9
p-I_{ON} (mA/μA)	1.04	1.31	0.28	3.03	2.15
p-SS (mV/dec)	70.6	70	70.4	66	75.2

Table 2.1. *Summary of the n- and p-type ON-state currents I_{ON} and of the subthreshold slopes SS extracted from the I_D-V_{GS} transfer characteristics in Figure 2.9*

The I_{ON} and SS of all simulated 2D TMD FETs are summarized in Table 2.1. Overall, under ideal conditions, only one 2D TMD offers an ON-state current significantly larger than 1 mA/μm, both in its n- and p-type configuration, WS_2: the n-type I_{ON} is 2.71 mA/μm, the p-type one 3.03 mA/μm. Due to the aforementioned issues with the subthreshold region of these FETs and the possibility that the WS_2 mobility might be lower than expected, it must be concluded that 2D TMDs are probably not the best candidates to replace Si as future, more-than-Moore transistors. Further challenges, as well as opportunities to improve their properties, will be discussed in section 2.4.

2.3.2. *Novel 2D materials*

Besides TMDs, other 2D materials suitable for logic applications have emerged over the years, starting with black phosphorus: a monolayer of phosphorus (BP) atoms with a buckled honeycomb lattice (Liu et al. 2014b). Its high hole mobility makes it particularly attractive as the *p*-type field-effect transistor, as demonstrated experimentally in Li et al. (2014). While BP has established itself as the most prominent alternative to TMDs, other novel 2D materials have earned themselves a place under the spotlight, for example, silicene in Tao et al. (2015), germanene in Derivaz et al. (2015), antimonene in Ji et al. (2016), InSe in Bandurin et al. (2017) or Bi_2O_2Se in Wu et al. (2017). Many others are also in the pipeline; see Franklin (2015) and Liu et al. (2018).

On the theoretical side, a high-throughput (HT) investigation by Mounet et al. (2018) revealed that more than 1,800 2D materials might exist, among them about 1,000 easily exfoliable monolayers. To come up with these numbers, the authors considered a large set of 3D parent crystals from the Inorganic Crystal Structure Database (n.d.) and the Crystallographic Open Database (COD, (Grazulis et al. 2011)). They then applied geometrical criteria to identify layered compounds and extract 2D children from them, tested the stability of the latter in vacuum by computing their phonon band structure, and finally classified them according to their inter-layer binding energy E_b and the influence of van der Waals forces. Those of interest have a low E_b and their layers are kept together by non-covalent, van der Waals bonds. All notable 2D materials (graphene, TMDs, BP, etc.) were recovered by the HT study of Mounet et al. (2018). They are accompanied by components with a huge variety of band gaps (from metals to oxides), effective masses (from isotropic to strongly anisotropic band structures) and monolayer thicknesses (from one to several repeatable atomic layers). A few examples are proposed in Figure 2.10.

Independent of this study, the potential of 2D materials beyond TMDs and BP as field-effect transistors has been evaluated through device simulations, with empirical and ab initio models. The related literature is abundant, and not all important contributions can be listed here. Nevertheless, the following examples are deemed representative: Bi_2O_2Se in Quhe et al. (2019), monochalcogenides in Das et al. (2019), group IV in Gaddemane et al. (2018), group V in Pizzi et al. (2016), as well as on more exotic 2D materials such as Tl_2O in Ma et al. (2017). To be able to compare the performance of transistors made of these very different 2D components, a uniform modeling approach would be preferable, for instance the one described in section 2.2. It is validated in Figure 2.11 for non-conventional 2D monolayers, Ag_2N_6, GeSe and O_6Sb_4. As for TMDs, the DFT and MLWF band structure results agree very well so that everything is in place to conduct a large-scale and systematic performance comparison of 2D FETs. These results were originally published in Klinkert et al. (2020). All approximations, simulation results and extracted material parameters are presented in its Supplementary Information.

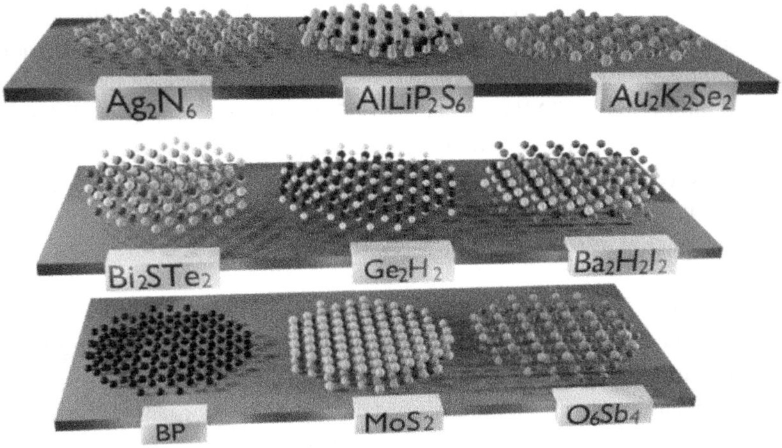

Figure 2.10. *Atomic unit cells of few selected 2D materials from Mounet et al. (2018)*

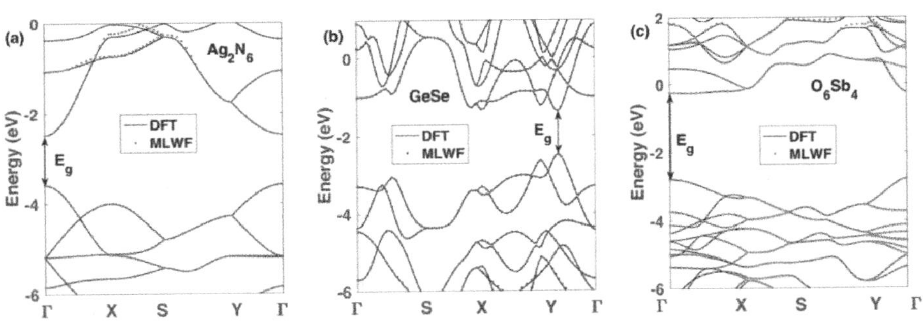

Figure 2.11. *(a) Band structure of Ag_2N_6 as computed with DFT using VASP (solid black lines) and after a transformation of the plane-wave results into MLWFs (orange dots). (b) Same as (a), but for a GeSe monolayer. (c) Same as (a) and (b), but for O_6Sb_4*

In this context, we designed a realistic 2D transistor structure, as in Figure 2.12, and defined a set of targeted figures of merits (FOM). The dimensions and specifications of this single-gate FET derive inspiration from the International Roadmap for Devices and Systems (IRDS) (2018) on their roadmap for the year 2025, i.e. a gate length L_G = 15 nm, a supply voltage V_{DD} = 0.7 V and an equivalent oxide thickness (EOT) of 0.6 nm. This EOT is achieved through a 3 nm HfO_2 dielectric layer with a relative permittivity ϵ_R = 20. The 2D materials are deposited onto a SiO_2 substrate with a thickness t_{box} = 20 nm and ϵ_R = 3.9. To ensure a satisfactory electrostatic control, the source and drain extensions of the FETs are doped with a donor/acceptor concentration $N_{D/A} = 5x10^{13}$ cm^{-2}. Such high values cannot be achieved experimentally for the moment; see Suh et al. (2014) and McClellan et al. (2021), but they could be in the future, for example, by combining different doping techniques. The SiO_2 and HfO_2 domains do not enter the NEGF equations; they are treated as perfectly insulating layers that only impact the electric field profile.

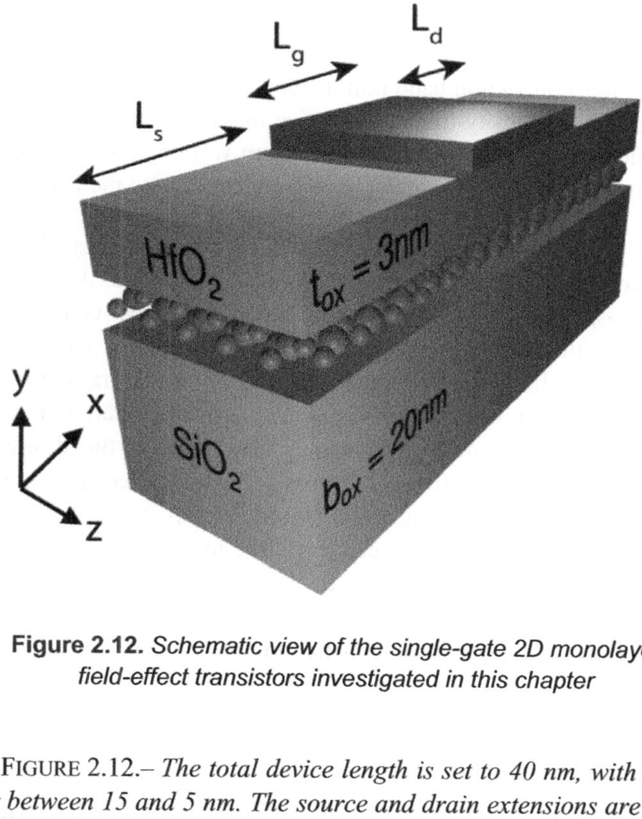

Figure 2.12. *Schematic view of the single-gate 2D monolayer field-effect transistors investigated in this chapter*

NOTES ON FIGURE 2.12.– *The total device length is set to 40 nm, with a gate length L_G varying between 15 and 5 nm. The source and drain extensions are doped with a donor N_D or acceptor N_A concentration of $5x10^{13}$ cm^{-2} for the n- and p-type configurations, respectively. A $t_{ox} = 3$ nm thick high-κ dielectric layer (HfO_2) separates the 2D channel, which is deposited on a SiO_2 box with $t_{box} = 20$ nm from the gate contact. Transport occurs along the x-axis; y is a direction of confinement, whereas z is assumed to be periodic and gives rise to the k_z-dependence in equation [2.4]. Adapted from Klinkert et al. (2020).*

In terms of FOM, we are looking for 2D materials offering an ON-state current larger than 3 mA/μm, both in their *n*- and *p*-type configurations, at an OFF-state current $I_{OFF} = 0.1$ μA/μm. At the same time, the subthreshold slope *SS* should not exceed 80 mV/dec at a gate length L_G = 15 nm. We selected 100 different 2D materials from the database of Mounet et al. (2018) that could potentially reach these objectives. As a first criterion, we singled out thin semiconductor monolayers with a small number of atoms in their primitive unit cell (PUC). No compounds thicker than 1.5 nm and with more than 30 atoms in their PUC were considered. Secondly, using the band structure calculation results of Mounet et al. (2018), we further restricted ourselves to 2D materials with a band gap larger than 1 eV and, if possible, anisotropic conduction band minima and/or valence band maxima so that a low transport and high-density-of-state effective mass are simultaneously obtained. Finally, only monolayers that are stable in vacuum were retained, i.e. those whose phonon band structure does not have negative branches around the Γ-point.

Figure 2.13 reports the *n*- and *p*-type transfer characteristics (I_D-V_{GS}) of six promising 2D materials that satisfy the conditions listed above: Ag_2N_6, As_2, GeSe, O_6Sb_4, black phosphorus (BP) and SiH (silicane). Under ideal conditions, they all deliver $I_{ON} \geq 3$ mA/μm, which is about 3x larger than MoS_2, BP even reaching ON-state currents of the order of 5 mA/μm. The *SS* of all transistors is around 70 mV/dec at $L_G = 15$ nm, which is 10 mV/dec lower than the target that was set. In total, out of the 100 examined 2D materials, 13 arrive at the desired level of performance. Their FOM and effective masses are summarized in Table 2.2. None of the conventional TMDs from Figure 2.4 (MoS_2, $MoSe_2$, $MoTe_2$, WS_2 and WSe_2) belongs to that group, although WS_2 gets very close to it (*n*-type $I_{ON} = 2.71$ mA/μm and *p*-type $I_{ON} = 3.03$ mA/μm), as can be seen in Table 2.1. Two less common TMDs, HfS_2 and ZrS_2, seem to have a higher potential, their *n*- and *p*-type ON-state currents being above 3 mA/μm in the ballistic limit of transport.

To give an overview of all 2D materials that have been simulated, we put together their "*n*-type I_{ON} versus *p*-type I_{ON}" characteristics in Figure 2.14(a). This plot allows us to rapidly identify the 13 best-performing candidates as they are situated in the upper right corner delimited by black dashed lines. Overall, several 2D materials (39) exhibit an ON-state current larger than 3 mA/μm in their *n*-type form, for example, P_8Si_4 (5.43 mA/μm), As_8Si_4 (5.23 mA/μm) or Tl_2O (5.09 mA/μm), much less (17) as *p*-FET, for example, $C_2N_4Pb_2$ (4.08 mA/μm) or $I_2Nb_2O_4$ (3.41 mA/μm). As for the Si-based CMOS technology, the fact that more 2D compounds have a high *n*-type rather than *p*-type ON-state current indicates that fabricating high-performance pFETs might be a challenging task in the flat land as well.

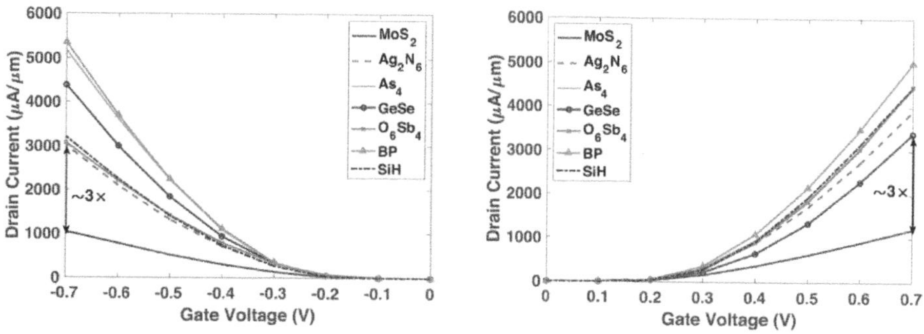

Figure 2.13. *Transfer characteristics of transistors based on selected novel 2D materials and comparison to MoS2*

NOTES ON FIGURE 2.13.– *(left) Transfer characteristics I_D-V_{GS} at V_{DS} = -0.7 V of p-type monolayer FETs with a gate length L_G = 15 nm. They are made of promising 2D materials (Ag_2N_6, As_2, GeSe, O_6Sb_4, black phosphorus, SiH) and compared to a MoS_2 device. (Right) Same as on the left, but for n-type monolayer FETs at V_{DS} = 0.7 V. For both transistor configurations, the ON-state current of the promising 2D materials increases by a factor of ~3x (at the same I_{OFF} = 0.1 μA/μm), as compared to MoS_2.*

Among all 2D materials that were simulated, black phosphorus stands out as it displays the largest "*n*-type I_{ON} versus *p*-type I_{ON}" combination, when the transport direction of the FET is aligned with the Γ–*X* crystal axis of BP. The conduction and valence band anisotropy of this monolayer (see Figure 2.15) is at the origin of the high current densities. Its transport effective mass, m_{trans}, is equal to 0.16 m_0 for electrons (0.14 for holes), whereas its density-of-states counterpart, m_{DOS}, amounts to 0.42 m_0 (0.82 for holes). Other 2D materials benefit from anisotropic band structures, which is the reason why they deliver I_{ON} larger than 3 mA/μm, as it can be generally seen in Figure 2.14(b) and (c). This is the case of the electrons and holes in Ag_2N_6, As_8Ge_4 or As_8Si_4, and of the holes in $I_4O_4Sc_4$, for example. Their band extrema have an ellipsoid shape. In fact, almost all 13 best components are characterized by an m_{trans} lower than their m_{DOS}, i.e. they are situated above the dashed black lines that correspond to materials with an isotropic band structure in Figure 2.14(b) and (c). It should be noted that m_{trans} and m_{DOS} were not directly extracted from the band structure of the 2D materials, but by calculating their charge and current densities with analytical equations, as in Klinkert et al. (2020).

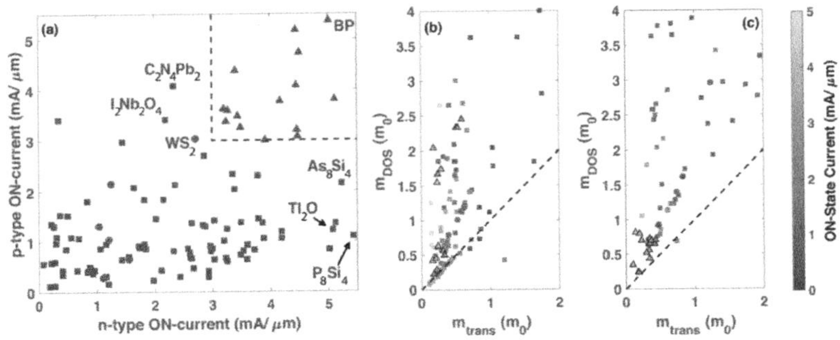

Figure 2.14. *Key transistor metrics of the 100 selected 2D materials*

NOTES ON FIGURE 2.14.– *(a) "n-type ON-state current versus p-type ON-state current" for the 100 selected 2D materials. The red circles refer to the usual TMDs, while the green triangles encompass the best-performing compounds, i.e. those with an n- and p-type I_{ON} larger than 3 mA/μm. (b) ON-current of the 100 simulated n-type 2D FETs in (a) as a function of their transport m_{trans} and density-of-states m_{DOS} effective masses. The same marker shapes as in (a) are used. The current magnitude is encoded in the marker color (see the color bar on the right). The black dashed line indicates 2D materials with an isotropic band structure ($m_{trans} = m_{DOS}$). (c) Same as (b), but for p-type 2D FETs. Adapted from Klinkert et al. (2020).*

The question that arises with materials having an anisotropic conduction band minimum and/or valence maximum is "what happens if the direction along which the electrical current flows is not perfectly aligned with the most suitable crystal axis?". For example, it is well known in BP that transport along the Γ–Y axis is (much) less efficient than along Γ–X (Xia et al. 2014). Using the proposed MLWF+NEGF approach, we found in Klinkert et al. (2021) that orientation misalignments up to 50° from the ideal case do not significantly alter the ON-state current of BP transistors, with almost negligible performance loss up to a misalignment angle of 20°. This means that I_{ON} does not linearly decrease as a function of the misalignment angle, but rather first stays constant up to 20°, slightly decreases up to 50°, and finally rapidly drops. This behavior, which occurs both in the ballistic limit of transport and in the presence of electron/hole-phonon and charged impurity scattering, can be explained by considering the angle-dependent value of the m_{trans} and m_{DOS} effective masses.

It is important to realize that the impact of the misalignment angles that was discussed above for black phosphorus transistors can be generalized to any 2D materials with an anisotropic band structure. If the $m_{\Gamma\text{-}X}$ /$m_{\Gamma\text{-}Y}$ ratio of the effective masses extracted from the band structure along the two main axes of the Brillouin zone is smaller than 0.1, i.e. if $m_{\Gamma\text{-}Y} \geq 10\ m_{\Gamma\text{-}X}$ in Figure 2.15, the ON-state current only marginally decreases up to $\delta \leq 20°$, by 25% if the misalignment is pushed to 40°. It is clear that the magnitude of the ON-state current still depends on the DOS of each 2D material. Nevertheless, a region where the current is almost insensitive to the misalignment angle can be expected in all cases, as suggested in Klinkert et al. (2021).

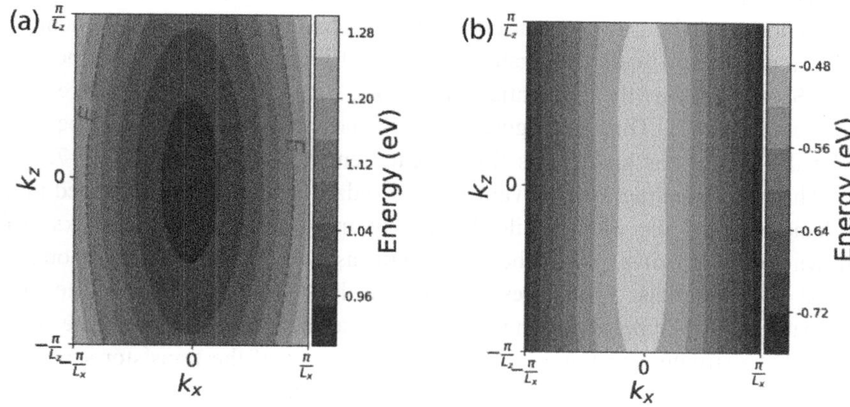

Figure 2.15. *Contour plot of the band edges of black phosphorus*

NOTES ON FIGURE 3.1.– *(a) Contour plot of the bottom conduction band of a black phosphorus monolayer corresponding to an orthorhombic (rectangular) unit cell of dimensions L_x = 2.77 nm and L_z = 1.98 nm. In the transistor configuration, k_x is aligned with the transport direction, k_z with the direction assumed periodic. The dashed red line indicates the Fermi level iso-energy. The effective mass along k_x, $m_{\Gamma\text{-}X}$, is equal to 0.16 m_0, along k_z, $m_{\Gamma\text{-}Y}$ = 1.2 m_0. (b) Same as (a), but for the top valence band of black phosphorus ($m_{\Gamma\text{-}X}$ = 0.14 m_0, $m_{\Gamma\text{-}Y}$ = 4.46 m_0). Adapted from Klinkert et al. (2020).*

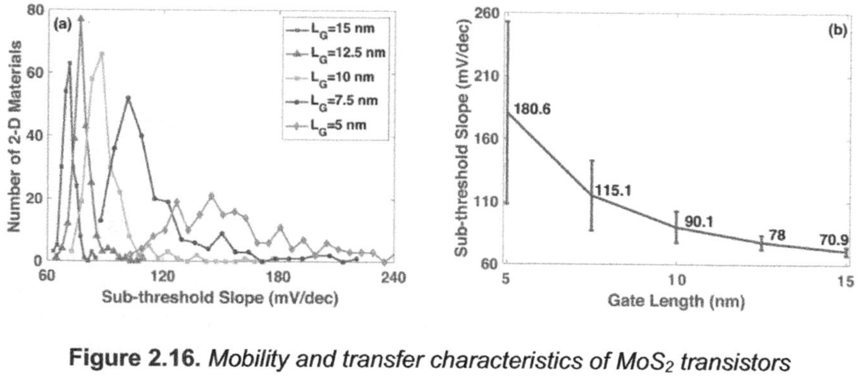

Figure 2.16. *Mobility and transfer characteristics of MoS_2 transistors*

NOTES ON FIGURE 2.16.– *(a) Histogram of the subthreshold slope of all investigated n- and p-type FETs at a gate length of L_G = 15, 12.5, 10, 7.5 and 5 nm. (b) Average SS as a function of L_G. The error bars indicate the standard deviation. Adapted from Klinkert et al. (2020).*

If 2D materials should replace Si as the channel of future field-effect transistors, they should be used during several consecutive technology nodes. Consequently, it should be possible to scale their gate length and still get a performance superior to that of Si FinFETs or their potential successors, gate-all-around nano-sheets; see Loubet et al. (2017). Thus, in Figure 2.16, we show the subthreshold slope of all simulated 2D FETs as histograms for the gate lengths L_G = 15, 12.5, 10, 7.5 and 5 nm. The corresponding average values and standard deviations are provided in the right-hand side of the plot. While *SS* slowly increases when L_G shrinks from 15 down to 10 nm, it explodes below 10 nm, as does the variations among the different 2D materials. From these results, it does not seem feasible to scale single-gate 2D FETs below 10 nm. By adding a second, symmetric gate at the bottom of the structure in Figure 2.12, an improvement of the transistor scalability can be expected. The double-gate architecture is probably the only viable path for 2D materials to compete with Si FETs. Still, other technology issues remain to be solved, as detailed in the next section. At the same time, new logic switch opportunities could emerge for 2D materials.

	n-FET				*p*-FET			
	I_{ON} (mA/µm)	m_{tran} (m_0)	m_{DOS} (m_0)	*SS* (mV/dec)	I_{ON} (mA/µm)	m_{tran} (m_0)	m_{DOS} (m_0)	*SS* (mV/dec)
Ag_2N_6	3.91	0.31	0.56	67	3	0.42	0.65	66.6
As_2	4.17	0.24	1.67	67.4	3.77	0.34	0.56	69.7
As_4	4.44	0.25	0.61	68.4	5.18	0.1	0.35	73.8
Ge_2S_2	4.45	0.24	0.63	67.8	4.08	0.23	0.7	65.8
Ge_2Se_2	3.4	0.33	0.5	69.1	4.38	0.35	0.44	68
HfS_2	3.27	0.51	2.33	69.1	3.59	0.35	0.66	68.8
O_6Sb_4	4.47	0.18	2.04	67.6	3.07	0.35	0.73	62.6
BP	5.01	0.16	0.42	69.5	5.36	0.14	0.82	70.4
Sb_2	5.11	0.23	0.55	67.5	3.8	0.3	0.52	70.1
Si_2H_2	4.46	0.29	1.74	69.5	3.2	0.43	0.71	69.5
$Ti_2Br_2N_2$	3.49	0.18	0.23	71.6	3.25	0.18	0.24	75.8
$Ti_2N_2Cl_2$	3.44	0.22	0.29	69.9	3.48	0.19	0.24	70.4
ZrS_2	3.24	0.59	2.45	66.4	3.37	0.37	0.7	68.5

Table 2.2. *ON-state current (I_{ON}), transport effective mass (m_{trans}), density-of-states effective mass (m_{DOS}) and subthreshold slope (SS) of the 13 2D single-gate FETs with I_{ON} >3 mA/µm in both their n- and p-type configurations at L_G = 15 nm. If more than one transport direction satisfies the ON-state current condition for a given 2D material, only the best one is reported here. Adapted from Klinkert et al. (2020)*

2.4. Challenges and opportunities

2.4.1. *Electrical contacts between metals and 2D monolayers*

One of the key challenges 2D materials are facing is their contact with metallic electrodes. Different approaches can be used to inject electrons into monolayers, the most common ones being top (Radisavljevic et al. 2011), side (Guimarães et al. 2016) and phase-engineered contacts (Kappera et al. 2014). The former two are plotted in Figure 2.17. Besides these geometries, the choice of the metal (Das et al. 2013), the introduction of an interfacial layer between the metal and semiconductor (Cui et al. 2017) or the doping of the channel (McDonnell et al. 2014) represent additional design options. Typically, all these contact configurations are characterized by resistances in the kΩ·µm range (Cheng et al. 2018) instead of 150–200 Ω·µm as in Si FinFETs; see International Roadmap for Devices and Systems (IRDS) (2018). It should nevertheless be mentioned that phase-engineered or nickel-etched graphene electrodes, as in Leong et al. (2015), can give lower resistance values, of the order of 200 Ω·µm, but for multilayer, not monolayer MoS_2.

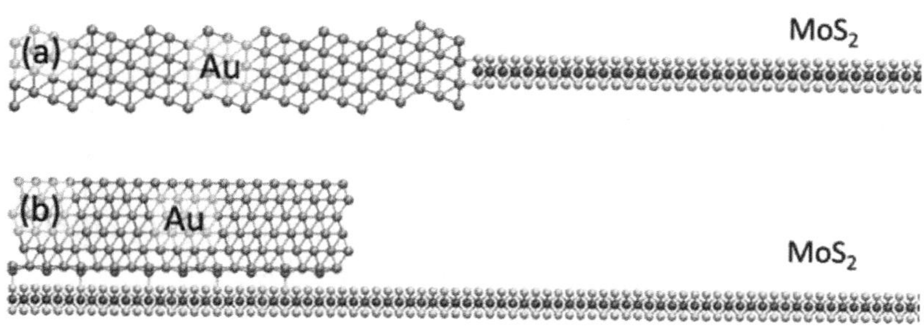

Figure 2.17. *(a) Schematic view of an Au-MoS2 structure with a side contact configuration, as simulated with the MLWF+NEGF approach introduced in section 2.2. (b) Same as (a), but for an Au top contact*

A question that remains open concerns the transfer of electrons from the metallic contact to the semiconductor channel, especially in the case of top-contact architectures. The transfer length L_T measures the average distance that an electron needs in order to completely leave the metal electrode and enter the semiconducting channel. Two different scenarios are possible, one labeled “edge process” (electrons flow through the metal up to the edge of the metal–semiconductor interface; L_T is close to 0), the other one "area-dependent process" (electrons are gradually transferred from the metal to the semiconductor, L_T>>0). For example, in Liu et al. (2014b), a transfer length of $L_T \approx$ 600 nm was found for a monolayer MoS_2 with titanium contacts, while in English et al. (2016) Au electrodes on top of bilayer MoS_2 led to low contact resistances R_C=740 Ω·µm and a short L_T of roughly 30 nm, i.e. a near-edge process.

On the theoretical side, different explanations for such a behavior have been proposed. Relying on DFT (Kang et al. 2014) came to the conclusion that the MoS_2 layer below the metallic contact metallizes, which creates an area-dependent injection of electrons. Through device simulations performed in the effective mass approximation, Arutchelvan et al. (2017) determined that the transfer length depends on the number of layers composing the 2D material, going from an edge process in monolayers to an area-dependent one in multi-layer compounds. Using our ab initio quantum transport approach, we managed to reconcile these theories. If the metal–semiconductor interface is perfectly clean, the transmission of electrons from the electrode to the channel tends to be edge-dependent in monolayers, while the transfer length increases if an interfacial layer is present between both materials (Szabó et al. 2019). While the exact nature of the transfer process is not yet fully understood, it is clear that L_T should be as small as possible to make the 2D technology fully scalable. Recently, Smets et al. (2019) demonstrated few-layer MoS_2 FETs with 13-nm-long top contacts and still very good performance in terms of ON-state current, subthreshold slope and contact resistance. These devices therefore have an ultra-short L_T, which goes in exactly the right direction. Note that side contacts are intrinsically more scalable, but they do not currently offer the same performance as top ones, as illustrated in Jain et al. (2019).

2.4.2. *2D mobility limiting factors*

Ab initio calculations have been widely used to predict the phonon-limited mobility of 2D materials, with a strong focus on TMDs. One of the first such calculations was done in 2012 by Kaasbjerg et al. (2012) for monolayer MoS_2. The electron–phonon scattering rates obtained from DFT served as inputs to fit deformation potentials that were then used in a standard linearized Boltzmann transport equation (LBTE) solver. A mobility of 410 cm^2/Vs at room temperature was returned by this approach. It is much larger than what has been so far measured experimentally, for example, 63 cm^2/Vs at 240 K in Radisavljevic and Kis (2013) or 35.7±2.6 cm^2/Vs at 300 K in Smithe et al. (2018), depending on the dielectric environment. More recent MoS_2 mobility calculations based on fully ab initio electron–phonon scattering rates revealed values much closer to experiments, 150 cm^2/Vs in Li (2015) or 144 cm^2/Vs in Sohier et al. (2018). However, other effects such as charged impurity scattering (CIS) (Ong and Fischetti 2013) or surface optical phonons (SOP) (Ma and Jena 2014) could also play a critical role and bring the mobility closer to measurements.

In Figure 2.18(a), we show mobility results for single-layer MoS_2 that were computed with our in-house LBTE solver described in Rhyner and Luisier (2013), including ab initio electron–phonon interactions as well as CIS and SOP. The only parameter that was neither calculated nor taken from the literature is the concentration of charged impurities, n_{imp}. This quantity was used as the fitting parameter to best reproduce the experimental data of Radisavljevic and Kis (2013), which was achieved with $n_{imp} = 2.5\text{x}10^{12}$ cm^{-2}. An excellent agreement between simulations and experiments is obtained for two different electron concentrations, $7.6\text{x}10^{12}$ and $1.15\text{x}10^{13}$ cm^{-2}.

Getting the mobility gives a lot of information about the transport properties of a material, but being able to determine the influence of this figure of merit on the "current versus voltage" characteristics of a device is equally important. To do that, we used our LBTE solver to calibrate the magnitude of the scattering mechanisms in our MLWF+NEGF QT simulator, OMEN (Lee et al. 2019). This step is necessary as several approximations to the scattering self-energies must be applied in the NEGF formalism. To compensate for them, each scattering rate (electron–phonon, CIS and SOP) can be scaled by a different factor so that both LBTE and QT produce the same mobility. This is what we did for a MoS_2 monolayer before simulating the transfer characteristics of a device similar to the one in Figure 2.8(a). Results are presented in Figure 2.18(b), where the currents with electron–phonon scattering only and with additionally CIS and SOP are compared to each other. It can be seen that CIS and SOP, together, are responsible for a reduction of the ON-state current by a factor 2, as compared to the case without them so that I_{ON} does not exceed 500 μA/μm at a gate length L_G =15 nm. This finding indicates that there is room for improvement. The impurity concentration might indeed be decreased by improving the 2D crystal quality and its interface with the dielectric environment, while SOP might be minimized through substrate engineering. Furthermore, strain might have a beneficial impact on the electronic properties of MoS_2, as theoretically demonstrated in Sohier et al. (2019) for other 2D materials. Overall, there are certainly multiple paths to push the ON-state current of MoS_2 above 1 mA/μm.

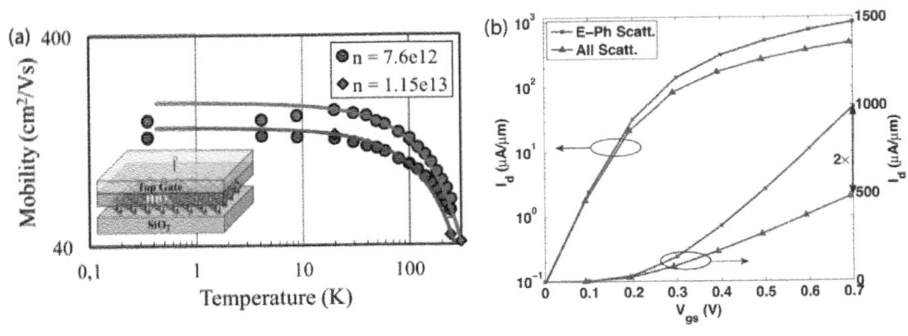

Figure 2.18. *Mobility and transfer characteristics of MoS2 transistors*

NOTES ON FIGURE 2.18.– *(a) Temperature-dependent mobility of monolayer* MoS_2 *embedded between two oxide layers (see the inset) at two different carrier densities. Symbols refer to the experimental data of Radisavljevic, Kis (2013), lines to simulation results. Electron–phonon, charged impurity (CI) and surface optical phonon (SOP) scattering are included. (b) Transfer characteristics* I_D-V_{GS} *at* $V_{DS} = 0.7$ *V of a* MoS_2 *2D FET with the same single-gate structure as in Figure 2.8 and* $L_G = 15$ *nm. The current with electron–phonon interactions only (blue line with crosses) and with additionally CI and SOP scattering (red lines with triangles) are reported. Adapted from Lee et al. (2019).*

2.4.3. *2D oxides*

What has made silicon such an attractive channel material for transistors is the existence of a native oxide, SiO_2, which can be used to separate it from gate contacts or from another dielectric layer. Although not perfect, the Si–SiO_2 interface contains a low defect density, contrary to what is found if SiO_2, Al_2O_3 or HfO_2 is deposited on a TMD monolayer; see Illarionov et al. (2020). There are few exceptions such as ZrO_2 on $ZrSe_2$ or HfO_2 on $HfSe_2$: although large ON/OFF current ratios were demonstrated, the extracted carrier mobility remained small and the *I–V* characteristics still exhibited an hysteresis (Mleczko et al. 2017) when sweeping the gate voltage back and forth. Such a behavior is usually induced by the presence of interface traps that are charged and discharged as the gate potential varies.

Another approach consists of placing a 2D oxide on a 2D channel material. The most common 2D insulator is hexagonal boron nitride (hBN). For example, encapsulating MoS_2 between hBN layers has been shown to produce higher mobility values, both in mono- and few-layer configurations (Cui et al. 2015). Using hBN in transistor applications might, however, not be optimal because the relative dielectric permittivity, ϵ_R, of this materials is of the order of 5. To reach an EOT of 1 nm or less, a hBN thickness of <1.3 nm is needed, which corresponds to three to four layers. We used our ab initio QT simulator to determine what the implications of such thin oxides might be for gate leakage currents (Knobloch et al. 2021). As testbed, a Au-hBN-Si metal–oxide–semiconductor (MOS) capacitor was constructed at the atomic level and the current that flows through it computed as a function of the applied voltage. It was found that three and four layers of hBN are not sufficient to satisfy the IRDS requirements. The situation dramatically worsens if a defect is present in the hBN dielectric. Bridges can form between adjacent hBN layers, which locally increases the leakage current, as can be seen in Figure 2.19. All current trajectories converge towards the defect location.

Other 2D oxides have been deposited on TMD channels, for example, CaF_2 in Illarionov et al. (2019). The advantages of such crystals are that they have a perfectly ordered structure, contrary to SiO_2 or HfO_2, which are amorphous, and that they form a quasi van der Waals interface with 2D channels. Furthermore, the dielectric constant of CaF_2 is higher than that of hBN such that 2 nm of this oxide yields an EOT of roughly 1 nm. Given the fact that the 2D material space includes more than 1,800 compounds (Mounet et al. 2018), it can be expected that several other 2D oxides might be exfoliated in the future and that some of them are competitive with HfO_2 in terms of ϵ_R and band offsets, while preserving a clean oxide–semiconductor interface.

Figure 2.19. *Leakage current trajectories through a metal (Au)–oxide (hexagonal boron nitride)–semiconductor (Si) structure. The oxide layer is made of three h-BN layers, which corresponds to a physical thickness of ~1 nm and an equivalent oxide thickness EOT=0.76 nm (ϵ_R = 5). The h-BN insulator contains a defect in its central layer. Credit: Dr. Jean Favre (CSCS)*

2.4.4. *Advanced logic concepts*

2D materials do not only face challenges, they also offer opportunities in advanced logic applications. Their excellent electrostatic control properties are particularly appealing for the realization of band-to-band tunneling field-effect transistors (TFETs). An entire chapter of this book is dedicated to these logic concepts that could theoretically exhibit a subthreshold slope below 60 mV/dec at room temperature (Ionescu and Riel 2011). Such a feature enables a drastic reduction of the supply voltage and therefore of the power consumption of integrated circuits. However, due to the tunneling nature of the injection mechanism, TFETs tend to suffer from very low ON-state currents. Typically, a large band gap is needed to ensure a steep *SS* and a low I_{OFF}, while a small band gap is necessary to boost I_{ON}. This dichotomy can be partly addressed through the usage of heterojunctions, as demonstrated in Memisevic et al. (2017).

Through device simulation, it has been shown in Ghosh and Mahapatra (2013) and Ilatikhameneh et al. (2015) that conventional TMD monolayers are probably not the best TFET candidates as their large band gaps do not allow the ON-state current to reach large values. This behavior is confirmed in Figure 2.20(a): if the OFF-state current is fixed to ~0.1 nA/μm and the supply voltage V_{DD} to 0.5 V, I_{ON} does not exceed 1 μA/μm, except for WTe_2, a TMD that is rather difficult to stabilize in the 2H phase. Even worse, the *SS* is superior to 60 mV/dec for several of the considered TMDs. This can be explained by the fact that even in the OFF-state, the tunneling channel is already open, i.e. the conduction band below the gate contact has already been pushed below the valence band of the source; see Szabó et al. (2018). Once this condition is satisfied, the I_{ON} increase with respect V_{GS} slows down and becomes almost linear instead of exponential.

There are different solutions to obtain a better TFET performance, for example, by combining different TMDs and forming van der Waals heterostructures. The benefit of such approaches has been highlighted both theoretically in Lam et al. (2013), Szabó et al. (2015a) or Cao et al. (2016) and experimentally in Sarkar et al. (2015), Balaji et al. (2020) or Oliva et al. (2020), combining different TMDs together or one TMD with another material, for example, germanium. Alternatively, the huge variety of properties encountered in 2D materials (see Mounet et al. 2018) can be taken advantage of to identify compounds with low band gaps, compatible with a high ON-state current. With the help of our MLWF+NEGF solver, we have simulated the electrical behavior of relevant examples. Their transfer characteristics are displayed in Figure 2.20(b). All have I_{ON}'s of the order of 100 μA/μm at V_{DD} = 0.5 V and I_{OFF} = 0.1 nA/μm, which is 100x larger than for most TMDs. At the same time, a steep *SS* is obtained. Further investigations would be needed to screen the available design space, which may contain many more 2D materials with band gaps comprised between 0.5 and 1 eV.

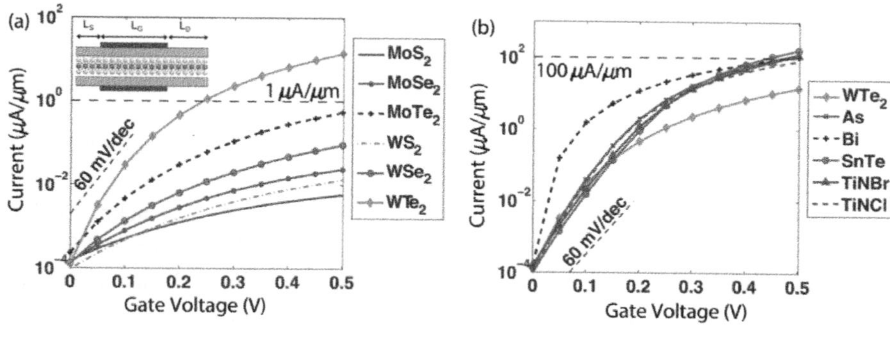

Figure 2.20. *Transfer characteristics of tunneling transistors based on selected 2D materials*

NOTES ON FIGURE 2.20.– *(a) Transfer characteristics I_D-V_{GS} at $V_{DS} = 0.5$ V of band-to-band tunneling field-effect transistors (TFETs) made of conventional TMDs (+WTe_2) with a double-gate configuration (see the inset). The gate length L_G is set to 40 nm, the source and drain extensions to $L_S = 13$ nm (acceptor doping concentration $N_A = 5x10^{13}$ cm^{-2}) and $L_D = 26$ nm (donor doping concentration $N_D = 5x10^{12}$ cm^{-2}), respectively. The 60 mV/dec slope and the 1 μA/μm current level are indicated by dashed black lines. (b) Same as (a), but for novel 2D materials from the database of Mounet et al. (2018). Adapted from Szabó et al. (2018).*

2.5. Conclusion and outlook

In this chapter, the potential of 2D materials as field-effect transistors has been discussed from a modeling perspective, starting from the key features of monolayers. The importance of being able to simulate their electrical characteristics has been introduced. Among all possible approaches, the combination of plane-wave density functional theory, maximally localized Wannier functions and quantum transport has been selected for its versatility and accuracy. By applying it, it has been revealed that transition metal dichalcogenides cannot currently provide a switching performance that is comparable to that of Si FinFETs. They suffer from technical difficulties that will probably disappear with time, such as their electrical contacting, as well as from inherent deficiencies, for example, relatively large effective masses that negatively impact their carrier mobility. We have shown theoretically that novel 2D materials represent an attractive alternative to TMDs with excellent figures of merits as logic switches. It remains to experimentally confirm the predicted performance, which first requires isolating the desired mono- or few-layer components.

2.6. Acknowledgments

The work presented in this chapter was supported by ETH Zurich (grant ETH-32 15-1) and by the Swiss National Science Foundation (SNSF) under grant no. 200021_175479 (ABIME) and under the NCCR MARVEL. We acknowledge PRACE for awarding us access to Piz Daint at CSCS under Project pr28, PRACE for the allocated computational resources on Marconi at CINECA under Project 2016163963 and CSCS for Project s876.

2.7. References

Afzalian, A. (2021). Ab initio perspective of ultra-scaled CMOS from 2D-material fundamentals to dynamically doped transistors. *NPJ 2D Materials and Applications*, 5(1), 5.

Akinwande, D., Huyghebaert, C., Wang, C.-H., Serna, M.I., Goossens, S., Li, L.-J., Wong, H.-S.P., Koppens, F.H.L. (2019). Graphene and two-dimensional materials for silicon technology. *Nature*, 573(7775), 507–518.

Allain, A. and Kis, A. (2014). Electron and hole mobilities in single-layer WSe_2. *ACS Nano*, 8(7), 7180–7185.

AlMutairi, A., Yin, D., Yoon, Y. (2018). $PtSe_2$ field-effect transistors: New opportunities for electronic devices. *IEEE Electron Device Letters*, 39(1), 151–154.

Appenzeller, J., Knoch, J., Bjork, M.T., Riel, H., Schmid, H., Riess, W. (2008). Toward nanowire electronics. *IEEE Transactions on Electron Devices*, 55(11), 2827–2845.

Arutchelvan, G., Lockhart de la Rosa, C.J., Matagne, P., Sutar, S., Radu, I., Huyghebaert, C., De Gendt, S., Heyns, M. (2017). From the metal to the channel: A study of carrier injection through the metal/2D MoS_2 interface. *Nanoscale*, 9(30), 10869–10879.

Balaji, Y., Smets, Q., Szabó, Á., Mascaro, M., Lin, D., Asselberghs, I., Radu, I., Luisier, M., Groeseneken, G. (2020). MoS_2/$MoTe_2$ heterostructure tunnel FETs using gated schottky contacts. *Advanced Functional Materials*, 30(4), 1905970.

Balasubramanyam, S., Merkx, M.J.M., Verheijen, M.A., Kessels, W.M.M., Mackus, A.J.M., Bol, A.A. (2020). Area-selective atomic layer deposition of two-dimensional WS_2 nanolayers. *ACS Materials Letters*, 2(5), 511–518.

Bandurin, D.A., Tyurnina, A.V., Yu, G.L., Mishchenko, A., Zólyomi, V., Morozov, S.V., Krishna Kumar, R., Gorbachev, R.V., Kudrynskyi, Z.R., Pezzini S., et al. (2017). High electron mobility, quantum Hall effect and anomalous optical response in atomically thin InSe. *Nature Nanotechnology*, 12(3), 223–227.

Beyn, W.-J. (2012). An integral method for solving nonlinear eigenvalue problems. *Linear Algebra and its Applications*, 436(10), 3839–3863.

Brück, S., Calderara, M., Bani-Hashemian, M.H., VandeVondele, J., Luisier, M. (2017). Efficient algorithms for large-scale quantum transport calculations. *The Journal of Chemical Physics*, 147(7), 074116.

Bruzzone, S., Iannaccone, G., Marzari, N., Fiori, G. (2014). An open-source multiscale framework for the simulation of nanoscale devices. *IEEE Transactions on Electron Devices*, 61(1), 48–53.

Cai, J., Ruffieux, P., Jaafar, R., Bieri, M., Braun, T., Blankenburg, S., Muoth, M., Seitsonen, A.P., Saleh, M., Feng, X. et al. (2010). Atomically precise bottom-up fabrication of graphene nanoribbons. *Nature*, 466(7305), 470–473.

Calderara, M., Brück, S., Pedersen, A., Bani-Hashemian, M.H., VandeVondele, J., Luisier, M. (2015). Pushing back the limit of ab-initio quantum transport simulations on hybrid supercomputers. *SC '15: Proceedings of the International Conference for High Performance Computing, Networking, Storage and Analysis*. Association for Computing Machinery, New York [Online]. DOI: 10.1145/2807591.2807673.

Cao, J., Cresti, A., Esseni, D., Pala, M. (2016). Quantum simulation of a heterojunction vertical tunnel FET based on 2D transition metal dichalcogenides. *Solid-State Electronics*, 116, 1–7.

Chang, J., Register, L.F., Banerjee, S.K. (2014). Ballistic performance comparison of monolayer transition metal dichalcogenide MX_2 (M = Mo, W; X = S, Se, Te) metal-oxide-semiconductor field effect transistors. *Journal of Applied Physics*, 115(8), 084506.

Cheng, Z., Price, K., Franklin, A.D. (2018). Contacting and gating 2-D nanomaterials. *IEEE Transactions on Electron Devices*, 65(10), 4073–4083.

Cui, X., Lee, G.-H., Kim, Y.D., Arefe, G., Huang, P.Y., Lee, C.-H., Chenet, D.A., Zhang, X., Wang, L., Ye, F. et al. (2015). Multi-terminal transport measurements of MoS_2 using a van der Waals heterostructure device platform. *Nature Nanotechnology*, 10(6), 534–540.

Cui, X., Shih, E.-M., Jauregui, L.A., Chae, S.H., Kim, Y.D., Li, B., Seo, D., Pistunova, K., Yin, J., Park, J.-H. et al. (2017). Low-temperature ohmic contact to monolayer MoS_2 by van der Waals bonded Co/h-BN Electrodes. *Nano Letters*, 17(8), 4781–4786.

Das, S., Chen, H.-Y., Penumatcha, A.V., Appenzeller, J. (2013). High performance multilayer MoS_2 transistors with scandium contacts. *Nano Letters*, 13(1), 100–105.

Das, P., Wickramaratne, D., Debnath, B., Yin, G., Lake, R.K. (2019). Charged impurity scattering in two-dimensional materials with ring-shaped valence bands: GaS, GaSe, InS, and InSe. *Physical Review B*, 99(8), 085409.

Datta, S. (1995). *Electronic Transport in Mesoscopic Systems*. Cambridge University Press, Cambridge.

Dennard, R.H., Gaensslen, F.H., Yu, H., Rideout, V.L., Bassous, E., LeBlanc, A.R. (1974). Design of ion-implanted MOSFET's with very small physical dimensions. *IEEE Journal of Solid-State Circuits*, 9(5), 256–268.

Derivaz, M., Dentel, D., Stephan, R., Hanf, M.-C., Mehdaoui, A., Sonnet, P., Pirri, C. (2015). Continuous germanene layer on Al(111). *Nano Letters*, 15(4), 2510–2516.

Dong, W. and Littlewood, P.B. (2019). Quantum electron transport in ohmic edge contacts between two-dimensional materials. *ACS Applied Electronic Materials*, 1(6), 799–803.

Dongarra, J.J., Luszczek, P., Petitet, A. (2003). The LINPACK benchmark: Past, present and future. *Concurrency and Computation: Practice and Experience*, 15(9), 803–820.

Duerloo, K.-A.N., Li, Y., Reed, E.J. (2014). Structural phase transitions in two-dimensional Mo- and W-dichalcogenide monolayers. *Nature Communications*, 5(1), 4214.

English, C.D., Shine, G., Dorgan, V.E., Saraswat, K.C., Pop, E. (2016). Improved contacts to MoS_2 transistors by ultra-high vacuum metal deposition. *Nano Letters*, 16(6), 3824–3830.

Fang, H., Chuang, S., Chang, T.C., Takei, K., Takahashi, T., Javey, A. (2012). High-performance single layered WSe_2 p-FETs with chemically doped contacts. *Nano Letters*, 12(7), 3788–3792.

Fathipour, S., Ma, N., Hwang, W.S., Protasenko, V., Vishwanath, S., Xing, H.G., Xu, H., Jena, D., Appenzeller, J., Seabaugh, A. (2014). Exfoliated multilayer $MoTe_2$ field-effect transistors. *Applied Physics Letters*, 105(19), 192101.

Fiori, G., Bonaccorso, F., Iannaccone, G., Palacios, T., Neumaier, D., Seabaugh, A., Banerjee, S.K., Colombo, L. (2014). Electronics based on two-dimensional materials. *Nature Nanotechnology*, 9(10), 768–779.

Fischetti, M.V. (1998). Theory of electron transport in small semiconductor devices using the Pauli master equation. *Journal of Applied Physics*, 83(1), 270–291.

Franklin, A.D. (2015). Nanomaterials in transistors: From high-performance to thin-film applications. *Science*, 349(6249) [Online]. Available at: https://science.sciencemag.org/content/349/6249/aab2750.

Frindt, R.F. and Yoffe, A.D. (1963). Physical properties of layer structures: Optical properties and photoconductivity of thin crystals of molybdenum disulphide. *Proceedings of the Royal Society of London*, 273(1352), A27369-83.

Gaddemane, G., Vandenberghe, W.G., de Put, M.L.V., Chen, E., Fischetti, M.V. (2018). Monte-Carlo study of electronic transport in non-σ_h-symmetric two-dimensional materials: Silicene and germanene. *Journal of Applied Physics*, 124, 044306.

Ghosh, R.K. and Mahapatra, S. (2013). Monolayer transition metal dichalcogenide channel-based tunnel transistor. *IEEE Journal of the Electron Devices Society*, 1, 175.

Giannozzi, P., Baroni, S., Bonini, N., Calandra, M., Car, R., Cavazzoni, C., Ceresoli, D., Chiarotti, G.L., Cococcioni, M., Dabo, I., et al. (2009). QUANTUM ESPRESSO: A modular and open-source software project for quantum simulations of materials. *Journal of Physics: Condensed Matter*, 21(39), 395502.

Gnani, E., Gnudi, A., Reggiani, S., Luisier, M., Baccarani, G. (2008). Band effects on the transport characteristics of ultrascaled SNW-FETs. *IEEE Transactions on Nanotechnology*, 7(6), 700–709.

Grazulis, S., Daskevic, A., Merkys, A., Chateigner, D., Lutterotti, L., Quiros, M., Serebryanaya, N.R., Moeck, P., Downs, R.T., Le Bail, A. (2011). Crystallography open database (COD): An open-access collection of crystal structures and platform for world-wide collaboration. *Nucleic Acids Research*, 40(D1), D420–D427.

Guimarães, M.H.D., Gao, H., Han, Y., Kang, K., Xie, S., Kim, C.-J., Muller, D.A., Ralph, D.C., Park, J. (2016). Atomically thin ohmic edge contacts between two-dimensional materials. *ACS Nano*, 10(6), 6392–6399.

Han, M.Y., Özyilmaz, B., Zhang, Y., Kim, P. (2007). Energy band-gap engineering of graphene nanoribbons. *Phys. Rev. Lett.*, 98(20), 206805.

Heyd, J., Scuseria, G.E., Ernzerhof, M. (2003). Hybrid functionals based on a screened Coulomb potential. *The Journal of Chemical Physics*, 118(18), 8207–8215.

Hisamoto, D., Lee, W.-C., Kedzierski, J., Anderson, E., Takeuchi, H., Asano, K., King, T.-J., Bokor, J., Hu, C. (1998). A folded-channel MOSFET for deep-sub-tenth micron era. *International Electron Devices Meeting 1998. Technical Digest*, 1032–1034.

Hybertsen, M.S. and Louie, S.G. (1986). Electron correlation in semiconductors and insulators: Band gaps and quasiparticle energies. *Physical Review B*, 34(8), 5390–5413.

Ilatikhameneh, H., Tan, Y., Novakovic, B., Klimeck, G., Rahman, R., Appenzeller, J. (2015). Tunnel field-effect transistors in 2-D transition metal dichalcogenide materials. *IEEE Journal on Exploratory Solid-State Computational Devices and Circuits*, 1, 12–18.

Illarionov, Y.Y., Banshchikov, A.G., Polyushkin, D.K., Wachter, S., Knobloch, T., Thesberg, M., Mennel, L., Paur, M., Stöger-Pollach, M., Steiger-Thirsfeld, A. et al. (2019). Ultrathin calcium fluoride insulators for two-dimensional field-effect transistors. *Nature Electronics*, 2(6), 230–235.

Illarionov, Y.Y., Knobloch, T., Jech, M., Lanza, M., Akinwande, D., Vexler, M.I., Mueller, T., Lemme, M.C., Fiori, G., Schwierz, F. et al. (2020). Insulators for 2D nanoelectronics: The gap to bridge. *Nature Communications*, 11(1), 3385.

Inorganic Crystal Structure Database (n.d.). Inorganic Crystal Structure Database – ICSD [Online]. Available at: https://www.fiz-karlsruhe.de/de/produkte-und-dienstleistungen/inorganic-crystal-structure-database-icsd [Accessed 19 February 2021].

International Roadmap for Devices and Systems (IRDS) (2018). International Roadmap for Devices and Systems (IRDS™) 2018 Edition [Online]. Available at: https://irds.ieee.org/editions/2018 [Accessed 22 February 2021].

Ionescu, A.M. and Riel, H. (2011). Tunnel field-effect transistors as energy-efficient electronic switches. *Nature*, 479, 329.

Jain, A., Szabó, Á., Parzefall, M., Bonvin, E., Taniguchi, T., Watanabe, K., Bharadwaj, P., Luisier, M., Novotny, L. (2019). One-dimensional edge contacts to a monolayer semiconductor. *Nano Letters*, 19(10), 6914–6923.

Ji, J., Song, X., Liu, J., Yan, Z., Huo, C., Zhang, S., Su, M., Liao, L., Wang, W., Ni, Z. et al. (2016). Two-dimensional antimonene single crystals grown by van der Waals epitaxy. *Nature Communications*, 7(1), 13352.

Joensen, P., Frindt, R.F., Morrison, S.R. (1986). Single-layer MoS_2. *Materials Research Bulletin*, 21(4), 457–461.

Kaasbjerg, K., Thygesen, K.S., Jacobsen, K.W. (2012). Phonon-limited mobility in *n*-type single-layer MoS_2 from first principles. *Physical Review B*, 85(11), 115317.

Kang, J., Liu, W., Sarkar, D., Jena, D., Banerjee, K. (2014). Computational study of metal contacts to monolayer transition-metal dichalcogenide semiconductors. *Physical Review X*, 4(3), 031005.

Kang, K., Xie, S., Huang, L., Han, Y., Huang, P.Y., Mak, K.F., Kim, C.-J., Muller, D., Park, J. (2015). High-mobility three-atom-thick semiconducting films with wafer-scale homogeneity. *Nature*, 520(7549), 656–660.

Kappera, R., Voiry, D., Yalcin, S.E., Branch, B., Gupta, G., Mohite, A.D., Chhowalla, M. (2014). Phase-engineered low-resistance contacts for ultrathin MoS_2 transistors. *Nature Materials*, 13(12), 1128–1134.

Khan, H.N., Hounshell, D.A., Fuchs, E.R.H. (2018). Science and research policy at the end of Moore's law. *Nature Electronics*, 1(1), 14–21.

Klinkert, C., Szabó, Á., Stieger, C., Campi, D., Marzari, N., Luisier, M. (2020). 2-D materials for ultrascaled field-effect transistors: One hundred candidates under the ab initio microscope. *ACS Nano*, 14(7), 8605–8615.

Klinkert, C., Fiore, S., Backman, J., Lee, Y., Luisier, M. (2021). Impact of orientation misalignments on black phosphorus ultrascaled field-effect transistors. *IEEE Electron Device Letters*, 42(3), 434–437.

Knobloch, T., Rzepa, G., Illarionov, Y.Y., Waltl, M., Schanovsky, F., Jech, M., Stampfer, B., Furchi, M.M., Müller, T., Grasser, T. (2017). Physical modeling of the hysteresis in MoS_2 transistors. *2017 47th European Solid-State Device Research Conference (ESSDERC)*, 284–287.

Knobloch, T., Illarionov, Y.Y., Ducry, F., Schleich, C., Wachter, S., Watanabe, K., Taniguchi, T., Mueller, T., Waltl, M., Lanza, M. et al. (2021). The performance limits of hexagonal boron nitride as an insulator for scaled CMOS devices based on two-dimensional materials. *Nature Electronics*, 4(2), 98–108.

Kohn, W. and Sham, L.J. (1965). Self-consistent equations including exchange and correlation effects. *Physical Review*, 140(4A), A1133–A1138.

Kormányos, A., Burkard, G., Gmitra, M., Fabian, J., Zólyomi, V., Drummond, N.D., Fal'ko, V. (2015). k·p theory for two-dimensional transition metal dichalcogenide semiconductors. *2D Materials*, 2(2), 022001.

Kresse, G. and Furthmüller, J. (1996). Efficiency of ab-initio total energy calculations for metals and semiconductors using a plane-wave basis set. *Computational Materials Science*, 6(1), 15–50.

Kuhn, K.J. (2012). Considerations for ultimate CMOS scaling. *IEEE Transactions on Electron Devices*, 59(7), 1813–1828.

Lake, R., Klimeck, G., Bowen, R.C., Jovanovic, D. (1997). Single and multiband modeling of quantum electron transport through layered semiconductor devices. *Journal of Applied Physics*, 81(12), 7845–7869.

Lam, K.-T., Cao, X., Guo, J. (2013). Device performance of heterojunction tunneling field-effect transistors based on transition metal dichalcogenide monolayer. *IEEE Electron Device Letters*, 34, 1331.

Lee, Y., Fiore, S., Luisier, M. (2019). Ab initio mobility of single-layer MoS_2 and WS_2: Comparison to experiments and impact on the device characteristics. *2019 IEEE International Electron Devices Meeting (IEDM)*, 24.4.1–24.4.4.

Lemme, M.C., Echtermeyer, T.J., Baus, M., Kurz, H. (2007). A graphene field-effect device. *IEEE Electron Device Letters*, 28(4), 282–284.

Lent, C.S. and Kirkner, D.J. (1990). The quantum transmitting boundary method. *Journal of Applied Physics*, 67(10), 6353–6359.

Leong, W.S., Luo, X., Li, Y., Khoo, K.H., Quek, S.Y., Thong, J.T.L. (2015). Low resistance metal contacts to MoS_2 devices with nickel-etched-graphene electrodes. *ACS Nano*, 9(1), 869–877.

Li, W. (2015). Electrical transport limited by electron–phonon coupling from Boltzmann transport equation: An ab initio study of Si, Al, and MoS_2. *Physical Review B*, 92(7), 075405.

Li, L., Yu, Y., Ye, G.J., Ge, Q., Ou, X., Wu, H., Feng, D., Chen, X.H., Zhang, Y. (2014). Black phosphorus field-effect transistors. *Nature Nanotechnology*, 9(5), 372–377.

Liu, L., Kumar, S.B., Ouyang, Y., Guo, J. (2011). Performance limits of monolayer transition metal dichalcogenide transistors. *IEEE Transactions on Electron Devices*, 58(9), 3042–3047.

Liu, H., Neal, A.T., Zhu, Z., Luo, Z., Xu, X., Tománek, D., Ye, P.D. (2014a). Phosphorene: An unexplored 2D semiconductor with a high hole mobility. *ACS Nano*, 8(4), 4033–4041.

Liu, H., Si, M., Deng, Y., Neal, A.T., Du, Y., Najmaei, S., Ajayan, P.M., Lou, J., Ye, P.D. (2014b). Switching mechanism in single-layer molybdenum disulfide transistors: An insight into current flow across Schottky barriers. *ACS Nano*, 8(1), 1031–1038.

Liu, E., Fu, Y., Wang, Y., Feng, Y., Liu, H., Wan, X., Zhou, W., Wang, B., Shao, L., Ho, C.-H. et al. (2015). Integrated digital inverters based on two-dimensional anisotropic ReS2 field-effect transistors. *Nature Communications*, 6(1), 6991.

Liu, Y., Duan, X., Huang, Y., Duan, X. (2018). Two-dimensional transistors beyond graphene and TMDCs. *Chemical Society Reviews*, 47(16), 6388–6409.

Loubet, N., Hook, T., Montanini, P., Yeung, C.-W., Kanakasabapathy, S., Guillom, M., Yamashita, T., Zhang, J., Miao, X., Wang, J. et al. (2017). Stacked nanosheet gate-all-around transistor to enable scaling beyond FinFET. *2017 Symposium on VLSI Technology*, T230–T231.

Lu, A.K.A., Pourtois, G., Luisier, M., Radu, I.P., Houssa, M. (2017). On the electrostatic control achieved in transistors based on multilayered MoS_2: A first-principles study. *Journal of Applied Physics*, 121(4), 044505.

Luisier, M., Schenk, A., Fichtner, W., Klimeck, G. (2006). Atomistic simulation of nanowires in the $sp^3d^5s^*$ tight-binding formalism: From boundary conditions to strain calculations. *Physical Review B*, 74(20), 205323.

Luisier, M., Szabó, Á., Stieger, C., Klinkert, C., Brück, S., Jain, A., Novotny, L. (2016). First-principles simulations of 2-D semiconductor devices: Mobility, I-V characteristics, and contact resistance. *2016 IEEE International Electron Devices Meeting (IEDM)*, 5.4.1–5.4.4.

Ma, N. and Jena, D. (2014). Charge scattering and mobility in atomically thin semiconductors. *Physical Review X*, 4(1), 011043.

Ma, Y., Kuc, A., Heine, T. (2017). Single-layer Tl_2O: A metal-shrouded 2D semiconductor with high electronic mobility. *Journal of the American Chemical Society*, 139(34), 11694–11697.

Marzari, N. and Vanderbilt, D. (1997). Maximally localized generalized Wannier functions for composite energy bands. *Physical Review B*, 56(20), 12847–12865.

McClellan, C.J., Yalon, E., Smithe, K.K.H., Suryavanshi, S.V., Pop, E. (2021). High current density in monolayer MoS_2 doped by AlOx. *ACS Nano*, 15(1), 1587–1596.

McDonnell, S., Addou, R., Buie, C., Wallace, R.M., Hinkle, C.L. (2014). Defect-dominated doping and contact resistance in MoS_2. *ACS Nano*, 8(3), 2880–2888.

Memisevic, E., Hellenbrand, M., Lind, E., Persson, A.R., Sant, S., Schenk, A., Svensson, J., Wallenberg, R., Wernersson, L.-E. (2017). Individual defects in InAs/InGaAsSb/GaSb nanowire tunnel field-effect transistors operating below 60 mV/decade. *Nano Letters*, 17, 4373–4380.

Meng, Y., Ling, C., Xin, R., Wang, P., Song, Y., Bu, H., Gao, S., Wang, X., Song, F., Wang, J., Wang, X., Wang, B., Wang, G. (2017). Repairing atomic vacancies in single-layer $MoSe_2$ field-effect transistor and its defect dynamics. *NPJ Quantum Materials*, 2(1), 16.

Mistry, K., Allen, C., Auth, C., Beattie, B., Bergstrom, D., Bost, M., Brazier, M., Buehler, M., Cappellani, A., Chau, R. et al. (2007). A 45nm logic technology with high-k+Metal gate transistors, strained silicon, 9 Cu interconnect layers, 193nm dry patterning, and 100% Pb-free Packaging. *2007 IEEE International Electron Devices Meeting*, 247–250.

Mleczko, M.J., Zhang, C., Lee, H.R., Kuo, H.-H., Magyari-Köpe, B., Moore, R.G., Shen, Z.-X., Fisher, I.R., Nishi, Y., Pop, E. (2017). $HfSe_2$ and $ZrSe_2$: Two-dimensional semiconductors with native high-κ oxides. *Science Advances*, 3(8) [Online]. Available at: https://advances.sciencemag.org/content/3/8/e1700481.

Moore, G.E. (2006). Cramming more components onto integrated circuits. Reprinted from *Electronics*, volume 38, number 8, April 19, 1965, 114 ff. *IEEE Solid-State Circuits Society Newsletter*, 11(3), 33–35.

Mostofi, A.A., Yates, J.R., Lee, Y.-S., Souza, I., Vanderbilt, D., Marzari, N. (2008). wannier90: A tool for obtaining maximally-localised Wannier functions. *Computer Physics Communications*, 178(9), 685–699.

Mounet, N., Gibertini, M., Schwaller, P., Campi, D., Merkys, A., Marrazzo, A., Sohier, T., Castelli, I.E., Cepellotti, A., Pizzi, G. et al. (2018). Two-dimensional materials from high-throughput computational exfoliation of experimentally known compounds. *Nature Nanotechnology*, 13(3), 246–252.

Najmaei, S., Liu, Z., Zhou, W., Zou, X., Shi, G., Lei, S., Yakobson, B.I., Idrobo, J.-C., Ajayan, P.M., Lou, J. (2013). Vapour phase growth and grain boundary structure of molybdenum disulphide atomic layers. *Nature Materials*, 12(8), 754–759.

Nedjalkov, M., Dimov, I., Rossi, F., Jacoboni, C. (1996). Convergency of the Monte Carlo algorithm for the solution of the Wigner quantum-transport equation. *Mathematical and Computer Modelling*, 23(8), 159–166.

Novoselov, K.S., Geim, A.K., Morozov, S.V., Jiang, D., Zhang, Y., Dubonos, S.V., Grigorieva, I.V., Firsov, A.A. (2004). Electric field effect in atomically thin carbon films. *Science*, 306(5696), 666–669.

Oliva, N., Backman, J., Capua, L., Cavalieri, M., Luisier, M., Ionescu, A.M. (2020). $WSe_2/SnSe_2$ vdW heterojunction tunnel FET with subthermionic characteristic and MOSFET co-integrated on same WSe_2 flake. *NPJ 2D Materials and Applications*, 4(1), 5.

Ong, Z.-Y. and Fischetti, M.V. (2013). Mobility enhancement and temperature dependence in top-gated single-layer MoS_2. *Physical Review B*, 88(16), 165316.

Ovchinnikov, D., Allain, A., Huang, Y.-S., Dumcenco, D., Kis, A. (2014). Electrical transport properties of single-layer WS_2. *ACS Nano*, 8(8), 8174–8181.

Perdew, J.P., Burke, K., Ernzerhof, M. (1996). Generalized gradient approximation made simple. *Physical Review Letters*, 77(18), 3865–3868.

Pizzi, G., Gibertini, M., Dib, E., Marzari, N., Iannaccone, G., Fiori, G. (2016). Performance of arsenene and antimonene double-gate MOSFETs from first principles. *Nature Communications*, 7(1), 12585.

Pop, E., Sinha, S., Goodson, K.E. (2006). Heat generation and transport in nanometer-scale transistors. *Proceedings of the IEEE*, 94(8), 1587–1601.

Quhe, R., Liu, J., Wu, J., Yang, J., Wang, Y., Li, Q., Li, T., Guo, Y., Yang, J, Peng, H. et al. (2019). High-performance sub-10 nm monolayer Bi_2O_2Se transistors. *Nanoscale*, 11(2), 532–540.

Radisavljevic, B. and Kis, A. (2013). Mobility engineering and a metal–insulator transition in monolayer MoS_2. *Nature Materials*, 12(9), 815–820.

Radisavljevic, B., Radenovic, A., Brivio, J., Giacometti, V., Kis, A. (2011). Single-layer MoS_2 transistors. *Nature Nanotechnology*, 6(3), 147–150.

Rhyner, R. and Luisier, M. (2013). Phonon-limited low-field mobility in silicon: Quantum transport vs. linearized Boltzmann transport equation. *Journal of Applied Physics*, 114(22), 223708.

Sancho, M.P.L., Sancho, J.M.L., Rubio, J. (1984). Quick iterative scheme for the calculation of transfer matrices: Application to Mo (100). *Journal of Physics F: Metal Physics*, 14(5), 1205–1215.

Sarkar, D., Xie, X., Liu, W., Cao, W., Kang, J., Gong, Y., Kraemer, S., Ajayan, P.M., Banerjee, K. (2015). A subthermionic tunnel field-effect transistor with an atomically thin channel. *Nature*, 526, 91.

Schwierz, F., Pezoldt, J., Granzner, R. (2015). Two-dimensional materials and their prospects in transistor electronics. *Nanoscale*, 7(18), 8261–8283.

Smets, Q., Arutchelvan, G., Jussot, J., Verreck, D., Asselberghs, I., Mehta, A.N., Gaur, A., Lin, D., Kazzi, S.E., Groven, B. et al. (2019). Ultra-scaled MOCVD MoS_2 MOSFETs with 42nm contact pitch and 250μA/μm drain current. *2019 IEEE International Electron Devices Meeting (IEDM)*, 23.2.1–23.2.4.

Smithe, K.K.H., English, C.D., Suryavanshi, S.V., Pop, E. (2018). High-field transport and velocity saturation in synthetic monolayer MoS_2. *Nano Letters*, 18(7), 4516–4522.

Sohier, T., Campi, D., Marzari, N., Gibertini, M. (2018). Mobility of two-dimensional materials from first principles in an accurate and automated framework. *Physical Review Materials*, 2(11), 114010.

Sohier, T., Gibertini, M., Campi, D., Pizzi, G., Marzari, N. (2019). Valley-engineering mobilities in two-dimensional materials. *Nano Letters*, 19(6), 3723–3729.

Stieger, C. and Luisier, M. (2020). *Winterface: An Interface from Wannier90 to Quantum Transport*. DOI: 10.48550/arXiv.2007.04268.

Stieger, C., Szabó, Á., Bunjaku, T., Luisier, M. (2017). Ab-initio quantum transport simulation of self-heating in single-layer 2-D materials. *Journal of Applied Physics*, 122(4), 045708.

Suh, J., Park, T.-E., Lin, D.-Y., Fu, D., Park, J., Jung, H.J., Chen, Y., Ko, C., Jang, C., Sun, Y. et al. (2014). Doping against the native propensity of MoS_2: Degenerate hole doping by cation substitution. *Nano Letters*, 14(12), 6976–6982.

Suh, J., Tan, T.L., Zhao, W., Park, J., Lin, D.-Y., Park, T.-E., Kim, J., Jin, C., Saigal, N., Ghosh, S. et al. (2018). Reconfiguring crystal and electronic structures of MoS_2 by substitutional doping. *Nature Communications*, 9(1), 199.

Svizhenko, A., Anantram, M.P., Govindan, T.R., Biegel, B., Venugopal, R. (2002). Two-dimensional quantum mechanical modeling of nanotransistors. *Journal of Applied Physics*, 91(4), 2343–2354.

Szabó, Á., Koester, S.J., Luisier, M. (2015a). Ab-initio simulation of van der Waals $MoTe_2$-SnS_2 heterotunneling FETs for low-power electronics. *IEEE Electron Device Letters*, 36(5), 514–516.

Szabó, Á., Rhyner, R., Luisier, M. (2015b). Ab initio simulation of single- and few-layer MoS_2 transistors: Effect of electron–phonon scattering. *Physical Review B*, 92(3), 035435.

Szabó, Á., Klinkert, C., Campi, D., Stieger, C., Marzari, N., Luisier, M. (2018). Ab initio simulation of band-to-band tunneling FETs with single- and few-layer 2-D materials as channels. *IEEE Transactions on Electron Devices*, 65(10), 4180–4187.

Szabó, Á., Jain, A., Parzefall, M., Novotny, L., Luisier, M. (2019). Electron transport through Metal/MoS_2 interfaces: Edge- or area-dependent process? *Nano Letters*, 19(6), 3641–3647.

Tao, L., Cinquanta, E., Chiappe, D., Grazianetti, C., Fanciulli, M., Dubey, M., Molle, A., Akinwande, D. (2015). Silicene field-effect transistors operating at room temperature. *Nature Nanotechnology*, 10(3), 227–231.

Thompson, S., Anand, N., Armstrong, M., Auth, C., Arcot, B., Alavi, M., Bai, P., Bielefeld, J., Bigwood, R., Brandenburg, J. et al. (2002). A 90 nm logic technology featuring 50 nm strained silicon channel transistors, 7 layers of Cu interconnects, low k ILD, and 1 /spl mu/m/sup 2/ SRAM cell. *Digest. International Electron Devices Meeting*, 61–64.

Top500 (2020). Top500: The List [Online]. Available at: http://www.top500.org [Accessed 10 February 2021].

Tosun, M., Chuang, S., Fang, H., Sachid, A.B., Hettick, M., Lin, Y., Zeng, Y., Javey, A. (2014). High-gain inverters based on WSe_2 complementary field-effect transistors. *ACS Nano*, 8(5), 4948–4953.

Uchida, K., Koga, J., Takagi, S. (2007). Experimental study on electron mobility in ultrathin-body silicon-on-insulator metal-oxide-semiconductor field-effect transistors. *Journal of Applied Physics*, 102(7), 074510.

Wallace, P.R. (1947). The band theory of graphite. *Physical Review*, 71(9), 622–634.

Wang, X., Ouyang, Y., Li, X., Wang, H., Guo, J., Dai, H. (2008). Room-temperature all-semiconducting sub-10-nm graphene nanoribbon field-effect transistors. *Physical Review Letters*, 100(20), 206803.

Wang, K., Valencia, D., Charles, J., He, Y., Povolotskyi, M., Klimeck, G., Maassen, J., Lundstrom, M., Kubis, T. (2016). NEMO5: Predicting MoS_2 heterojunctions. *2016 International Conference on Simulation of Semiconductor Processes and Devices (SISPAD)*, 221–224.

Wu, J., Yuan, H., Meng, M., Chen, C., Sun, Y., Chen, Z., Dang, W., Tan, C., Liu, Y., Yin, J. et al. (2017). High electron mobility and quantum oscillations in non-encapsulated ultrathin semiconducting Bi_2O_2Se. *Nature Nanotechnology*, 12(6), 530–534.

Xia, F., Wang, H., Jia, Y. (2014). Rediscovering black phosphorus as an anisotropic layered material for optoelectronics and electronics. *Nature Communications*, 5(1), 4458.

Yoon, Y., Ganapathi, K., Salahuddin, S. (2011). How good can monolayer MoS_2 transistors be? *Nano Letters*, 11(9), 3768–3773.

Zahid, F., Liu, L., Zhu, Y., Wang, J., Guo, H. (2013). A generic tight-binding model for monolayer, bilayer and bulk MoS_2. *AIP Advances*, 3(5), 052111.

3

Negative Capacitance Field-Effect Transistors

Wei CAO and Kaustav BANERJEE
Department of Electrical and Computer Engineering, University of California, Santa Barbara, USA

3.1. Introduction

The continuous dimensional scaling of the metal oxide semiconductor field-effect transistors (MOSFETs) (Sakurai 2004; Lundstrom 2006) in the past few decades has enabled unprecedentedly faster and cheaper chips, which play an essential role in driving forward important modern technologies such as artificial intelligence (AI) and Internet of Things (IoT). Unfortunately, limited by the inability of supply voltage scaling constrained by the thermionic emission-based transport mechanism of MOSFETs, the energy/power efficiency of integrated circuit (IC) did not improve at the same pace as integration density and operation speed. To overcome this issue, many beyond-MOSFET transistor concepts with novel device physics have been proposed to allow effective lowering of the supply voltage, without sacrificing operation speed or increasing fabrication cost to an unaffordable level. Among them, negative-capacitance (NC) FET (Salahuddin and Datta 2008), as self-explained by the name, employs a negative capacitance in the gate stack to save switching voltage, and is emerging as one of the most promising and widely studied devices, not only due to its supply voltage scalability, but also thanks to its simple structure that involves minimal penalty in added (with respect to MOSFETs) manufacturing complexity.

For a color version of all figures in this chapter, see www.iste.co.uk/cresti/beyondcmos.zip.

Beyond-CMOS, coordinated by Alessandro CRESTI.

Currently, the fundamentals of NC-FETs, as well as potential challenges and opportunities, have been well established and recognized by the device physicists (Cao and Banerjee 2020). However, it remains controversial and debatable on the experimental front, particularly on the characterization and proof of authentic negative capacitance under different test conditions (DC vs. AC, discrete vs. integrated gate stack) and different ferroelectric material systems, and the interpretation of measured unexpected NC-FET behavior. Thus, to properly serve the purpose of this book, this chapter will be focused on introducing its fundamentals, and uncovering its challenges and opportunities from a device physics point of view, after which recent experimental progress will be benchmarked and commented upon based on the presented NC-FET device physics, as complementary information.

3.2. The rise of NC-FETs

The inability to effectively scale the supply voltage of MOSFETs stems from their thermionic emission-based transport mechanism that sets a lower bound of their subthreshold swing (*SS*) to 60mV of gate voltage swing per decade of drain current swing. For a general requirement of switching current on/off ratio of at least three to four orders of magnitudes for low-power digital application, the threshold voltage of MOSFETs cannot be lower than 0.18–0.24 V, no matter how perfect the device structure is designed to be. Small *SS* can help FETs save supply voltage (V_{dd}) and hence dynamic (switching) power that has a quadratic dependence on the V_{dd}, with fixed on-current (I_{on}) and off-current (I_{off}), as illustrated in Figure 3.1(a). Thus, pursuing novel device physics that can break the 60 mV/dec thermionic limit for device *SS* represents the most important research topic for scaling supply voltage. So far, there have been numerous types of beyond-thermionic transistors proposed (Baba 1992; Gopalakrishnan et al. 2005; Sarkar et al. 2015), such as nano-electro-mechanical switch, phase change material gated FET, band-to-band tunneling FET, impact ionization MOSFET, charge feedback FET and NC-FET. Despite these steep-slope device concepts showing promising *SS*, most of them are not feasible for mass production or suffer from major limitations such as low operation speed, drive current, requirement of large source-drain voltage, etc. Since the concept of NC-FET was proposed in 2007/2008 by Salahuddin and Datta (2008), it quickly rose to a notable position, thanks to its fabrication-friendly structure as illustrated in Figure 3.1(b), which is basically a MOSFET with an additional NC (typically ferroelectric material, FE) layer inserted between the gate oxide and gate electrode. Moreover, doping Hafnium oxide (HfOx), which has been widely used by industry as high-k dielectric material in MOSFETs, has been found to own robust ferroelectricity even when its physical thickness is scaled to nanometer scale, which further fuels the NC-FET research surge in both academia and industry.

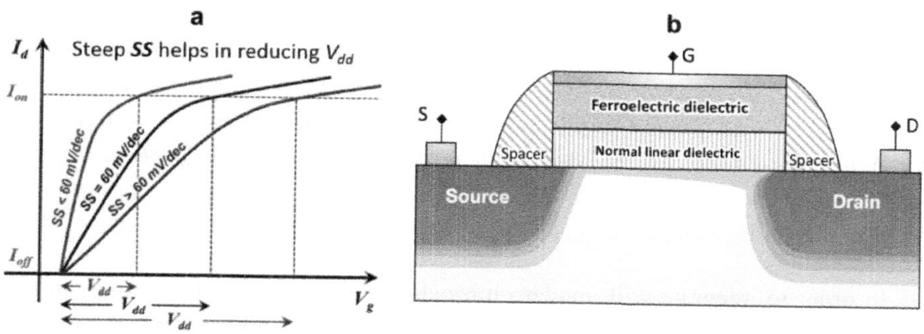

Figure 3.1. *The importance of steep slope and the simple structure of NC-FETs*

NOTES ON FIGURE 3.1.– *(a) The advantage of steep subthreshold slope (steep SS) or small subthreshold swing in terms of saving supply voltage V_{dd}. (b) Compared to MOSFET, NC-FET simply adds one additional ferroelectric dielectric layer.*

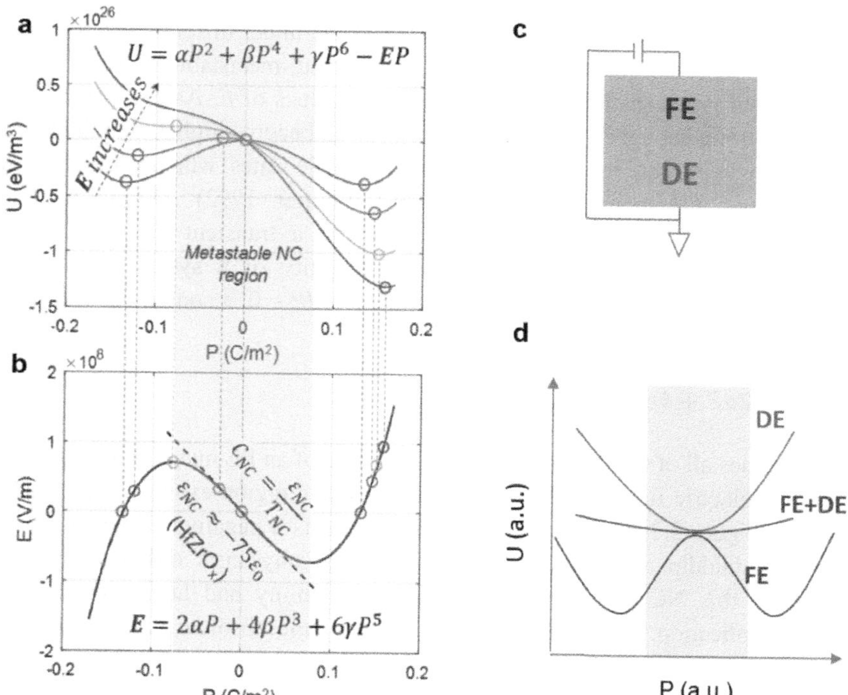

Figure 3.2. *Energy landscape of ferroelectric polarization*

NOTES ON FIGURE 3.2.– *(a) Energy landscape (U) of a ferroelectric (FE) material in terms of polarization (P) and external electric field E. (b) Stable (including metastable) P–E relation derived from a.* α*=-7×10*^{8} *mF*$^{-1}$*,* β*=2×10*^{10} *m*^{5}*F*$^{-1}$*C*$^{-2}$*,* γ*=0 m*^{9}*F*$^{-1}$*C*$^{-4}$*. The shaded region represents the negative capacitance (NC) state. (c) Schematic of an FE – normal dielectric (DE) stack. (d) Schematic energy landscape of an individual FE, an individual DE, and an FE-DE stack (absolute value of the FE capacitance is made larger than the DE capacitance, i.e.* $|C_{FE}| \geq C_{DE}$*).*

In order to integrate with modern ultra-short-channel MOSFETs, single-domain ferroelectric (FE) dielectrics are desired, in order to guarantee acceptable uniformity and performance robustness. Thus, the discussion in this chapter is restricted to single-domain FE NC-FETs. According to the Landau–Devonshire theory, the Gibbs free energy density (U) of a single-domain FE material is given by Tolédano and Tolédano (1987).

$$U = \alpha P^2 + \beta P^4 + \gamma P^6 - E \cdot P \qquad [3.1]$$

where P is the polarization and E is the applied electric field across the FE layer. Figure 3.2(a) shows the Gibbs free energy landscape under different electric fields. The circles represent bistable (valley regions) and metastable (barrier region, indicated with green shade) P states for different values of E. As external bias and hence E is strong enough, one of the two valleys can become higher than the barrier, thereby realizing a state transition between the bistable states, which is the operation mechanism of FE memory (Miller and McWhorter 1992). During the state transition, the barrier region is considered to be a transient phase due to its metastable nature in energy landscape. Metastability of a system can also be interpreted as a negative capacitance. By letting $dU/dP = 0$, a continuous P–E curve can be obtained

$$E = 2\alpha P + 4\beta P^3 + 6\gamma P^5 \qquad [3.2]$$

which includes all of the possible polarization states of an FE material in response to an external electric field. In this curve, the metastable region exhibits negative P–E slope (see Figure 3.2(b)), which in physics represents a negative capacitance (NC) region (Salahuddin and Datta 2008). Historically, physicists did not pay much attention to this NC region due to its uncontrollability and hence feasibility to practical application. It is worthwhile to note that the absolute value of the P–E slope is generally very large, i.e. the absolute NC dielectric constant $|\varepsilon_{NC}|$ is very large. For example, $|\varepsilon_{NC}|$ of the CMOS compatible $Hf_{0.5}Zr_{0.5}O_2$, can reach ~ $75\varepsilon_0$, which is more than three times that of HfO_2.

While the physics of the above FE material has been known for a long time, the really interesting part of the narrative began with a prediction by Salahuddin and Datta (2008) in 2007/2008. By connecting an FE in series with a normal dielectric (DE) (see Figure 3.2(c)) with proper capacitance matching between FE and DE, i.e. $|C_{FE}| \geq C_{DE}$, to eliminate the metastable state, the stack energy can be stable, as schematically shown in Figure 3.2(d). By improving the FE material and FE/DE interface quality and designing interesting experiments, experimentalists (Hoffmann et al. 2019; Yadav et al. 2019) have recently shown preliminary evidence to justify this model. Note that there remain many debates on this model such as the atomistic interaction between FE and DE at the FE/DE interface, and the role of interface traps in the energy stabilization process, as well as if this model applies to generic FE/DE stack and broad range of frequency up to GHz, etc. Therefore, more comprehensive experimental investigation on the structurally simple FE-DE stack is strongly suggested to gain deeper understanding on the FE switching dynamics and interplay between FE and DE layers, and achieve robust FE/DE performance before diving into the much more complex NC-FET experiment (Hoffman et al. 2020).

Although experimental demonstrations of a controllable NC state in an FE/DE stack have not become fully convincing, it does not prevent the exploration of the device physics of NC-FETs, especially for ultra-short-channel MOSFETs, thereby uncovering the opportunities and challenges of NC-FETs in modern MOSFET application. Early attempts at understanding the workings of NC-FETs either assumed the total FET gate capacitance to be a constant (Salahuddin and Datta 2008; Hoffmann et al. 2018), i.e. the bias-dependent channel capacitance was ignored, or were focused on performing intensive numerical simulations (Khan et al. 2011; Kobayashi and Hiramoto 2016; Ota et al. 2016; You and Su 2017; Rollo et al. 2018) or complex analytical modeling (Duarte et al. 2016; Pahwa et al. 2016) of a specific NC-FET, such as NC-FinFET and NC-2DFET, in which the generic device physics, design rule and major challenges of NC-FETs get obscured. Due to the lack of a good understanding of NC-FETs, but simply driven by the enthusiasm towards this novel "steep-slope" device, experimentalists have been indiscriminately trying to incorporate FE materials in the gate of various "X-FETs", where X can be SOI, Fin, nanowire (NW), 2D (such as MoS_2 and WSe_2) or 1D (such as carbon nanotube (CNT) and graphene nanoribbon (GNR)), generating numerous "NC-FETs" while ending up with numerous ambiguous or un-interpretable results in the last several years (Rusu et al. 2010; Cheng and Chin 2014; Krivokapic et al. 2017; McGuire et al. 2017; Saeidi et al. 2017; Si et al. 2018a, 2018b; Wang et al. 2019). In a recent work, Cao and Banerjee (2020) analytically formulated the slope of the transfer characteristics of a generic NC-FET structure, from which a physics-rich but concise *picture* is innovatively extracted to "*visualize*" the intriguing device physics and design space of NC-FETs (Cao and Banerjee 2020). This *picture* can not only serve as a convenient design platform for NC-FET experimentalists, but also leads to the discovery of a novel functionality of NC in modern MOSFET applications.

3.3. Understanding NC-FETs from scratch

3.3.1. *Electrostatics in a generic NC-FET*

The switching of FETs is realized by electrostatic (or capacitive) modulation of the potential of the channel through which current is conducted. Figure 3.3(a) depicts all of the relevant capacitors in a generic NC-FET structure. The discussion throughout this work is based on n-type devices, if not specified otherwise. C_{NC} and C_{OX} are the capacitances of the NC layer and gate oxide, respectively. C_Q, C_{trap}, C_{dep} and $C_{s/d,geo}$ represent the quantum, interface trap-induced, depletion, and source/drain geometrical capacitances, respectively, and can be grouped as gate voltage divider capacitance C_{div},

$$C_{div} = C_Q + C_{trap} + C_{dep} + C_{s,geo} + C_{d,geo} \qquad [3.3]$$

C_Q in FETs describes the change of electron charge density (Q_e) in response to channel potential (φ_{Ch}) modulation, and can be expressed as

$$C_Q = \frac{dQ_e}{d\varphi_{Ch}} = q^2 \cdot \sum_i \frac{DOS_i}{1+exp(\frac{E_{c,i}-E_f}{kT})} \qquad [3.4]$$

where Q_e is

$$Q_e = q \cdot \sum_i \int_{E_{c,i}}^{+\infty} \frac{DOS_i}{1+exp(\frac{E-E_f}{kT})} dE = q \cdot kT \cdot \sum_i DOS_i \cdot ln\left(1 + e^{\frac{E_f-E_{c,i}}{kT}}\right) \qquad [3.5]$$

where q is the elementary charge, i is the index of conduction modes, DOS_i is the density-of-states of the ith mode, $E_{c,i}$ is the energy level of the ith mode, and E_f is the Fermi level. C_Q approaches zero below threshold, while approaching $q^2 \cdot \Sigma_i DOS_i$ above threshold. C_{trap} is the trap state filling induced capacitance (~$dQ_{trap}/d\varphi_{Ch}$). C_{dep} is the geometrical capacitance of the depletion layer. $C_{s/d,geo}$ stem from the capacitive coupling from the source and drain. This capacitance also captures the effect of the fringing field from the source/drain towards the gate through the path inside the device in the short-channel condition (Pahwa et al. 2018). As indicated in the small-signal diagram in Figure 3.3(b), C_{div} prevents the channel potential φ_{Ch} from being efficiently modulated by the gate voltage, and thus, its unnecessary components (C_{trap}, C_{dep} and $C_{s/d,geo}$) are normally minimized in modern MOSFET design. Note that the source/drain-to-gate fringing capacitance (including overlap capacitance) $C_{fring,s/d\text{-}g}$ is directly connected between power rails (gate to source/drain), and hence it is irrelevant to central-channel potential modulation (that determines device performance) in any FETs with long-channel or excellent electrostatic integrity. Therefore, any steep slope of fabricated NC-FETs (Si et al. 2018a) in the long-channel condition should not be misattributed to $C_{fring,s/d\text{-}g}$. In the operation of NC-FETs, the negative C_{NC} induces a negative voltage drop (V_{NC}) across the NC layer, as illustrated in the conduction band diagram in Figure 3.3(c), to compensate for the positive voltage (V_{MOS}) on the underlying FET, thereby potentially reducing the total gate voltage (V_g=V_{NC}+V_{MOS}).

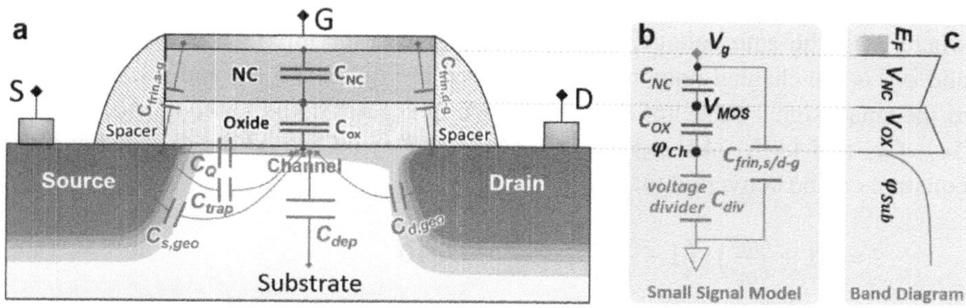

Figure 3.3. *Capacitors in a generic NC-FET structure and its small-signal model*

NOTES ON FIGURE 3.3.– *(a) Capacitors potentially involved in determining the channel potential of a generic NC-FET. C_Q, C_{dep}, $C_{s/d,geo}$ and C_{trap} are the quantum-, depletion-, s/d geometrical and trap-induced capacitors, respectively. C_{NC} and C_{OX} represent the capacitance of the NC and oxide layers, respectively. (b) Simplified small-signal model for NC-FETs. Note that in the small-signal model, the change in the values of the capacitances near the vicinity of a given DC bias point is negligible. C_{div} is the voltage divider capacitance that includes C_Q, C_{dep}, $C_{s/d,geo}$ and C_{trap}. The effect of $C_{frin,s/d\text{-}g}$ is screened by power rails, and hence it is irrelevant to channel potential control. (c) The conduction band diagram of NC-FETs under a positive gate bias. Note that due to negative C_{NC}, V_{NC} is negative.*

3.3.2. *Formulating switching slope of a generic NC-FET*

To quantify the operation mechanism of NC-FET, the inverse of slope or swing (S) of the entire I_d-V_g curve of a generic NC-FET is formulated in the form of decoupled contributions from electrostatics and the transport mechanism.

The drain current of an (n-type) FET can be expressed as

$$I_d = W \cdot v \cdot Q_e \qquad [3.6]$$

where W is the device width and v is the carrier velocity.

The slope of the I_d-V_g curve of an NC-FET is,

$$S = \left(\frac{d[logI_d]}{dV_g}\right)^{-1} = \frac{dV_{MOS}}{d\varphi_{Ch}} \cdot \frac{dV_g}{dV_{MOS}} \cdot \frac{d\varphi_{Ch}}{d[logI_d]} \qquad [3.7]$$

where V_g is the gate voltage, V_{MOS} is the voltage at the interface of NC and oxide, and φ_{Ch} is the channel potential, as illustrated in Figures 3.3 (b) and (c). According to the small-signal capacitor network shown in Figure 3.3(b), $(\Delta V_{MOS} - \Delta\varphi_{Ch})\cdot C_{OX} = \Delta\varphi_{Ch}\cdot C_{div}$ and $(\Delta V_g - \Delta V_{MOS})\cdot C_{NC} = -(\Delta V_g - \Delta V_{MOS})\cdot|C_{NC}|=\Delta V_{MOS}\cdot C_{MOS}$, above the equation can be converted to:

$$S = \left(1 + \frac{C_{div}}{C_{OX}}\right) \cdot \left(1 - \frac{C_{MOS}}{|C_{NC}|}\right) \cdot \frac{I_d d\varphi_{Ch}}{dI_d} \cdot ln10 \qquad [3.8]$$

C_{MOS} is the total gate capacitance of the FET under the NC layer

$$C_{MOS} = \frac{C_{div} \cdot C_{OX}}{C_{div} + C_{OX}} \qquad [3.9]$$

Substituting equations [3.3], [3.6] and [3.9] into equation [3.8], and considering the lowest conduction mode (i.e. i=1, DOS=DOS_1, E_c=$E_{c,1}$) only,

$$S = \left(1 + \frac{C_{div}}{C_{OX}}\right) \cdot \left(1 - \frac{1}{|C_{NC}|}\frac{C_{div}C_{OX}}{C_{div}+C_{OX}}\right) \cdot \frac{q \cdot kT \cdot DOS \cdot ln\left(1 + e^{\frac{E_f - E_c}{kT}}\right) d(-E_c/q)}{q \cdot kT \cdot DOS \cdot dln\left(1 + e^{\frac{E_f - E_c}{kT}}\right)} \cdot ln10$$

$$= \left(1 - C_{div}\left[\frac{1}{|C_{NC}|} - \frac{1}{C_{OX}}\right]\right) \cdot S_{TP} \qquad [3.10]$$

where

$$S_{TP} = \left(1 + e^{\frac{E_c - E_f}{kT}}\right) \cdot ln\left(1 + e^{\frac{E_f - E_c}{kT}}\right) \cdot \frac{kT}{q} \cdot ln10 \qquad [3.11]$$

S_{TP} represents the thermionic transport mechanism limited S. The expression of S_{TP} only takes into account the lowest conduction mode for conciseness of analytical expression, which does not affect the conclusions. Note that all numerical simulations shown in subsequent sections have rigorously included all conduction modes.

The battle field for any steep-slope device to pursue small SS is the subthreshold regime, where E_f is well below E_c. In this regime, $S_{NC\text{-}FET}$ is reduced to

$$SS = SS_{TP} \cdot \frac{m}{A_V} = \left(1 - C_{div,<Vth}\left[\frac{1}{|C_{NC}|} - \frac{1}{C_{OX}}\right]\right) \cdot \frac{kT}{q} \cdot ln10 \qquad [3.12]$$

where

$$SS_{TP} = \frac{kT}{q} \cdot ln10 \qquad [3.13]$$

$$m = \frac{dV_{MOS}}{d\varphi_{Ch}} = 1 + \frac{C_{div,<Vth}}{C_{OX}} \qquad [3.14]$$

$$A_V = \frac{dV_{MOS}}{dV_g} = \frac{|C_{NC}|}{|C_{NC}| - C_{MOS,<Vth}} \qquad [3.15]$$

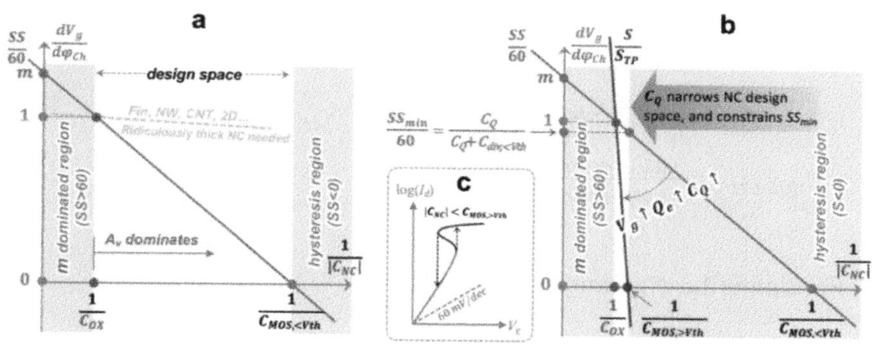

Figure 3.4. *Impact of electrostatics and quantum capacitance on NC-FET design space*

NOTES ON FIGURE 3.4.– *(a) Visualizing the SS formula and its physics. In the shaded region on the LHS, m dominates, thus SS>60 mV per decade. Room temperature is assumed if not specified otherwise. In the shaded region on the RHS, SS<0, represents the appearance of hysteresis. The design space here is between the two-shaded regions, where A_V dominates and 0<SS<60. The dashed line illustrates the ultra-small negative slope of SS/60 line. (b) A picture that captures all of the essential quasi-static device physics of NC-FETs. Note that in the near- and above-threshold region, the SS/60 line transforms into the S/S_{TP} line, which rotates clockwise due to the rapidly increasing quantum capacitance C_Q with gate bias until* *a nearly vertical position is reached when the device is fully turned on. Since S in the entire range of I_d-V_g curve should be guaranteed to be positive, in order to prevent hysteresis in both sub- and super-threshold regimes, the lower bound of $1/|C_{NC}|$ is extended to $1/C_{MOS,>Vth}$, leading to a narrow NC design space and a minimum hysteresis-free SS ($=C_Q/(C_Q+C_{div,<Vth})$) at $1/|C_{NC}|=1/C_{MOS,>Vth}$. (c) I_d-V_g curve of an example NC-FET, in which $1/|C_{NC}|$ is designed to be within the shaded region on the RHS of b. Although SS (brown colored portion) is small, negative S (i.e. hysteresis) appears in the near- or above-threshold region (blue colored portion).*

SS_{TP} is exclusively determined by the transport mechanism, specifically thermionic emission, which has no difference compared to MOSFETs. At room temperature, SS_{TP} gives the famous thermionic *SS* lower limit of *60 mV per decade*, often referred to as "Boltzmann tyranny". m is called the body factor that captures the gate efficiency of the FET under the NC layer, and A_V is the voltage gain introduced by the NC layer, which is expected to be larger than 1 (Lundstrom 2006). In other words, m and A_V compete with each other in determining the final *SS*. $C_{div,<Vth}$ and $C_{MOS,<Vth}$ represent C_{div} and C_{MOS}, respectively, in the subthreshold regime. Note that C_Q is negligibly small in the subthreshold regime (see equation [3.4]), thus

$$C_{div,<Vth} = C_{trap} + C_{dep} + C_{s,geo} + C_{d,geo} \quad [3.16]$$

$$C_{MOS,<Vth} = \frac{C_{div,<Vth} \cdot C_{OX}}{C_{div,<Vth} + C_{OX}} \quad [3.17]$$

3.4. Fundamental challenges of NC-FET

3.4.1. *NC does not help good FETs*

The detailed physics of equations [3.12]–[3.17] can be "visualized" in Figure 3.4(a). As shown, when $|C_{NC}|$ is larger than C_{OX}, m wins the competition, and $SS > 60$ mV per decade (shaded region on the left). When $|C_{NC}|$ is smaller than C_{OX}, A_V dominates *SS*, and enables sub-60 *SS*.

However, $|C_{NC}|$ should not be smaller than $C_{MOS,<Vth}$, otherwise A_V and *SS* become smaller than zero (see equation [3.15] and the shaded region on the right in Figure 3.4(a)), and the total gate capacitance $C_g=|C_{NC}|\cdot C_{MOS,<Vth}/(|C_{NC}|-C_{MOS,<Vth})$ (note $C_{NC}= -|C_{NC}|$) is negative, indicating that the metastable state of the NC layer is not well stabilized by the FET structure underneath. In the charge-voltage or current-voltage characteristics, the metastable negative C_g or negative *SS* portion is reflected as hysteresis (Salahuddin and Datta 2008), resembling that in a ferroelectric memory

device (Miller and McWhorter 1992). As indicated in equation [3.12], $C_{div,<Vth}$ represents the slope of the $SS/60$ line. As long as $C_{div,<Vth}$ is not too large compared to C_{OX}, the design space of $1/|C_{NC}|$ (normally spanned by designing the NC layer thickness, T_{NC}) for sub-60 SS, i.e. between $1/C_{MOS,<Vth}$ (= $1/C_{OX}$+$1/C_{div,<Vth}$) and $1/C_{OX}$, is reasonably spacious. Meanwhile, $C_{div,<Vth}$ should not be too small either, otherwise $|C_{NC}|$ is required to be equally small to achieve steep slope, as illustrated by the dashed line in Figure 3.4(a). Considering the generally large absolute value of the dielectric constant of ferroelectric materials ($|\varepsilon_{NC}|$) (Miller and McWhorter 1992; Salahuddin and Datta 2008; Saeidi et al. 2017), small $|C_{NC}|$ translates to a very thick NC layer, which is impractical for state-of-the-art very-large-scale-integration (VLSI) technology. Modern MOSFETs have evolved into the ultrathin-body (UTB) era, in the form of semiconductor-on-insulator (SOI) (Mistry et al. 1997), Fin-FET (Auth et al. 2012), NW-FET (Appenzeller et al. 2008), CNT-FET (Appenzeller et al. 2008) and 2D-FET (Cao et al. 2015), accompanied with the requirement of high-quality material/interface and low channel doping (to suppress impurity scattering and performance variation), all of which are the constituents of a "good" FET for optimal gate efficiency and current drive. Unfortunately, these designs result in small $C_{s/d,geo}$, C_{dep} and C_{trap}, resulting in very small $C_{div,<Vth}$ and hence small $C_{MOS,<Vth}$ (see equation [3.12]). In other words, NC does not help these "good" MOSFET platforms to achieve steep slope. Note that many researchers may intuitively arrive at the conclusion that the UTB of modern MOSFETs can serve as a large channel capacitance or a large effective C_{dep} (see Figure 3.3), which, however, is a fundamental misunderstanding of FET physics. In UTB MOSFET, the body terminal is absent, i.e. the other side (with respect to the gate) of UTB is electrically floating. Therefore, the thickness of UTB does not determine the channel capacitance or effective C_{dep}. In fact, UTB suppresses $C_{div,<Vth}$ and hence $C_{MOS,<Vth}$ (=C_{OX} $C_{div,<Vth}$/(C_{OX}+$C_{div,<Vth}$)). Numerical simulation, by solving Poisson's and drift-diffusion equations in the entire device region, is performed to obtain the total gate capacitance C_{MOS} of a Si SOI MOSFET versus gate bias V_{MOS}.

The doping density in the channel is set to be ~1×10^{16} cm^{-3}. As shown in Figure 3.5(a), C_{MOS} of a trap-free device in the subthreshold regime (-0.9V – -0.2 V) is negligibly small, with respect to gate oxide capacitance C_{OX}. As a result, it is difficult to achieve sub-60 SS, unless an unfeasibly thick NC layer is used. From an NC operation point of view, subthreshold charge density $Q_{MOS,<Vth}$ (=$C_{MOS,<Vth}\cdot V_g$) in these "good" FETs induces (through electric displacement field D (=$C_{MOS,<Vth}\cdot V_g$), which is continuous across oxide/NC interface) negligible polarization in NC, and thus miniscule voltage gain A_V(=$|C_{NC}|$/($|C_{NC}|$-$C_{MOS,<Vth}$)) (see equation [3.15]). Therefore, contrary to MOSFET design, a voltage divider (i.e. finite $C_{div,<Vth}$) should be

intentionally introduced into NC-FETs to exploit A_v benefits. It is worthwhile to note that a mid-gap trap density of ~1 ×10^{11} cm^{-2} can introduce appreciable capacitance into $C_{MOS,<Vth}$. In other words, NC benefit and steep slope may be more readily observable in devices in which defective materials such as 2D, organic, amorphous and polycrystalline materials, and poor interfaces such as III–V/Oxide and Ge/Oxide are involved, compared to high-quality Si devices.

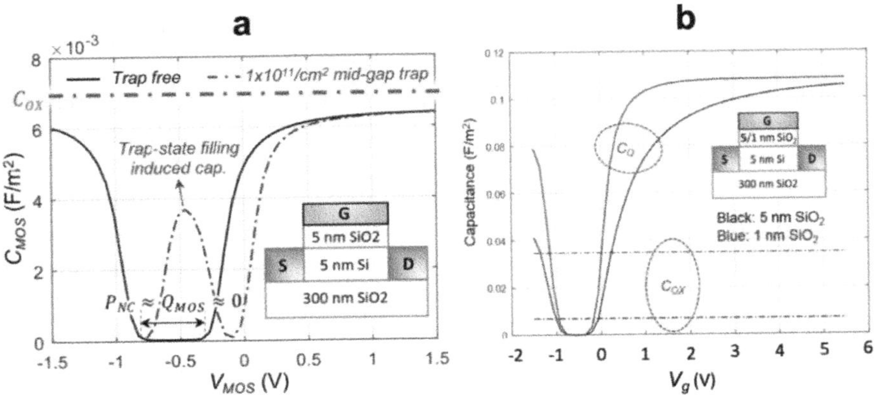

Figure 3.5. *The limitation of good FET platform for NC-FET construction*

NOTES ON FIGURE 3.5.– *(a) Simulated C_{MOS} in an ultra-thin Si SOI MOSFET (see inset sketch) with and without traps. Note that trap states can introduce appreciable C_{trap}, and hence $C_{div,<Vth}$. (b) Simulated C_Q in an ultra-thin Si SOI MOSFET. Even in such a thin film, C_Q is still much larger with respect to the 1 nm SiO_2 gate oxide capacitance C_{OX}. Note that C_Q for 1 nm and 5 nm SiO_2 gate oxide will converge to the same value when V_g keeps increasing, i.e. the maximum C_Q is independent of gate oxide thickness.*

Although required $C_{div,<Vth}$ (as discussed above) is not feasible in state-of-the-art MOSFET platforms, it can be introduced in unconventional ways. For example, in traditional bulk MOSFET technology, $C_{div,<Vth}$ can be realized in the form of high doping induced C_{dep}; in SOI structure, the buried oxide (BOX) can be designed to be comparable with gate oxide in terms of capacitance value, thereby serving as an effective C_{dep}; in poorly designed ultra-short-channel devices and/or devices with poor interface, $C_{s/d,geo}$ and/or C_{trap} can, although in undesired ways, supply an appreciable $C_{div,<Vth}$.

3.4.2. *Quantum capacitance may "kill" NC-FETs*

Although, as discussed above, the tuning of $C_{div,<Vth}$ is non-trivial, the greatest challenge of NC-FETs arises from the rapidly increasing channel electron density (or C_Q in equation [3.4]) with gate bias near and above the threshold.

As shown in the simulation results in Figure 3.5(b), even in the ultra-thin channel of a 5-nm UTB SOI MOSFET with EOT of 1 nm, C_Q is still much larger than C_{OX}. Since C_{div} is the slope of S/S_{TP} versus $1/|C_{NC}|$ (see equation [3.10]), increased C_Q, and hence, C_{div} is equivalent to rotating the S/S_{TP} line clockwise with respect to the *SS*/60 line, as illustrated in Figure 3.4(b). In order to guarantee hysteresis-free *I-V* characteristics, or positive *S*, not only in the subthreshold regime, but also near- and above-threshold regimes, the lower bound of the forbidden hysteresis region (shaded region on the RHS of Figure 3.4(b)) of $1/|C_{NC}|$ gets extended from $1/C_{MOS,<Vth}$ $(=1/C_{div,<Vth}+1/C_{OX})$ to $1/C_{MOS,>Vth}= (1/C_{div,>Vth}+1/C_{OX}\approx 1/C_Q+1/C_{OX})$, which is very close to $1/C_{OX}$, since C_Q in near- and above-threshold regimes generally becomes much larger than C_{OX} in typical Si or Ge MOSFETs. In other words, the design space of NC is significantly narrowed by C_Q. If $|C_{NC}|$ is designed to be smaller than $C_{MOS,>Vth}$, although *SS* can be noticeably lower than 60 mV per decade, a hysteresis appears at the near- or above-threshold regime, as schematically illustrated in Figure 3.4(c), ending up with a memory device rather than a logic device. In the design of an FET structured FE memory (Miller and McWhorter 1992), in which hysteresis is a desired property, the oxide layer is normally made very thin, or completely removed if the employed FE material is a good insulator. Interestingly, the reason for these devices always exhibiting hysteresis becomes very clear from Figure 3.4(b). With finite oxide layer thickness, if $1/|C_{NC}|$ is designed to be within the "*m* dominated region" (shaded region on the LHS of Figure 3.4(b)), *S* in the entire range of I_d-V_g curve is guaranteed to be positive, i.e. the I_d-V_g curve is free of hysteresis. On the contrary, in the absence of an oxide layer (i.e. $1/C_{OX} \approx 0$), the "*m* dominated region" disappears, leaving negligible $1/|C_{NC}|$ design space (between 0 and $1/C_{MOS,>Vth} \approx 1/C_Q$), in which hysteresis is absent. Hence, in the feasible $1/|C_{NC}|$ range, *S* is always smaller than 0, which guarantees hysteresis.

As indicated in Figure 3.4(b), minimum *SS* (SS_{min}) that does not suffer from hysteresis can be obtained by matching $|C_{NC}|$ to $C_{MOS,>Vth}$, i.e. by substituting $|C_{NC}| = C_{MOS,>Vth}$ into the *SS* formula in equation [3.12],

$$SS_{min} = \left(1 - C_{div,<Vth}\left[\frac{1}{C_{MOS,>Vth}} - \frac{1}{C_{OX}}\right]\right)\cdot\frac{kT}{q}\cdot ln10 = \frac{C_Q}{C_Q+C_{div,<Vth}}\cdot\frac{kT}{q}\cdot ln10 \quad [3.18]$$

3.5. Design and optimization of NC-FET

3.5.1. *Designing NC-FET in the quantum capacitance limit*

Inspired by equation [3.18], lowering C_Q, desirably into the quantum capacitance limit (C_Q<C_{OX}), seems to be an effective direction in enlarging the NC design space and reducing SS_{min}. According to the dependence of C_Q on DOS (equation [3.4]), low-DOS material systems could help. In order to illustrate this, SS_{min} is evaluated for Si (as reference), $Ga_{0.47}In_{0.53}As$ and monolayer MoS_2, as shown in Figure 3.6(a), versus $C_{div,<Vth}$, which is normalized with respect to the equivalent oxide thickness (EOT) for the convenience of comparison. If not specified otherwise, $Hf_{0.5}Zr_{0.5}O_2$ is used as the NC material in the calculation. As shown, the low-DOS $Ga_{0.47}In_{0.53}As$ can provide much smaller SS_{min}, with respect to Si and MoS_2. However, it is technologically rather difficult to implement large $C_{div,<Vth}$ with EOT less than 3 nm (i.e. close to C_{OX} in state-of-the-art MOSFETs); thus, hysteresis-free sub-60 *SS* and sub-50 *SS* can hardly be achieved in Si/MoS_2 and $Ga_{0.47}In_{0.53}As$ based NC-FETs, respectively.

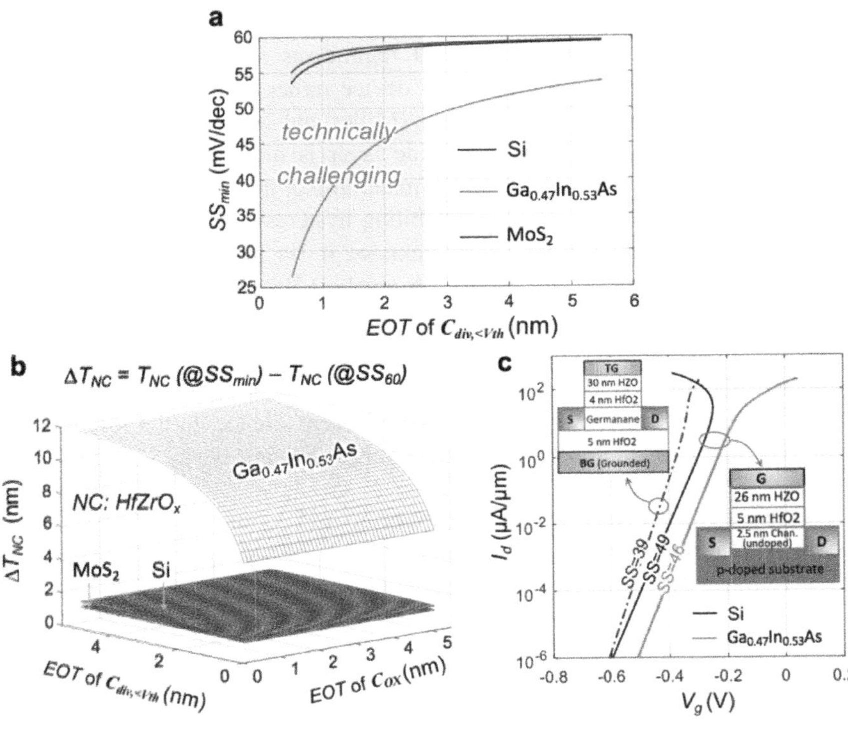

Figure 3.6. *Impact of density-of-states on SS_{min} and NC design space*

NOTES ON FIGURE 3.6.– *(a) SS_{min} at room temperature and (b) NC thickness design space (ΔT_{NC}) for Si, $Ga_{0.47}In_{0.53}As$ and monolayer MoS_2. Note that typical DOS (conduction band) values are as follows: Si: 3.2×10^{19} per cm^3; $Ga_{0.47}In_{0.53}As$: 2.1×10^{17} per cm^3; MoS_2: 2.4×10^{14} per cm^2; Germanane: 2.9×10^{13} per cm^2. The low-DOS $Ga_{0.47}In_{0.53}As$ exhibits significant advantages. $C_{div,<Vth}$ and C_{OX} are normalized with respect to the equivalent oxide thickness (EOT). (c) With the same device structure/size, $Ga_{0.47}In_{0.53}As$ device can (in principle) achieve smaller SS, with respect to Si device, without hysteresis. SS of monolayer Germanane device can reach 39 mV per decade due to its low-DOS.*

Figure 3.6(b) shows the NC thickness design space (ΔT_{NC}, essentially the width of the design space in Figure 3.4(b)) in terms of $C_{div,<Vth}$ and C_{OX}. ΔT_{NC} of Si and MoS_2-based NC-FETs almost vanish due to their large DOS, which is consistent with their near-60 SS_{min} shown in Figure 3.6(a). The low-DOS of $Ga_{0.47}In_{0.53}As$ can help derive a feasible ΔT_{NC} of ~12 nm due to its low C_Q. Further reduction of the DOS of GaInAs system by increasing the In content is possible, but will result in lowered bandgap that degrades on–off current ratio. To further confirm the effect of DOS and hence C_Q, numerical simulations are carried out to obtain the I_d-V_g curves of Si and $Ga_{0.47}In_{0.53}As$-based NC-FETs, as shown in Figure 3.6(c). $C_{div,<Vth}$ in bulk Si and $Ga_{0.47}In_{0.53}As$ devices are intentionally implemented as C_{dep} with an engineered doping profile, as illustrated in the inset figure on the RHS. With exactly the same device structure/size, the $Ga_{0.47}In_{0.53}As$ device shows a hysteresis-free *SS* of ~46 mV per decade, while the Si device shows an *SS* of ~49 mV per decade suffering from hysteresis, which the is consistent with the predicted hysteresis-free SS_{min} in Figure 3.6(a). The thinness of 2D material is intrinsically beneficial in terms of achieving low DOS. However, the most widely studied MoS_2, although is promising for short-channel MOSFETs (Cao et al. 2015), suffers due to its large electron effective mass (~$0.6m_0$) that results in high DOS. In contrast, Germanane (also a single-layer 2D semiconductor) possesses a much smaller electron effective mass (~$0.07m_0$) and a finite band gap of ~1.5 eV (Ghosh et al. 2014), and can thus be an ideal channel material for NC-FET design. Figure 3.6(c) also shows the simulated I_d-V_g curve of a Germanane NC-FET. The atomic scale thickness of Germanane, as well as the lack of effective doping techniques, prevents Germanane-based NC-FETs from exploiting C_{dep}. Instead, an HfO_2-based BOX layer is employed to act as an effective C_{dep}, as illustrated in the inset on the LHS of Figure 3.6(c). By optimizing the oxide and NC thicknesses, *SS* of Germanane-based NC-FETs can be as low as 39 mV per decade.

3.5.2. *The role of NC nonlinearity*

It is worthwhile to mention that the *P–E* characteristics in the NC region of an FE material are not strictly linear, as shown in Figure 3.2(b), i.e. $|C_{NC}|$ is not a constant with bias or charge density. Using the analysis developed above, it is found that this secondary effect can slightly enlarge the NC design space, and may help a bit in deriving smaller SS_{min}.

The linear term (with coefficient α) in equation [3.2] gives a constant (bias independent)

$$\varepsilon_{NC,0} = \left(\varepsilon_0 - \frac{1}{2|\alpha|}\right) \qquad [3.19]$$

(and hence C_{NC}) as the *0*th order approximation, which is valid in low-to-medium electric-field range, as indicated in Figure 3.2(b). Note that α is negative, which is the source of negative capacitance. When the electric field approaches the coercive field of FE (the two turning points in Figure 3.2(b)), the nonlinearity terms (with β and γ coefficients) begin to deviate *P–E* relation from linearity. To simplify the analysis, the γ coefficient is set to be zero in this work. As a result, $|\varepsilon_{NC}|$ (and hence $|C_{NC}|$) increases with bias or charge density.

$$\varepsilon_{NC} = \varepsilon_{NC,0} / \left(1 - \frac{6\beta}{|\alpha|} Q_{MOS}^2\right) \qquad [3.20]$$

This nonlinearity introduces a damping term to the slopes of both *SS*/60 and S/S_{TP} lines versus NC layer thickness T_{NC},

$$\frac{SS}{60} = 1 + \frac{C_{div,<Vth}}{C_{OX}} - \frac{C_{div,<Vth}}{|\varepsilon_{NC,0}|}\left(1 - \frac{6\beta}{|\alpha|} Q_{MOS}^2\right) T_{NC} \qquad [3.21]$$

$$\frac{S}{S_{TP}} = \frac{C_{div}}{C_{MOS}} - \left(\frac{C_{div,<Vth} + C_Q}{|\varepsilon_{NC,0}|}\right)\left(1 - \frac{6\beta}{|\alpha|} Q_{MOS}^2\right) T_{NC} \qquad [3.22]$$

which can be depicted as a counterclockwise rotation of both S/S_{TP} and *SS*/60 lines, leading to enlarged NC design space, at the expense of increased *SS*, as illustrated in Figures 3.7 (a), (b) and (c) show simulated I_d-V_g and *S*-I_d characteristics of a Si SOI NC-FET (inset), respectively, with different nonlinearity term β. With increased β, hysteresis is effectively eliminated, at the expense of increased *SS*, which is consistent with the uncovered role of NC nonlinearity in Figure 3.7(a).

Note that the simulation here is performed on a fixed device structure/size. With larger β, the NC design space is wider, SS_{min} can be lower if the device structure/size is optimized. Note that although the effect of NC nonlinearity on the design space of hysteresis-free *SS* has been discussed in Jain and Alam (2014), the clear physics picture developed in Cao and Banerjee (2020) allows us to explain this effect much more effectively, in terms of clarity, depth and generality.

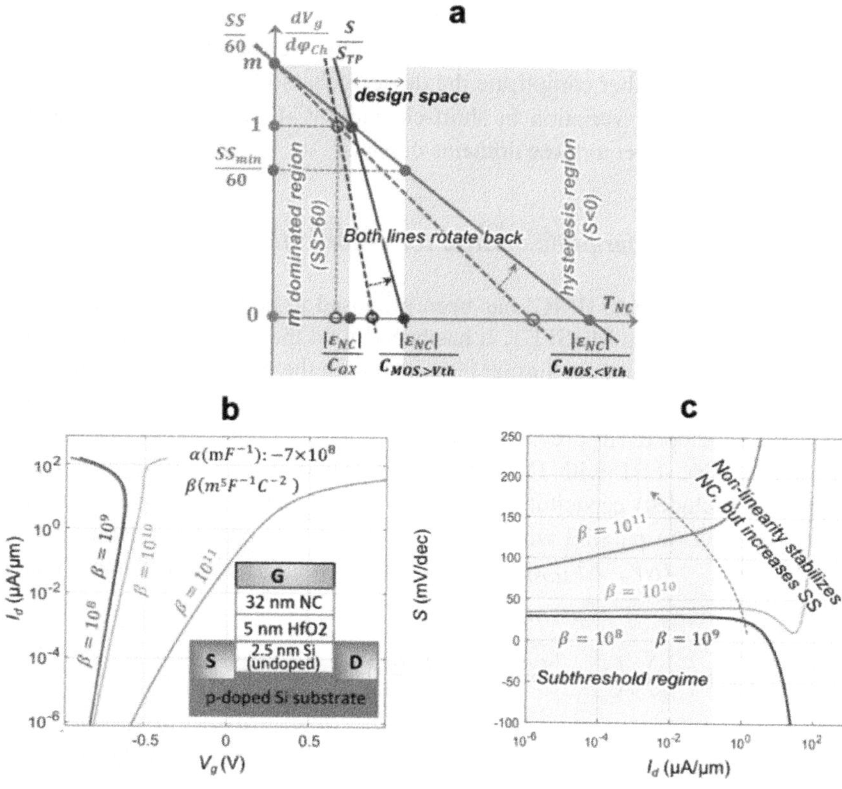

Figure 3.7. *Nonlinearity of NC*

NOTES ON FIGURE 3.7.– *(a) The nonlinearity of NC dynamically increases* $|\varepsilon_{NC}|$ *(or* $|C_{NC}|$*) with the charge density on the MOSFET gate,* Q_{MOS} *or gate bias, and rotates back both* S/S_{TP} *and SS/60 lines, which enlarges the NC design space constrained by* C_Q*, but increases SS. (b), (c) Numerical simulation of an NC-FET (inset) shows that with stronger NC nonlinearity, i.e. larger* β*, the hysteresis is suppressed, but SS gets degraded.*

It is worth mentioning that the appearance of multi-domains and/or poly-crystals, with respect to the ideal single-domain case considered in this work, in the ferroelectric layer may introduce additional and unpredictable nonlinearity, because these discrete domains/crystals are unlikely to have uniform properties (*P–E* curves). As well known, today's scaled transistors already get severely loaded by "bias-dependent" capacitances such as gate-to-source/drain capacitances $C_{g,s/d}$ and Miller capacitance, which makes the device/circuit behavior difficult to be precisely evaluated. The multi-domain and poly-crystalline ferroelectric layer added in NC-FETs, not only further complicate the device behavior, but also serve as a major source of performance variation in short-channel NC-FETs that have comparable feature sizes with respect to these domains/crystals.

3.5.3. *IMG: borrow parasitic charge for polarization in NC*

An internal metal gate (IMG) has been proposed to be inserted between the NC layer and the underlying MOSFET. It has been found that the IMG-induced fringing capacitance (that helps induce further polarization in the NC layer) can help improve *SS* (Pahwa et al. 2018). This section clarifies the underlying physics in the framework developed in previous sections. Figure 3.8(a) shows the capacitor network in an NC-FET with IMG. It can be found that two additional fringing (overlapping included) capacitors, $C_{frin,s/d\text{-}IMG}$, appear between the source/drain and the floating IMG, in parallel with C_{MOS}, the charge conservation relation gives us $(\Delta V_g\text{-}\Delta V_{MOS})\cdot C_{NC}=\text{-}(\Delta V_g\text{-}\ \Delta V_{MOS})\cdot|C_{NC}|=\Delta V_{MOS}\cdot(C_{MOS}+\ C_{frin,s/d\text{-}IMG})$. *S* and *SS* should be updated to,

$$S=\left(1+\frac{C_{div}}{C_{OX}}\right)\cdot\left(1-\frac{C_{MOS}+C_{frin,s/d-IMG}}{|C_{NC}|}\right)\cdot S_{TP} \quad [3.23]$$

$$SS=SS_{TP}\cdot\frac{m}{A_V}=\left(1+\frac{C_{div,<Vth}}{C_{OX}}\right)\cdot\left(1-\frac{C_{MOS,<Vth}+C_{frin,\frac{s}{d}-IMG}}{|C_{NC}|}\right)\cdot\frac{kT}{q}\cdot ln10 \quad [3.24]$$

where

$$A_V=\frac{dV_{MOS}}{dV_g}=\frac{|C_{NC}|}{|C_{NC}|-C_{frin,s/d-IMG}-C_{MOS,<Vth}} \quad [3.25]$$

Compared to equation [3.15], A_V here does not rely on $C_{MOS,<Vth}$ anymore, $C_{frin,s/d\text{-}IMG}$ can be used to effectively increase the amplification due to the presence of IMG. S/S_{TP} and *SS/60* are plotted in Figure 3.8(b). The minimum *SS* (SS_{min}) that does not suffer from hysteresis can be obtained at $|C_{NC}|=C_{MOS,>Vth}+\ C_{frin,s/d\text{-}IMG}$,

$$SS_{min}=\frac{C_Q}{C_Q+C_{div,<Vth}}\cdot\frac{C_{MOS,>Vth}}{C_{MOS,>Vth}+C_{frin,s/d-IMG}}\cdot\frac{kT}{q}\cdot ln10 \quad [3.26]$$

The first term is limited by the quantum capacitance, while the second term reflects the contribution of IMG. To the first-order approximation for "good" FETs, $C_Q >> C_{div,<Vth}$, and $C_{MOS,>Vth} \sim C_{OX}$, thus, at room temperature,

$$SS_{min} \sim \left(\frac{C_{OX}}{C_{OX}+C_{frin,s/d-IMG}}\right) \cdot 60 \quad [3.27]$$

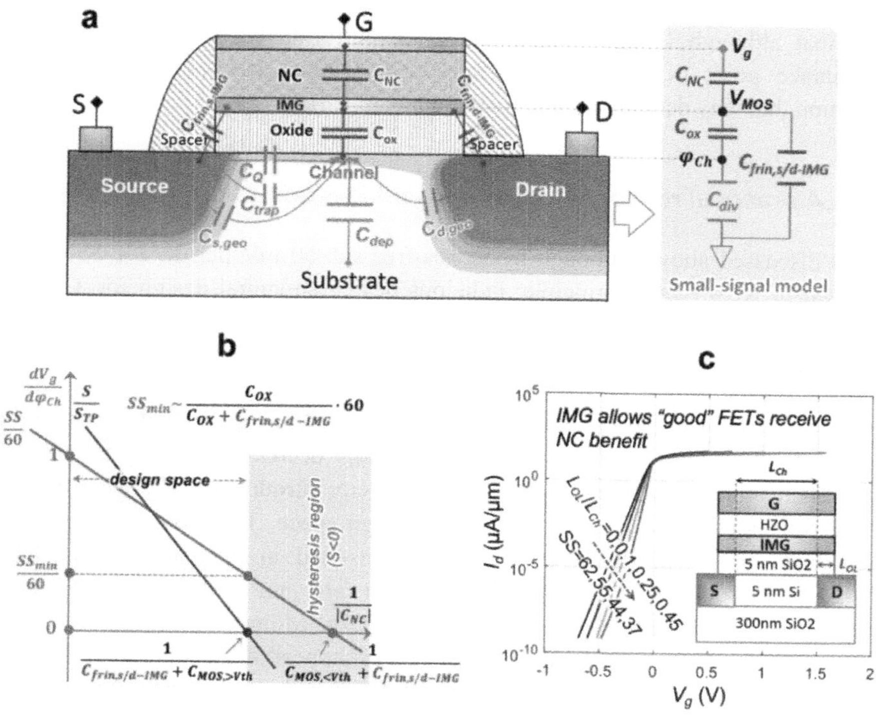

Figure 3.8. *The physics of internal metal gate-based NC-FETs*

NOTES ON FIGURE 3.8.– *(a) Capacitors in an NC-FET with internal metal gate (IMG). (b) IMG can borrow charge from fringing/overlap capacitance in the subthreshold regime, thereby unlocking polarization from C_Q, and enlarging the NC design space for small SS_{min}. (c) Simulation results based on a Si SOI MOSFET verify the uncovered IMG mechanism. The larger the L_{OL}/L_{Ch}, the more the NC benefits.*

Interestingly, the NC design space (between 0 and $1/(C_{OX}+C_{frin,s/d-IMG})$) and SS_{min} becomes irrelevant with respect to C_Q, and only dependent on C_{OX} and $C_{frin,s/d-IMG}$. This is because $C_{frin,s/d-IMG}$ provide parasitic charge in both subthreshold and above-threshold regimes to induce polarization in NC, which unlocks the dependence of polarization on mobile charge (or C_Q) in "good" FETs. As indicated by the SS_{min} formula, SS_{min} can be effectively reduced by increasing the ratio of $C_{frin,s/d\text{-}IMG}$ to C_{OX}, which perfectly explains the previous report that short-channel NC FinFETs with IMG can achieve small *SS*. For long-channel NC-FETs, large overlap between source/drain and IMG, with respect to channel length, can be designed to reduce *SS*, which is confirmed by simulation of a Si SOI NC-FET, as shown in Figure 3.8(c). Note that although IMG is beneficial to the performance of NC-FETs from a capacitance point of view, it will introduce floating gate effect and domain formation that may destabilize NC (Hoffmann et al. 2018).

3.5.4. *A practical role of NC for FETs: voltage-loss saver*

As discussed above, although hysteresis-free sub-60 mV per decade *SS* can be achieved in NC-FETs in principle, judicious device structural design for a proper $C_{div,<Vth}$, rigorous matching between NC and the total gate capacitance (C_{MOS}), which involves a strongly bias-dependent C_Q, and non-trivial fabrication efforts are required to demonstrate a prototype NC-FET. Moreover, there is a critical thickness for ferroelectric materials (~5 nm for $BiFeO_3$) (Steffes et al. 2019), below which the ferroelectricity begins to degrade, and hence $|\varepsilon_{NC}|$ decreases, leading to reduced $|C_{NC}|$. Such a thickness dependence of ferroelectricity introduces additional difficulty to precisely match $C_{MOS,>Vth}$ in the narrow design space. Therefore, all fabricated "NC-FETs" in which FE layers are simply inserted in the gate of Fin-FETs, NW-FETs, Tunnel-FETs or 2D-FETs, which are not suitable for NC-FET construction, are unlikely to be genuine NC-FETs. Moreover, almost all of the observed steep *SS*s in those fabricated "NC-FETs" appear at low drain current level (close to the off-current and/or current noise floor) only, as reflected by the data collected (Chung et al. 2017; Fan et al. 2017; Lee et al. 2017; Yu et al. 2017; Zhou et al. 2017; Si et al. 2018a) in Figure 3.9, which not only obscures their prospects, but also, again, casts doubt on their claims of being NC-FETs, since those observations are not consistent with or supported by the fundamental physics-based predictions of the subthreshold behavior of NC-FETs in this work and other relevant work in the literature. In fact, more and more studies (Jin et al. 2018; Li and Toriumi 2018; Saha et al. 2018; Wang et al. 2018) have indicated that the steep *SS*s observed in many of those fabricated "NC-FETs" can be attributed to the transient effects (not captured in this work because of the ideal steady-state model employed, aimed at uncovering the essential physics and evaluating the upper-limit of NC-FET performance) during the measurement and/or ferroelectric polarization dynamics, instead of the negative capacitance effect through capacitance matching, which provide further support to the analysis and the insightful conclusions of this work, albeit from a different perspective. Here, based on the *SS* physics of NC-FETs developed above, we also provide our insight/suggestion to those who are trying to interpret the observed steep *SS*s in various reported "NC-FETs" from a transient effect point of view. Essentially, to achieve subthermionic *SS*, the polarization dynamics should allow the FE layer in the subthreshold regime to be greatly depolarized and allow the weak residual polarization to be self-sustained (i.e. $D \approx Q_{MOS,<Vth} \approx 0$ or $P \approx -\varepsilon_0 E$), instead of relying on the charge compensation (i.e. $\varepsilon_0 E \approx 0$ or $P \approx D \approx Q_{MOS,<Vth}$) from the FET underneath, since in modern FETs, $C_{div,<Vth}$ and $C_{MOS,<Vth}$ are negligibly small. It is worth noting that such transient effects induced steep *SS*s, imposed strict constraints on bias sweeping rate/direction, and operating frequency (Jin et al. 2018; Wang et al. 2018), and are therefore of limited practical use in logic transistor applications, and hence, irrelevant to the discussion in this work. It is also worthwhile to mention that the polarization response of ferroelectric materials must be enhanced (Kobayashi et al. 2016; Yuan et al. 2016) to allow the NC to function effectively at high frequencies (GHz range).

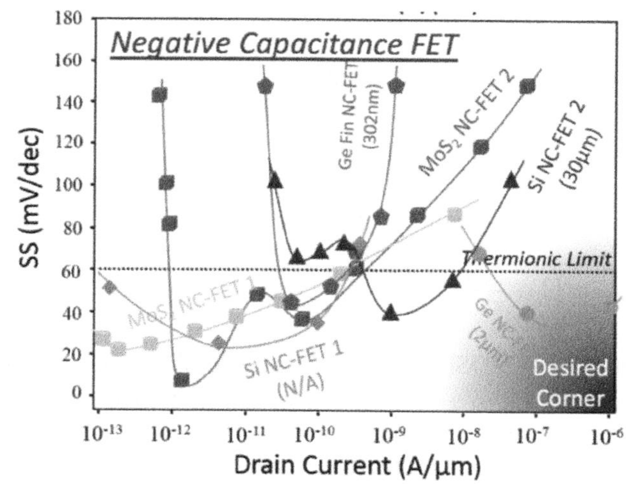

Figure 3.9. *State-of-the-art SS data from reported NC-FET experiments*

In fact, in terms of saving switching energy, NC-FET does not have to be designed as a steep-slope device. As indicated in Figure 3.4(b), the S/S_{TP} line rotates with bias about a constant (bias and NC material independent) point ($1/C_{OX}$, 1), which provides an important implication for a new role NC can play. To make it clearer, *SS* formula for short-channel MOSFETs and NC-FETs, and *S* formula in the near- and above-threshold regimes for MOSFETs and NC-FETs of any scale, are reorganized below. In the near- and above-threshold regimes, C_Q dominates C_{div}, i.e. $C_{div} \approx C_Q$,

$$S = \left(1 - C_Q\left[\frac{1}{|C_{NC}|} - \frac{1}{C_{OX}}\right]\right) \cdot S_{TP} \quad [3.28]$$

As discussed above, in the subthreshold regime of a poorly designed short-channel NC-FET, short-channel capacitance C_{SCE}, which is mainly composed of source/drain geometry capacitance $C_{s/d,geo}$, dominates C_{div}, i.e. $C_{div} \approx C_{SCE}$, $S_{TP}=60$,

$$SS = \left(1 - C_{SCE}\left[\frac{1}{|C_{NC}|} - \frac{1}{C_{OX}}\right]\right) \cdot 60 \quad [3.29]$$

For MOSFETs in which the NC layer is absent, the slope of the I_d-V_g curve can be derived by making $|C_{NC}|$ infinitely large in the above equations,

$$S_{MOSFET} = \left(\frac{d[\log I_d]}{dV_g}\right)^{-1} = \frac{dV_g}{d\varphi_{Ch}} \cdot \frac{d\varphi_{Ch}}{d[\log I_d]} = \left(1 + \frac{C_{div}}{C_{OX}}\right) \cdot S_{TP} \quad [3.30]$$

In the near- and above-threshold regimes, $C_{div} \approx C_Q$, then,

$$S_{MOSFET} = \left(1 + \frac{C_Q}{C_{OX}}\right) \cdot S_{TP} \quad [3.31]$$

In the subthreshold regime of a short-channel MOSFET, $C_{div} \approx C_{SCE}$, $S_{TP}=60$,

$$SS_{MOSFET} = \left(1 + \frac{C_{SCE}}{C_{OX}}\right) \cdot 60 \quad [3.32]$$

In the short-channel MOSFETs, C_{div} is mainly attributed to short-channel effect (SCE), specifically, $C_{div}=C_{SCE}=C_{s,geo}+C_{d,geo}$. C_{SCE} prevents gate voltage in the subthreshold regime from dropping entirely in the channel, i.e. it causes a subthreshold gate voltage loss on the gate oxide, which is reflected by a poor SS, $SS_{MOSFET}=60(1+C_{SCE}/C_{OX})>60$. On the contrary, in the near- and above-threshold regimes of MOSFETs, large C_Q, or electron charge screening effect, forces V_g-V_{th}, which is usually referred to as the overdrive voltage, to drop on the gate oxide, instead of in the channel, leading to a poor gate efficiency, or equivalently a large S, $S_{MOSFET}=S_{TP}\cdot(1+C_Q/C_{OX})>>S_{TP}$. In this sense, overdrive voltage is a type of voltage loss. Interestingly, by adding an NC layer on MOSFETs, and simply matching $|C_{NC}|$ to the constant C_{OX}, C_{SCE} and C_Q can be absorbed. In other words, SS in short-channel devices can be restored to 60 mV per decade, and S in the near- and above-threshold regimes can be restored to transport mechanism limited S_{TP}, as illustrated in Figure 3.10(a). The significance of the latter is that it provides an alternative solution to reduce the supply voltage and hence the switching energy, i.e. by saving overdrive voltage, which also occupies a big portion of the entire supply voltage, instead of struggling to save subthreshold voltage, as summarized in Figure 3.10(b). This idea is supported by the numerical simulation results shown in Figures 3.10(c) and (d). Note that small NC nonlinearity is desired to reduce S in the near- and above-threshold regimes. Compared to matching $|C_{NC}|$ to the variable C_{MOS} for sub-60 mV per decade SS, which is rather difficult to achieve and inevitably introduces variation issues, matching $|C_{NC}|$ to the constant C_{OX} is obviously much more practical in terms of saving switching energy. In several reported NC-FET experiments (Krivokapic et al. 2017; Saeidi et al. 2017; Kwon et al. 2018; Zhang et al. 2019; Cheema et al. 2022), a common observation is that the use of the FE layer helps improve SS (instead of achieving steep slope, i.e. < 60 mV per decade), and enhance the on-current, which very likely stems from the voltage loss saving role of NC uncovered in this section.

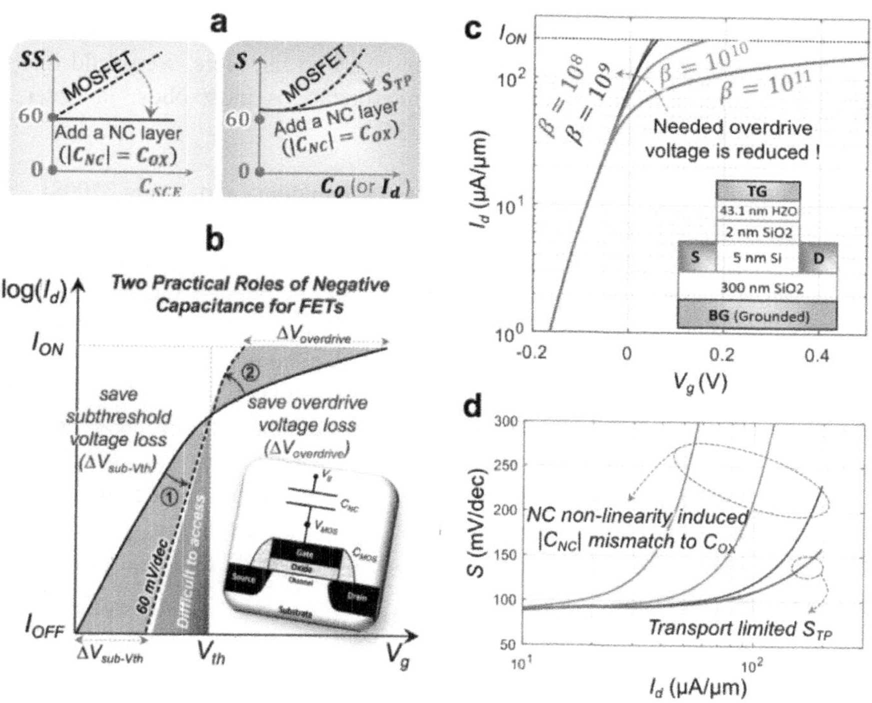

Figure 3.10. *Role for NC in MOSFETs as a voltage loss saver*

NOTES ON FIGURE 3.10.– *(a) The NC layer can help suppress the short-channel effect induced subthreshold voltage loss ($\Delta V_{sub\text{-}Vth}$), and inversion charge screening* *induced overdrive voltage loss ($\Delta V_{overdrive}$). It is difficult for NC to help obtain hysteresis-free steep slope. (b) A summary of the two practical roles of NC for MOSFETs. (c) Simulated I_d-V_g and (d) S versus I_d curve of a Si SOI NC-FET, in which $|C_{NC}|$ is matched to C_{OX}. Lower NC nonlinearity (quantified by the factor of β) is desired, in terms of restoring S back to S_{TP}.*

3.6. Appendix: A rule for polarization dynamics-based interpretation of the subthermionic SS

As proved above, it is very difficult to obtain steep slope in quasi-static conditions from those NC-FETs made on "good" FET platforms without the help of IMG. There have been many studies that interpret the observed subthermionic SS in various reported "NC-FETs" as the transient effects during measurement (Kobayashi et al. 2016; Saha et al. 2018) and/or intricate FE polarization dynamics $P(E, t)$ (Jin et al. 2018; Li and Toriumi 2018; Wang et al. 2018). Because of the great complexity, there has not been any consensus so far. Here, we would like to point out a constraint or rule that these interpretations must obey, in order to correctly capture the observed subthermionic SS.

Ever since the proposal of NC-FETs (Salahuddin and Datta 2008), an approximation, $Q_{MOS} = D_{FE} = \varepsilon_0 E_{FE} + P_{FE} \approx P_{FE}$, has been widely used to describe the relation between the polarization and charge in the MOSFET underneath. This approximation is obviously inconsistent in physics, as pointed out by Liu et al. (2018), but remains mathematically valid in most FE memories and FE-DE double-layer capacitors, as well as NC-FETs in the turn-on condition. On the one hand, in those conditions, sufficient charge from the electrode plates or the channel of the MOSFET underneath can be supplied to compensate/induce strong polarization in the FE layer. On the other hand, for most FEs, $|\varepsilon_{FE}| >> \varepsilon_0$, thus $P_{FE} >> \varepsilon_0 E_{FE}$ and $Q_{MOS} \approx P_{FE}$. However, this approximation can be misleading in the subthreshold regime of NC-FETs. As discussed above, Q_{MOS} is negligibly small in the subthreshold regime of NC-FETs that are made on a "good" FET platform, where and IMG is not employed, i.e.,

$$Q_{MOS} = D_{FE} = \varepsilon_0 E_{FE}(t) + P_{FE}(t) \to 0 \qquad [3.33]$$

where t is the additional time dimension that determines the FE polarization. Note that this equation stays valid in any transient process. In this condition, the approximation of $Q_{MOS} \approx P_{FE}$ will result in a trivial solution, i.e. $Q_{MOS} \approx E_{FE} \approx P_{FE} \to 0$. In fact, equation [3.33] describes a non-conventional self-sustaining polarization scheme, in the absence of compensating charge for FE polarization. Simply speaking, the polarization in the FE can be sustained by the development of a non-trivial electric field E_{FE}, instead of relying on the compensating charge from external circuitry. As shown in Figure 3.11(a), the green and blue dotted curves represent the quasi-static or stable P–E traces based on NC theory and conventional FE polarization switching theory, respectively, and equation [3.33] is essentially a self-sustaining line with a small ($|\varepsilon_{FE}| >> \varepsilon_0$) and negative slope, which translates to two pieces of important physics. On the one hand, the FE polarization in the subthreshold regime is weak, most likely a result of significant depolarization. On the other hand, the residual weak polarization can be exploited to build up a negative electric field in the FE, and hence a negative V_{FE}, (see Figure 3.11(b)) in the subthreshold regime, as long as the polarization dynamics $P(E, t)$ allow the transient P–E operation point to evolve (with t and V_g) on the self-sustaining line and away from the origin (grey circle), as shown in Figure 3.11(a). This finding sets a rule for the future efforts in exploring potential polarization dynamics to capture the subthermionic SSs of various reported "NC-FETs".

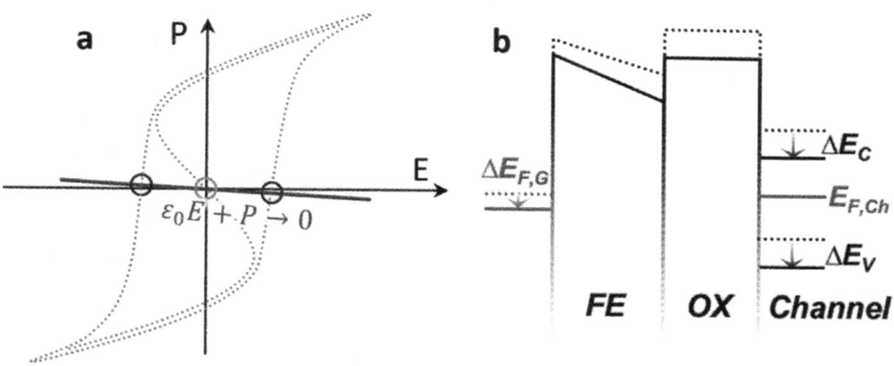

Figure 3.11. *The expected P–E relation in the subthreshold regime*

NOTES ON FIGURE 3.11.– *(a) To receive NC benefit in the subthreshold regime, i.e. negative voltage on the FE layer, the transient FE operation points predicted by any polarization dynamics models should be along the self-sustaining line ($\varepsilon_0 E_{FE} + P_{FE} \to 0$) and away from the origin (grey circle). Dotted curves represent possible polarization traces. (b) Band diagrams of an NC-FET in the subthreshold regime before (dotted lines) and after (solid lines) a gate voltage tuning.*

3.7. References

Appenzeller, J., Knoch, J., Bjork, M.T., Riel, H., Schmid, H., Riess, W. (2008). Toward nanowire electronics. *IEEE Trans. Electron Devices*, 55, 2827–2845.

Auth, C., Allen, C., Blattner, A., Bergstrom, D., Brazier, M., Bost, M., Buehler, M., Chikarmane, V., Ghani, T., Glassman, T. et al. (2012). A 22 nm high performance and low-power CMOS technology featuring fully-depleted tri-gate transistors, self-aligned contacts and high density MIM capacitors. *IEEE VLSI Tech. Dig.*, 131–132.

Baba, T. (1992). Proposal for surface tunnel transistors. *Jpn J. Appl. Phys.*, 31, 455–457.

Cao, W. and Banerjee, K. (2020). Is negative capacitance FET a steep-slope logic switch? *Nat. Comm.*, 11, 1–8.

Cao, W., Kang, J., Sarkar, D., Liu, W., Banerjee, K. (2015). 2D semiconductor FETs-projections and design for sub-10 nm VLSI. *IEEE Trans. Electron Devices*, 62, 3459–3469.

Cheema, S.S., Shanker, N., Wang, L.C., Hsu, C.-H., Hsu, S.-L., Liao, Y.-H., Jose, M.S., Gomez, J., Chakraborty, W., Li, W. et al. (2022). Ultrathin ferroic HfO_2–ZrO_2 superlattice gate stack for advanced transistors. *Nature*, 604, 65–71.

Cheng, C.H. and Chin, A. (2014). Low-voltage steep turn-on pMOSFET using ferroelectric high-κ gate dielectric. *IEEE Electron Device Lett.*, 35, 274–276.

Chung, W., Si, M., Ye, P.D. (2017). Hysteresis-free negative capacitance germanium CMOS FinFETs with bi-directional sub-60 mV/dec. *IEDM Tech. Dig.*, 365–368.

Duarte, J.P., Khandelwal, S., Khan, A.I., Sachid, A., Lin, Y.-K., Chang, H.-L., Salahuddin, S., Hu, C. (2016). Compact models of negative-capacitance FinFETs: Lumped and distributed charge models. *IEEE Int. Elec. Dev. Meeting*, 754–757.

Fan, C.-C., Cheng, C.-H., Chen, Y.-R., Liu, C., Chang, C.-Y. (2017). Energy-efficient HfAlOx NCFET: Using gate strain and defect passivation to realize nearly hysteresis-free sub-25mV/dec switch with ultralow leakage. *IEDM Tech. Dig.*, 561–564.

Ghosh, R.K., Brahma, M., Mahapatra, S. (2014). Germanane: A low effective mass and bandgap 2-D channel material for future FETs. *IEEE Trans. Electron Devices*, 61, 2309–2315.

Gopalakrishnan, K., Griffin, P., Plummer, J. (2005). Impact ionization MOS (I-MOS)–Part I: Device and circuit simulations. *IEEE Tran. Elec. Dev.*, 52, 69–76.

Hoffmann, M., Pesic, M., Slesazeck, S., Schroeder, U., Mikolajick, T. (2018). On the stabilization of ferroelectric negative capacitance in nanoscale devices. *Nanoscale*, 10, 10891–10899.

Hoffmann, M., Fengler, F.P.G., Herzig, M., Mittmann, T., Max, B., Schroeder, U., Negrea, R., Lucian, P., Slesazeck, S., Mikolajick, T. (2019). Unveiling the double-well energy landscape in a ferroelectric layer. *Nature*, 565, 464–467.

Hoffmann, M., Slesazeck, S., Schroeder, U., Mikolajick, T. (2020). What's next for negative capacitance electronics? *Nat. Electro.*, 3, 504–506.

Jain, A. and Alam, M. (2014). Stability constraints define the minimum subthreshold swing of a negative capacitance field-effect transistor. *IEEE Trans. Electron Devices*, 61, 2235–2242.

Jin, C., Jang, K., Saraya, T., Hiramoto, T., Kobayashi, M. (2018). Experimental study on the role of polarization switching in subthreshold characteristics of HfO_2-based ferroelectric and anti-ferroelectric FET. *IEEE Int. Elec. Dev. Meeting*, 31.5.1–31.5.4.

Khan, A.I., Yeung, C.W., Hu, C., Salahuddin, S. (2011). Ferroelectric negative capacitance MOSFET: Capacitance tuning & antiferroelectric operation. *IEEE Int. Elec. Dev. Meeting*, 255–258.

Kobayashi, M. and Hiramoto, T. (2016). On device design for steep-slope negative-capacitance field-effect-transistor operating at sub-0.2V supply voltage with ferroelectric HfO_2 thin film. *AIP Advances*, 6, 025113.

Kobayashi, M., Ueyama, N., Jang, K., Hiramoto, T. (2016). Experimental study on polarization-limited operation speed of negative capacitance FET with ferroelectric HfO_2. *IEEE Int. Elec. Dev. Meeting*, 314–317.

Krivokapic, Z., Rana, U., Galatage, R., Razavieh, A., Aziz, A., Liu, J., Shi, J., Kim, H.J. Sporer, R., Serrao, C. et al. (2017). 14 nm ferroelectric FinFET technology with steep subthreshold slope for ultra low power applications. *IEEE Int. Elec. Dev. Meeting*, 357–360.

Kwon, D., Chatterjee, K., Tan, A.J., Yadav, A.K., Zhou, H., Sachid, A.B., Reis, R.D., Hu, C., Salahuddin, S. (2018). Improved subthreshold swing and short channel effect in FDSOI n-channel negative capacitance field effect transistors. *IEEE Electron Device Lett.*, 39, 300–303.

Lee, M.H., Chen, P.-G., Fan, S.-T., Chou, Y.-C., Kuo, C.-Y., Tang, C.-H., Chen, H.-H., Gu, S.-S., Hong, R.-C., Wang, Z.-Y. et al. (2017). Ferroelectric Al:HfO2 negative capacitance FETs. *IEDM Tech. Dig.*, 565–568.

Li, X. and Toriumi, A. (2018). Direct relationship between sub-60 mV/dec subthreshold swing and internal potential instability in MOSFET externally connected to ferroelectric capacitor. *IEEE Int. Elec. Dev. Meeting*, 31.3.1–31.3.4.

Liu, Z., Bhuiyan, M., Ma, T. (2018). A critical examination of "quasi-static negative capacitance" (QSNC) theory. *IEEE Int. Elec. Dev. Meeting*, 31.2.1–31.2.4.

Lundstrom, M.S. (2006). The MOSFET revisited: Device physics and modeling at the nanoscale. *IEEE Int. SOI Conf. Proceedings*, 17–19.

McGuire, F.A., Lin, Y-C., Price, K., Rayner, G.B., Khandelwal||, S., Salahuddin, S., Franklin, A.D. (2017). Sustained sub-60 mV/decade switching via the negative capacitance effect in MoS_2 transistors. *Nano Lett.*, 17, 4801−4806.

Miller, S.L. and McWhorter, P.J. (1992). Physics of the ferroelectric nonvolatile memory field effect transistor. *J. Appl. Phys.*, 72, 5999–6010.

Mistry, K., Grula, G., Sleight, J., Bai, L., Stephany, R., Flatley, R., Skerry, P. (1997). A 2.0V, 0.35μm partially depleted SOI-CMOS Technology. *IEEE Int. Elec. Dev. Meeting*, 583–586.

Ota, H., Ikegami, T., Hattori, J., Fukuda, K., Migita, S., Toriumi, A. (2016). Fully coupled 3-D device simulation of negative capacitance FinFETs for sub 10 nm integration. *IEEE Int. Elec. Dev. Meeting*, 318–321.

Pahwa, G., Dutta, T., Agarwal, A., Khandelwal, S., Salahuddin, S., Hu, C., Chauhan, Y,S. (2016). Analysis and compact modeling of negative capacitance transistor with high ON-current and negative output differential resistance, Part II: Model validation. *IEEE Trans. Electron Devices*, 63, 4986–4992.

Pahwa, G., Agarwal, A., Chauhan, Y.S. (2018). Numerical investigation of short-channel effects in negative capacitance MFIS and MFMIS transistors: Subthreshold behavior. *IEEE Trans. Electron Devices*, 65, 5130–5136.

Rollo, T., Wang, H., Han, G., Esseni, D. (2018). A simulation-based study of NC-FETs design: Off-state versus on-state perspective. *IEEE Int. Elec. Dev. Meeting*, 213–216.

Rusu, A., Salvatore, G.A., Jimenez, D., Ionescu, A.M. (2010). Metal-ferroelectric-meta-oxide-semiconductor field effect transistor with sub-60 mV/decade subthreshold swing and internal voltage amplification. *IEEE Int. Elec. Dev. Meeting*, 16.3.1–16.3.4.

Saeidi, A., Jazaeri, F., Bellando, F., Stolichnov, I., Luong, G.V., Zhao, Q.-T., Mantl, S., Enz, C.C., Ionescu, A.M. (2017). Negative capacitance as performance booster for tunnel FETs and MOSFETs: An experimental study. *IEEE Electron Device Lett.*, 38, 1485–1488.

Saha, A., Datta, S., Gupta, S. (2018). Negative capacitance in resistor-ferroelectric and ferroelectric-dielectric networks: Apparent or intrinsic? *J. Appl. Phys.*, 123, 105102-1.

Sakurai, T. (2004). Perspectives of low-power VLSI's. *IEICE Trans. Electron*, E87-C, 429–436.

Salahuddin, S. and Datta, S. (2008). Use of negative capacitance to provide voltage amplification for low power nanoscale devices. *Nano Lett.*, 8, 405–410.

Sarkar, D., Xie, X., Liu, W., Cao, W., Kang, J., Gong, Y., Kraemer, S., Ajayan, P.M., Banerjee, K. (2015). A subthermionic tunnel field-effect transistor with an atomically thin channel. *Nature*, 526, 91–95.

Si, M., Su, C.-J., Jiang, C., Conrad, N.J., Zhou, H., Maize, K.D., Qiu, G., Wu, C.-T., Shakouri, A., Alam, M.A., Ye, P.D. (2018a). Steep-slope hysteresis-free negative capacitance MoS_2 transistors. *Nat. Nano. Technol.*, 13, 24–28.

Si, M., Jiang, C., Chung, W., Du, Y., Alam, M.A., Ye, P.D. (2018b). Steep-slope WSe_2 negative capacitance field-effect transistor. *Nano Lett.*, 18, 3682–3687.

Steffes, J.J., Ristau, R.A., Ramesh, R., Huey, B.D. (2019). Thickness scaling of ferroelectricity in $BiFeO_3$ by tomographic atomic force microscopy. *PNAS*, 116, 2413–2418.

Sze, S.M. and Ng, K.K. (2006). *Physics of Semiconductor Devices*, 3rd edition. John Wiley & Sons, Inc., Hoboken.

Tolédano, J.C. and Tolédano, P. (1987). *The Landau Theory of Phase Transitions*. World Scientific, Singapore.

Wang, H., Yang, M., Huang, Q., Zhu, K., Zhao, Y., Liang, Z., Chen, C., Wang, Z., Zhong, Y., Zhang, X., Huang, R. (2018). New insights into the physical origin of negative capacitance and hysteresis in NCFETs. *IEEE Int. Elec. Dev. Meeting*, 31.1.1–31.1.4.

Wang, X., Yu, P., Lei, Z., Zhu, C., Cao, X., Liu, F., You, L., Zeng, Q., Deng, Y., Zhu, C. et al. (2019). Van der Waals negative capacitance transistors. *Nat. Comm.*, 10, 1–8.

Yadav, A.K., Nguyen, K.X., Hong, Z., García-Fernández, P., Aguado-Puente, P., Nelson, C.T., Das, S., Prasad, B., Kwon, D., Cheema, S. et al. (2019). Spatially resolved steady-state negative capacitance. *Nature*, 565, 468–471.

You, W. and Su, P. (2017). Design space exploration considering back-gate biasing effects for 2D negative-capacitance field-effect transistors. *IEEE Trans. Electron Devices*, 64, 3476–3481.

Yu, Z., Wang, H., Li, W., Xu, S., Song, X., Wang, S., Wang, P., Zhou, P., Shi, Y., Chai, Y., Wang, X. (2017). Negative capacitance 2D MoS_2 transistors with sub-60 mV/dec subthreshold swing over 6 orders, 250 μA/μm current density, and nearly-hysteresis-free. *IEDM Tech. Dig.*, 577–580.

Yuan, Z.C., Rizwan, S., Wong, M., Holland, K., Anderson, S., Hook, T.B., Kienle, D., Gadelrab, S., Gudem, P.S., Vaidyanathan, M. (2016). Switching-speed limitations of ferroelectric negative-capacitance FETs. *IEEE Trans. Electron Devices*, 63, 4046–4052.

Zhang, Z., Xu, G., Zhang, Q., Hou, Z., Li, J., Kong, Z.Z., Zhang, Y., Xiang, J., Xu, Q., Wu, Z. et al. (2019). FinFET with improved subthreshold swing and drain current using 3-nm ferroelectric $Hf_{0.5}Zr_{0.5}O_2$. *IEEE Electron Device Lett.*, 40, 367–370.

Zhou, J., Wu, J., Han, G., Kanyang, R., Peng, Y., Li, J., Wang, H., Liu, Y., Zhang, J., Sun, Q. et al. (2017). Frequency dependence of performance in Ge negative capacitance PFETs achieving sub-30 mV/decade swing and 110 mV hysteresis. *IEDM Tech. Dig.*, 373–376.

4

Z2 Field-Effect Transistors

Joris Lacord
CEA, LETI, Université Grenoble Alpes, France

4.1. Introduction

The standard CMOS technology widely used for a large panel of microelectronic applications faces an unavoidable switching limitation, usually named the "60 mV/dec" or the "kT/q" limit. This means that below the threshold voltage, an increase of the current of one decade requires at least 60 mV of gate polarization. This limitation is due to the nature of the transport in the MOS transistor below the threshold. Indeed, the current in the MOS transistor subthreshold regime is a diffusion current, which depends exponentially on the gate voltage (Sze 1981; Taur and Ning 1998). This leads to a slope value that cannot be lower than "ln(10)xkT/q", so 60 mV/dec at room temperature.

The "60 mV/dec limit" is consequently intrinsic to MOS transistor operation. A solution to overcome this is to develop new devices with an operation different from that MOS transistor. These kinds of devices are part of the "steep slope" device family. The sharp switch can allow ultra-low voltage operations of the logical switch to replace the standard CMOS technology in digital circuit for ultra-low power applications. The sharp switch associated with low leakage and the ability to handle high current could be interesting for ElectroStatic Discharge (ESD) protection (Fonteneau et al. 2015). Finally, the last potential application is memory, because the sharp switch guarantees a large memory window associated with a large ratio of current between the two memory states. This justifies the proliferation of new steep

Beyond-CMOS, coordinated by Alessandro Cresti.

slope device structures. A part of them are based on the PIN diode structure with different locations of the gate and different polarization schemes, leading to different operations, performances and intrinsic behaviors. Some examples are listed as follows:

– The tunnel field-effect transistor (TFET):

This is a reverse-biased PIN diode with a fully gated intrinsic body (Figure 4.1-a). In this device, the current is due to band-to-band tunneling at the source/intrinsic region. The gate polarization modulates the band's bending at the source/intrinsic region junction (Figure 4.1-b), allowing the carrier injection by band-to-band tunneling and leading to the possibility of a steeper slope than 60 mV/dec (Mayer et al. 2008a). This device is ambilpolar (so it can be used as nMOS or pMOS), compliant with ultra-low power operation, but the main drawback is the very low drive current compared to standard CMOS technology (Jeon et al. 2015).

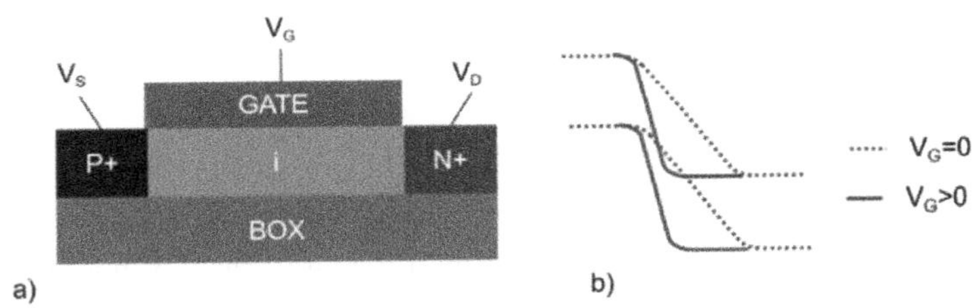

Figure 4.1. *(a) n-type TFET structure schematic: source is the P+ region, drain is the N+ region and (b) band bending at source/intrinsic region at V_G=0 V and V_G>0 V. For a color version of this figure, see www.iste.co.uk/cresti/beyondcmos.zip*

– The impact ionization MOS (I-MOS):

This is a reverse-biased partially gated PIN diode. The gated part of the intrinsic region is close to the drain (Figure 4.2-a), so the N+ area in the case of nMOS (Gopalakrishnan et al. 2005a). At a sufficient drain-to-source positive voltage, increasing the gate voltage forms and then increases the inversion layer below the gate. This reduces the intrinsic layer length of the PIN diode and produces an increase of the electric field in this region until reaching breakdown. The breakdown on the PIN diode corresponds to the I-MOS switching, which is very sharp. The I-MOS drive current corresponds to the avalanche current of the reverse-biased PIN diode, and is consequently very high (Figure 4.2-b). However, because of the avalanche phenomena, it needs high operating voltage and its reliability is very limited (Mayer et al. 2007, 2008b).

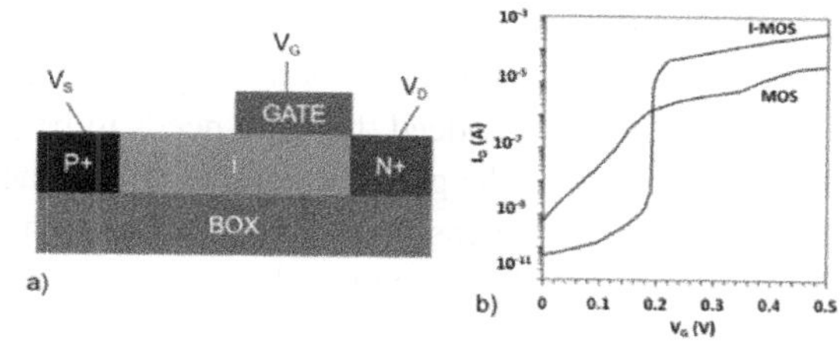

Figure 4.2. *(a) n-type I-MOS structure schematic: source is the P+ region, drain is the N+ region and (b) I-MOS and MOS I_D-V_G at V_D=1V trend comparison. For a color version of this figure, see www.iste.co.uk/cresti/beyondcmos.zip*

– The feedback field-effect transistor (FB-FET):

This is also a partially gated PIN diode, with the gate in middle of the intrinsic region, and two ungated portions of the intrinsic layer on N+ and P+ area side (Figure 4.3-a). By increasing the gate voltage, some carriers flowing from source to drain accumulate in the intrinsic ungated regions (so below the spacers). This produces a reduction of the source/intrinsic region's and drain/intrinsic region potential barriers leading to a dynamic threshold of voltage lowering. This behavior is similar to the floating body effect in the partially depleted SOI transistor, but it occurs in the subthreshold regime in the FB-FET, leading to the sharp switch (Figure 4.3-b) (an experimental demonstration of this can be found in Padilla et al. 2008).

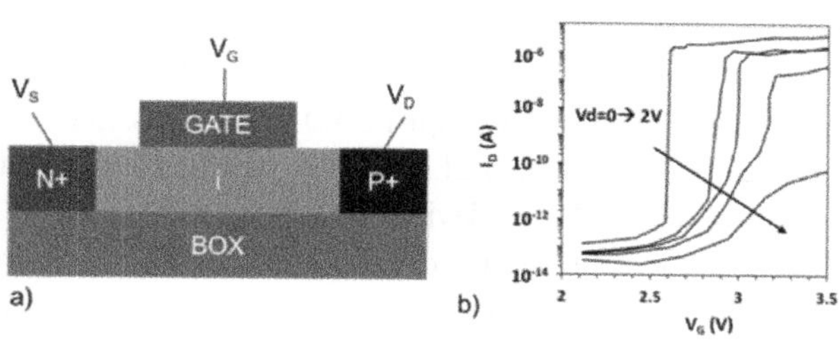

Figure 4.3. *(a) n-type FB-FET structure schematic: source is the N+ region, drain is the P+ region and (b) FeedbackFET I_D-V_G trend for V_D=0 V to 2 V. For a color version of this figure, see www.iste.co.uk/cresti/beyondcmos.zip*

– The zero subthreshold swing and zero impact ionization field-effect transistor (Z2FET):

Z2FET is a forward biased partially gated PIN diode on SOI. Its structure is similar to I-MOS, but the drain of the device is the P+ region (Figure 4.4-a) and the back gate needs to be polarized (V_B). Indeed, the gate needs to be positively biased to increase electron charge density in the gated region, while the back gate has to be negatively biased to increase the hole density in the ungated region. This transforms the device into an electrostatic "NPNP" structure. Therefore, it behaves almost like a thyristor, but it does not require impact ionization for switching, allowing low voltage application (Wan et al. 2012a, 2012b). The drawback of the other devices briefly described above does not affect Z2FET. Indeed, as it does not involve impact ionization, it is compliant with a low voltage operation and should not suffer from a reliability issue. Z2FET presents sharp switch (Figure 4.4-b) and its drive current is high, because it corresponds to a PIN diode forward current.

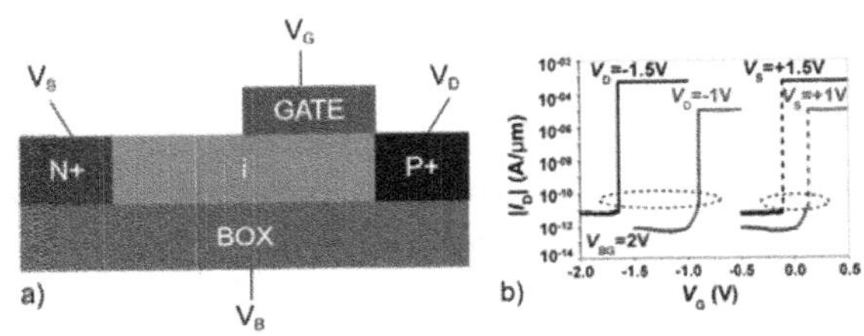

Figure 4.4. *(a) n-type Z2FET structure schematic: source is the N+ region, drain is the P+ region and (b) experimental Z2FET I_D-V_G curves for V_D=0 V to 1.8 V (Wan et al. 2012b). For a color version of this figure, see www.iste.co.uk/cresti/beyondcmos.zip*

The following sections of this chapter is dedicated to Z2FET. Section 4.2 describes the Z2FET operation in steady state, based on TCAD simulations. Section 4.3 details the analytical and compact modeling of the Z2FET steady-state operation. Section 4.4 shows experimental evidence of Z2FET devices. Section 4.5 presents the Z2FET 1T-DRAM operation. Section 4.6 briefly describes Z2FET device optimizations, and finally, section 4.7 concludes the chapter with other Z2FET potential applications.

4.2. Z2FET steady-state operation

Z2FET is a partially gated PIN diode with four electrodes. For the n-type Z2FET, the source electrode is the N+ region, the drain electrode is the P+ region and the backgate is the doped layer below the buried oxide (BOX). Source and drain regions are doped around 1e20 cm^{-3}, while the backgate is doped around 1e18 cm^{-3}. Note that in the literature, the source is also called the cathode and the drain the anode by analogy with the PIN diode. The gate length is noted as L_G and the length of the ungated region is noted as L_{UN} (Figure 4.5). Furthermore, source voltage is V_S, drain voltage is V_D, gate voltage is V_G, and back gate voltage is V_B.

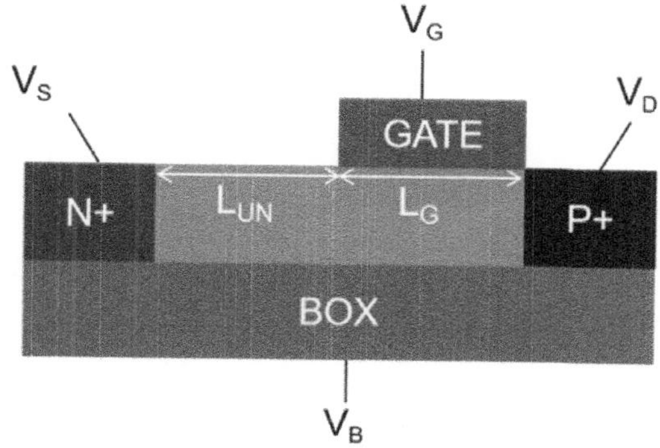

Figure 4.5. *n-type Z2FET structure schematic: source is the N+ region, drain is the P+ region. L_G is the gate length and L_{UN} is the ungated region length. For a color version of this figure, see www.iste.co.uk/cresti/beyondcmos.zip*

4.2.1. *Z2FET sharp switch evidence*

A convenient way to describe a steady-state operation, including switching, is to use the drain current (I_D) versus drain voltage (V_D) characteristic. To perform a deep explanation of the Z2FET behavior, we need to monitor the electrostatic potential and the charge densities in the device during its operation. Therefore, the use of TCAD simulation (Synopsys) is an adequate tool. To capture the intrinsic behavior, basic physics is included in the simulation: Poisson's equation, drift diffusion, SRH generation/recombination and constant mobility. Note that impact ionization model is not activated in the simulations, in agreement with the literature (Wan et al. 2012b). Figure 4.6 shows the I_D–V_D characteristic of the Z2FET device (L_G=L_{UN}=200 nm) obtained by steady-state simulation. The gate and the back gate voltages are initially set, respectively, to 2 V and -2 V to build the electrostatic field-induced NPNP structure. Then, the drain is swept to 2 V and then back to 0 V. We first observe on the forward bias characteristic (V_D swept from 0 to 2 V), the sharp switch between off and on state at V_D=V_{ON}: eight decades of drain current increase in a few mV. On the reverse bias characteristic, sharp switch also occurs, but at a lower drain voltage V_{OFF}. This simulation's results demonstrate the presence of a steady-state hysteresis in the Z2FET behavior.

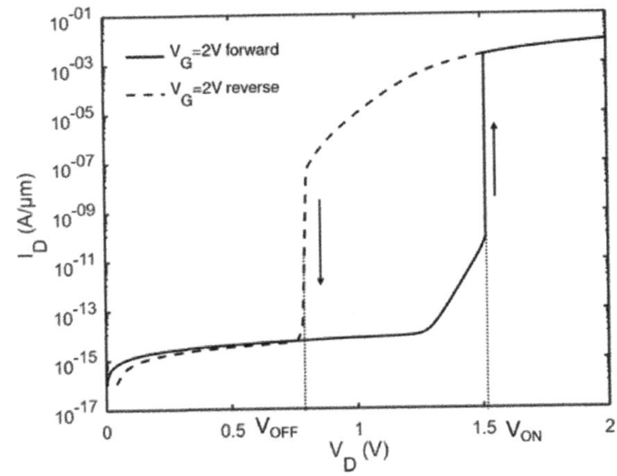

Figure 4.6. *Z2FET I_D–V_D under forward (solid line) and reverse (dashed line) for V_G=2 V and V_B=-2 V (L_G=L_{UN}=200 nm)*

To highlight the effect of gate voltage (V_G) and back gate voltage (V_B), similar simulations are performed with various values of V_G (Figure 4.7) and V_B (Figure 4.8), where forward and reverse characteristics are plotted on different graphs. The sharp switch voltage in forward bias V_{ON} is strongly related to the gate voltage, while its dependence on the back gate voltage is weaker. If the gate voltage is not high enough, sharp switch disappears and Z2FET behaves as a PIN diode. Concerning the sharp switch voltage under reverse bias V_{OFF}, gate voltage has a low impact on its value, while the effect of the back gate is stronger.

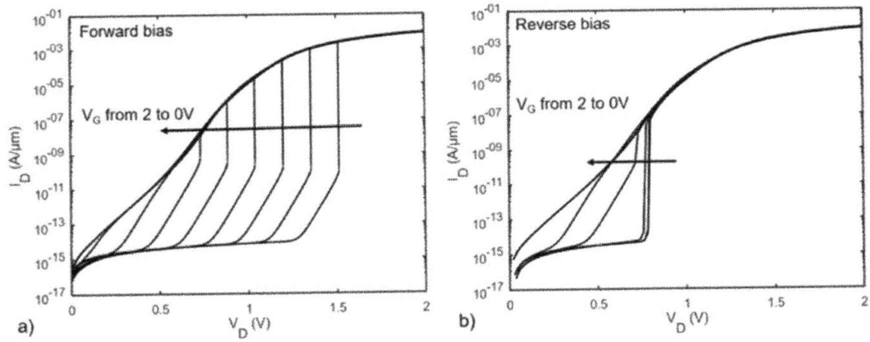

Figure 4.7. *Z2FET I_D–V_D under direct (a) and reverse (b) bias for V_G varying from 0 to 2 V*

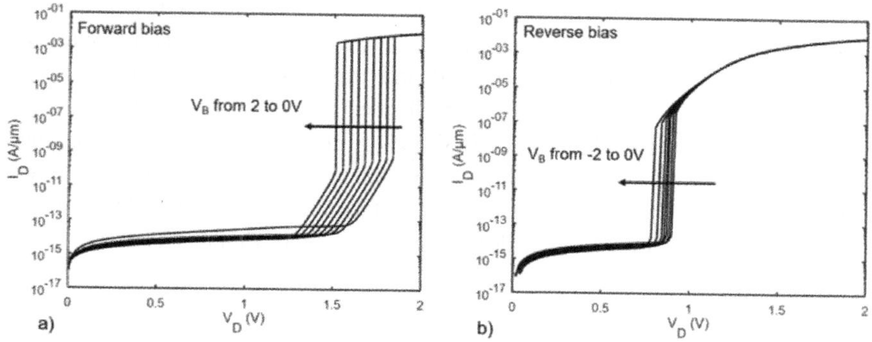

Figure 4.8. *Z2FET I_D–V_D under direct (a) and reverse (b) bias for V_B varying from -2 V to 0 V*

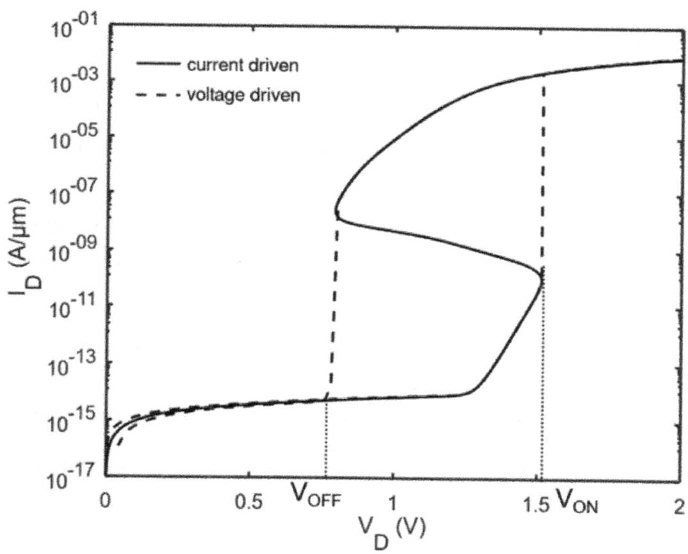

Figure 4.9. *Z2FET I_D–V_D obtained by current-driven TCAD simulation (line) and voltage-driven TCAD simulation (dotted line)*

4.2.2. *Z2FET "S-shape" characteristic*

Z2FET could be seen as a "field-induced thyristor" even though it does not involve impact ionization. Thyristor is known to present an S-shape I_D–V_D

steady-state characteristic (Sze 1981). To obtain this shape of curve, it is necessary to characterize the device by imposing the current (and not the voltage as usual). In other words, the input is the drain current and the output is the drain voltage. Current-driven TCAD simulations are performed in the same conditions as Figure 4.6 and the corresponding I_D–V_D is shown in Figure 4.9. It demonstrates that Z2FET presents, as a thyristor device, S-shape I_D–V_D in a current-driven characteristic. We can remark that a negative resistance region links the sharp switch voltages V_{ON} and V_{OFF} of the voltage driven characteristic of Figure 4.6.

4.2.3. *Z2FET detailed description*

To describe Z2FET steady-state operation, we use the S-shape I_D–V_D curve, because it is the most complete characteristic. Five operating regimes are identified as follows (Figure 4.10) (Lacord et al. 2017):

– "*Initialization*": VD=0 and VG and VB are, respectively, positively and negatively polarized to transform the chemical PIN structure into an electrostatic NPNP structure.

– "*OFF-region*": the NPNP electrostatic structure is built and the device is blocked. The drain voltage increase has almost no impact on the drain current, which stays very low.

– "*Barriers collapsing*": the drain current increases with the drain voltage with a slope of around 60 mV/dec. The end of this regime corresponds to the sharp switch under the forward bias (V_D=V_{ON}).

– "*Negative resistance*": in this regime, the drain voltage decreases while the drain current increases. Therefore, the device behaves as a negative resistance. This occurs between both Z2FET sharp switches, so between V_D=V_{ON} and V_D=V_{OFF}.

– "*PIN diode*": in this regime, the device behaves as a forward bias PIN diode and delivers high drain current.

Figure 4.10 shows an illustration of each regime on the Z2FET S-Shape I_D–V_D curve. In the following figure, each regime is described based on potential and charge densities plotted from source to drain in the middle of the silicon film.

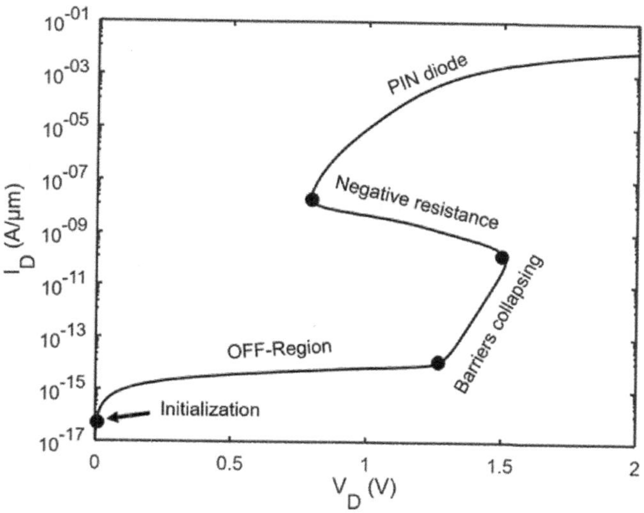

Figure 4.10. *S-shape I_D–V_D with the definition of the five steady-state operating regimes*

4.2.2.1. *Initialization regime*

In this regime, V_D is kept equal to 0 V, V_G is increased from 0 to a positive value (here 2 V) and V_B is decreased from 0 V to a negative value (here -2 V). In Figure 4.11, we can see the electrostatic potential in the middle of the channel from the source (S) to the drain (D), which shows the action of the top and back gate polarizations. At V_G=V_B=0 V, potential in gated and ungated region is constant: there is no potential barrier between gated and ungated regions (dashed line in Figure 4.11-a). But, under positive V_G, potential in gated regions increases, while under negative back gate polarization, potential in ungated region decreases. It produces the potential barrier between gated and ungated regions (continuous line in Figure 4.11-a). If we look at carrier densities in the same way (Figure 4.11-b), we can see that electron and hole concentration are close to the intrinsic value (n_i~1e10 cm^{-3}) at V_G=V_B=0 V (respectively circle and crosshair in Figure 4.11-b). Then, for biased top and back gates, electron density (n) is high under the gate, while hole density (p) is high in ungated regions with the product $np=n_i^2$ kept constant all along the device. This indicates that the field-induced PN junction formed by the gated and ungated silicon region behaves as a grounded PN junction. The observations on carrier densities are consistent with the one done on electrostatic potential. It also demonstrates that a positive polarization of the gate produces an N-type electrostatic region, while negative polarization produces a P-type electrostatic ungated region. Finally, we can remark that the electron concentration in the gated region (~1e18 cm^{-3}) is higher than the hole concentration in the ungated region (~1e17 cm^{-3}). This is due to the thicker oxide in the ungated (BOX) than in the gated region (front gate oxide, EOT). With such carrier densities, we can conclude that both regions are in (or close to) the inversion regime. The potential under the gate reaches the inversion potential value of MOS capacitor, noted as ψGinv. Hence, during the initialization regime, Z2FET behavior is fully described by the standard MOS operation.

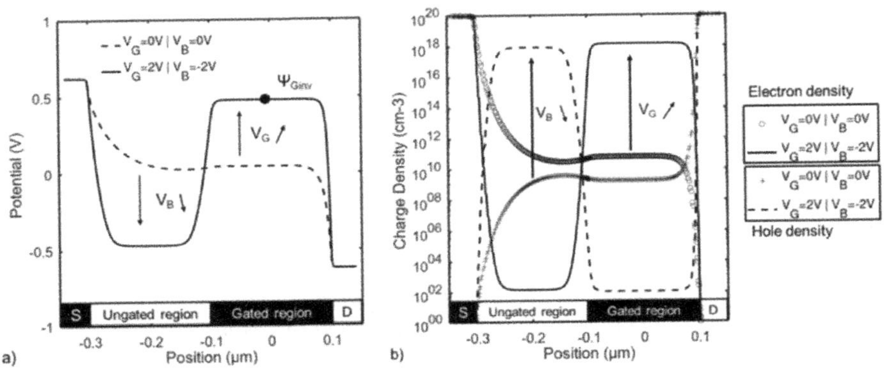

Figure 4.11. *Initialization regime potential (a) and charge densities (b) from source to drain*

4.2.2.2. *OFF-region*

In the "*OFF-region*" drain voltage, V_D increases from 0 to ~1.2 V and the drain current I_D is very low and almost constant, with only a slight increase of one decade. Similar to the previous section, we plot the electrostatic potential and the charges densities (electron and hole on separated graphs) in the middle of the film from source to drain at the start and at the end of the "*OFF-region*" operating regime (Figures 4.12 and 4.13). We observe that the electrostatic potential increases in the gated region while it stays constant in the ungated region, leading to the growth of the potential barrier between the gated and ungated regions. The potential in the gated region reaches a value corresponding to the surface potential in the depletion regime in a classical MOS capacitor, noted as Ψ_{Gdep}. This indicates that the drain voltage increase in the "*OFF-region*" produces the depletion of the gated region. This is consistent with the electron density (n) decrease observed in the gated region in Figure 4.13. As the hole density (p) increases in gated region and remains constant in ungated, the "np" product remains constant in the complete intrinsic region and is equal to the square of the intrinsic value ($np=n_i^2$).

Z2FET behavior in the "*OFF-region*" can also be described with the analogy of the PN junction operation, where the P region is the ungated region with a doping equal to the hole concentration and the N area is the gated region with a doping equal to the electron concentration. Therefore, the intrinsic region of Z2FET can be seen as a PN junction with doping depending on the polarization. Indeed, the increase of drain polarization V_D produces the electron density diminution until it reaches a value close to the intrinsic concentration (n_i). Therefore, in this regime, the V_D increase leads to a lowering of the N-type field-induced doping in the gated region. It also produces an increase of the hole density in the gated region that enables it to keep the product $np{\sim}n_i^2$ constant. At the device level, due to the product $np{\sim}n_i^2$ in the complete device, the PN junctions included in Z2FET device are in an unpolarized regime. This explained why the drain current I_D is very low and very slightly increases with V_D during the *OFF-region*.

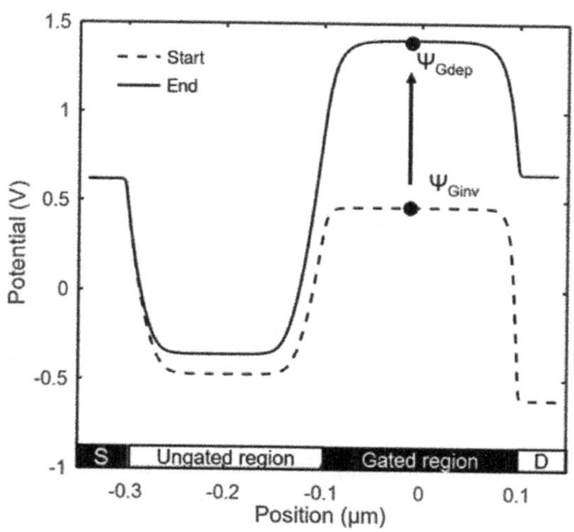

Figure 4.12. *Electrostatic potential from source to drain at the start (line) and at the end (dotted line) of the "OFF-regime"*

To summarize, in the "*OFF-region*", increasing V_D produces a lowering of the field-induced doping in the gated region, allowing the side PN junctions at source and drain to remain in an unpolarized state while the middle junction is heavily reverse-biased. Note that the ungated region is not relevantly impacted by the V_D increase during *OFF-region*.

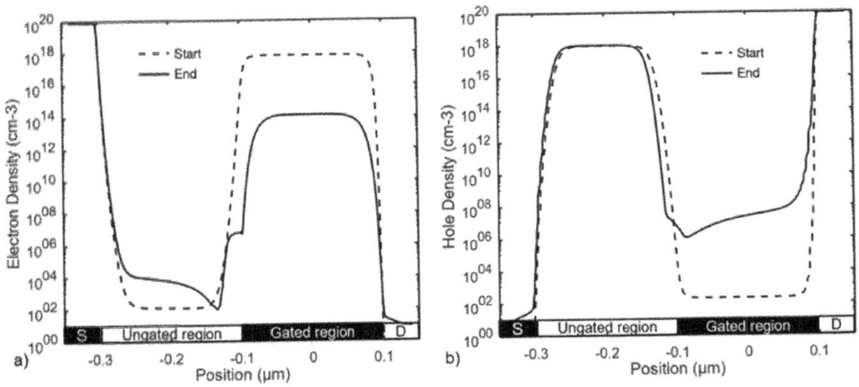

Figure 4.13. *Electron (a) and hole (b) densities from source to drain at the start (line) and at the end (dotted line) of the "OFF-regime"*

4.2.2.3. *Barriers collapsing*

In the "*barriers collapsing*" regime, the increase of current is stronger: ~4 decades in ~300 mV in the example in Figure 4.10. The electrostatic potential (Figure 4.14) increases in the ungated region, while it stays constant in the gated region. The electron density (Figure 4.15-a) continues to decrease in the gated region until it reaches the intrinsic concentration (n_i) and then increases. In the ungated region, it regularly increases. The hole density (p) (Figure 4.15-b) continues to increase in the gated region, while it is constant and still equal to the initialization value in the ungated region.

Therefore, in the "*barriers collapsing*" regime, as the potential in the gated region has reached its maximum value during the "*OFF-region*" regime, any additional drain voltage increase provokes the direct polarization of the ungated region/cathode and the gated region/anode PN junctions. This is demonstrated by the values of np product in gated and ungated region, superior to n_i^2. As a result, as the PN junctions are forward biased, this produces (and justifies) the strong drain current increase. A last consequence of the PN junction polarization is that, because $np>n_i^2$, reasoning based on standard MOS behavior cannot be used anymore, because the nature of the current is now ambipolar (also in the case for the "*negative resistance*" and "*PIN diode*" regimes).

Finally, the end of the "*barriers collapsing*" regime corresponds to the V_D value, where electron density in the ungated region is equal to hole density in the gated region (Figure 4.15). This condition corresponds to PIN diode direct polarization condition, and is consistent with the observation done on the S-shape I_D–V_D: the end of the "*barriers collapsing*" regime corresponds to sharp switch location (V_{ON}).

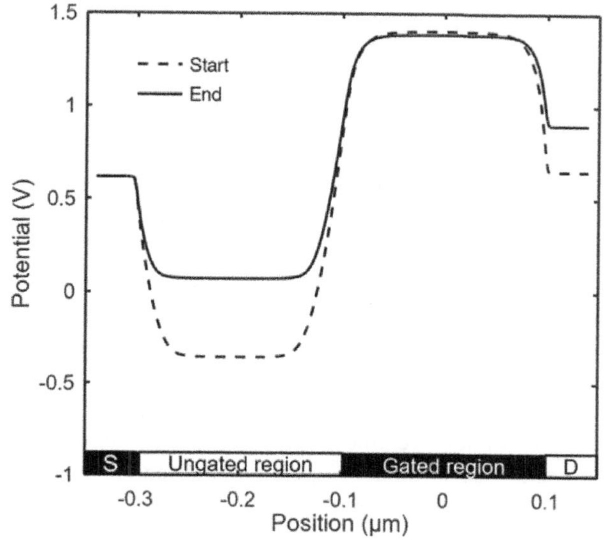

Figure 4.14. *Electrostatic potential from source to drain at the start (line) and at the end (dotted line) of the "barriers collapsing" regime*

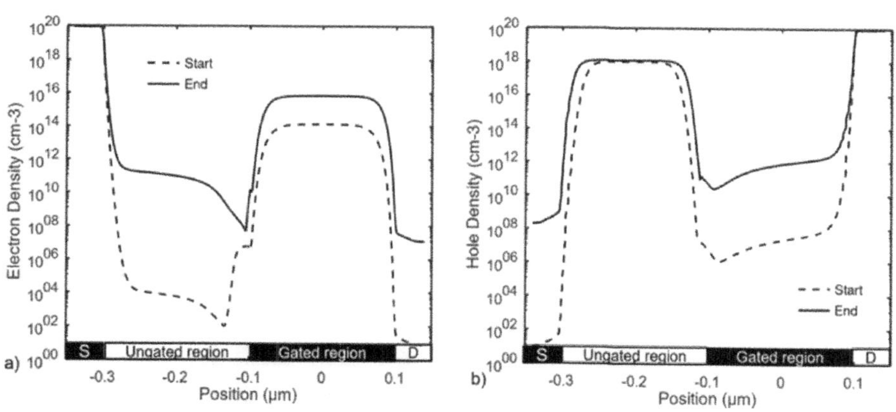

Figure 4.15. *Electron (a) and hole (b) densities at from source to drain at the start (line) and at the end (dotted line) of the "barriers collapsing" regime*

4.2.2.4. *Negative resistance*

The "*negative resistance*" regime corresponds to drain voltage V_D decrease between the sharp switching voltages (V_{ON} and V_{OFF}) for an increase of drain current I_D. The electrostatic potential decreases strongly in the gated region until it reaches the value imposed by the gate only: ψGinv. In the ungated region, the potential increases slightly. It leads to a strong lowering of potential barrier between gated and ungated regions (Figure 4.16). The electron density (n) (Figure 4.17-a) continues to increase in the gated region until it reaches a maximum value. This maximum corresponds to the electron density at the end of the initialization, so it corresponds to the density imposed by the gate. The hole density (p) (Figure 4.17-b) continues to increase in the gated region, while it is still constant and equal to the initialization value. Finally, when the electron density in the ungated region is equal to the hole density in the gated region and the hole density in the ungated region is equal to the electron density in the gated region, Z2FET enters the "*PIN diode*" regime.

The "*negative resistance*" regime is an intermediate regime between classical MOS operation and "*PIN diode*" operation: potential and minority carrier concentration (electron in gated region, hole in ungated region) are driven by the MOS capacitors, while the majority carrier concentration and the current are driven by the PIN diode.

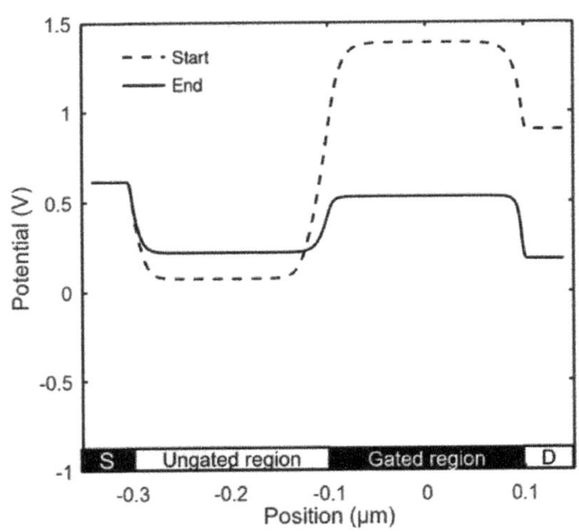

Figure 4.16. *Electrostatic potential from source to drain at the start (line) and at the end (dotted line) of the "negative resistance" regime*

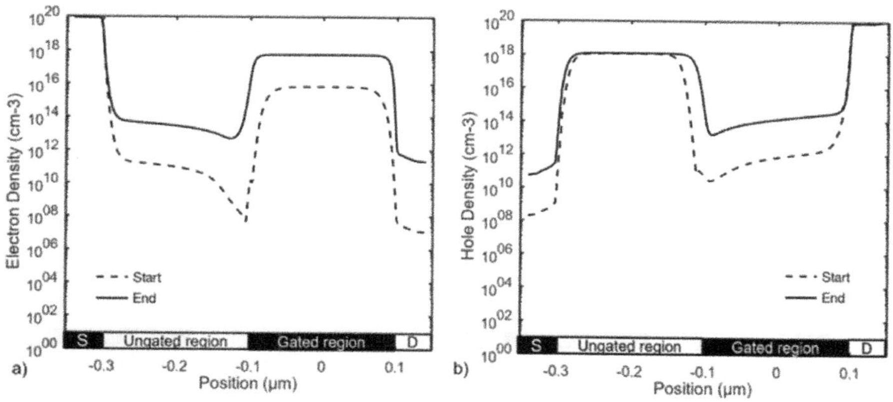

Figure 4.17. *Electron (a) and hole (b) densities at from source to drain at the start (line) and at the end (dotted line) of the "negative resistance" regime*

4.2.2.5. *PIN diode*

During the "*PIN diode*" regime, the electrostatic potential increases in both regions and is similar to the standard potential variation in a forward bias PIN diode (Figure 4.18). Electron and hole are present in the whole silicon film (Figure 4.19): current is driven by volume conduction (ambipolar transport) and the current saturates for high drain voltage V_D, because of the intrinsic resistance of the silicon film.

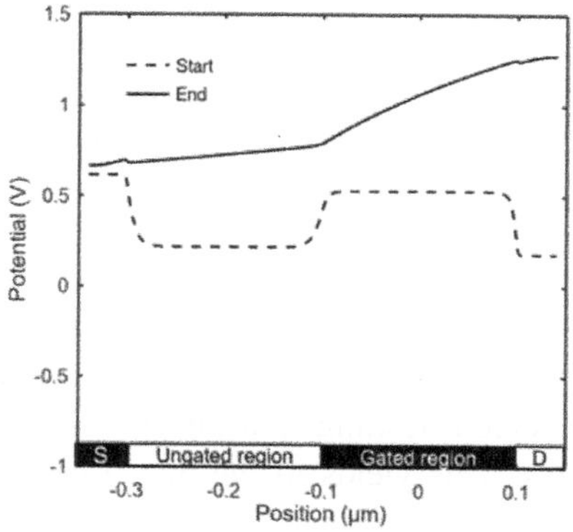

Figure 4.18. *Electrostatic potential from source to drain at the start (line) and at the end (dotted line) of the "PIN diode" regime*

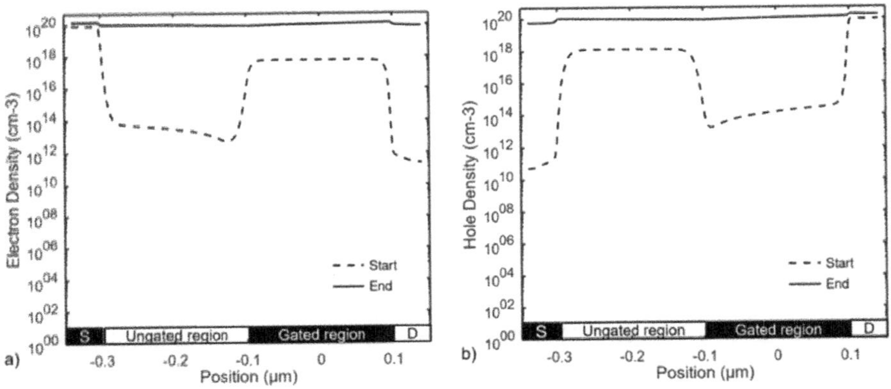

Figure 4.19. *Electron (a) and hole (b) densities at from source to drain at the start (line) and at the end (dotted line) of the "PIN diode" regime*

4.2.2.6. *Summary of Z2FET steady-state operation*

With all polarizations at 0 V, Z2FET is a PIN diode from chemical and electrostatic points of view.

During the "*initialization*" regime, V_G and V_B are biased to produce the NPNP electrostatic structure and transform the chemical PIN diode into an electrostatic thyristor structure. Potential and carrier densities are only driven by the MOS capacitors formed by the gate and the intrinsic region and by the back gate, i.e. the BOX the intrinsic region.

Through the "*OFF-region*", the drain voltage increase produces the increase of the potential in gated region associated with the lowering of the electron density. The Z2FET current is the leakage current of the PN junctions, which are in an unpolarized state (as shown by $np=n_i^2$).

During the "*barriers collapsing*" regime, the potential in gated region has reached its maximum, so the drain voltage increase produces a lowering of the potential in the ungated region, leading to a lowering of potential barrier between gated and ungated regions. This yields to the increase of carriers densities in the device, resulting to $np>n_i^2$ this means that PN junctions are now forward biased and the current increase is stronger. The end of this regime corresponds to the sharp switch under direct bias V_{ON}.

Through the "*negative resistance*" regime, the potential in gated region goes from MOS depletion to inversion value. The end of this regime corresponds to the sharp switch under reverse bias V_{OFF}.

During the "*PIN diode*" regime, Z2FET acts as a forward biased PIN diode.

This short summary highlights that the Z2FET behavior is due to the competition between MOS and PIN diode operations and that the sharp switch is due to the sudden pass from one to the other one.

4.3. Z2FET steady-state analytical and compact model

4.3.1. *Z2FET steady-state analytical drain current model*

As Z2FET is a partially gated PIN diode, its model can be based on a standard PIN diode anode current model ($I_A(V_{APP})$), given by (Gatard 2006):

$$I_A = I_{La}\left(e^{\frac{V_{APP}-R_{PIN}I_A}{N.ut}} - 1\right) + I_{Rec}\left(e^{\frac{V_{APP}-R_{PIN}I_A}{N.ut}} - 1\right)^2 \quad [4.1]$$

where I_{La} and I_{Rec}, respectively, stand for diffusion and recombination currents, R_{PIN} represents the intrinsic region resistance, V_{APP} is the applied voltage on the PIN diode and $ut=kT/q$ and N the diode ideality factor equal to 2. In the "*PIN diode*" regime, this equation is used natively ($V_{APP}=V_D$), while in the regimes "*OFF-region*" and "*Barriers collapsing*", V_{APP} has to account for change in potential reference. Indeed, the applied voltage can be defined as the difference of electrostatic potential in the drain ψ_D at $V_D=0$ and $V_D>0$. From the illustration of Figure 4.20, it leads to:

$$V_{APP} = \Psi_D(V_D) - \Psi_D(0\ V) \quad [4.2]$$

$$V_{APP} = V_D + \Psi_{Ginv} - \Psi_G \quad [4.3]$$

where ψ_D is the potential in the drain, ψ_G is the potential in the gated region, ψ_{Ginv} is the potential in the gated region in the inversion regime, and Ψ_d is the built-in potential of the drain. Note that ψ_G and ψ_{Ginv} can be evaluated analytically through an adaptation of surface potential evaluation in standard FDSOI device (Martinie et al. 2021).

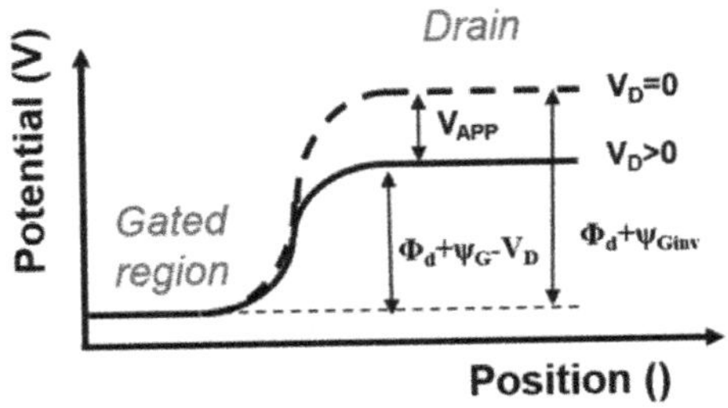

Figure 4.20. *Illustration of the applied voltage evaluation on the potential distribution from gated region to drain at V_D=0 and V_D>0*

Z2FET drain current can be obtained in injecting the applied voltage equation [4.3] into the PIN diode current equation [4.1]. To simplify the expression and make it explicit, we plot in Figure 4.21, the variation of ψ_G and of the sum $V_D + \psi_{Ginv}$ versus drain voltage (so the difference between the two curves is the applied voltage V_{APP} of equation [4.3]). We can see that, in the "*OFF-region*" regime, the applied voltage V_{APP} is close to zero, so the square term of equation [4.1] can be neglected. Moreover, as the drain current in this regime is low, the term $R_{PIN}I_D$ can also be neglected. It leads to this equation of drain current in the "*OFF-region*" $I_{D\text{-}OFF}$:

$$I_{D-OFF} = I_{LA-OFF}(e^{\frac{V_{APP}}{2ut}} - 1) \quad [4.4]$$

where $I_{LA\text{-}BC}$ is a fitting parameter related to diffusion current.

In the "*barriers collapsing*" regime, we observe in Figure 4.20 that V_{APP} cannot be neglected, but as the drain current is still low, the term $R_{PIN}I_D$ is still negligible. Moreover, the first term of equation [4.1] can be neglected. It leads to this expression of drain current in the "*barriers collapsing*" regime, $I_{D\text{-}BC}$:

$$\mathrm{I_{D-BC}} = \mathrm{I_{REC-BC}}\left(\mathrm{e}^{\frac{\mathrm{V_{APP}}}{2.\mathrm{ut}}} - 1\right)^2 \quad [4.5]$$

where $I_{REC\text{-}BC}$ is a fitting parameter related to recombination current.

Concerning the "*negative resistance*" regime, the behavior is too complex for a physics-based model description. However, we can remark on the S-shape I_D–V_D that the variation of the logarithm of the drain current is linear with drain voltage. As a result, we can use the following behavioral equation to describe drain current in the "*negative resistance*" regime $I_{D\text{-}NEG}$:

$$I_{D_{NEG}} = \frac{I_{ON}-I_{OFF}}{V_{ON}-V_{OFF}} V_D + \frac{V_{ON}I_{OFF}-V_{OFF}I_{ON}}{V_{ON}-V_{OFF}} \quad [4.6]$$

where V_{ON} (V_{OFF}) is the switching voltage under forward (reverse) bias and I_{ON} (I_{OFF}) is the corresponding drain current just before the switch.

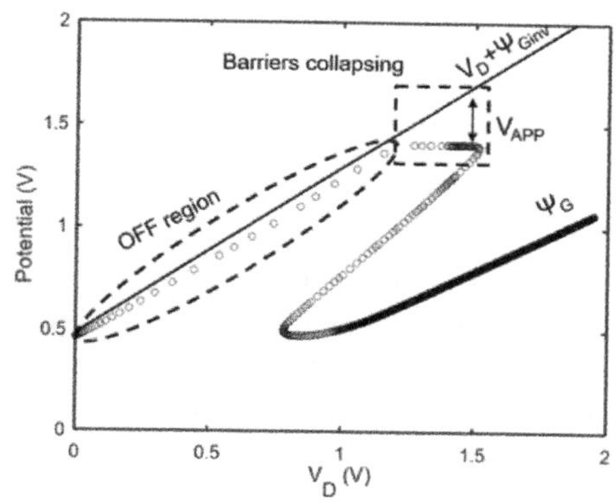

Figure 4.21. *Variation of the potential in gated region (Ψ_G, open circle) and of the sum of drain voltage and potential in gated region in the inversion regime (V_D+Ψ_{Ginv}) versus drain voltage obtained by the same TCAD simulations as shown in Figure 4.7*

As can be seen in Figure 4.22, the analytical expressions provide a nice agreement with TCAD simulation.

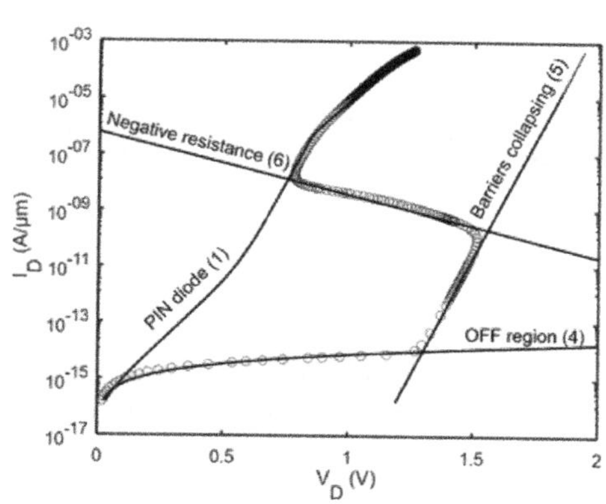

Figure 4.22. *Z2FET S-Shape I_D–V_D from TCAD (open circle) compare to analytical expressions [4.1]–[4.6] of drain current versus drain voltage in all the steady-state regime*

4.3.2. *Z2FET analytical evaluation of switching voltage*

Previous part provides analytical expressions of drain current in all the steady-state Z2FET regime. In order to go towards a unified description of the Z2FET drain current, so to compact model, analytical expressions of drain voltage limiting each regime is necessary. From TCAD simulations analysis, we know that:

– "*OFF-region*" ends at $\psi_G=\psi_{Ginv}$;

– "*Barriers collapsing*" ends at $V_D=V_{ON}$;

– "*Negative resistance*" ends at $V_D=V_{OFF}$.

The condition $\psi_G=\psi_{Ginv}$ is obtained analytically in the work of Martinie et al. (2021). The two other conditions require analytical expression of switching voltages. The evaluation of MOS capacitor potential in the depletion and inversion regimes ψ_{Gdep} and ψ_{Ginv} are also detailed in Martinie et al. (2021).

4.3.2.1. V_{ON} *analytical expression*

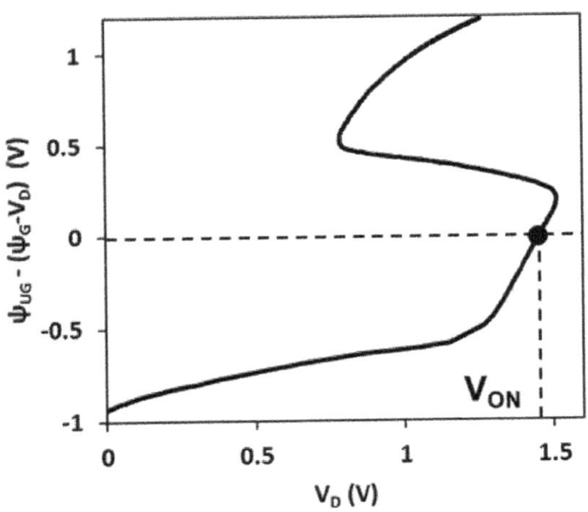

Figure 4.23. *Variation of Ψ_{UG}-(Ψ_G-V_D) versus drain voltage V_D from TCAD simulation to determine a criterion for V_{ON} evaluation*

To evaluate V_{ON}, the first step is to determine a criterion. The variation of Ψ_{UG}-(Ψ_G-V_D) versus drain voltage V_D from TCAD simulation is plotted in Figure 4.23, where Ψ_{UG} is the potential in ungated region and Ψ_G is the potential in gated region. We remark that V_{ON} corresponds to Ψ_{UG}-(Ψ_G-V_D)=0. Thanks to a step-by-step method detailed in Martinie et al. (2021), we obtained the following analytical expression for V_{ON}:

$$V_{ON} = \psi_{G_{dep}} - 0.5\psi_{UG}\left(\psi_{G_{dep}}\right) - 0.5\psi_{UG}\left(\psi_{G_{dep}} - 0.5\psi_{UG}\left(\psi_{G_{dep}}\right)\right) \quad [4.7]$$

where Ψ_{Gdep} is the potential in the gated region in the depletion regime and $\Psi_{UG}(V_D)$ is the potential in the ungated region as a function of drain voltage. Note that the analytical expression of $\Psi_{UG}(V_D)$ is given by Martinie et al. (2021).

4.3.2.2. *V_{OFF} analytical expression*

With a similar methodology, V_{OFF} criterion is determined from the variation of (Ψ_G-Ψ_{Ginv}) versus the drain voltage (Figure 4.24). We can see that V_{OFF} is the drain voltage giving Ψ_G-Ψ_{Ginv}=0. First, we consider that transport is ohmic in gated and ungated regions and is equal in both regions:

$$I_{gated} = I_{ungated} = \frac{V_D - V_M}{R_{Gated}} = \frac{V_M - V_S}{R_{Ungated}} \quad [4.8]$$

where I_{Gated} and $I_{Ungated}$ are, respectively, the current through the gated region and the ungated region, R_{Gated} and $R_{Ungated}$ are, respectively, the resistance of the gated and the ungated regions, and V_D, V_S and V_M are, respectively, the potential at the drain, the source and the limit between gated and ungated regions. Then, we assume a linear variation of the potential in the gated region, so from V_M to V_D and consider that the potential in gated region Ψ_G is the value of the potential in the middle of the gated region:

$$\psi_G = \frac{1}{2}(V_D - V_M) \quad [4.9]$$

Finally, considering that the resistance in the gated and the ungated regions is given by the ratio of the length over the surface of the corresponding region, multiplied by its resistivity (ρxL/S) and injecting equation [4.8] into equation [4.9], we obtain the analytical expression of V_{OFF}:

$$V_{OFF} = \frac{2t_{si-UG} + 2t_{si-G}}{t_{si-UG} + 2t_{si-G}} \psi_{Ginv} \quad [4.10]$$

where $t_{si\text{-}G}$ is the gated region silicon thickness, and $t_{si\text{-}UG}$ is the ungated region silicon thickness.

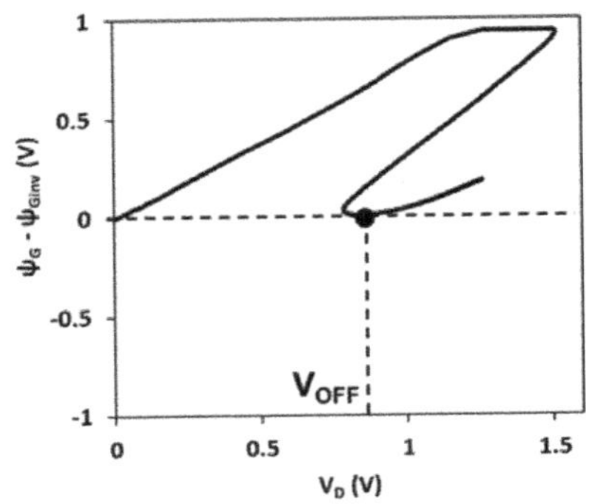

Figure 4.24. *Variation of Ψ_G-Ψ_{Ginv} versus drain voltage V_D from TCAD simulation to determine a criterion for V_{OFF} evaluation*

4.3.2.3. *Discussion around sharp switches voltage V_{ON} and V_{OFF}*

From the analytical expressions of V_{ON} and V_{OFF}, we can remark that they are first-order dependent, respectively, on the depleted and inverted surface potential in gated region. The variation of V_{ON} is considerable, because it is first-order dependent on the depleted surface potential (Ψ_{Gdep}), which is strongly impacted by the gate voltage. Concerning V_{OFF}, its evolution is smaller with gate voltage, because of its first-order dependency on the inverted surface potential (Ψ_{Ginv}). Similar remarks can be done on the back gate voltage effect, which is weaker than the gate voltage because of the larger thickness of the buried oxide compared to the gate oxide. Figure 4.25 illustrates the variation of switching voltages with back gate voltage and also demonstrates the nice agreement between TCAD and analytical expressions.

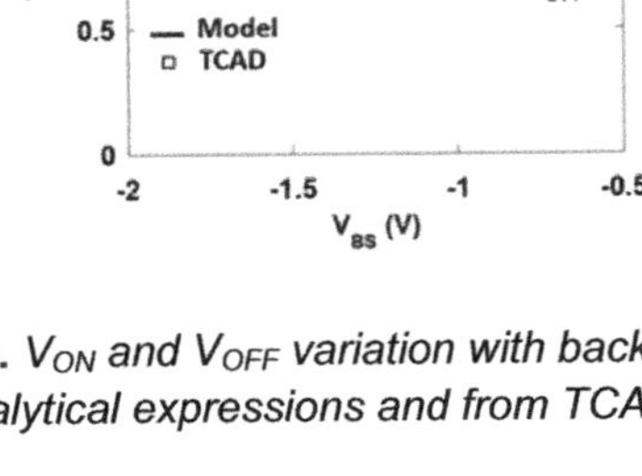

Figure 4.25. *V_{ON} and V_{OFF} variation with back gate voltage V_{BS} from analytical expressions and from TCAD simulations*

4.3.3. *Z2FET compact model*

To be able to perform the Z2FET circuit design, a compact model compatible with conventional CAD tools is necessary. The main Z2FET compact modeling challenge is a unified description of the steady-state hysteresis. This means that same equations need to be used regardless of the polarization scheme (here forward or direct bias). This cannot be managed by a current versus voltage I(V) as described in section 4.3.1, but it can be described by voltage versus current V(I) formulation, where there is no more hysteresis, as illustrated in Figure 4.26.

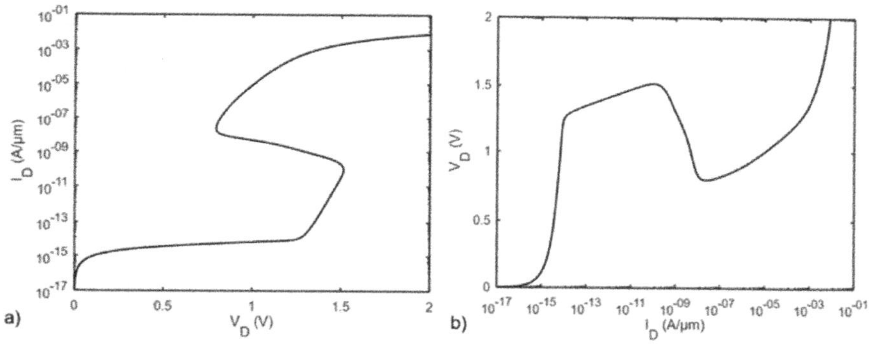

Figure 4.26. *Z2FET I_D–V_D (a) and V_D–I_D (b) curves illustrating that there is no V(I) hysteresis*

Therefore, drain current analytical expressions of section 4.3.1 need to be reversed, thus leading to:

$$V_{D-OFF} = N_{OFF}.\,ut.\ln(1 + \frac{I_D}{I_{La-OFF}}) \quad [4.11]$$

$$V_{D-BC} = N_{BC}.\,ut.\ln\left(1 + \sqrt{\frac{I_D}{I_{Rec-BC}}}\right) + N_{BC}(\psi_{G_{dep}} - \psi_{G_{inv}}) \quad [4.12]$$

$$V_{D-NEG} = \frac{V_{ON} - V_{OFF}}{I_{ON} - I_{OFF}} I_D + \frac{V_{ON} I_{OFF} - V_{OFF} I_{ON}}{I_{ON} - I_{OFF}} \quad [4.13]$$

$$V_{D-PIN} = N_{PIN}.\,ut.\ln\left(\frac{2.I_{Rec-PIN} - I_{La-PIN} + \sqrt{I_{La}^2 + 4 I_D I_{Rec-PIN}}}{I_{Rec-PI}}\right) + R_{PIN} I_D \quad [4.14]$$

Then, V_{ON}, V_{OFF}, gated potential Ψ_G and ungated potential Ψ_{UG} analytical expression of sections 4.3.1 and 4.3.2 can be used directly into a V(I) formulation.

The final step is to guarantee a unified description with only a continuous variation of the output of the model, so here the voltage. This is achieved through linking continuous functions, such as the Fermi function. Finally, Z2FET I_D–V_D curve can be plotted through SPICE simulation with current as input to obtain S-Shape I_D–V_D (Figure 4.27-a), and also with voltage as input to obtain the sharp switches under forward and reverse biases (Figure 4.27-b). Indeed, the SPICE simulator is able to provide the current as output with the voltage as input, even though the model uses a V(I) formulation. Figure 4.27 shows the very nice agreement between I_D–V_D from TCAD and from experimental measurements.

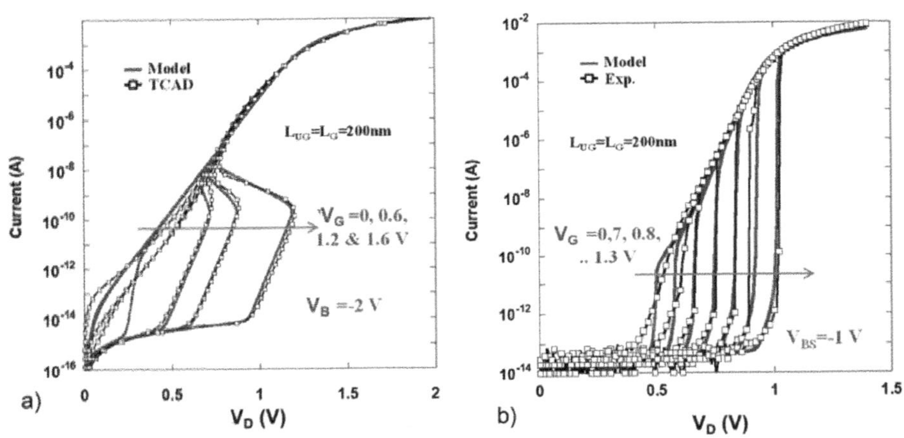

Figure 4.27. *Z2FET I_D–V_D current driven from model and TCAD (a) and voltage driven from model and experimental measurement (b). For a color version of this figure, see www.iste.co.uk/cresti/beyondcmos.zip*

4.4. Z2FET experimental evidence

4.4.1. *Z2FET fabrication*

Z2FET fabrication is fully compatible with standard FDSOI process and does not require any additional or different fabrication steps. Z2FETs have already been fabricated in 28FDSOI (Planes et al. 2012; Solaro et al. 2014) and 14FDSOI (Weber et al. 2014; El Dirani et al. 2015) technologies. TEM cross-section of Z2FET is shown in Figure 4.28.

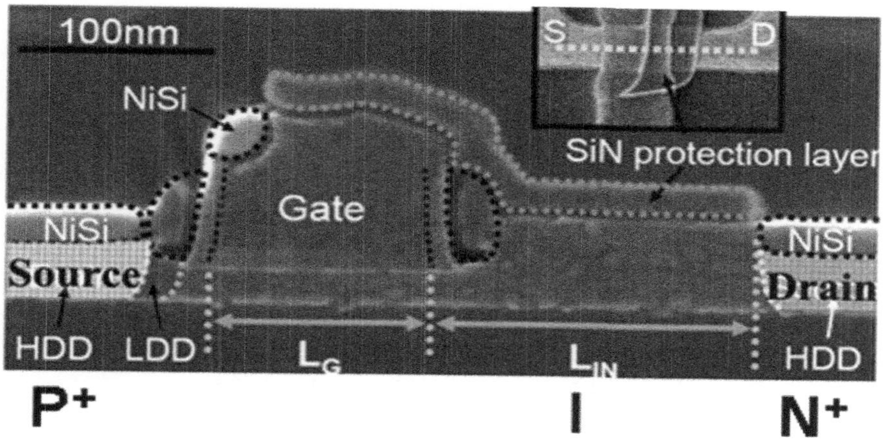

Figure 4.28. *Z2FET TEM cross-section fabricated with 28FDSOI FDSOI technology. For a color version of this figure, see www.iste.co.uk/cresti/beyondcmos.zip*

4.4.3. *Z2FET switching characteristic under gate sweep*

To obtain a Z2FET sharp switch with gate sweep, the first step is to negatively bias the back gate to increase the hole density in the intrinsic region and get an electrostatic PN structure. Then, the source is grounded and the drain positively biased to obtain large current. At this step, the device behaves as a PIN diode under forward bias. When the gate voltage is increased, the electron density in the intrinsic region below the gate increases, while the hole density decreases. If the gate voltage is sufficient, Z2FET becomes equivalent to an NPNP structure and is blocked. This sudden blockage leads to a sharp switch. This is experimentally evidenced in Lee et al. (2019) on n-type Z2FET (L_G=L_{UN}=500 nm) fabricated in 28FDSOI technology (Figure 4.29). However, Z2FET I_D-V_G curves of Figure 4.29 performed at drain voltage varying from 925 mV to 1 V also highlight the strong impact of drain voltage on the gate voltage needed to block Z2FET. Indeed, for the drain voltage increase of 75 mV, the gate voltage needs to be increased to 1.5 V to turn off Z2FET. It also shows that Z2FET cannot be blocked if the drain voltage is too high (i.e. >1 V).

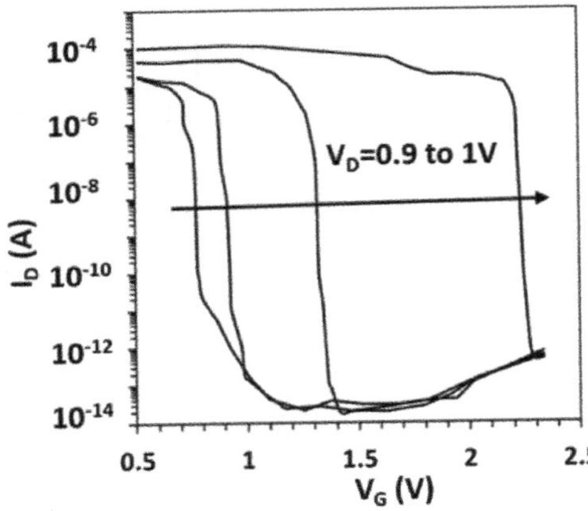

Figure 4.29. *Experimental evidence of Z2FET sharp switching for gate voltage variation*

4.4.4. *Z2FET switching characteristic under drain sweep*

Z2FET switching under drain sweep is experimentally demonstrated by many papers, for instance, Wan et al. (2012b). Figure 4.30 shows experimental I_D–V_D of Z2FET fabricated with 28FDSOI technology. Experimental I_D–V_D shape is the same as the simulated one, as well as the gate voltage effect on switching voltage.

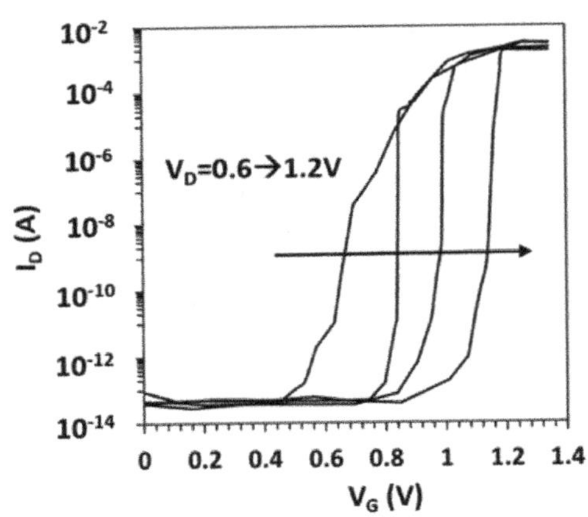

Figure 4.30. *Experimental evidence of Z2FET I_D–V_D sharp switching for gate voltage variation*

4.5. Z2FET as 1T-DRAM

The classical DRAM (Dynamic Random Access Memory) cell is composed of a transistor and a capacitor (1T-1C) and needs to be billion-times cloned, leading to mandatory aggressive scaling. As in CMOS technology, scaling is still feasible, but is more and more complex and costly. A disruptive solution could be the single-transistor DRAM (1T-DRAM), proposed more than 25 years ago (Wann and Hu 1993). In general, 1T-DRAMs take advantage of the floating-body effects usually considered in CMOS technology as parasitic and detrimental. "1" state is defined by the presence of charge stored in the floating body and provides high current. Z2FET capacitorless 1T-DRAM (Z2-RAM) has emerged as a possible candidate for designing low power, low cost efficient embedded DRAM for IoT applications (Wan et al. 2013; El Dirani et al. 2016b; Parihar et al. 2017). As in other 1T-DRAMs, information is stored by charge collection under the gate.

4.5.1. *Z2FET 1T-DRAM operation description*

For "1" state programming (write operation (W)), the device is turned on (V_G=0 at high drain voltage), then the gate polarization is increased to build potential barriers and so carriers are retained under the gate. The presence of charge, corresponding to "1" state produces a shift of the switching voltage V_{ON}, used to read information. The device turns on in "1" state and remains blocked in "0" state. Therefore, the read operation (R) is performed with high V_G and V_D chosen higher than "1" state V_{ON} and lower than "0" state V_{ON}. In this way, the "1" state read current I_1 corresponds to forward PIN diode current and is largely higher than the "0" state read current I_0, which relies on blocked Z2FET. "0" state programming (or erasing (E)) is performed with V_G=V_D=0, thus enabling the carriers stored below the gate to be removed, while the device goes back to the equilibrium. Hold operation is performed with high V_G to preserve the potential barriers to keep the information in the device and with V_D=0 V to guarantee no current flow. Note that "0" state hold is the worst case for retention (Wan et al. 2012a). The parasitic generation of electrons and their accumulation under the gate during "0" state holding is the limiting mechanism (El Dirani et al. 2016a).

The Z2FET 1T-DRAM operation memory sequence (Erase(E)-Write(W)-Read(R)-Erase(E)-Read(R)) is shown in Figure 4.31, and the detailed analysis of each operation based on TCAD simulations can be found in Cristoloveanu et al. (2018).

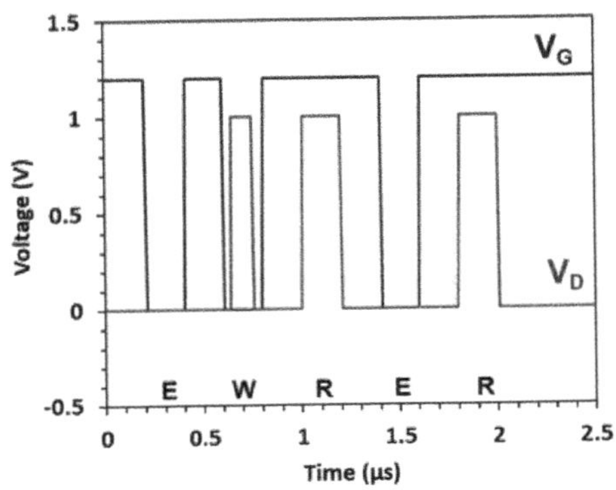

Figure 4.31. *Z2FET 1T-DRAM operation E-W-R-E-R memory sequence gate and drain voltages chronogram. For a color version of this figure, see www.iste.co.uk/cresti/beyondcmos.zip*

4.5.2. Z2FET 1T-DRAM operation experimental evidence

Z2FET 1T-DRAM operation is first experimentally evidenced by Wan et al. (2012a). Figure 4.32 shows the measurement of the memory sequence Erase (E) – Read "0" (R) – Program "1" (W) – Read "1" (R) performed on Z2FET fabricated with 28FDSOI technology. For voltages below 1.2 V, a large memory window is demonstrated with "1" state read current I_1 is 46 µA/µm, while "0" state read current I_0 is around 1 nA/µm.

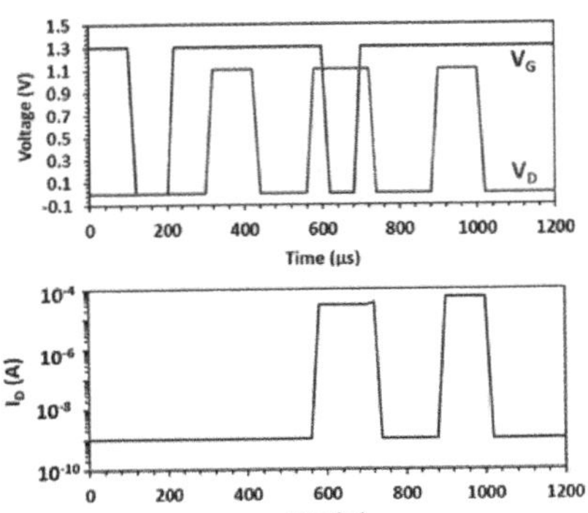

Figure 4.32. *Experimental evidence of Z2FET 1TDRAM operation with the sequence E-R-W-R. voltages. "1" state and "0" state are successfully programmed and read with I1>>I0. For a color version of this figure, see www.iste.co.uk/cresti/beyondcmos.zip*

Thanks to a programming scheme study, Parihar et al. (2017) showed that "1" state can be programmed at very low drain voltage, up to 500 mV to reduce power consumption without degradation of "1" state read current I_1.

It is also shown that the "1" state read current and the sense margin largely depend on the drain voltage applied during reading, while they are almost insensitive to the gate voltage V_G. Indeed, as described in section 4.2.2.5, gate voltage impacts the switching voltage value, but not the drain current in the "PIN" regime.

The Z2FET retention time is experimentally evaluated in the worst case, so during "0" state holding. Figure 4.33 shows its variation with drain voltage applied during reading and gate voltage applied during holding at room temperature. It demonstrates retention time up to 5 s in the best bias condition, while the worst case reaches 200 ms.

As standard DRAM usually operates around 85°C, the Z2FET memory operation has also been experimentally explored at high temperatures. Lee (2019) shows that "1" state read current is almost not affected by the temperature increase, while "0" state read current rises after 120°C (Figure 4.34-a). Retention time is expected to decrease for higher temperatures, due to the stronger carrier generation. It is measured by Wan et al. (2013) up to 80°C. Figure 4.34-b shows that, in the best bias condition, retention time is 5 s at room temperature and decreases to 200 ms at 80°C, which is still compatible with the DRAM operation.

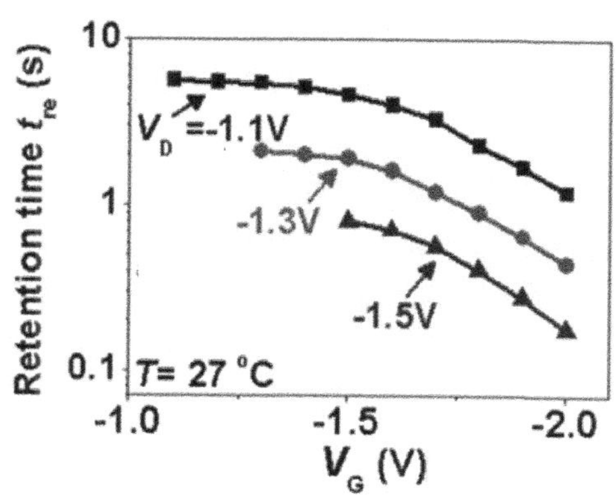

Figure 4.33. *Z2FET retention time evaluated in the worst-case condition ("0" state holding) as a function of gate voltage applied during hold operation and drain voltage applied during read operation. For a color version of this figure, see www.iste.co.uk/cresti/beyondcmos.zip*

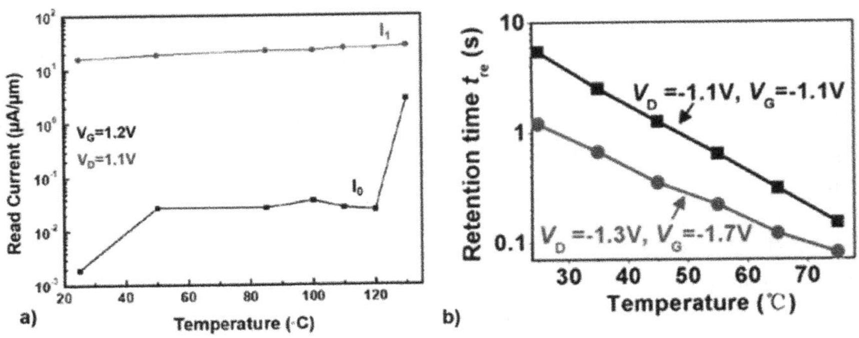

Figure 4.34. *(a) Z2FET I_1 and I_0 variation with temperature and (b) Z2FET retention time with temperature. For a color version of this figure, see www.iste.co.uk/cresti/beyondcmos.zip*

1T-DRAM cells are intended to be used in matrix environments where specific behaviors, such as parasitic writing, can occur and be detrimental for industrial use. 2x2 Z2FET matrix has first been simulated by TCAD in Parihar et al. (2018). It is shown that the word "1010" can be written and read in the matrix.

2x2 Z2FET matrixes have also been fabricated with 28FDSOI technology and characterized (Navarro et al. 2018). With *i* being the number of the rows and *j* being the number of the columns in the matrix, each Z2EFT Z_{ij} is connected as follow (Figure 4.35):

– the gate (G) is connected to the wordline WL_i;

– the drain (D) is connected to the bitline BL_j. The source (S) is connected to the selector word access M_i. The back gate (B) is common to all Z2FET of the matrix and is plugged to BG.

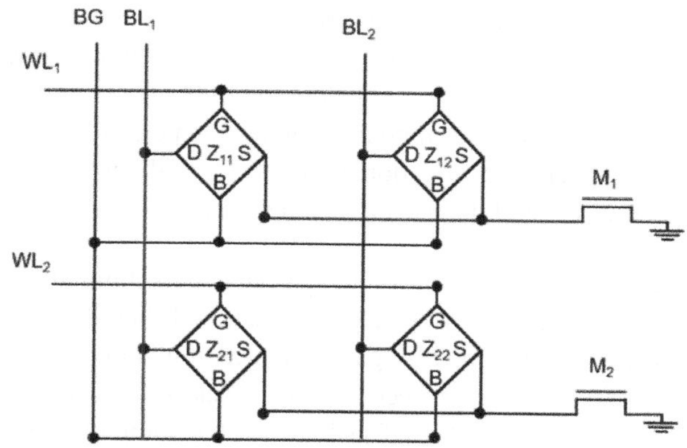

Figure 4.35. *2X2 Z2FET matrix schematic where the connection of each Z2FET (Z_{ij}) electrode (gate (G), drain (D), source (S) and backgate (B)) to the circuit is represented*

NOTES ON FIGURE 4.35.– *WL_i is the wordline of the ith row and BL_j is the bitline of jth column. All back gates are connected together and one selector per row is used, connected to the source of each Z2FET of the row.*

First, the objective is to write and read the words "00", "01", "10" and "11" only in the first row of the matrix. For each program or read operation, the selector M1 needs to be activated. To program the "0" state, the BL (drain) is at 0 V while for the "1" state, it is at 0.5 V while WL (the gate) is at 0 V in both cases. To read the memory state, WL and BL are set to 1.2 V. Note that it is the same program/read scheme as single cell characterization (Parihar et al. 2017). Navarro et al. (2018) demonstrates by measurement that the programmed word is well read in each case.

As each raw can be programmed and read simultaneously, only four access cycles are necessary to program and read the word "1001". The corresponding sequence is the following: "10" word is programmed on the first row, then "01" word is programmed on the second row, and finally, reading is performed row by row. Navarro et al. (2018) demonstrates experimentally that "1001" word is programmed and read in only four access cycles.

4.6. Z2FET structure optimization

In the previous sections, we showed that Z2FET presents a sharp switch whose position depends on the gate and back gate voltages, low leakage, high drive current and interesting memory performance. Standard Z2FET is usually fabricated with P-type doped silicon below the buried oxide (i.e. the Ground Plane, GP). In this case, the back gate voltage V_B is applied below the BOX all along the Z2FET device from source to drain and is used to tune the ungated area electrostatic P-type doping. Therefore, the N-type electrostatic doping of the gated region is related to the coupling between the gate and the back gate polarizations.

However, we saw that the control of the switch is limited by the coupling between gate and back gates. It was also demonstrated that for high drain voltage (>1 V), Z2FET cannot be blocked by gate voltage. In addition, we highlighted the strong sensitivity of gate voltage necessary to block Z2FET to drain voltage (if V_D<1 V). This makes the use of Z2FET as logic switch complex. To overcome those issues, two optimizations of Z2FET device structures have been proposed: the Z2FET with dual ground plane (Z2FET DGP) and the Z3FET.

4.6.1. *Z2FET DGP*

The Z2FET limitation described above is due to the P-type single GP, which implies that the carrier density in the gated region is due to the coupling between gate and back gate voltages. This coupling limits the Z2FET switching ability at high drain voltage. To go through it, N-type ground plane below the gated region is introduced to make a Z2FET with two ground planes: Z2FET DGP (El Dirani et al. 2016b). The device is sketched in Figure 4.36 and V_{BP} is the P-type ground plane (GP-P) voltage, while V_{BN} is the N-type ground plane (GP-N) voltage. With Z2FET DGP, the carrier concentration of the gated region is only due to front gate bias V_G and GP-N bias V_{BN}. Potential barriers are better controlled, so V_{ON} can be better managed and can also reach higher values. To prevent leakage, the NP diode formed by GP-P and GP-N is not to be forward biased: this is the only limitation of the flexibility enhancement of Z2FET DGP compared to standard Z2FET.

Z2FETs DGP has been fabricated with 14FDSOI and 28FDSOI technologies. Figure 4.37 shows measured I_D–V_D curves with V_{BN}=0 and V_{BN}=V_G. Z2FET DGP operation is the same as standard Z2FET but switching occurs at higher drain voltage in the case V_G=V_{BN}, because potential barriers are stronger and require higher drain voltage to allow switching (Dirani 2016b). This feature can be interesting for ESD applications (see section 4.7.1).

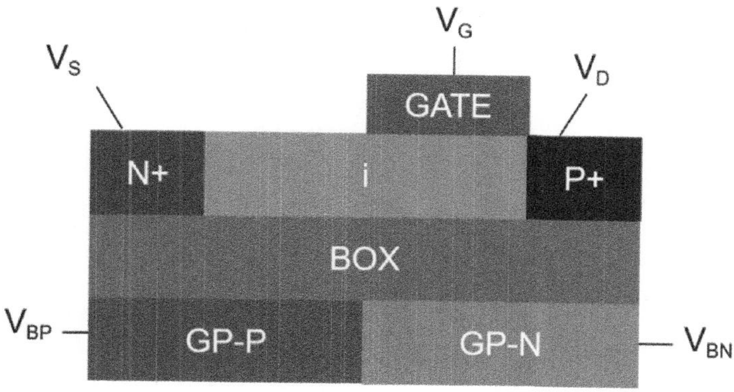

Figure 4.36. *Z2FET DGP device structure. For a color version of this figure, see www.iste.co.uk/cresti/beyondcmos.zip*

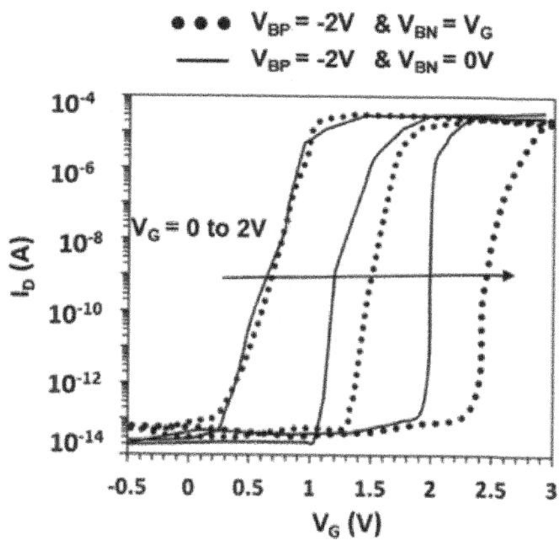

Figure 4.37. *Experimental I_D–V_D curves with V_{BN}=0 (a) and V_{BN}=V_G (b)*

Standard Z2FET cannot be blocked if drain voltage is too high (V_D>1 V, see Figure 4.29) and the switching gate voltage sensitivity to drain voltage is strong. With the introduction of the GP-N in Z2FET DGP, this sensitivity is largely reduced but switching at high drain current is still not possible (Figure 4.38). However, applying 1 V on the GP-N enables to block Z2FET, even at V_D=1.1 V (Figure 4.38). This result makes possible the use of Z2FET DGP as a logic switch (see section 4.7.2). Finally, the stronger potential barriers of Z2FET DGP allow the use of shorter device for the 1T-DRAM application (Lee et al. 2018).

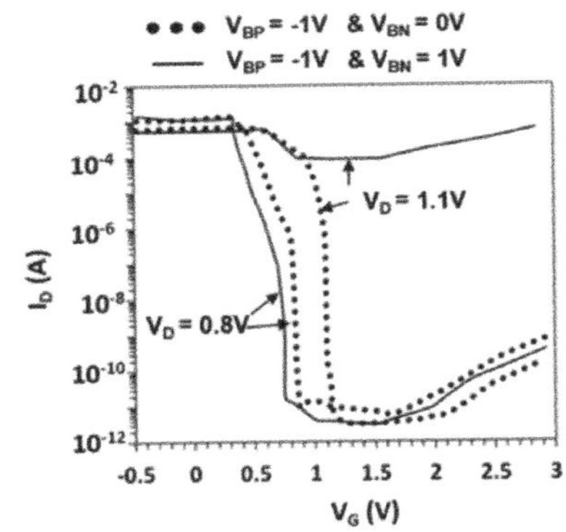

Figure 4.38. *Z2FET DGP I_D-V_G measurement with (a) V_{BN}=0 and (b) V_{BN}=1 V*

4.6.2. *Z3FET*

Z2FET DGP already improved Z2FET performance, especially for the high voltage application. However, Z2FET DGP high voltage can be limited by gate stack reliability. This observation is the basis of the definition of Z3FET, which is a Z2FET DGP with a free top surface, i.e. without top gate, so the third Z of Z3FET means "zero gate" (Solaro et al. 2016) (Figure 4.39). The absence of the conventional ultrathin gate dielectric is a remarkable advantage in terms of reliability: it transfers the reliability concern to the BOX, which is sufficiently thick to support a higher voltage.

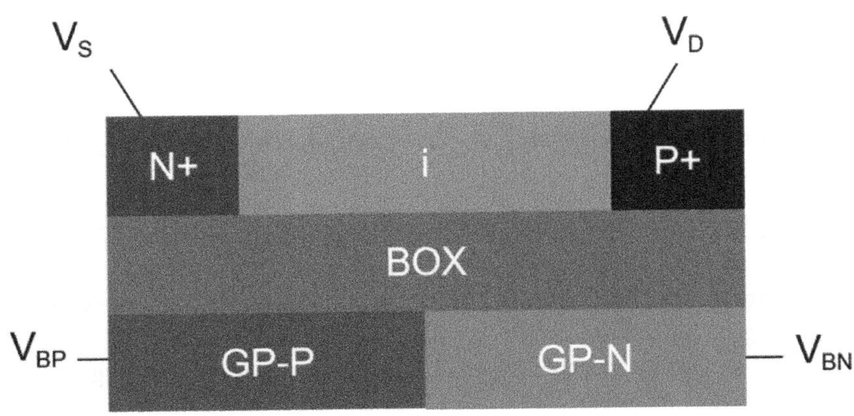

Figure 4.39. *Z3FET device structure. For a color version of this figure, see www.iste.co.uk/cresti/beyondcmos.zip*

Z3FET operation is the same as Z2FET, but the electrostatic "NPNP" device is formed thanks to the GP-P and GP-N polarizations, respectively, to build P-type and N-type electrostatic regions in the silicon film. The sharp switch is experimentally evidenced on Z3FET fabricated with 14FDOI technology (Solaro et al. 2016). As Z2FET, switching voltage is adjustable by varying ground plane voltages. As Z2FET DGP, the only limitation is due to the PN diode formed by both ground planes which are not to be forward biased. Z3FET can be blocked even at a high drain voltage like Z2FET DGP and contrary to standard Z2FET. As Z3FET includes no top gate, it requires slightly higher V_{BN} to be blocked than Z2FET DGP.

The high voltage operation ability is evidenced by measurement on longer Z3FET (500 nm long GP-N and GP-P areas), where I_D–V_D are measured up to V_D=7 V. This measurement also demonstrates that the switching voltage is efficiently managed by GP-N polarization (Solaro et al. 2016).

Z3FET is interesting for ESD application, because it is able to handle high voltages. Moreover, the free surface of Z3FET can be functionalized for sensing applications.

4.7. Z2FET advanced applications

This final section aims to describe the possible use of Z2FET in other applications, such as ESD, logic switch or photo-detector.

4.7.1. *Z2FET as ESD*

ElectroStatic Discharge (ESD) events are generally short (<100 ns) and involve high current and high voltage. They are responsible for electronic circuit damage or destruction. To be used as ESD protection, a device needs to be:

– efficient: the device needs to evacuate large current before the circuit reaches its breakdown voltage. The device needs to switch at a voltage between the power supply of the circuit and its breakdown voltage;

– transparent: the device leakage needs to be as low as possible;

– robust: the device needs to handle high voltage and high current without breaking;

– fast: ESD events are very short, so the device needs to switch very fast.

It is demonstrated by Solaro et al. (2013) that Z2FET fulfills all ESD protection requirements: it is as robust as standard ESD protection like gated diode or field-effect diode, but it presents lower leakage. As it is fully compatible with FDSOI process, it is a very interesting candidate for FDSOI technology ESD protection. Thanks to its better switching of voltage control, Z2FET DGP is also interesting for ESD application (El Dirani et al. 2016b). Finally, thanks to the absence of top gate, Z3FET is demonstrated as an ESD protection efficient even at a high voltage (Solaro et al. 2016).

4.7.2. *Z2FET as logic switch*

In steady-state operations, Z2FET blockage by gate voltage requires high gate voltage or can even be impossible if drain voltage is too high (i.e. >1 V, see section 4.4.2, Figure 4.27). Moreover, the presence of hysteresis in Z2FET prevents its use for logic applications. To solve both issues, the use of fast pulses on the gate is proposed (Lee et al. 2019). The experimental protocol is the following:

– drain voltage is low (V_D=0.5 V) and gate voltage is high (V_G=1.8 V): Z2FET is blocked;

– drain voltage is increased up to the operation value (V_D=0.8 V) with 50 ns rise time. Z2FET is still blocked, but leakage increases;

– gate voltage is lowered to 500 mV (ideally to V_G=0 V, but it was not possible with measurement tool used by the authors) with 50 ns fall time. Z2FET turns on and delivers a high current;

– gate voltage is increased back to its initial value with 50 ns rise time. Z2FET turns off and delivers only a weak leakage current.

Figure 4.40 shows the corresponding voltage and current versus time variation measured for several maximum value of gate bias (from 1.2 to 1.5V V). It demonstrates that the Z2FETs switch on for low gate voltages and switch off for high gate voltages and also suggests that the hysteresis is canceled thanks to the fast pulse applied on the gate. Indeed, when V_G is sharply increased, the electron concentration underneath the gate cannot be adjusted instantly, thus leading to a potential barrier higher than in the steady-state mode. This non-equilibrium mechanism is also used to program “1” state in the Z2FET 1T-DRAM operation (see section 4.5). This explained why the gate voltage needed to block the Z2FET is lower in pulsed mode than in steady-state mode.

To enhance the switching performance, gate control needs to be improved, so Z2FET DGP should present a higher performance. This is evidenced by Figure 4.41

with $V_{BN}>0$: OFF state current is fully suppressed for V_{BN}=2V even at gate voltage V_G as low as 0.5 V. However, these conditions are no longer sufficient if the drain voltage is increased to deliver higher drive current. Applying negative bias on the second ground plane allows us to recover the operation with higher drain voltage (Figure 4.41).

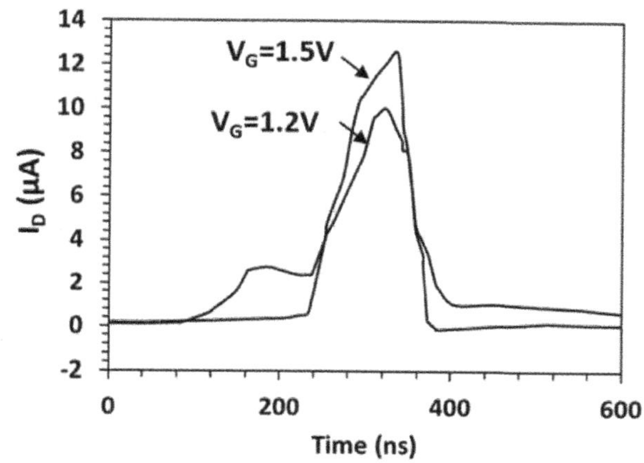

Figure 4.40. *Z2FET measurement with short gate pulse for logic switch application*

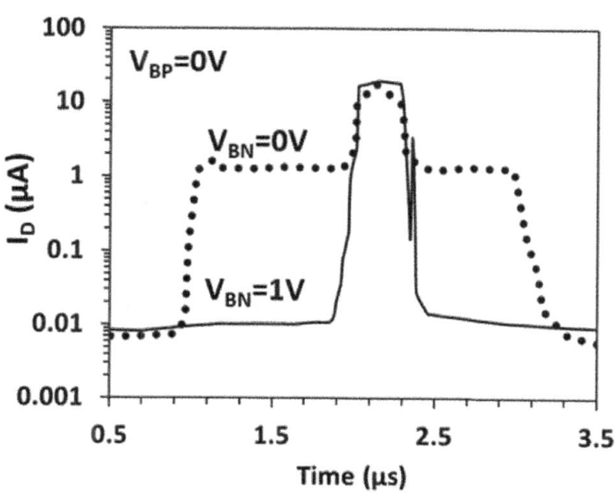

Figure 4.41. *Z2FET DGP short gate pulse measurement for logic switch application V_{BN} variation*

4.7.3. *Z2FET as photodetector*

The free surface of Z2FET (the ungated region) can be used to detect light. Z2FET needs to be blocked, so positive V_G and negative V_B are necessary with intermediate V_D. Z2FET is then illuminated after 1 ms. If the Z2FET is not illuminated, it remains blocked, while it turns on after light exposure. The higher the light intensity, the shorter the time necessary for Z2FET to turn on (Liu et al. 2019). In this operation, Z2FET switching is due to photo-generated carriers in the ungated region, which reduce the potential barrier until switching. With higher light intensity, the amount of photo-generated carrier is bigger, explaining why Z2FET switching occurs faster.

The light sensitive area of Z2FET is the ungated region of the device. As a result, Z3FET photo-detection performance should be better, but it still has to be investigated.

4.8. Conclusion

Z2FET is a device presenting low leakage, high drive current and a sharp switch without involving impact ionization and fully compatible with existing FDSOI technologies. Through TCAD simulations and modeling, we demonstrated that Z2FET device operation is a combination of MOS capacitors and PIN diode operations. The MOS capacitors allow the control of the electrostatic doping in the "PNPN" electrostatic structure to tune switching voltages. The PIN diode enables the ultra-low leakage and the high drive current. Finally, we saw that sharp switches are due to the sudden from the MOS to the PIN diode operation and vice versa. Then, Z2FET analytical and compact models are presented to allow circuit design studies.

1T-DRAM is the most studied Z2FET application in the literature with extensive TCAD simulation, fabrication and electrical characterization from single cell to 2x2 matrix. It shows a very interesting performance: large memory window and read margin and high retention current even at high temperatures. However, Z2FET 1T-DRAM performances are still not high enough to replace the standard 1T-1C DRAM.

Z2FET as ESD protection was also experimentally studied. It demonstrated that Z2FET fulfills all ESD requirements (efficiency, transparency, robustness and speed) and outperforms standard ESD protection as gated diode.

Z2FET is also considered as a possible solution for a logical switch and an image sensor with preliminary studies in the literature based on experimental data and TCAD simulation.

The Z2FET structure is optimized to improve the switching ability, the switching voltage adaptability with gate and back gate biases and the reliability with the Z2FET DGP (Dual Ground Plane) and the Z3FET. With independent ground planes, Z2FET DGP presents better a switching ability and flexibility, because the gated region is not impacted by the back gate voltage required to build the P concentration in the ungated region. Z3FET has the same benefit as Z2FET DGP, but with a higher reliability thanks to the absence of gate.

As a conclusion, Z2FET as ESD protection seems to be the closer to an industrial application, even though the Z2FET 1T-DRAM performance can certainly be improved to reach the DRAM standard performance target.

4.9. References

Cristoloveanu, S., Lee, K.H., Parihar, M.S., El Dirani, H., Lacord, J., Martinie, S., Le Royer, C., Barbe, J.-C., Mescot, X., Fonteneau, P. et al. (2018). A review of the Z2-FET 1T-DRAM memory: Operation mechanisms and key parameters. *Solid State Electronics*, 143, 10–19. DOI: 10.1016/j.sse.2017.11.012.

El Dirani, H., Solaro, Y., Fonteneau, P., Ferrari, P., Cristoloveanu, S. (2015). Sharp-switching Z2-FET device in 14 nm FDSOI technology. *45th European Solid State Device Research Conference (ESSDERC)*. DOI: 10.1109/ESSDERC.2015.7324761.

El Dirani, H., Bawedin, M., Lee, K., Parihar, M., Mescot, X., Fonteneau, P., Galy, P., Gamiz, F., Kim, Y-T., Ferrari, P. (2016a). Competitive 1T-DRAM in 28 nm FDSOI technology for low-power embedded memory. *IEEE SOI-3D-Subthreshold Microelectronics Technology Unified Conference (S3S)*. DOI: 10.1109/S3S.2016.7804402 and DOI: 10.1109/ESSDERC.2016.7599623.

El Dirani, H., Fonteneau, P., Solaro, Y., Ferrari, P., Cristoloveanu, S. (2016b). Novel FDSOI band-modulation device: Z2-FET with dual ground planes. *46th European Solid-State Device Research Conference (ESSDERC)*. DOI: 10.1109/ESSDERC.2016.7599623.

Fonteneau, P., Solaro, Y., Marin-Cudraz, D., Legrand, C.-A., Fenouillet-Beranger, C. (2015). Innovative high-density ESD protection device in state of the art UTBB FDSOI technologies. *EOS/ESD Symposium*. DOI: 10.1109/EOSESD.2015.7314808.

Gatard, E. (2006). Analyse des phénomènes physiques dans les diodes p-i-n : contribution à la modélisation électrothermique pour les applications de puissance RF et hyperfréquences. PhD Dissertation, Université de Limoges.

Gopalakrishnan, K., Woo, R., Jungemann, C., Griffin, P.B., Plummer, J.D. (2005a). Impact ionization MOS (I-MOS), Part II: Experimental results. *IEEE Transactions on Electron Devices*, 52(1). DOI: 10.1109/TED.2004.841345.

Gopalakrishnan, K., Griffin, P.B., Plummer, J.D. (2005b). Impact ionization MOS (I-MOS), Part I: Device and circuit simulation. *IEEE Transactions on Electron Devices*, 52(1). DOI: 10.1109/TED.2004.841344.

Jeon, K., Loh, W.-Y., Patel, P., Yong Kang, C., Oh, J., Bowonder, A., Park, C., Park, C.S., Smith, C., Majhi, P. et al. (2015). Si tunnel transistors with a novel silicided source and 46mV/dec swing. *Digest of Technical Papers – Symposium on VLSI Technology*. DOI: 10.1109/VLSIT.2010.5556195.

Lacord, J., Martinie, S., Parihar, M.-S., Lee, K., Bawedin, M., Cristoloveanu, S., Taur, Y., Barbé, J.-C. (2017). Z2-FET DC hysteresis: Deep understanding and preliminary model. *International Conference on Simulation of Semiconductor Processes and Devices (SISPAD)*. DOI: 10.23919/SISPAD.2017.8085329.

Lee, K. (2019). Etude des dispositifs à dopage électrostatique et des applications dans les technologies FD-SOI. PhD Dissertation, Université Grenoble Alpes.

Lee, K., El Dirani, H., Fonteneau, P., Bawedin, M., Sato, S., Cristoloveanu, S. (2018). Sharp switching, hysteresis-free characteristics of Z2–FET for fast logic applications. *48th European Solid-State Device Research Conference (ESSDERC)*. DOI: 10.1109/ESSDERC.2018.8486915.

Lee, K., El Dirani, H., Fonteneau, P., Bawedin, M., Cristoloveanu,S. (2019). Sharp logic switch based on band modulation. *IEEE Electron Device Letters*, 40(11). DOI: 10.1109/LED.2019.2939665.

Liu, J., Cao, X.Y., Lu, B.R., Chen, Y.F., Zaslavsky, A., Cristoloveanu, S., Wan, J. (2019). Dynamic coupling effect in Z2-FET and its application for photodetection. *IEEE Journal of the Electron Devices Society*, 7. DOI: 10.1109/JEDS.2019.2918203.

Martinie, S., Lacord, J., Lee, K., Bawedin, M., Cristoloveanu, S. (2021). Pragmatic Z^2-FET compact model including DC and 1T-DRAM memory operation. *Solid-State Electronics*, 179, 107960.

Mayer, F., Le Royer, C., Le Carval, G., Clavelier, L., Deleonibus, S. (2007). Experimental and TCAD investigation of the two components of the impact ionization MOSFET (IMOS) switching. *IEEE Electron Device Letters*, 28(7). DOI: 10.1109/LED.2007.897876.

Mayer, F., Le Royer, C., Blachier, D., Clavelier, L., Deleonibus, S. (2008a). Avalanche breakdown due to 3-D effects in the impact-ionization MOS (I-MOS) on SOI: Reliability issues. *IEEE Transactions on Electron Devices*, 55(6). DOI: 10.1109/TED.2008.922797.

Mayer, F., Le Royer, C., Damlencourt, J.F., Romanjek, K., Andrieu, F., Tabone, C., Previtali, B., Deleonibus, S. (2008b). Impact of SOI, Si1-xGexOI and GeOI substrates on CMOS compatible tunnel FET performance. *Proc. Int. Electron Devices Meeting*. DOI: 10.1109/IEDM.2008.4796641.

Navarro, S., Navarro, C., Marquez, C., El Dirani, H., Galy, P., Bawedin, M., Pickering, A., Cristoloveanu, S., Gamiz, F. (2018). Experimental demonstration of operational Z2-FET memory matrix. *IEEE Electron Device Letters*, 39(5). DOI: 10.1109/LED.2018.2819801.

Padilla, C., Wing Yeung, C., Shin, C., Hu, T.-J., King, L. (2008). Feedback FET: A novel transistor exhibiting steep switching behavior at low bias voltages. *IEEE International Electron Devices Meeting*. DOI: 10.1109/IEDM.2008.4796643.

Parihar, M.-S., Lee, K.H., Dirani, H.E., Navarro, C., Lacord, J., Martinie, S., Barbe, J.-C., Fonteneau, P., Galy, P., Le Royer, C. et al. (2017). Low-power Z2-FET capacitorless 1T-DRAM. *IEEE International Memory Workshop (IMW)*. DOI: 10.1109/IMW.2017.7939093.

Parihar, M.-S., Lee, K., Park, H.-J., Navarro, C., Lacord, J., Gamiz, F., Galy, P., Cristoloveanu, S., Bawedin, M. (2018). Z2-FET memory matrix in 28 nm FDSOI technology. *2018 Joint International EUROSOI Workshop and International Conference on Ultimate Integration on Silicon (EUROSOI-ULIS)*, 1–4. DOI: 10.1109/ULIS.2018.8354341.

Planes, N., Weber, O., Barral, V., Haendler, S., Noblet, D., Croain, D., Bocat, M., Sassoulas, P.O., Federspiel, X., Cros, A. et al. (2012). 28nm FDSOI technology platform for high-speed low-voltage digital applications. *VLSI Tech. Dig.*, 133–134. DOI: 10.1109/VLSIT.2012.6242497.

Solaro, Y., Fonteneau, P., Legrand, C.-A., Marin-Cudraz, D., Passieux, J., Guyader, P., Clement, L.-R., Fenouillet-Beranger, C., Ferrari, P., Cristoloveanu, S. (2013). Innovative ESD protections for UTBB FD-SOI technology. *IEEE International Electron Devices Meetingin*. DOI: 10.1109/IEDM.2013.6724580.

Solaro, Y., Wan, J., Fonteneau, P., Fenouillet-Beranger, C., Le Royer, C., Zaslavsky, A., Ferrari, P., Cristoloveanu, S. (2014). Z2-FET: A promising FDSOI device for ESD protection. *Solid-State Electronics*, 97, 23–29. DOI: 10.1016/j.sse.2014.04.032.

Solaro, Y., Fonteneau, P., Legrand, C.A., Fenouillet-Beranger, C., Ferrari, P., Cristoloveanu, S. (2016). A sharp-switching device with free surface and buried gates based on band modulation and feedback mechanisms. *Solid-State Electronics*, 116, 8–11. DOI: 10.1016/j.sse.2015.10.010.

Sze, S.M. (1981). *Physics of Semiconductor Devices*, 2nd edition. Wiley, New York.

Taur, Y. and Ning, T.H. (1998). *Fundamentals of Modern VLSI Devices*. Cambridge University Press.

Wan, J., Le Royer, C., Zaslavsky, A., Cristoloveanu, S. (2012a). A compact capacitor-less high-speed DRAM using field effect-controlled charge regeneration. *IEEE Electron Device Letters*, 33(2). DOI: 10.1109/LED.2011.2176908.

Wan, J., Le Royer, C., Zaslavsky, A., Cristoloveanu, S. (2012b). A feedback silicon-on-insulator steep switching device with gate-controlled carrier injection. *Solid-State Electronics*. DOI: 10.1016/j.sse.2012.05.061.

Wan, J., Le Royer, C., Zaslavsky, A., Cristoloveanu, S. (2013). Progress in Z2-FET 1T-DRAM: Retention time, writing modes, selective array operation, and dual bit storage. *Solid-State Electronics*, 84, 147–154. DOI: 10.1016/j.sse.2013.02.010.

Wann, H.J. and Hu, C. (1993). A capacitorless DRAM cell on SOI substrate. *IEDM'93 Technical Digest International*, 635–638. DOI: 10.1109/IEDM.1993.347280.

Weber, O., Josse, E., Andrieu, F., Cros, A., Richard, E., Perreau, P., Baylac, E., Degors, N., Gallon, C., Perrin, E. et al. (2014). 14nm FDSOI technology for high speed and energy efficient applications. *Symposium on VLSI Technology (VLSI-Technology): Digest of Technical Papers*. DOI: 10.1109/VLSIT.2014.6894343.

5

Two-Dimensional Spintronics

Matthieu JAMET[1], Diogo C. VAZ[2], Juan F. SIERRA[3], Josef SVĚTLÍK[3], Sergio O. VALENZUELA[3,4], Bruno DLUBAK[5], Pierre SENEOR[5], Frédéric BONELL[1] and Thomas GUILLET[3]

[1] CNRS, CEA, Grenoble INP, IRIG-Spintec, Université Grenoble Alpes, France
[2] CIC nanoGUNE BRTA, Donostia-San Sebastián, Spain
[3] Catalan Institute of Nanoscience and Nanotechnology (ICN2), CSIC and The Barcelona Institute of Science and Technology (BIST), Spain
[4] Institució Catalana de Recerca i Estudis Avançats (ICREA), Barcelona, Spain
[5] Unité Mixte de Physique, CNRS, Thales, Université Paris-Sud, Université Paris-Saclay, Palaiseau, France

5.1. Introduction[1]

In this chapter, we review recent advances in two-dimensional (2D) spintronics. With the extraordinary progress in material growth and synthesis, it is now possible to stabilize 2D systems such as two-dimensional electron gas (2DEG) at the interface between oxide layers or use two-dimensional materials that have emerged in the last two decades, like graphene or transition metal dichalcogenides, to develop spin-based devices. Moreover, another class of quantum materials has been discovered in 2007, namely topological insulators, exhibiting insulating bulk and

1. This section written by Matthieu JAMET.

Beyond-CMOS, coordinated by Alessandro CRESTI.

conducting topologically protected surface states. These surface states, as well as 2DEG and 2D materials, possess very specific spin properties that could advantageously be used in spintronic devices to boost their performance and lower their power consumption. The introduction of 2D systems in spintronics has been at the origin of several major discoveries. Indeed, their low-dimensional character allowed for the experimental evidence of original spin-dependent phenomena, such as the (inverse) Rashba–Edelstein effect, long spin diffusion lengths or spin-momentum locking, as well as very specific spin filtering effects. This chapter thus reviews these advances and is divided into four distinct parts. In the first one, we review the contribution to the field of 2D spintronics of 2DEG at the surface/interface of oxide layers. Their specific electronic structure leads to exceptionally large spin-to-charge conversion by the inverse Rashba–Edelstein effect, which is electrically tunable. The second part deals with spin transport in 2D materials and, in particular, in graphene. This material exhibits remarkably long spin diffusion lengths thanks to its low spin–orbit coupling and high carrier mobility. Moreover, its two-dimensional character makes it an ideal platform to study the influence of proximity effects with other 2D materials on spin dynamics. The third part is dedicated to the integration of 2D materials into magnetic tunnel junctions, which is the prototypical spintronic device. 2D materials used as the tunnel barrier can act as efficient spin filters or enhance perpendicular magnetic anisotropy, which are key ingredients for the development of magnetic random access memories (MRAMs). Finally, in the last part, we review the spin properties of topological insulators. Spin-momentum locking at the topologically protected surface states of topological insulators such as bismuth chalcogenides (BiSe, BiTe or BiSbTe) appears to be a key asset for the conversion of a charge current into a spin current for the manipulation and switching of the magnetization of an adjacent ferromagnet thanks to the Rashba–Edelstein effect.

5.2. Spintronics in 2D Rashba gases at oxide surfaces–interfaces[2]

In the last decade, complex oxides have emerged as an important class of materials for spintronics due to their multifunctionality and exotic properties. From a material perspective, the field saw a large boost in popularity in the late 1980s with advances in material growth techniques, such as pulsed laser deposition and molecular beam epitaxy, which allowed for the precise control of heterostructures at

2. This section written by Diogo C. VAZ.

For a color version of all figures in this section, see www.iste.co.uk/cresti/beyondcmos.zip.

the atomic level. By carefully constructing them, these heterostructures can attain a multitude of properties, such as superconductivity, ferromagnetism or ferroelectricity. Meanwhile, spintronics saw its first breakthrough with the discovery of giant magnetoresistance in metallic multilayers (Baibich et al. 1988), followed by major advances and rediscoveries of spin-based effects, such as tunneling magnetoresistance, spin Hall effect, and spin-transfer torque, all of which led to important technological progress for device applications.

Given the limitations of metallic systems in terms of scaling, energy consumption and overall tunability, the combination of complex oxides and spintronics seemed almost inevitable. The first trigger came with the report of colossal magnetoresistance in manganite films, which allowed for tunneling magnetoresistance (TMR) ratios that were one order of magnitude larger than observed in metal-based systems (Lu et al. 1996), and that were later expanded with the concept of spin filtering through ferromagnetic insulator and ferroelectric tunneling barriers (Gajek et al. 2005; Tsymbal and Kohlstedt 2006).

Concomitant with the first experimental demonstration of spin-transfer torque magnetization switching in the Co/Cu/Co multilayer (Myers et al. 1999), magnetoelectric multiferroics resurfaced, with $BiFeO_3$ in the spotlight (Ramesh and Spaldin 2007). Given that multiferroics are insulating oxides that possess two (or more) ferroic order parameters that are coupled, a pathway to voltage control of magnetism opened up, which could in turn profoundly reduce the energy consumption of logic and memory devices.

Roughly at the same time, the discovery of a two-dimensional electron gas (2DEG) at the interface between two insulating oxides propelled the field to new heights, with a massive body of research being dedicated to unraveling its intriguing properties. In the following pages, we will focus on the recent advances in 2DEGs at different oxide surfaces–interfaces, the rebirth of Rashba physics, and the new paradigm that these concepts open for spintronic devices.

5.2.1. *Emergent 2D conductivity at oxide interfaces*

Following extensive studies on the epitaxial growth by pulsed laser deposition of oxide thin films (Ohtomo et al. 2002a), and the discovery of charge-transfer-based conductivity in $LaTiO_3/SrTiO_3$ (LTO/STO) heterostructures (Ohtomo et al. 2002b), Ohtomo et al. (2004) reported the emergence of a high mobility electron gas between $LaAlO_3$ (LAO) and $SrTiO_3$ (Ohtomo and Hwang 2004). Unlike LTO, which is a Mott insulator and can become conductive via Ti valence changes, LAO is a band insulator with a large band gap of 5.6 eV, thus making the conductivity in LAO/STO especially intriguing. Not only that, but the conducting behavior was only observed when at least 4 unit cells (u.c.) of LAO (roughly 1.5 nm, as shown by a transmission electron microscopy image of the interface in Figure 5.1 (a)) were deposited on TiO_2-terminated STO (Thiel et al. 2006). These findings triggered an intense discussion within the oxide community regarding the exact mechanisms at play, with initial theories pointing towards an electronic reconstruction scenario, where charge is transferred from the surface of the LAO to the interface, to release an electrostatic build-up – a scenario famously called *polar catastrophe*. The next 10 years saw a rampant interest in the topic, with several works investigating the validity of the polar catastrophe (Reinle-Schmitt et al. 2012), the role of cation interdiffusion at the interface (Willmott et al. 2007), polar-induced defects (Yu and Zunger 2014) and the role of chemical and electrostatic reactions at the LAO surface. Regarding this last approach, the critical thickness of 4 u.c. could even be tuned by the usage of a conductive tip AFM (Cen et al. 2008), by changing the environmental conditions where the LAO/STO samples were measured (Scheiderer et al. 2015), or by depositing different metallic capping layers (Vaz et al. 2017). Altogether, these studies established an unofficial consensus that interplay between electronic and chemical interactions was responsible for the unforeseen conductivity.

An important challenge was also to clarify where exactly the conduction was located. It was well known up to this point that, while STO is an insulator with a band gap of 3.25 eV, Sr and Ti substitution, or oxygen removal can lead to electron doping, inducing slight Fermi level shifts, and Ti^{4+} to Ti^{3+} valence changes that render the material conductive. As shown in Figure 5.1(b), while the polar catastrophe scenario predicted that the charge transfer from the LAO surface would render the STO conductive at the interface only, oxygen vacancies generated during the high-temperature growth process were still on the table. The issue was resolved with experimental imaging of the interfacial conduction, using conductive-tip atomic force microscopy in samples with and without a high-temperature oxygen annealing treatment (Basletic et al. 2008). As shown in Figure 5.1(c), by measuring the cross-section of oxygen-annealed LAO/STO samples, the conductivity was confirmed to be quasi-two-dimensional, fueling a movement of 2DEGs at oxide interfaces.

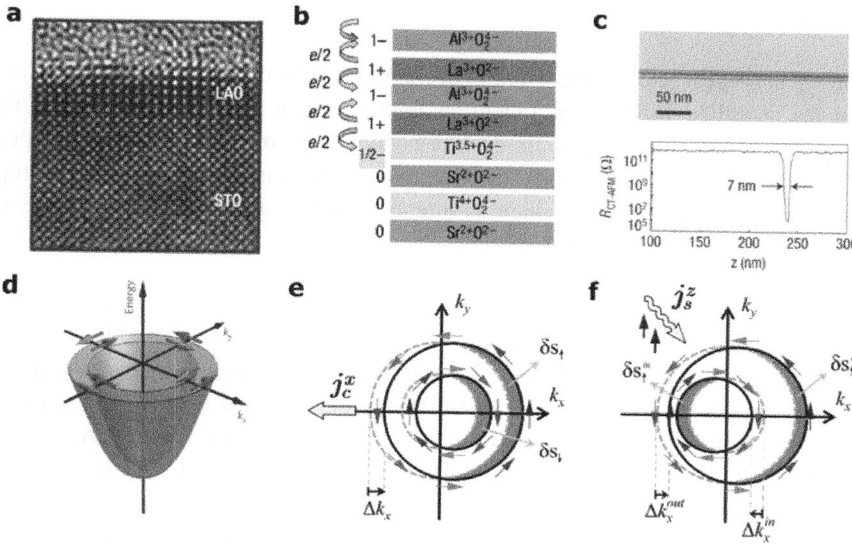

Figure 5.1. *Atomic, electronic and spin structures of LAO/STO interface*

NOTES ON FIGURE 5.1.– *(a) High-resolution transmission electron microscopy image of an LAO/STO heterostructure. (b) Illustration of the polar catastrophe, where half an electron is transferred to the first* TiO_2 *layer, avoiding a potential build-up in the LAO (Nakagawa et al. 2006). (c) Conductive-tip atomic force microscopy resistance mapping at the LAO/STO interface. Bottom: the resistance profile, showing a 2DEG thickness of about 7 nm (Basletic et al. 2008). (d) Energy dispersion for Rashba-split bands (Lesne et al. 2016). (e) Edelstein effect, and (f) inverse Edelstein effect (Vaz 2018).*

5.2.2. Rashba spin–orbit interactions

In systems that lack space-inversion symmetry, i.e. there is no inversion centre to which an atom in coordinates (x,y,z) as an indistinguishable atom in (-x,-y,-z), spin–orbit coupling splits the conduction bands into two branches, each with its own spin orientation, as shown in Figure 5.1(d). Dresselhaus (1954) reported that this effect occurs in what are called non-centrosymmetric semiconductors, such as GaAs or InSb. In the 1980s, Vas'ko, Bychkov and Rashba identified a similar effect in quantum wells and 2DEGs (Vas'ko 1979; Bychkov and Rashba 1984), where an interfacial electric field $\boldsymbol{E} = E_z \boldsymbol{z}$ results in the following spin–orbit Hamiltonian:

$$H_R = \alpha_R(\boldsymbol{z} \times \boldsymbol{k}) \cdot \boldsymbol{\sigma}$$

Here, $\boldsymbol{\sigma}$ is the vector of the Pauli spin matrices, $\boldsymbol{k}$ is the momentum vector and α_R is the so-called Rashba coefficient/parameter, which quantifies the size of the spin splitting. Indeed, besides bulk crystal asymmetries, one may intuitively expect that another way of breaking inversion symmetry is by having an abrupt discontinuity in the crystal, which is precisely what occurs at any surface or interface between two different materials.

The Rashba spin–orbit coupling opened a valuable platform to study the interaction between charge and spin. The field attracted considerable attention during the 1990s, with the proposal of a *spin transistor*, a device where the spins of conducting electrons would be controlled during their passage from source to drain, through the manipulation of α_R using gate voltages (Datta and Das 1990). Around the same time, another effect called the *spin galvanic effect* (later the *Edelstein effect*) described a direct way of generating and manipulating spin currents in two-dimensional systems (Edelstein 1990). Akin to the spin Hall effect in bulk materials with high spin–orbit coupling, a finite spin accumulation can be generated by flowing a charge current in a 2DEG. Conversely, a charge current can also be generated by injecting a non-equilibrium spin density, an effect usually called the *inverse Edelstein effect*. The nature of the effect can be intuitively visualized by looking at what happens to the band structure upon application of these two stimuli. In Figure 5.1(e), we observe that in the presence of a charge current in $-y$, electrons are accelerated in x, so that both Fermi contours experience an equal shift in Δk_x. From the Hamiltonian above, since k and σ are perpendicular (a property called *spin-momentum locking*), the shift in k slightly tilts the spins around the Fermi contour, thus leading to two uncompensated spin densities (with opposite spin) represented as blue- and red-striped areas. Because both densities are unequal, a finite non-equilibrium spin density is generated. In Figure 5.1(f) we can see the inverse effect. Upon injection of a spin current, both contours are shifted in the opposite direction to compensate for the excess spin population, leading to the generation of a finite 2D charge current. As we will see in the following, the Edelstein effect opened up a path to the experimental generation and manipulation of spin currents using conventional current/voltage probes, expanding the field of spintronics, yet again.

5.2.3. *Spin-to-charge current conversion in oxide 2DEGs*

After a few years of exploration of spin–charge interconversion in semiconductor quantum wells (Ganichev et al. 2002), metallic interfaces (Sánchez et al. 2013) and even topological insulators, the 2DEG in LAO/STO established an exciting new playground for the spintronics community. The 2DEG was shown to possess a modest Rashba spin–orbit coupling, which could be tuned using electrostatic gating. Here, the large dielectric constant of STO at low temperatures allowed for large modulation of the carrier density, which drastically affected the estimated α_R, and was even observed to shift the superconductor critical temperature (Caviglia et al. 2008, 2010). In 2016, the inverse Edelstein effect was for the first time demonstrated in LAO/STO at low temperatures (Lesne et al. 2016). The setup of the experiment is shown in Figure 5.2(a). Here, a ferromagnetic NiFe metallic layer was deposited on an LAO/STO heterostructure and was used to produce a pure spin current through *spin pumping*. The technique relies on the concept of ferromagnetic resonance, where the application of a well-defined external magnetic field $\boldsymbol{H}_{d.c.}$ and microwave radiation $\boldsymbol{h}_{rf}$ can sustain a precessional motion of the magnetization $\boldsymbol{M}$. While $\boldsymbol{M}$ is precessing, the ferromagnet will pump its own spin angular momentum into any adjacent layer, giving rise to a spin current j_s with spin $\boldsymbol{\sigma}$ parallel to $\boldsymbol{M}$. In this experiment, the LAO thin layer acts as a barrier, through which the spin current passes to be injected into the interfacial 2DEG. Through the inverse Edelstein effect (exemplified in Figure 5.1(f)), this out-of-equilibrium spin density induces a finite charge current j_c^{2D} in the 2DEG, which can be picked up using a two-point voltage probe. In Figure 5.2(b), we observe that by fixing the microwave radiation and sweeping the external magnetic field, at around 98 mT the resonance condition is met, precisely where a voltage peak is measured.

Just like in the spin Hall effect, where the spin Hall angle θ_{SH} quantifies the efficiency of the spin–charge conversion, for the inverse Edelstein effect, we can use:

$$\lambda_{IEE} = \frac{j_c^{2D}}{j_s^{3D}} = \frac{\alpha_R \tau}{\hbar},$$

where τ is the relaxation time of the out-of-equilibrium spin density distribution in the 2DEG states, and $\hbar$ is the reduced Planck constant. Note that, while θ_{SH} is given by the ratio of two 3D currents, λ_{IEE} is given instead by a 2D charge current, so that it is normally expressed in nm. Another key difference comes from the fact that since θ_{SH} is adimensional, it can be used reciprocally for the direct and inverse effect. Since the efficiency of the direct Edelstein effect implies a reversal of the ratio, given by q_{EE} in nm^{-1}, the Onsager reciprocity is experimentally less trivial, especially when different experiments are used to measure the direct and inverse effects.

The maximum λ_{IEE} in LAO/STO was estimated to be 6.4 nm at low temperatures, the largest value found in any system up to that point, dropping to about 0.15 nm at room temperature (Chauleau et al. 2016). Following these seminal works, several groups also demonstrated charge-to-spin conversion in LAO/STO using spin-torque ferromagnetic resonance experiments (Wang et al. 2017), as well as interconversion in ferromagnetic-free systems (Jin et al. 2017; Trier et al. 2020). The gate dependence of λ_{IEE} was another remarkable feature, with a sign change observed over a gate voltage range ±200V. The result was linked to a modulation of the Fermi level position, allowing the exploration of the complex 2DEG band structure, already hinted to be the reason for the large modulation of α_R.

Over time, the oxide community started to explore 2DEGs beyond the classic LAO/STO system, with important progress made in metal/STO systems (Rödel et al. 2016). These were particularly interesting for two reasons: first, the reduced complexity in terms of growth, since LAO still required pulsed laser deposition, high temperatures and a finely tuned process, while metals could be easily grown using sputtering; second, depending on the metal chosen, the properties of the 2DEG could also be modified, with added functionality through proximity effects between ferromagnetic or superconductor metals and the 2DEG in STO. The first spintronics-oriented result was obtained for Al/STO (Vaz et al. 2019). Here, instead of relying on charge transfer from the epitaxially grown LAO layer, the sputtered Al chemically reacts with STO by pulling oxygen and becoming oxidized. As mentioned above, oxygen vacancies created in STO release electrons in the lattice. Since the Al layer is ultra-thin (~0.3 nm), oxygen removal only occurs at the top layers of STO, thus leading to the formation of a 2DEG. A transmission electron microscopy image of this system is shown in Figure 5.2(c). Spin–charge conversion experiments revealed a maximum λ_{IEE} of about 30 nm, almost five times larger than in LAO/STO, with several sign changes over a gate voltage range. This peculiar behavior was mapped to the band structure of the 2DEG, using a combination of transport, angle-resolved photoemission spectroscopy (ARPES) and semiclassical Boltzmann calculations, with the latter showing remarkable similarity with the experimental data (see Figure 5.2(d)). Given the larger carrier density seen in Al/STO when compared to that in LAO/STO, additional features at higher energies were reachable. By carefully analyzing the spin structure in this region, it was found that some λ_{IEE} maxima were attributed not to regions of enhanced Rashba splitting, but to non-trivial spin textures rising from spin-polarized topological edge states, pointing towards unexplored topological features in oxide-based 2DEGs.

The last ingredient that explained the large λ_{IEE} was the Al itself. Since the Al layer is oxidative when deposited on STO, the barrier between the NiFe and the 2DEG was mostly composed of Al oxide. Upon spin injection, the spins can either relax in the 2DEG, contributing to the spin–charge conversion effects, or scatter back to the ferromagnet. In all-metal systems, since the whole structure is conductive, the second effect largely reduces the experimental λ_{IEE} values. By estimating the resistance across the barrier for LAO/STO and Al/STO, a difference of about five orders of magnitude is found, making the 2DEG in the latter much more encapsulated, which prevents undesired scattering back to the NiFe. It is, however, important to understand that if the barrier is too resistive or too thick, the spin injection may also be degraded, eventually leading to a reduction in λ_{IEE}.

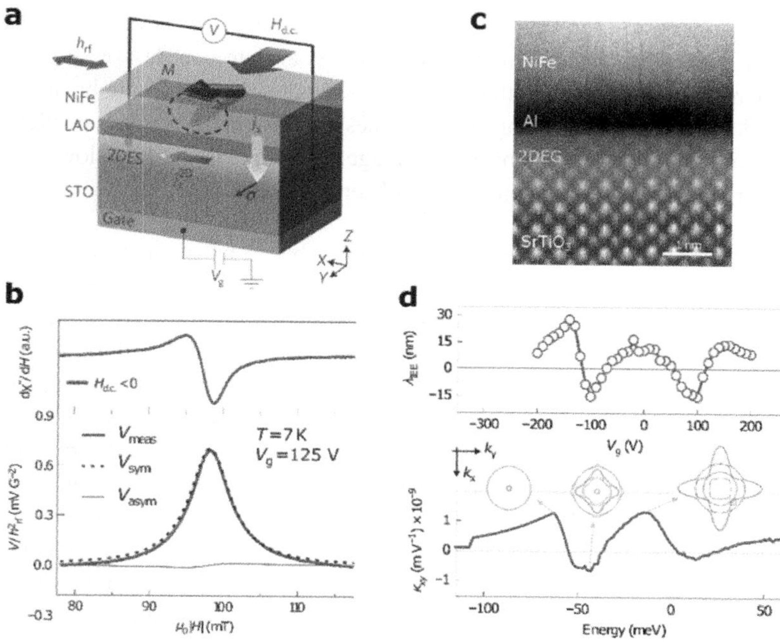

Figure 5.2. *Spin pumping at LAO/STO and Al/STO interfaces*

NOTES ON FIGURE 5.2.– *(a) Schematic of spin pumping experiments in LAO/STO. (b) Top: ferromagnetic resonance signal. Bottom: the measured voltage upon spin pumping (Lesne et al. 2016). (c) Scanning transmission electron microscopy of a NiFe/Al/STO structure, with false-colored 2DEG region. (d) Top: gate-dependence of spin–charge conversion efficiency at 15 K. Bottom: energy dependence of the sum of the Edelstein tensor κ_{xy} for all bands. Fermi lines at difference energies are also shown (Vaz et al. 2019).*

5.2.4. *Device applications and prospects*

One of the most important requirements for a functional spintronic device is its ability to handle spin information. Instead of using electric charge like in conventional transistors, these devices can store information as a magnetization direction in a ferromagnetic metal, such as Co-, Fe- or Ni-based alloys, and be modified by current-induced magnetic fields. While this concept led to a first generation of non-volatile magnetic random-access memories (MRAM), and later spin-transfer torque (STT-MRAM) and spin–orbit torque-based devices (SOT-MRAM), issues with scalability, endurance and high energy consumption prevent mass adoption, as well as their use in low-power applications. Even though Moore's law is prophesized to be nearing its limit, new materials and three-dimensional gate control are still enabling steady improvements. However, the bottleneck seems to come from the breakdown of the lesser-known Dennard scaling, which states that as transistors get smaller, their power density should stay constant. With voltage and frequency scaling slowing down, the need for a post-CMOS solution has never been greater.

In 2019, Intel proposed a new type of device based on magneto-electric spin–orbit (MESO) logic (Manipatruni et al. 2019). A sketch of the device is shown in Figure 5.3(a). The device uses the magnetization direction of a ferromagnetic metal to store information, as in previous spin-based devices, and two modules for writing and reading operations (Figure 5.3(b)). The first one uses a magnetoelectric component to switch the ferromagnet upon a voltage input. This can be achieved using a multiferroic such as $BiFeO_3$. Due to its ferroelectric and antiferromagnetic nature, electrically switching the polarization also drives a switch of the antiferromagnetic order. When coupled with a ferromagnetic layer, the $BiFeO_3$ can drive voltage-controlled magnetization switching, a major advantage when compared with the charge current-based STT and SOT devices. The reading module is based on a spin–orbit coupling component, which converts a vertical spin-polarized current in a positive or negative voltage output, depending on the magnetization orientation of the ferromagnet. These devices can also be cascaded, so that the output voltage of one device can be the input of another. For this to work, the main challenge lies in reducing the switching voltage of the magnetoelectric, and increasing the output voltage of the spin–orbit module, so that both values match. For spin–orbit module, high spin–orbit coupling metals such as Pt or Ta could be considered, due to their high spin–charge conversion efficiency, and industry-compatible deposition techniques. However, due to their small resistance, the effective voltage output is too small. To tackle this issue, STO-based 2DEGs are one of the strongest candidates. Despite the modest conversion efficiencies reported at room temperature, which are on par with metallic systems, resistances can be two to three orders of magnitude larger, pushing the output voltage closer to the required values.

Since low-voltage switching of magnetoelectric materials is still a challenge, other alternatives have been proposed with similar goals in mind, some of which use STO simultaneously as the writing and reading module. Based on several works from the late 1990s, it was reported in 2019 that with sufficiently large voltages, or sufficiently thin structures, a remanent ferroelectric state can be induced in STO (Noël et al. 2020). Since the physics of the 2DEG are governed by an electric-field-induced Rashba spin–orbit coupling, by switching the polarization of the STO the spin–charge conversion output signal can also be switched. A sketch of such a device is shown in Figure 5.3(c). Compared with the MESO proposal, in this device, the magnetization of the ferromagnet is fixed. Information is written by switching the polarization of the ferroelectric material. The polarization state is then read by injecting a spin-polarized current into the ferroelectric-embedded 2DEG, which outputs (through spin–charge conversion) a positive or negative voltage. The spin pumping experiments displayed in Figure 5.3(d) show the sign reversal of the output signal upon the application of gate voltages with different polarities.

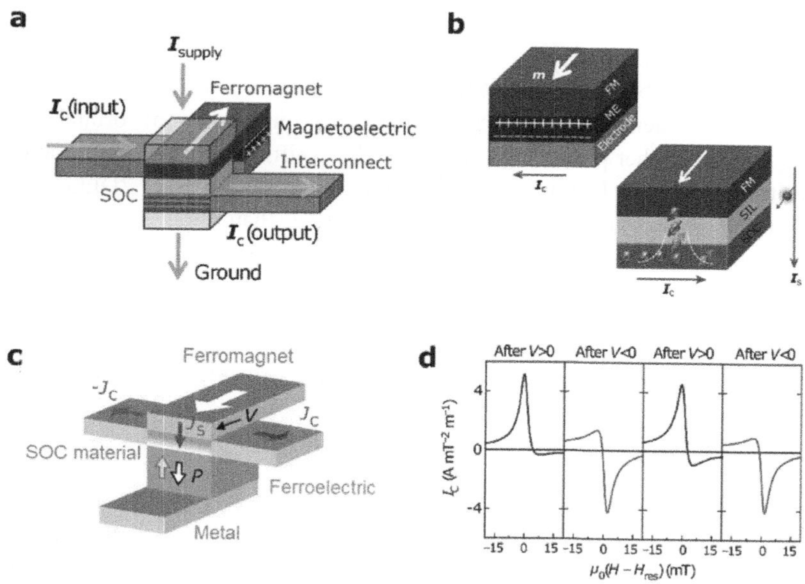

Figure 5.3. *(a) Sketch of the MESO device including (b) a magnetoelectric module for writing, and a spin–orbit module for reading (Manipatruni et al. 2019). (c) Non-volatile device based on a ferroelectric material with an embedded 2DEG. The polarization dictates the state, while the reading is performed through spin–charge conversion. Switchable charge current output in a ferroelectric STO-based 2DEG (Noël et al. 2020)*

In spite of the relatively small amount of work showing oxide-based 2DEGs with spintronic functionalities, the search for beyond-STO 2DEGs has been extremely active (Vaz et al. 2018; Kornblum 2019). $KTaO_3$ was shown to host a 2DEG with slight mobility improvements at room temperature (Zou et al. 2015), but with a predicted and experimentally observed Rashba splitting about one order of magnitude larger than in STO (Bruno et al. 2019). Although the larger splitting does not necessarily mean that larger spin–charge conversion efficiencies can be attained, since the doping may not be enough to reach the predicted hotspots of the band structure, it may translate into unexplored opportunities for spintronic applications. On another front, stannates such as $BaSnO_3$ have also been in the spotlight due to their high mobility at room temperature, with some works already exploring 2D conductivity (Kim et al. 2018). Given their transparency in visible light, applications for transparent spintronics may also gain traction in the future. Then, following the work on ferroelectric STO, 2DEGs embedded in substrates with additional properties can drastically expand their functionality. A switchable 2DEG has been demonstrated in ferroelectric Ca-doped STO, where the 2DEG properties can be modulated in a non-volatile way (Bréhin et al. 2020). Also, the band structure of a 2DEG at the surface of ferroelectric $BaTiO_3$, appearing upon the deposition of a thin Al layer, has been observed using ARPES (Rödel et al. 2016). Lastly, even within STO, new phenomena keep emerging, with the importance of orbital effects emerging as yet another promising ingredient, thus shifting the attention to potential spin-orbitronic devices. In this regard, it was reported that in STO 2DEGs, orbital effects are at least one order of magnitude larger than their spin counterpart (Johansson et al. 2020), opening up the discussion of whether the exploration of an orbital Hall effect may be a viable path for device optimization.

5.3. Spintronics in lateral spin devices in 2D materials[3]

5.3.1. *Introduction*

Since their implementation (Johnson and Silsbee 1985), lateral spin devices represent an ideal platform for studying spin-related phenomena. In contrast to vertical devices, the dominant spin current flows in a horizontal direction, along a spin channel into which spins are injected, and subsequently propagate, typically in a diffusive fashion. Lateral devices have been successfully developed with non-magnetic metals, semimetals, semiconductors and superconductors, sparking a deep understanding of their spin-related characteristics, such as their spin diffusion constant and the spin relaxation length and lifetime (Valenzuela 2009).

In the past few years, the focus has shifted towards 2D materials (Han et al. 2014; Avsar et al. 2020) because of their potential for spin propagation and manipulation. The paradigmatic and most studied example is grapheme, but increasing efforts are being directed towards related 2D materials such as black

3. This section written by Juan F. SIERRA, Josef SVĚTLÍK and Sergio O. VALENZUELA.

phosphorous and transition metal dichalcogenides (TMDCs). In particular, their atomically thin nature has enabled us to build 2D heterostructures, and further investigation has demonstrated that this leads to highly relevant proximity effects in which specific properties can be imprinted across interfaces. Such properties could be absent in the isolated materials and lead to dramatic changes in the spin dynamics.

Among all the 2D materials, graphene is particularly appealing. Graphene is a 2D semimetal offering a vast catalogue of technologically relevant properties for spintronics, including high carrier mobility, gate-tunable carrier density and low spin–orbit coupling (SOC), the latter enabling the propagation of spins over very long distances, in excess of tens of micrometers at room temperature. When combined with high spin–orbit materials, such as some families of TMDCs (e.g. MoS_2, $MoSe_2$, $MoTe_2$, WS_2, WSe_2 and WTe_2), the graphene SOC can be enhanced with high prospects for spin control and manipulation, not accessible in any other material platform. Beyond graphene, black phosphorous, a sizable bandgap 2D semiconductor, is also a low SOC material with relatively high carrier mobility at room temperature, which also transports spins efficiently.

All of these experimental realizations have been mostly performed using fully electrical means, where spin-sensitive planar contacts, typically ferromagnetic (FM) metals attached to the 2D spin-channel material, are used for spin injection and detection. These electrical contacts require the intercalation of tunnel barriers between the FM contacts and the spin channel, which enhances the spin injection efficiency by alleviating the conductivity mismatch problem. The latter refers to the low spin-injection and detection efficiency that is obtained with a direct contact between a high-conductivity FM metal and a relatively low-conductivity semimetallic or semiconducting spin channel material. Noteworthy, when combined with optically active materials such as TMDCs, optical spin injection from the TMDC to graphene can be achieved without a tunnel barrier (Avsar et al. 2017a; Luo et al. 2017). Indeed, such hybrids of graphene and TMDC lateral spin devices are currently at the forefront of spintronic and optospintronic research.

This chapter provides an overview of spin lateral devices based on 2D materials. In section 5.3.2, we first describe the essential aspects of spin injection and spin detection using non-local techniques. Next, in section 5.3.3, we introduce spin precession measurements, which allow the determination of all the spin-transport parameters of the system, with a single properly designed device. In section 5.3.4, we overview models of spin relaxation, which in the case of 2D materials, provide a rich variety of spin relaxation mechanisms. A brief survey of spin transport phenomena in 2D materials, particularly in graphene, when proximitized to high spin–orbit TMDCs, and the emergence of anisotropic spin dynamics, spin Hall (SHE) and spin galvanic effects (SGE) in such heterostructures is given in section 5.3.5.

5.3.2. *Spin injection and detection*

Spin transport in lateral devices relies on non-local transport schemes, in which no current circulates by the detection region. This avoids the contributions of, for example, the anisotropic magnetoresistance or the Hall effect, which are present in local measurements and can mimic spin transport signals. The detected signal is the voltage V_{nl} induced by the spin accumulation in the detector FM, away from the electrical current circuit used for spin injection. The typical non-local device is composed of a non-magnetic (NM) channel material through which spins diffuse, two (typically NM) metallic electrodes attached to the outer sides of the channel and two FM probes (FM1 and FM2) in the middle acting as spin injector and detector (Figure 5.4(a)). Due to the different populations, derived from the spin splitting, and mobilities of spin-up and spin-down electrons in the FM injector, a charge current I_{inj} passing through it is spin-polarized. When current I_{inj} runs through the FM injector towards the spin channel, it creates a non-equilibrium spin accumulation. The magnitude of the resulting spin accumulation is quantified by a splitting $\Delta s = (s_\uparrow - s_\downarrow)$ in the spin-dependent electrochemical potentials $s_\uparrow$ $(s_\downarrow)$ of the channel. These accumulated spins then diffuse in both directions away from the injection point over a characteristic distance λ_s, called the spin diffusion length. If the second FM contact is within this distance ($L<\lambda_s$, with L as the spin channel length), its Fermi level tends to equilibrate closer to the local $s_\uparrow$ or $s_\downarrow$ depending on its magnetization orientation. The electrochemical potential of the detector electrode is displaced by $P^d|\Delta s|/2$ relative to the mean potential of the channel (Figure 5.4(b)), where P^d is the effective polarization of the detector. In this way, pure spin currents can be detected as a voltage signal $V_{nl} = P^d|\Delta s|/2e$, where e is the elementary charge. The non-local signal is often converted in units of resistance defined as $R_{nl} = V_{nl}/I_{inj}$. As the injected spin current is given by $P^i I_{inj}$, with P^i the effective polarization of the injector, then $R_{nl} \propto P^i P^d$.

When the resistance of the spin channel is higher than that of the FM injector, the efficiency of the spin injection is low and the spins are partially lost due to the spin absorption in the FM, resulting in a low $P^i(P^d)$. This impedance mismatch problem can be circumvented by introducing a thin insulating barrier that facilitates tunneling spin injection, hence increasing the injector efficiency (that is now dominated by an effectively different density of states for spin up and spin down carriers at the Fermi level) and reducing contact-induced spin relaxation effects. Conventional tunnel barriers include oxides such as AlO_x, MgO and TiO_x, which, in the case of lateral spin devices based on 2D materials, can be spanned to other compatible barriers such as amorphous carbon interfacial layers, SrO and few layers of hexagonal boron nitride (hBN).

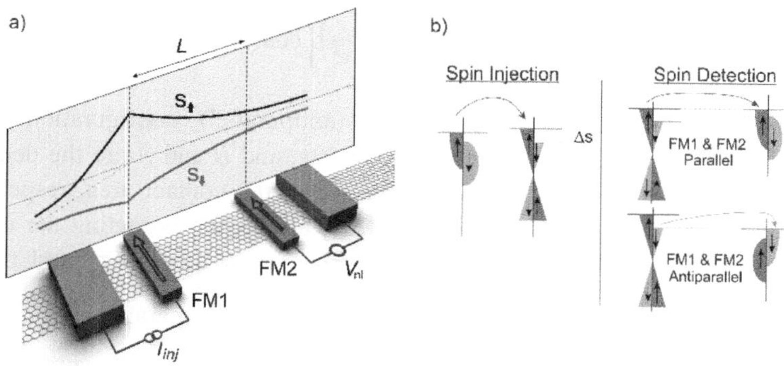

Figure 5.4. *a) Lateral spin device and spin diffusion. b) Spin injection and detection*

The spin density in the channel can be described by a diffusive transport model derived from the Boltzmann equation $\nabla^2 \Delta s = (1/\lambda_s)^2 \Delta s$. Assuming spin diffusion in one dimension only, the general solution to this equation is given by $\Delta s(x) = [s_\uparrow(0) - s_\downarrow(0)] \cdot \exp(-x/\lambda_S)$. Because R_{nl} depends on the spin polarization of both the injector (P^i) and detector (P^d), and the spin resistance R_S of the channel, its magnitude will be given by $R_{nl} = \frac{1}{2} P^i P^d R_s \cdot \exp(-x/\lambda_S)$.

5.3.3. *Spin precession*

When an external magnetic field B is applied, spins diffusing from the injector to the detector precess (unless the direction of the spins and the field are parallel or antiparallel). The precession frequency is given by the Larmor frequency $\omega_L = \gamma B$, where γ is the gyromagnetic ratio of the electron. To maximize the spin precession signal, in the conventional spin precession scheme, B is applied perpendicular to the plane containing the spin channel and the orientation of the spins. The latter is determined by the injector magnetization, which is typically along its length due to shape anisotropy. The combined effect of spin diffusion, precession and relaxation is described by the diffusive Bloch equations:

$$\frac{\partial \vec{s}}{\partial t} = \overline{D_S} \nabla^2 \vec{s} - \gamma(\vec{s} x \vec{B}) - \overline{\tau_S^{-1}} \cdot \vec{s} \quad [5.1]$$

with D_s as the spin diffusion constant and τ_s as the spin lifetime. The steady-state solution to equation [5.1] in the conventional scheme, considering highly resistive contacts, results in a non-local resistance given by the relation:

$$R_{nl} = \pm \frac{p^i p^d}{NAe^2} \int_0^\infty \frac{1}{\sqrt{4\pi D_S t}} \exp\left[-\left(\frac{L}{4D_S t} + \frac{t}{\tau_S^{\parallel}}\right)\right] \cos(\omega_L t) \mathrm{d}t \qquad [5.2]$$

where the +(-) sign is for the parallel (antiparallel) configuration of the magnetization of the injector and detector contacts, and N and A are the density of states of the spin channel at the Fermi energy and the contact area, respectively. Fitting the experimental data to expression [5.2] allows for extracting the in-plane spin-transport parameters, namely D_S, the in-plane spin lifetime $\tau_S^{\parallel}$ and the spin relaxation length $\lambda_S^{\parallel} = \sqrt{D_S \tau_S^{\parallel}}$. The determination of these parameters is of great importance in spintronics because they enable us to get insights into spin relaxation and spin dephasing processes.

5.3.4. *Mechanisms of spin relaxation*

Sources of spin relaxation in 2D materials have been one of the most fascinating puzzles, which still need to be fully solved. As a first approach, conventional mechanisms observed in metals and semiconductors such as the Elliot–Yafet (EY) and Dyakonov–Perel (DP) mechanisms have often been proposed to explain spin relaxation in 2D materials. In the EY mechanism, the spins relax via ordinary momentum scattering with phonons and impurities. Each scattering event changes the electron momentum and simultaneously induces a spin-flip event, resulting in a linear scaling between the momentum relaxation time τ_p and τ_S^{EY}, i.e. $\tau_S^{EY} \propto \tau_p$. In contrast, in the DP mechanism, the SOC generates an internal spin–orbit field about which electrons precess while diffusing. Here, the spin lifetime τ_S^{DP} is inversely proportional to τ_p, i.e., $\tau_S^{DP} \propto \tau_p^{-1}$.

Owing to its low SOC (~24 μeV), and the lack of hyperfine interaction (99% of ^{12}C nuclei with nuclear spin zero), graphene is highly promising for long spin communication at room temperature. Experimentally, it is the most studied and debated case in the literature. Spin relaxation in graphene has yielded vast experimental and theoretical efforts over the last few years. Early numerical calculations predicted spin lifetimes in the range of micro- to milliseconds (Ertler et al. 2009), in stark contrast with the first experimental observations, which measured $\tau_S^{\parallel}$ of the order of 100 ps (Tombros et al. 2007). Since then, experiments have concentrated on improving the quality of the lateral devices to enhance the spin relaxation parameters. Firstly, in reducing the effects of the substrate (primary silicon oxide) by suspending graphene, which resulted in spin relaxation lengths of up to 5 μm but with spin lifetimes still below 1 ns, likely due to short-channel effects. Later, by using graphene fully encapsulated with hBN, spin lifetimes exceeding 10 ns were found, still orders of magnitude below the original predictions. The sources of spin relaxation in graphene have also been thoroughly investigated. Early studies in single-layer and bilayer graphene have suggested predominant EY and DP mechanisms, respectively. Changes in spin lifetimes in single-layer graphene, when doped with Au adatoms, were attributed to the DP mechanism. However, later experiments in graphene on hBN showed that neither the EY nor the DY mechanism could describe the observed scaling between τ_p and $\tau_s^{\parallel}$ alone, suggesting the combination of both mechanisms as sources of spin relaxation.

The discrepancies between models and experiments and between experiments at different laboratories highlighted the need for new theoretical models of spin relaxation. Magnetic impurities, which can be induced either by vacancies in the graphene lattice, sp^3 bonding with, for example, hydrogen atoms, or artificially by introducing magnetic adatoms, which can be an efficient source of spin relaxation in graphene. First-principle calculations have shown that a very small amount of local magnetic moments (1 ppm) combined with resonant scattering strongly reduces the spin lifetime in graphene, which could explain the limited values observed in experiments (~100 ps) (Kochan et al. 2014). Another important source of spin relaxation arises from the spin–pseudospin entanglement (Tuan et al. 2014). The entanglement between the spin and pseudospin degrees of freedom is driven by SOC that can be generated by the presence of impurities, ripples and substrate-related effects. Such a source of SOC couples the spin to the orbital motion, and therefore to pseudospin. Spin relaxation driven by either magnetic impurities or spin–pseudospin entanglement predicts a decreasing spin lifetime when approaching the Dirac point, which is a common feature observed in the experiments with clean enough graphene (Raes et al. 2016).

Yet, beyond graphene, spin-related phenomena in lateral spin devices are scarce. Studies of spin dynamics in black phosphorus, a centrosymmetric material in which the DP mechanism is precluded, indicate that the EY mechanism is the main source for spin relaxation (Avsar et al. 2017b). At low temperatures, the observed $\tau_s^{\parallel}$ were as high as 4 ns, with $\lambda_s^{\parallel}$ exceeding 5 μm.

5.3.5. *Spin transport in van der Waals heterostructures*

Spin-related phenomena in 2D materials have evolved in the past few years from studying devices made of isolated layers to more sophisticated heterostructures, in which physical properties – absent in their isolated form – can be imprinted by proximity effect across interfaces. When stacked vertically in van der Waals heterostructures, the individual 2D layers are separated by a van der Waals gap of a few tenths of nanometers. Due to the weak interlayer hybridization, the properties of the individual 2D layers are largely preserved, whereas the small overlapping of the neighboring layers can induce subtle modifications in the electronic band structure, and, therefore, optical, electrical or spin-related properties can be modulated. Such proximity-induced effects enable us to engineer material properties, thus greatly enriching the variety of possible physical phenomena in 2D systems and making van der Waals heterostructures interesting for both fundamental research and technological applications. In the following sections, we provide a detailed explanation of the most relevant heterostructures for spintronics, which entail the combination of graphene, which holds high carrier mobilities, ambipolar transport and long spin lifetimes, with TMDCs.

5.3.5.1. *Anisotropic spin dynamics in graphene-TMDC lateral devices*

SOC describes the relativistic interaction between the electron spin and momentum degrees of freedom. It is a concept ubiquitous in spintronics, playing a central role for spin control and manipulation and spin-to-charge interconversion phenomena driven by the SHE and the SGE (Sinova et al. 2015). Graphene holds long spin lifetimes due to its low SOC. However, this property is also a major drawback for establishing spin manipulation protocols in graphene. Nevertheless, it is possible to enhance the SOC in graphene via proximity-induced effects. In particular, hybrid systems of graphene and TMDCs, which can have a sizable SOC, have been theoretically predicted and experimentally demonstrated to imprint SOC and spin-valley locking in graphene. First-principle calculations have shown that, when proximitized by a TMDC such as WS_2 or MoS_2, several changes occur in the band structure of graphene near the K point. First, an orbital gap in the meV range opens, while preserving the linear energy dispersion (Gmitra and Fabian 2015). This gap opening is exclusively due to the breaking of the pseudospin symmetry in graphene (carbon atoms at the sublattices A and B have a different landscape coming from the adjacent TMDC). In addition, the combined effect of the proximity-induced SOC and broken space inversion symmetry results in a spin splitting of the conduction and valence bands with a characteristic spin texture. In the K valley, at $k = 0$, the spins point out of the graphene plane and alternate between up and down polarization in the spin-split bands, with the order being reversed at the K and K' valleys. At $k \neq 0$, spins acquire a winding spin component with different helicity in the split bands, i.e. a Rashba spin texture, which is the same in both valleys. Such proximity-induced SOC has strong consequences for spin transport in graphene. A clear fingerprint is the anisotropic behavior in spin dynamics, with spin lifetimes depending on the spin orientation (Raes et al. 2017;

Benítez et al. 2019). These effects have been studied in lateral spin devices comprising a spin channel made of graphene partially covered by the TMDC layer. In graphene/WS_2 and graphene/$MoSe_2$ heterostructures (Ghiasi et al. 2017; Benítez et al. 2018), the spin lifetime for spins precessing out of the graphene plane $\tau_S^{\perp}$ = 30–40 ps, whereas the lifetime of spins precessing in the graphene plane $\tau_S^{\parallel}$= 3–4 ps. The spin anisotropy ratio $\zeta \equiv \tau_S^{\perp}/\tau_S^{\parallel} \sim 10$ when the Fermi energy in the TMDC is within its bandgap. Such anisotropic values are theoretically well reproduced considering strong intervalley scattering (Cummings et al. 2017). Within this scenario, the spin lifetimes are driven by the DP mechanism with approximately $\tau_S^{\perp} \propto \tau_p^{-1}$ and $\tau_S^{\parallel} \propto \tau_{iv}^{-1}$, where τ_{iv} is the intervalley scattering time. This anisotropic spin dynamic is in stark contrast with the isotropic behavior of spins in regular graphene ($\tau_S^{\parallel} = \tau_S^{\perp}$) (Raes et al. 2016) (see Figure 5.5). Note that, when the TMDC is conducting, the diffusing spins are absorbed by the TMDC and the devices acts as a spin-field switch (Yan et al. 2016).

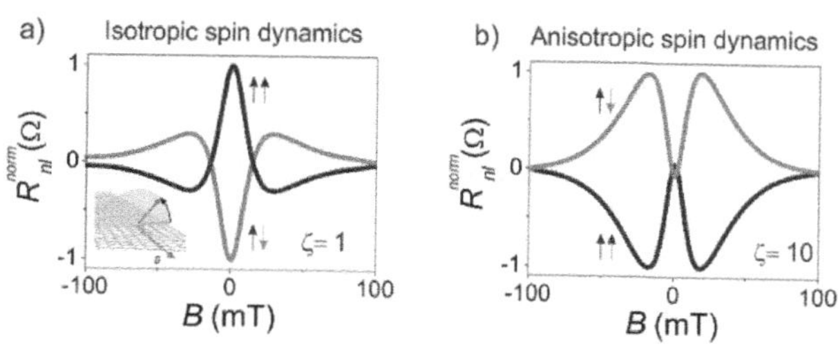

Figure 5.5. *R_{nl} (normalized by its maximum value) under B applied along the spin channel for: a) isotropic spin dynamics, b) anisotropic spin dynamics*

5.3.5.2. *Spin Hall and spin galvanic effects in graphene-TMDC lateral devices*

Another important consequence of enhancing the SOC in graphene is the emergence of spin–charge interconversion phenomena, which are absent in pristine graphene. In this case, the lateral devices are designed similarly to the spin Hall devices developed for metallic systems (Sinova et al. 2015). Here, graphene is patterned into a Hall cross with the TMDC along one of the Hall arms. Along this arm, the SOC in graphene is enhanced, thus defining the active region in which spin–charge interconversion occurs (Figure 5.6(a)). When a current runs along this arm, it induces two different effects: i) a transversal spin current with spins polarized out-of-plane due to the SHE and ii) an in-plane spin density with spin polarization transversal to the current at the interface driven by the inverse SGE. These two conventional mechanisms for spin–charge interconversion can be probed by spin-sensitive contacts, i.e. FMs, attached to the graphene spin channel. Alternatively, the same device can be used to study the reciprocal effects (inverse SHE and SGE), where spins are injected through the FM contact, diffusing towards the bilayer region, and generating a transversal voltage in the Hall arm.

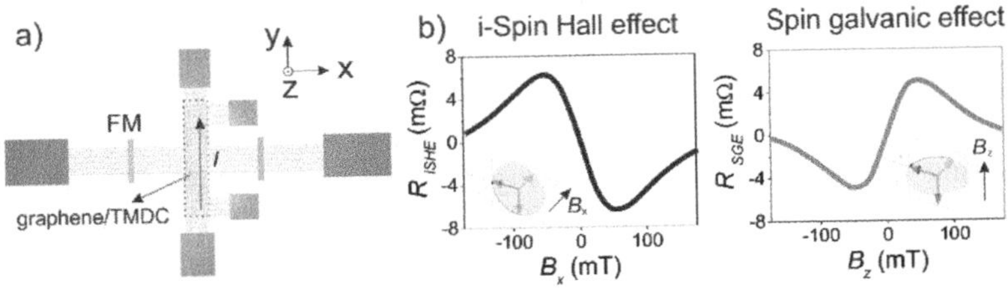

Figure 5.6. *a) Spin Hall devices for spin–charge interconversion. b) Inverse spin Hall and spin galvanic signals detected in the TMDC–graphene region*

The SHE and inverse SGE can be disentangled by studying the spin precession under selected magnetic field orientations (Figure 5.6(b)). Under an out-of-plane magnetic field, only the in-plane spins arise from the SGE precess, while under an in-plane magnetic field (applied along the spin channel), only the out-of-plane spins arise from the SHE precess. Experimental evidence of spin–charge interconversion in such spin Hall devices has been reported in different systems. The SHE has been reported in graphene combined with MoS_2, WSe_2 (Erling et al. 2019; Safeer et al. 2019a), while the SGE has been probed in graphene combined with WS_2 (Ghiasi et al. 2019; Benítez et al. 2020). Remarkably, gate-tunable SHE and SGE have been experimentally demonstrated in graphene/WS_2 heterostructures at room temperature (Benítez et al. 2020). Recently, unconventional mechanisms for spin–charge interconversion have been reported in conducting 1T′-MoTe and WTe_2, using similar device structures (Safeer et al. 2019b; Zhao et al. 2020).

5.4. 2D materials in magnetic tunnel junctions[4]

5.4.1. *Introduction*

Spintronics is a new paradigm for the future of electronics focusing on spin as the information vector. It naturally offers properties ranging from the spin's intrinsic quantum variable to the ultra-low-power oriented magnetic non-volatility, while also withstanding coupling to photonics through circular polarization of light.

4. This section written by Bruno DLUBAK and Pierre SENEOR.
For a color version of all figures in this section, see www.iste.co.uk/cresti/beyondcmos.zip.

Spintronics has already provided acclaimed applications such as hard disk drive read heads for the first storage revolution and is now engaging in the upcoming next revolution of embedded memories with MRAMs. Here, the spin degree of freedom could naturally unify memory functions with logic ones (Behin-Aein et al. 2010), (Manipatruni et al. 2019) and further provide solutions for beyond CMOS technologies such as stochastic, neuromorphic and quantum architectures (Borders et al. 2019; Grollier et al. 2020). As such, spintronics is already a promising candidate for the future of high performance and low-power electronics and beyond.

In this regard, the emergence of two-dimensional materials has attracted considerable interest in recent years due to their ability to provide promising new functionalities and the wealth of materials available to unveil. Sparkled by the discovery of graphene and its potential for spin information transport (Tombros et al. 2007; Dlubak et al. 2012a; Seneor et al. 2012), this has opened up novel exciting opportunities in terms of functionalities and performances for spintronics devices, fueled by the arrival of new 2D materials such as hexagonal boron nitride (h-BN) and more recently, 2D semiconductors such as transition metal dichalcogenides (TMDCs).

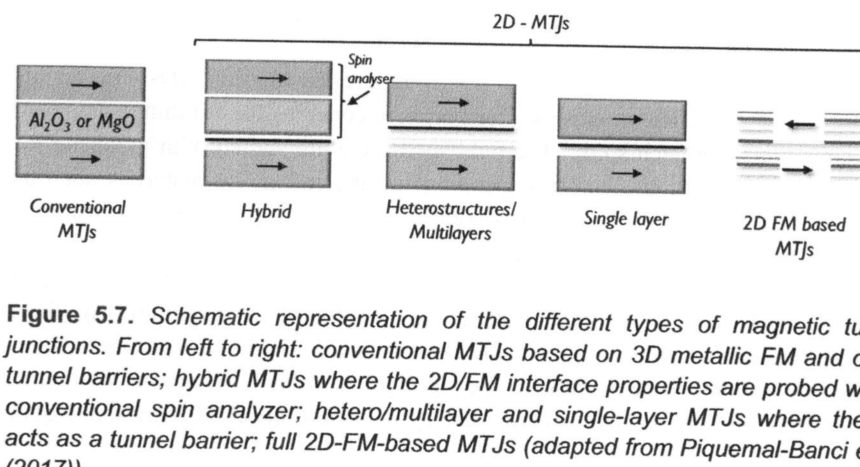

Figure 5.7. *Schematic representation of the different types of magnetic tunnel junctions. From left to right: conventional MTJs based on 3D metallic FM and oxide tunnel barriers; hybrid MTJs where the 2D/FM interface properties are probed with a conventional spin analyzer; hetero/multilayer and single-layer MTJs where the 2D acts as a tunnel barrier; full 2D-FM-based MTJs (adapted from Piquemal-Banci et al. (2017))*

In this chapter, we present an emerging and promising venue for 2D materials in spintronics: their integration into what is now the prototypical spintronics device, the spin valve. This will be illustrated in the following by the magnetic tunnel junction (MTJ), the core element of MRAMs (see Figure 5.7). An MTJ consists of an ultrathin insulating layer intercalated between two ferromagnetic electrodes. Here, the insulating layer serves as a tunnel barrier, while the two ferromagnetic (FM) electrodes act as a spin polarizer and analyzer (in analogy with optics), thus defining a spin valve. The most relevant property of an MTJ is that its resistance can be modulated by the relative orientation of the magnetization of the FM electrodes. The amplitude of this modulation is defined as tunneling magnetoresistance (TMR) and is usually expressed as percentage in the form: $TMR = \frac{R_{AP}-R_P}{R_P}$, where R_P and R_{AP} correspond to the parallel and antiparallel configurations of magnetizations. Trying to relate it to material properties, a simple model, known as Jullière's model, was introduced earlier (Jullière 1975) to give a quantitative estimation of the TMR as a function of the spin-polarizations P_1 and P_2 of both FM electrodes: $TMR = \frac{2P_1P_2}{1-P_1P_2}$. Nowadays, the spin polarizations are defined as effective ones, as it was later shown that the involved electrode spin polarizations were not intrinsic properties of the FMs, but more effective ones relating to the structural and electronic properties of the interfaces (Tsymbal et al. 2003; Žutić et al. 2004; Tsymbal and Zutic 2011; Galbiati et al. 2014). While this model does not fully capture the new richness of the underlying mechanisms as we will see later on, it is still a very useful tool to compare the performances of devices and electrodes (Tsymbal and Zutic 2011; Galbiati et al. 2014; Piquemal-Banci et al. 2017). Appropriately used, in many cases, it can account for the key features and highlight the impact of the relevant physics at play.

In the following, we summarize the literature concerning 2D-MTJ results published until 2021. The metric we focus on here concerns the maximum achieved TMR spin signal for each study. The published literature started with graphene for magnetic tunnel junctions, before expending towards other 2D materials such as h-BN and TMDCs 2D semiconductors (see Figure 5.8). In the following sections, we made the choice of introducing the progress of the field according to the historical evolution of the fabrication, from flake exfoliation to CVD transfer and then direct CVD growth. We detail these results, showing how the progression from exfoliated material to directly grown 2D layers allowed the interfacial degradation issue to be overcome to be overcome. We highlight how this CVD approach, in turn, unveiled spin-filtering mechanisms in these systems and unlocked large spin signal observations.

5.4.2. *First steps towards 2D material integration in magnetic tunnel junctions*

In an early computational study, Karpan et al. (2007) exposed the strong potential of graphene for magnetic tunnel junctions (MTJs). It was unveiled in particular that strong magnetoresistance (MR) spin signals could be expected for FM/graphene/FM systems reaching up to 10^{13} %, well in excess of what is achieved and expected for MgO-based structures. These expectations were based on the fundamental matching of the Dirac band structure of graphene with the spin-polarized fermi-surface of nickel: at the Dirac point, only minority spins would be available in nickel. This initial prediction ignited a strong interest for the integration of graphene – and later other 2D materials – in vertical spin valve heterostructures to exploit their performance and lever on their functionality benefits (Piquemal-Banci et al. 2017).

Following this early discussion, and relatively shortly after pioneering demonstrations of graphene exfoliation, the first experimental approaches were presented in the literature. An early study including spin and angle-resolved photoemission spectroscopy (ARPES) of graphene-covered Ni surface showed hints of spin-polarized current extraction (Dedkov et al. 2008). The same year, Mohiuddin et al. (2008) demonstrated the first fabrication of a complete spin valve encompassing a graphene layer. In this work, a regular SiO_2(100 nm)/Si substrate was covered with a dense array of 1 μm×20 μm magnetic NiFe electrodes. Graphene flakes were then derived by exfoliation using the scotch tape method (Novoselov et al. 2004) on top of the patterned silicon substrate, selected with a SiO_2 thickness providing enough optical contrast to spot micron-sized graphene flakes on the surface (Blake et al. 2007). Several multilayer flakes covering the prepatterned magnetic elements were thus detected. The relevant flakes were further contacted by a top magnetic contact. This fabrication approach required an adapted electronic-beam lithography step in which the design was carefully crafted to avoid shorts with the surrounding NiFe electrodes and other exfoliated graphene layers. This work led to the fabrication of the first reported NiFe/graphene/NiFe MTJ. The authors also investigated the spin-dependent transport across the graphene interlayer and reported an MR of 0.4%: the spin polarization thus appeared very limited and yet, there was no sign of spin filtering effect. It was at the time already discussed that various interfacial behaviors, either detrimental oxidation or complex hybridization effects, could jeopardize the exploitation of 2D materials in spin valve heterostructures (Karpan et al. 2007; 2008; Mohiuddin et al. 2008).

Hence, while the initial theoretical results triggered the hope of harnessing spin filtering effects to achieve large spin signals using the recently isolated 2D material family, the experimental studies promptly highlighted the associated difficulty to exploit these properties. As such, integration issues and complex interfacial physics remain the cornerstone of 2D-based MTJ exploration: many parameters impact the spin properties, and the integration of 2D materials in-between oxidation-prone metallic ferromagnets is far from trivial.

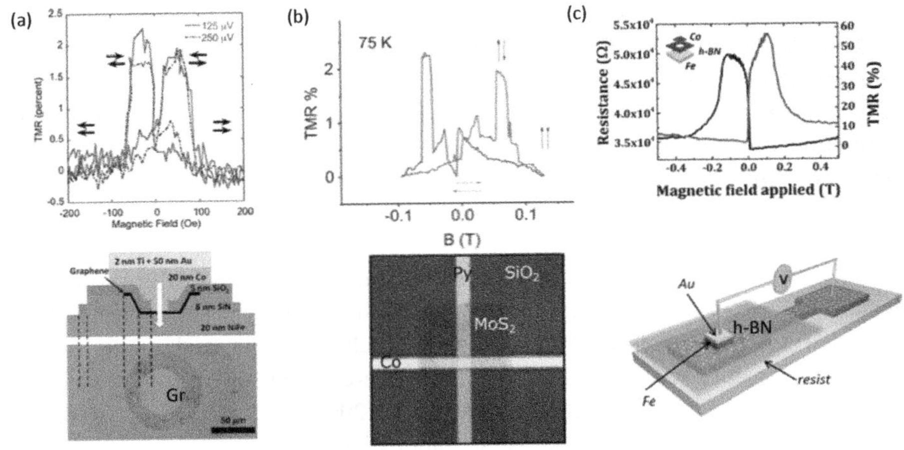

Figure 5.8. *Tunneling magnetoresistance and magnetic tunnel junctions incorporating 2D materials*

NOTES ON FIGURE 5.8.– *(a) Bottom: a NiFe/graphene/Co spin valve fabricated using CVD graphene by the wet-transfer method. Top: MR signal reaching 2%. Adapted from Cobas et al. (2012). (b) Bottom: a NiFe/MoS2/Co spin valve fabricated using CVD graphene by the wet-transfer method. Top: MR signal reaching 2%. Adapted from Dankert et al. (2017) (c) Bottom: a Fe/h-BN/Co spin valve fabricated using CVD graphene by the direct growth method. Top: MR signal reaching >50%. Adapted from Piquemal-Banci et al. (2018).*

5.4.3. *Exfoliated and transferred devices: early results*

The earliest studies were based on the pioneering method of graphene exfoliation (extended later with larger wet transferred CVD layers) in order to access 2D crystals and integrate them in spin valve structures. This approach offered the advantage of the easiness of the scotch tape method and gave immediate access to the sharply increasing number of available 2D materials. However, it was not directly adapted to the stringent constraints of spintronics, in particular for the preservation of the ferromagnet surfaces. Indeed, ambient exfoliation or transfers lead to oxidized interfaces. In the following, we describe the initial works and workarounds that have been proposed to allow the exfoliated/transferred material integration without interfacial degradations.

Mohiuddin et al. (2008) earlier highlighted the difficulty of integrating graphene in spin valve heterostructures because of oxidation issues. The bottom FM spin source was indeed exposed to air during exfoliation. However, since the inception of spintronics, it had been a constant requirement to rely on high vacuum and purity deposition conditions during fabrication in order to reach large spin signals. Graphene integration thus became problematic: a suggested solution at the time consisted of covering the NiFe electrode with a 2 nm thick Au layer leading to the fabrication of NiFe/Au/graphene/NiFe devices. Later, another early study by Banerjee et al. (2010) presented similar MTJ structures with graphene using a thin NiFe magnetic layer protected by an Au capping layer. The authors could measure a spin signal for NiFe/Au/graphite/Au/Co. But while the authors derived a minimal spin diffusion length (>100 nm) for spin transport perpendicular to graphite planes from these measurements, they did not, however, discuss the effect of the Au layer. As such, in the of both Mohiuddin et al. and Banerjee et al. an Au layer was inserted between the graphene/graphite layer and the spin source in an effort to circumvent the oxidation issue. However, while it could help to prevent the FM oxidation issue, the insertion of a thin non-magnetic layer (such as Au) at the interfaces is known to quench the spin properties of MTJ structures (Leclair et al. 2000).

Following this trend, the issue of oxidation has been a strong roadblock for 2D-based MTJ exploration. Illustratively, Cobas et al. (2012) presented a similar device structure, but using transferred graphene layers grown by CVD on Cu. They deposited the CVD layer by wet transfer on top of the bottom ferromagnetic NiFe electrodes. After defining a small window opening, a top ferromagnetic layer was evaporated on the graphene layer. This approach highlighted the benefit of large graphene layers for the systematic device fabrication. However, as underlined by the authors, the oxidation induced by the ambient/wet transfer step again probably caused the spin signal to be limited to MR=2%. Following these initial studies, many groups fabricated and characterized similar systems by exfoliation or transferred CVD graphene through a wet process. The measured spin signals still appeared in the same MR=1% range. For instance, Iqbal et al. (2013) presented structures with mono or bilayer graphene transferred on top of NiFe electrodes and made further contact with another NiFe electrode with MR below 1%. Studies by Chen et al. (2013) and Meng et al. (2013) presented data for Co/graphene/Co vertical structures. The MTJs based on monolayer CVD graphene showed an MR of 0.7% versus 1% for bilayers. Interestingly, in 2015, by using an Al_2O_3 interface (according to Dlubak et al. (2012b), see below), Iqbal et al. (2015) were able to improve their results and reach MR=-1.6%. Using h-BN similarly as a tunnel barrier on top of graphene, they reached MR=-1.88% (Iqbal et al. 2016). Note that the discussion on the sign of the MR will be further developed below. Nevertheless, this problem of low spin signal remains when the interfacial degradation is not addressed: Iqbal et al. (2017) showed that MR = 0.39%, Zhou et al. (2017) showed that MR=-1.2% and Sharma et al. (2020) showed that MR=1.2% for graphene bilayers.

These initial works focused mostly on graphene, but similar problems have been observed with the tentative integration of other 2D materials such as h-BN and TMDC 2D semiconductors. In the first study on a MoS_2-based MTJ, Wang et al. (2015a) used an Au capping layer to protect bottom NiFe electrodes, and report MR=0.73%. Subsequent studies have all shown similar fabrication approach as graphene and today the MR still culminates at around 3% for TMDCs with these methods (Dankert et al. 2017; Zhao et al. 2017; Khan et al. 2018) and similarly for h-BN (Dankert et al. 2015; Asshoff et al. 2017, 2018). While the conventional exfoliation and transfer methods quickly allowed for the first fabrication and exploration of devices, it has faced some interfacial material issues (well-known in spintronics) that severely impeded its progress.

These different pioneering studies highlight the practical difficulties to integrate 2D materials in spin valve structures – despite their strong interest – and to explore related fundamental spin transport phenomena. Spin transport quenching, because of interfacial issues, may have prevented expectations from being reached and hampered the study of spin transport mechanisms of these systems, severely limiting the opportunity to explore and exploit the vast library of 2D materials available.

5.4.4. *Exfoliated and transferred devices: improvement through in situ definition*

To go beyond this limitation, alternative processes have been proposed for exfoliated 2D materials resulting in larger spin signals. Park et al. (2014) presented the "flip-transfer" method where the FM is deposited on graphene, and the substrate is dissolved to establish contact on the other side of graphene. For a four-layer graphene device, the authors reported a measured MR of 4.58%, which was well above previous results. This highlights that better control of the oxidation clearly led to an improvement of the MR signal. This can also be acknowledged by the very low Resistance*Area product of 0.2 $\Omega\cdot\mu m^2$ observed in these devices. Another original approach has been to transfer the graphene layer over a perforated membrane and flip the sample in situ during growth, allowing Li et al. (2014b) to present a work with CVD graphene contacted on both sides by ferromagnets. This led to MRs of up to 3.4%, the largest for wet transferred CVD graphene. This was followed by Asshoff et al. (2017) who reported on the exfoliation of graphene on membranes with holes but with a measured MR of about 1%.

More recently, Galbiati et al. (2019) exposed another approach for exfoliation. They were able to prepare a 2D semiconductor – ferromagnet heterostructure in a glove-box inert atmosphere devoid of water and oxygen coupled with a high-vacuum deposition cluster. The in situ fabrication process flow allowed the preservation of the FM electrode interface integrity. Fabrication of spin valves with this approach has led to the largest measured MR signal up until now, with an exfoliated 2D material (above 5%). While this result further underlined the need for protection of the interfaces, it also unveiled that other physics mechanisms arising at 2D/FM interfaces may have precluded the progress in MR measurements so far. In particular, a lateral diffusion of spins was pinpointed as the main origin of spin flips in these TMDC-based systems. The authors projected the MR to reach about 100% in their specific 2D semiconductor-based heterostructure once corrected for lateral diffusion and loss effects.

By achieving the largest values for exfoliated and transferred materials, these works demonstrated the importance of preserving the quality of interfaces. They also unveiled that beyond material science, an underlying physics issue was at play. This holds strong promise for next steps of 2D-based spintronics…

5.4.5. *Direct CVD growth: the rise of large scale and high quality*

In parallel to these works integrating exfoliated or transferred 2D materials, another venue was proposed in 2012 using a direct graphene CVD growth process on top of the FM spin sources (Dlubak et al. 2012b). This process allowed us to define the metal/graphene interface in situ, fully avoiding the oxidation issue. XPS chemical characterizations showed the passivation effect of graphene after the heterostructure was left for an extended period in ambient conditions (Martin et al. 2015). This readily demonstrated the pertinence of the direct in situ growth approach to preserve the integrity of the 2D/FM interface. Remarkably, using a known analyzer, a strong spin signal above state of the art in the order of MR=-10% – corresponding to a spin polarization of P=-16% – was measured in this initial 2012 study. Beyond confirming that graphene could efficiently protect the FM surface from oxidation during fabrication steps, it showed that direct CVD growth would preserve a spin polarization for this ferromagnetic electrode. Interestingly, beyond its large amplitude, the sign of the measured MR in these devices was found to point to a spin filtering effect arising at the graphene/nickel interface (and reversal of the spin polarization). This will be discussed in the following. The result even increased a few years later in 2014 by using high-quality ALD-grown tunnel barriers that were specifically developed, with MR reaching -30% pointing to spin polarization P=-42% for the graphene/Ni interface (Martin et al. 2014).

Similar direct integration of graphene by CVD and significant spin signals were reported by Godel et al. (2014), and later by Cobas et al. (2016) and Entani et al. (2016), where the authors integrated multilayer graphene covered nickel electrodes in spintronics devices. These works confirmed the observation of a spin filtering effect at this interface and discussed the underlying physical mechanisms based on experimental observations. The direct growth approach was then extended to other 2D materials. In parallel, h-BN was grown by CVD on Fe and Co, with MR signals reaching 50% (Piquemal-Banci et al. 2016; Piquemal-Banci et al. 2018). As discussed below, the specificities of the CVD growth process led to different coupling being accessed at the 2D hBN-FM interfaces with strong modulation of the spin signals. Later, the direct graphene growth approach was even further engineered on epitaxial crystalline Ni(111) spin sources (Weatherup et al. 2014) (Piquemal-Banci et al. 2020). This led in particular to the measurement of very large magneto-resistance of the order of 80% for Co/monolayer graphene/Ni heterostructures. As described in the following section, this was made possible through several spin filtering mechanisms arising in these systems.

Up until now, this direct CVD-grown Co/Gr/Ni remains, with 80%, the largest reported MR spin signal for a 2D-based MTJ. The direct growth approach, by preserving pristine interfaces, was thus shown to unlock the potential of graphene/ferromagnet interfaces for spintronics. The only drawback would be that the CVD of graphene requires detailed investigations for each targeted ferromagnetic substrate (Reina et al. 2009; Weatherup et al. 2012; Dahal et al. 2014; Naganuma et al. 2019). Indeed, FM catalysts are exposed to carbon-based precursors towards the nucleation and growth of the graphene layers. Careful understanding of the processes was necessary to control the growth of graphene from the multilayer to the monolayer limit and further bring down the growth window to CMOS compatible temperatures around 450°C (Weatherup et al. 2012, 2014). Very recently, this type of large-scale integration directly on the FM spin source has been extended beyond graphene and h-BN thanks to techniques such as pulsed laser deposition. It is now also possible to grow large families of 2D semiconductors (WS_2, WSe_2, etc.) and their heterostructures on spin sources (Godel et al. 2020; Zatko et al. 2021), thus opening up uncharted territory for 2D-MTJ exploration.

5.4.6. *Experimental evidences of 2D-based spin filtering in hybrid 2D-MTJs*

Two main origins of spin filtering have been experimentally proposed since the original 2007–2008 works (see Figure 5.9): (a) a layer-dependent effect, corresponding to a "bulk" band-structure spin filtering arising from the matching of the spin-polarized Fermi level pockets of FM with the unique 2D material bands and (b) an interfacial hybridization effect, which can in some cases lead to an engineered increase of the spin polarization of the extracted current (Piquemal-Banci et al. 2017). At first, the literature's studies did not report on the observation of spin filtering or high MR. In fact, the problem was twofold. First, the degradation of interfaces – as discussed above – was a key issue. Second, the symmetric FM/Gr/FM heterostructures that were studied could not allow for the observation of negative spin polarization through MR due to its sign being even in spin polarization ($MR{\sim}P^2$) (Julliere 1975).

Hence, the first experimental evidence of the initial prediction of minority spin filtering at FM/graphene interfaces was only achieved once devices included clean interfaces such as those defined by direct CVD (Dlubak et al. 2012b; Godel et al. 2014; Martin et al. 2014; Cobas et al. 2016; Entani et al. 2016). Here, a key point was also to rely on a well-known Al_2O_3/Co spin analyzer to deduce the spin polarization of the extracted spin currents (Dlubak et al. 2012b; Martin et al. 2014). In these studies, the "bulk" spin filtering effect was shown through the measurement of inversed magnetoresistance spin signals, as expected by theory. Indeed, as exposed by Karpan et al. (2007), the majority of spin carriers should tunnel through graphene layers in the junction and hence were expected to be exponentially attenuated (as in MgO MTJs), while minority spin carriers would be transmitted in a metallic-like transport behavior. From this simple image of majority-tunneling versus minority-metallic conduction, the expected spin filtering effect had to increase exponentially with the thickness of the graphene stack. With thickness, the minority carrier current would become the dominant one. This prediction was observed experimentally: going from monolayer graphene with measured P=-9.8% (Martin et al. 2015) to multilayers with P=-42% (Martin et al. 2014), the observed spin filtering effect was shown to increase with the number of layers. Strikingly, the very strong spin filtering observed for graphene multilayers/Ni (P=-42%) already corresponded to a full reversal of the spin polarization compared to the best Ni/Al_2O_3 interfaces in the literature (P>30%) (Kim et al. 2004). This type of analysis was extended to the broad family of 2D semiconductors, where the band structure can be more specifically tailored (Chaves et al. 2020). A similar band spin filtering effect was described by Zatko et al. (2019) by following the same spin analyzer strategy in the case of WS_2. Interestingly, the band gap of WS_2 is expected to change with thickness switching from direct (one layer) to indirect (two layers and above). This was shown to have a direct impact on the extracted spin polarization from the WS_2/FM interface: these different gap arrangements would select different FM spin pockets at the Fermi level. Zatko et al. showed that this intrinsic band structure evolution of the 2D semiconductor led to a switch between K and Q points for the transport and to the dramatic inversion of the spin polarization with the WS_2 thickness. By providing insight at an additional lever compared to graphene devices and allowing a fundamental tailoring of the MTJ spin properties, this overall highlights that we are just touching on the richness of the band structure spin filtering opportunity offered by 2D material families.

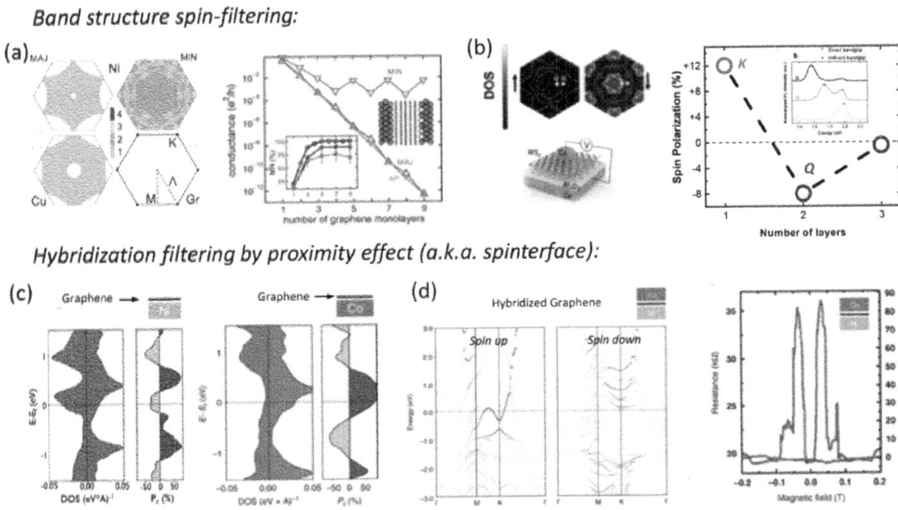

Figure 5.9. *Spin resolved band structure and resulting filtering effects at 2D materials/ferromagnet interfaces*

NOTES ON FIGURE 5.9.– *Top: band structure spin filtering. (a) Left: calculated Fermi surfaces for Ni. The Ni minority spins appear to match the graphene Dirac points. Adapted from Karpan et al. (2007). Right: dependence of majority and minority spin conduction through the graphene layers at the Ni/graphene interface (the inset shows the calculated resulting TMR signal) highlighting different metallic and tunnel channels for the two spin directions. Adapted from Zatko et al. (2019). (b) Left: Calculated Fermi surfaces for Co highlighting the K and Q points. Illustration of the measured spin valve device based on Co/WS_2. Right: measured evolution of the Co/WS_2 spin polarization as a function of WS_2 thickness showing inversion altogether with transition from K to Q point. In the inset, the measured photoluminescence showing the transition from K to Q. Bottom: hybridization filtering by the proximity effect. (c) Calculated density of states – and induced spin polarizations – of graphene for the Ni/gr and the Co/Gr interfaces. (d) Calculated band structure for the Ni/monolayer graphene/Co interface highlighting the impact* *of hybridization. Magnetoresistance signal achieving >80% in a Ni/monolayer Gr/Co spin valve device. Adapted from Piquemal-Banci et al. (2020).*

Remarkably, interfacial hybridizations between 2D layers and ferromagnetic electrodes were first seen as an issue to reach the perfect situation described by Karpan et al. (2007). Hence, hybridization was described as a setback preventing the maximum possible spin signals from being reached. Different approaches of interfacial engineering, for instance by the metallic dusting of Cu, have been suggested (Karpan et al. 2008). However, in stark contrast, Piquemal-Banci et al. (2018) showed that in some cases this hybridization could be leveraged to tailor the spin properties of the interface and reach even larger spin polarizations. Focusing first on the 2D insulator h-BN, the authors showed that in addition to the Resistance*Area control with h-BN evolving from an insulator to a conductor by proximity with the ferromagnet, the spin polarization extracted from hybridized interfaces could also be modulated. This opened up a large playground where the 2D nature of these materials would be harvested to enhance spin properties. This was further illustrated in the case of hybridization in epitaxial nickel–monolayer graphene interfaces (Piquemal-Banci et al. 2020). The hybridization occurring in a full Ni–graphene–Co stack was shown to play a pivotal role to reach large spin signals, above MR=80%, the largest reported for any 2D-based MTJ.

We note that we focused on discussing the specific interaction at the FM/graphene interface from the spin transport point-of-view, but it also has the potential for magnetic properties tailoring. It was discussed earlier how in situ intercalation of ultrathin FM below CVD graphene layers could lead to enhanced perpendicular magnetic anisotropies and allow tailoring of FM/graphene interfaces (Rougemaille et al. 2012; Yang et al. 2016). This was more recently followed by a study on the out-of-plane magnetization of graphene covered FePd $L1_0$ spin sources (Naganuma et al. 2019).

5.4.7. *Conclusion*

Overall, the recent experimental results for spin transport within 2D-based MTJs, especially those exploiting ultra-clean interfaces by direct growth, were discussed in light of two convoluted contributions. In particular for the Ni/graphene case, both of these contributions can give rise to a strong spin filtering effect: one arising from the exponential suppression of majority spin carriers through graphene's band structure selection at the Ni interface, and the second one based on the band structure modification through proximity effect leading to tailorable interfacial spin populations. While a deeper understanding remains to be achieved, especially for the many large families of 2D materials, these initial works already highlight a convergence of experiments and theory towards the very high potential of 2D material integration in MTJs. Beyond these works, we can only anticipate further potential in tailoring anisotropy, leveraging spin–orbit torque, controlling lateral spin diffusion or even further integrating the new families of arising 2D ferromagnets and ferroelectrics and combining them with gating ability.

5.5. Topological insulators in spintronics[5]

5.5.1. *Introduction*

Topological insulators (TIs) are bulk-insulating materials that host metallic states at their boundaries (Hasan and Kane 2010). These states derive from the specific electronic band topology of TIs, induced by the combination of large spin–orbit interaction (SOI) and time-reversal symmetry. Topological surface states (TSSs) display Dirac-like massless dispersion, spin-momentum locking and are robust against small disorder and non-magnetic impurities. These properties make TIs very promising materials for quantum computation and spintronics. In particular, the spin-momentum locking in TSSs leads to efficient spin–charge interconversion effects, which could be used in several electronic applications such as magnetoelectric spin–orbit (MESO) logic and magnetic random-access memories (MRAMs).

TIs were first predicted in 2005. The original model by Kane and Mele (2005) studies the effect of SOI on a graphene lattice. The SOI gaps the graphene bands and turns the system into the quantum spin Hall phase: graphene becomes an insulator, but it hosts dissipationless one-dimensional (1D) edge states with exactly quantized Hall conductance in the absence of a magnetic field. This quantum spin Hall insulator – or two-dimensional (2D) TI – can be thought of as two copies of the quantum Hall effect, where opposite spin components experience opposite magnetic fields as a result of SOI. The two spin channels propagate in opposite directions along the edges, forming a pure spin current that is potentially dissipationless due to the preserved time-reversal symmetry. In practice, the SOI in graphene turns out to be too small (< 1 meV) to realize the quantum spin Hall phase. The first experimental realization of a 2D TI was reported with strained HgTe in HgTe/(Hg,Cd)Te quantum wells (König et al. 2007). Three-dimensional (3D) TIs display equivalent properties, but in these materials, the boundary states are 2D

5. This section written by Frédéric BONELL and Thomas GUILLET.

For a color version of all figures in this section, see www.iste.co.uk/cresti/beyondcmos.zip.

(surface states) instead of 1D. Similarly to the edge states of 2D TIs, counter-propagating surface electrons in 3D TIs possess opposite spins, and the spin direction is locked perpendicularly to the momentum direction (see Figure 5.10(a)). Spin-momentum locking is the most interesting property of 3D TIs for spintronics, as detailed in the following section.

Several 3D TIs have been predicted from band theory before experimental confirmation, including Bi_xSb_{1-x}, strained HgTe, α-Sn, Bi_2Se_3, Bi_2Te_3 and Sb_2Te_3 (Fu and Kane 2007; Zhang et al. 2010). The key ingredients to drive a system into the TI phase are the SOI and the mechanism of band inversion. Band inversion refers to the crossing of the valence and conduction bands with opposite parities as a result of band shifting by SOI or lattice distortions (Fu and Kane 2007; Zhang et al. 2010; Zhu et al. 2012). The SOI is required to open a gap at the band crossing points. These two conditions imply that TIs are generally narrow gap semiconductors composed of heavy elements. Their inverted electronic structure possesses a non-trivial topology in the sense that it cannot be transformed continuously into one of the ordinary insulators without closing the insulating gap. At the interface between a TI and a trivial insulator such as a vacuum or most oxides, connecting the bands of equal polarities smoothly on each side necessarily implies electronic states that cross the Fermi level (the aforementioned gapless TSSs) (see Figure 5.10(b)). Their topological nature makes them robust against temperature, disorder and impurities that do not break the time reversal symmetry, including possible surface oxidation and non-magnetic impurities. However, these states require a minimal film thickness to develop, because in films thinner than a few nanometers, quantum tunneling couples the two surfaces and opens a trivial gap in the topological bands (Zhang et al. 2010). It is remarkable that many 3D TIs retain their unique properties above room temperature. This is for instance the case of the prototypal TI Bi_2Se_3, thanks to its rather large band gap of 0.3 eV.

5.5.2. *Spin-momentum locking and spin–charge interconversion*

5.5.2.1. *Spin-momentum locking*

The TSSs can be described by the low-energy Hamiltonian $\mathcal{H} = \hbar v_F(\mathbf{z} \times \mathbf{k}) \cdot \boldsymbol{\sigma}$, where $\hbar$ is the reduced Planck constant, v_F is the Fermi velocity, $\mathbf{z}$ is the unit vector normal to the surface, $\mathbf{k}$ is the 2D electron wave vector and $\boldsymbol{\sigma}$ is the vector of Pauli spin matrices. The corresponding graphene-like linear dispersion $E = \pm\, \hbar v_F k$ displays branches of opposite velocities $\pm v_F$ and associated spin components $\mathbf{s} = \pm\hbar/2\, (\mathbf{z} \times \boldsymbol{k}/|\boldsymbol{k}|)$. The latter expression means that in TSSs, the spin is in-plane and always perpendicular to the momentum. The opposite surface, described by considering $\boldsymbol{z} \rightarrow -\boldsymbol{z}$, displays an opposite spin helicity (see Figure 5.10(a)).

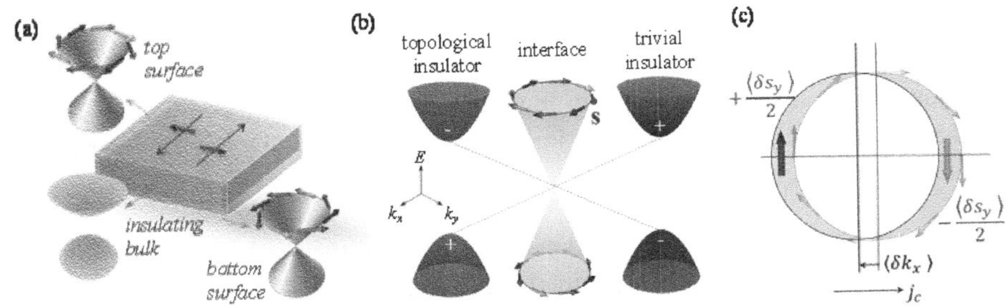

Figure 5.10. *(a) Spin-momentum locking in TSSs, with opposite helicities at opposite surfaces. (b) Electronic structure at a TI/trivial-insulator junction. In the TI, the bulk bands are inverted (+/- denote the band parity). Connecting the bands at the interface requires gap closure, resulting in spin-helical metallic states. (c) Edelstein effect. A charge current j_c shifts the Fermi contour by $\langle \delta k_x \rangle$, resulting in a net transverse spin density $\langle \delta s_y \rangle$*

5.5.2.2. *Edelstein effect*

The main transport phenomenon that exploits the spin-momentum locking property is the Edelstein effect (EE), also called the Rashba–Edelstein effect or inverse spin galvanic effect (Edelstein 1990). When a charge current density j_c flows at the surface of a 3D TI, say in the x direction, the carriers acquire a net momentum $\hbar\langle \delta k_x \rangle = -\hbar k_F\, v_d/v_F = -\,\hbar k_F/nev_F \times j_c$ and a non-equilibrium spin density $\langle \delta s_y \rangle = \hbar v_d/2v_F = \hbar/2nev_F \times j_c$ builds up (see Figure 5.10(c)). Here, k_F is the Fermi wave vector, v_d is the drift velocity, n is the electron density and e is the elementary charge. The EE is similar to the charge-to-spin conversion in Rashba-split 2D electron gases. However, in the case of Rashba bands, two Fermi contours with opposite spin helicities partly compensate and the resulting spin density is smaller by a factor $\Delta k_R/2k_F \ll 1$, with Δk_R as the Rashba splitting. Another difference with Rashba states is that the opposite surfaces of a 3D TI always host opposite spin densities. To exploit the EE in 3D TIs, it is therefore necessary to pick up the spin density at one of the two surfaces, which is easily done by creating a proper TI/ferromagnet (FM) interface.

The EE can be used to generate torques electrically on the magnetization of an FM in contact with the TI (Manchon et al. 2019; Han and Liu 2021), thus inducing sustained magnetization oscillations or magnetization switching (Miron et al. 2011). When injected into an FM, the current-induced spin density produces two different spin–orbit torques (SOTs): the field-like SOT (FL-SOT), due to direct exchange with the magnetization, and the damping-like SOT (DL-SOT), due to the spin transfer mechanism. Although Dirac and Rashba gas models predict a dominant

FL-SOT for the EE (Kim et al. 2017), a large DL-SOT has been reported in most experiments. This DL-SOT may originate from the spin relaxation in the FM, the spin Hall effect due to finite bulk conduction or the intrinsic contribution from the Berry curvature at hybridized interfaces (Ghosh and Manchon 2018). The torque efficiencies can be expressed as $B_{FL,DL}/j_c$, where $B_{FL,DL}$ are the SOT effective fields and j_c is either the 2D charge current j_c^{2D} flowing in the surface states or the 3D charge current j_c^{3D} flowing in the full TI layer. Since the DL-SOT is associated with a 3D spin current j_s^{3D} injected in the FM, other common figures of merit to quantify the charge-to-spin conversion efficiency are the Edelstein inverse length $q_{EE} = j_s^{3D}/j_c^{2D}$ (expressed in nm^{-1}), the dimensionless spin-torque efficiency $\theta_{EE} = j_s^{3D}/j_c^{3D}$, and the spin conductivity $\sigma_s = j_s^{3D}/E$, with E as the applied electric field. The DL-SOT is particularly interesting for applications, because it can deterministically switch the magnetization of a magnetic element without the use of an external magnetic field. This is the writing scheme of SOT-MRAMs: an in-plane charge current in the conversion layer is used to switch the free layer of an adjacent magnetic tunnel junction (Liu et al. 2012; Prenat et al. 2016). The great advantage is that the writing process does not require any current flow through the fragile tunnel barrier.

5.5.2.3. *Inverse Edelstein effect*

The Onsager reciprocal of the EE is called inverse EE (or direct spin galvanic effect) and refers to the spin-to-charge conversion process (Ganichev et al. 2002; Sánchez et al. 2013; Shen et al. 2014). A spin density $\langle \delta \boldsymbol{s} \rangle$ injected in the TSSs gives rise to a 2D charge current $\boldsymbol{j_c} = 2ev_F(\langle \delta \boldsymbol{s} \rangle \times \boldsymbol{z})/\hbar$. In practice, we inject a 3D spin current j_s^{3D} across a TI/FM interface. The resulting spin density is $\langle \delta s \rangle = \hbar j_s^{3D} \tau / 2e$ with τ the relaxation time of the TSSs. The figure of merit for the conversion is the inverse Edelstein length $\lambda_{iEE} = j_c^{2D}/j_s^{3D} = v_F \tau$ (expressed in nm) (Zhang and Fert 2016). Because of spin-momentum locking, λ_{iEE} is equal to the TI surface mean free path. Note that in general, $\lambda_{iEE} \neq 1/q_{EE}$, because λ_{iEE} includes the relaxation of a spin current into a spin density at the TI surface, whereas q_{EE} includes the relaxation of the spin density into a spin current in the FM layer, which are non-equivalent processes (Zhang and Fert 2016).

The inverse EE in TIs was proposed as a readout mechanism for non-volatile spin logic devices (Manipatruni et al. 2019). The so-called MESO device comprises a magnetoelectric element whose magnetic state is controlled by an input charge current. The state of the switch is detected by the inverse EE, which delivers an output charge current with a polarity depending on the magnetization direction. Compared to the CMOS technology, the MESO logic could reach 10–30 times lower switching energy and five times higher density. In order to achieve these performances and deliver a high voltage output, TIs with high resistivity (>10 mΩ.cm) and highly efficient inverse EE (λ_{iEE}>5 nm) are targeted.

5.5.2.4. *Nonlinear magnetoresistance*

Nonlinear contributions to the resistivity have been evidenced in systems with large SOI, and specifically in TIs. In Pt/Co bilayers, the spin accumulation induced by the spin Hall effect gives rise to spin-dependent scattering at the interface (Avci et al. 2015). The resistance of the bilayer thus depends on the relative alignment between the magnetization and the spin accumulation set by the current direction. Since the nonlinear resistance changes sign upon magnetization reversal, the effect is called unidirectional magnetoresistance (UMR). UMR could be applied as a magnetization readout mechanism in simple two-terminal devices. UMR with a much larger figure of merit was reported in TI/FM (Lv et al. 2018) and TI/magnetic-TI bilayers (Yasuda et al. 2016). With magnetic TIs, asymmetric electron–magnon scattering due to spin-momentum locking was identified as the main mechanism of UMR. Another kind of unidirectional effect, called bilinear magneto-electric resistance (BMER), was observed in TIs (He et al. 2018a, 2018b) and 2D electron gases (Guillet et al. 2020a, 2021; Vaz et al. 2020). In both cases, there is no ferromagnetic order and the resistance displays a component linear in both the current and magnetic field. Two mechanisms of BMER have been identified, both requiring spin-momentum locking, and involving either hexagonal warping of the TSSs (6-fold distortion, occurring in most Bi-based TIs) (He et al. 2018b) or scattering by spin–orbit impurities (Dyrdał et al. 2020).

5.5.3. *Materials, interfaces and fabrication methods*

Layered Bi and Sb chalcogenides are among the most studied 3D TIs. These compounds crystallize in the tetradymite structure, composed of covalently bonded X–M–X–M–X quintuple layers (X = Se, Te; M = Bi, Sb) separated by a van der Waals gap (see Figure 5.11(a)). The weak van der Waals interlayer bonding makes it possible to produce micrometer-size flakes by simple mechanical exfoliation of bulk crystals. Chemical vapor deposition, pulsed laser deposition, sputtering (Mahendra et al. 2018) and molecular beam epitaxy (MBE) have also been employed as bottom-up fabrication methods. MBE offers several advantages to grow high-quality TIs: the ultra-high vacuum environment, the accurate control of the thickness, the possibility to dope or alloy, and the possibility to form heterostructures with abrupt interfaces. Since the van der Waals bonding releases the constraint of lattice matching, the epitaxial growth of layered TIs can be achieved on a wide variety of substrates (e.g. graphene, Si(111), Al_2O_3(0001), GaAs(111), Ge(111), BaF_2(111), InP(111) and $SrTiO_3$(111)) (Zhang et al. 2011; Yoshimi et al. 2014; Tarakina et al.

2014; Kampmeier et al. 2015; Richardella et al. 2015; Bonell et al. 2017; Guillet et al. 2018). MBE is also a choice method to grow HgTe (König et al. 2007) and α-Sn (Rojas-Sánchez et al. 2016), in which strain engineering is required to stabilize the TI phase.

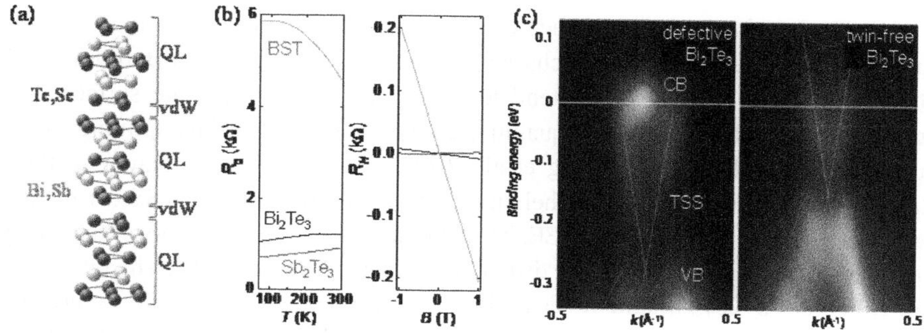

Figure 5.11. *Atomic, electrical and electronic properties of Bi,Sb-based TIs*

NOTES ON FIGURE 5.11. *(a) Crystal structure of $(Bi,Sb)_2(Se,Te)_3$ TIs, composed of quintuple layers (QL) separated by van der Waals (vdW) gaps. (b) Electrical characteristics of epitaxial TIs. (Left) Sheet resistance. (Right) Hall resistance at 77 K. Bi_2Te_3 and Sb_2Te_3 are degenerate semiconductors, whereas $Bi_{0.8}Sb_{1.2}Te_3$ (BST) displays insulating characteristics, with low carrier density and enhanced carrier mobility. (Adapted with permission from Bonell et al. (2020)) (c) Angle-resolved photoemission on Bi_2Te_3 grown by MBE. (Left) Defective films are highly n-doped, with the Fermi level in the conduction band (CB). (Right) Optimized single crystals are bulk insulating, with the Fermi level in the TSSs. (Adapted with permission from Bonell et al. (2017)).*

The particular interest of Bi and Sb chalcogenides lies in their rather large bulk band gap (0.1–0.3 eV), their simple surface electronic structure with a single topological Dirac cone (Zhang et al. 2010) and the possibility to engineer their electronic properties by composition tuning. Bi_2Se_3, Bi_2Te_3 and Sb_2Te_3 are generally highly doped due to native defects such as chalcogen vacancies or anti-site defects (Scanlon et al. 2012). The Fermi level is pinned in the conduction band for Bi_2Se_3 and Bi_2Te_3 and in the valence band for Sb_2Te_3. Harnessing the properties of TSSs and maximizing the spin–charge interconversion efficiency require reducing the bulk conduction, which can be done by adjusting the Bi/Sb ratio in $(Bi,Sb)_2Te_3$ (see Figure 5.11(b)). The Dirac point energy can further be tuned by Te/Se substitution. However, the drawback of chemical doping is a reduction of the electron mean free path and mobility (Beidenkopf et al. 2011). Therefore, important efforts are directed towards achieving crystals of higher quality, in which crystalline defects are suppressed. Suppressing rotational crystal twin boundaries in Bi_2Te_3 was shown to be an effective way to reduce the doping (see Figure 5.11(c)) (Bonell et al. 2017).

3D TIs are also prone to surface band bending due to electron doping, usually caused by surface oxidation or charge transfer at interfaces (ViolBarbosa et al. 2013). The band bending can extend to more than 10 nm from the surface. The downshift of the bands leads to quantum confinement and Rashba splitting of the bulk states near the surface (King et al. 2011). The outer Fermi contour of these Rashba bands possesses a spin helicity opposite to the one of TSSs, which is detrimental to the EE and inverse EE. TI/metal interfaces were also shown to display intermixing, principally Se and Te diffusion in the metallic layer (Walsh et al. 2017; Bonell et al. 2020). Magnetic dead layers were reported at TI/FM interfaces, probably due to material intermixing (Li et al. 2012). In theory, Dirac TSSs arise at the interface with a trivial insulator. At the metallic interface, multiple bands with different parities coexist, leading to complex hybridization effects. Ideally, spintronics devices based on 3D TIs should therefore incorporate FM insulators or sharp and weakly interacting van der Waals interfaces. Layered TIs have been grown on EuS, $BaFe_{12}O_{19}$, $Y_3Fe_5O_{12}$ and $Tm_3Fe_5O_{12}$ magnetic insulators. Although these films are generally defective (Reifsnyder Hickey et al. 2019), strong EE (Wang et al. 2016) and magnetic proximity effects (Tang et al. 2017) have been evidenced with them. More recently, high-quality epitaxial van der Waals heterostructures incorporating TIs have been reported. $(Bi,Sb)_2Te_3/Cr_2Ge_2Te_6$ and Bi_2Te_3/Fe_3GeTe_2 junctions were shown to display strong magnetic proximity effects, such as a strong enhancement of the anomalous Hall effect (Mogi et al. 2019) and Curie temperature (Wang et al. 2020).

5.5.4. *Spin–charge interconversion measurements*

5.5.4.1. *Lateral devices*

The earliest approach to probe the EE is the spin potentiometry on mechanically exfoliated flakes (Li et al. 2014; Tang et al. 2014). A charge current is passed in the TI plane, and the local spin electrochemical potential at the surface is probed with an FM/tunnel-barrier contact placed on top of the channel. The sign of the measured voltage reflects the relative orientation between the current-induced spin density at the TI surface and the magnetization of the FM. Several groups reported such a direct observation of the EE (Ando et al. 2014; Dankert et al. 2015; Lee et al. 2015; Liu et al. 2015; Li et al. 2016). However, it was pointed out that this method is prone to artifacts originating from the magnetic stray field of the FM and the ordinary Hall effect (de Vries et al. 2015). This difficulty is avoided in the non-local geometry, where the voltage detection is performed outside of the charge current path. In this case, a pure diffusive spin current reaches the detector, so that artifacts due to charge transport are suppressed. In conventional lateral spin valves, the spin injection/detection is performed with FM/tunnel-barrier contacts deposited on a channel with long spin-diffusion length (e.g. graphene, Ge, Cu). In order to study the spin–charge interconversion, we instead use TI contacts and a non-local Hall cross geometry (Benítez et al. 2020). Such measurements with graphene/$(Bi,Sb)_2(Se,Te)_3$ suggest a large gate-tunable spin–charge conversion efficiency (λ_{iEE}>5 nm) due to proximity-induced SOI and Rashba splitting in graphene (Vaklinova et al. 2016; Voerman et al. 2019; Khokhriakov et al. 2020). Lateral devices with a Ge channel and a Bi_2Se_3 detector have also been investigated (Guillet et al. 2020b). In these experiments, the spin accumulation in Ge was generated by optical spin orientation with circularly polarized light (see Figure 5.12(a) and (b)). An inverse Edelstein length λ_{iEE}= -0.026 nm was reported, originating from Rashba-split states at the hybridized Ge(111)/Bi_2Se_3 interface. These results show that the measured spin–charge conversion efficiency highly depends on the nature of the interface.

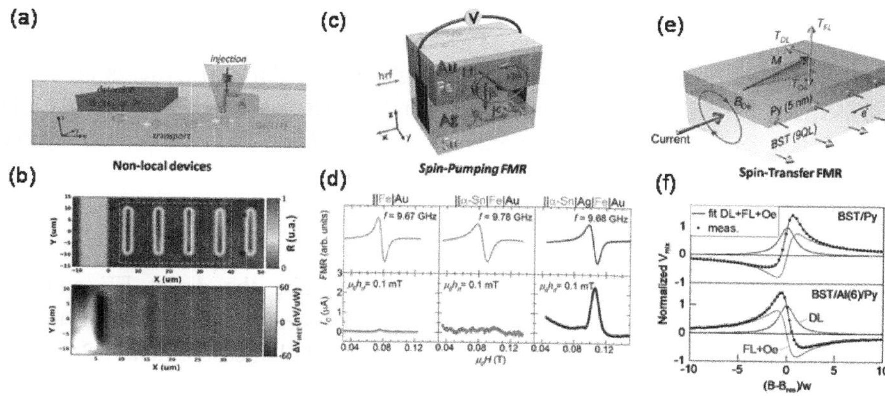

Figure 5.12. *Spin–charge interconversion experiments*

NOTES ON FIGURE 5.12.– *(a) Inverse EE spin detection in a lateral device using optical spin orientation in Ge. (b) Top: reflectivity map of the device. Bottom: voltage measured along the Bi_2Se_3 bar due to spin-to-charge conversion. (c) Spin-pumping experiment in α-Sn. The precessing magnetization injects a spin current in the TI, converted into a charge current by the inverse EE. (d) Top: FMR signal. Bottom: measured charge current. The inverse EE is detected in Sn/Ag/Fe. (e) ST-FMR measurements in a TI/FM bilayer. A charge current produces field-like (FL), damping-like (DL) and Oersted (Oe) torques on the magnetization. (f) Mixing* *voltage measured in (Bi,Sb)$_2$Te$_3$/spacer/permalloy multilayers and separation of the SOTs by line shape analysis. An Al spacer reverses the sum of the FL and Oe SOTs, which correlates with a reduction of material intermixing and indicates a strong EE. (a,b) Reprinted with permission from Guillet et al. (2020). (c,d) Reprinted with permission from Rojas-Sánchez et al. (2016). (e,f) Adapted with permission from Bonell et al. (2020).*

5.5.4.2. *Spin pumping*

Spin pumping is the standard technique to quantify the spin-to-charge conversion efficiency in heavy-metal/FM and TI/FM bilayers (Ando et al. 2011). The bilayer is placed in a microwave cavity immersed in a static magnetic field. Under ferromagnetic resonance (FMR) conditions, the precessing FM magnetization releases angular momentum and pumps a DC pure spin current across the interface (Tserkovnyak et al. 2002). The inverse EE in the TI gives rise to a directly measurable charge current. In prototypal Bi-based TIs such as Bi_2Se_3, the reported λ_{iEE} measured by spin pumping ranges between 0.01 nm and 0.3 nm (Jamali et al. 2015; Wang et al. 2016; Mendes et al. 2017). Enhancement of λ_{iEE} was found after insertion of a thin Ru spacer between (Bi,Sb)$_2$Te$_3$ and CoFeB, presumably due to the weak chemical bonding between the TI and Ru (He et al. 2021). The largest inverse Edelstein lengths were reported in strained HgTe (λ_{iEE}= 2.0 nm) (Noël et al. 2018) and α-Sn (λ_{iEE}= 2.1 nm; see Figures 5.12(c) and (d)) (Rojas-Sánchez et al. 2016). In both studies, a spacer was also inserted between the TI and FM in order to preserve the TI surface.

5.5.4.3. *SOT measurements*

Two techniques are commonly employed to probe the direct EE and are therefore complementary to spin pumping: spin-transfer ferromagnetic resonance (ST-FMR) (Liu et al. 2011) and second-harmonic Hall measurements (2ω-Hall) (Garello et al. 2013). In both methods, an oscillatory charge current along a TI/FM bilayer generates oscillatory SOTs and sustained magnetization precession. The SOTs and the charge-to-spin conversion efficiency are inferred from the amplitude, field and angular dependence of the DC and 2ω voltages arising from frequency mixing between the current and time-dependent magnetoresistance. ST-FMR and 2ω-Hall have been widely employed in TI/FM bilayers with Bi-based TIs (Mellnik et al. 2014; Wang et al. 2015; Kondou et al. 2016; Mahendra et al. 2018; Shi et al. 2018; Bonell et al. 2020; Han and Liu 2021; Xue et al. 2021). The reported DL-SOT efficiency typically reaches $\theta_{EE} = 0.1 - 10$, which is significantly larger than the spin Hall angle in heavy metals (Manchon et al. 2019). This corresponds to an Edelstein inverse length q_{EE} = 0.1-10 nm^{-1}. An efficiency $\theta_{EE} = 0.12$ was reported for α-Sn/FeCo, where α-Sn was grown with a Bi surfactant (Binda et al. 2021).

Thanks to the semiconducting nature of TIs, the SOTs can also be tuned with electrostatic gating (Fan et al. 2014). The large spread of DL-SOT efficiencies reflects the difficulty in determining accurate the SOTs due to large current shunting in the metallic FM layer accurately, as well as fluctuations in the material properties and TI/metal interfaces. It was shown that chalcogen diffusion in the FM, generally overlooked, strongly alters the SOTs due to spin memory loss or charge transfer (Figure 5.12(e) and (f)) (Bonell et al. 2020). Normal metal spacers inserted between the TI and FM layers turn to be effective in preserving the TI surface quality, but they may introduce hybridized states that modify the spin–charge conversion (Shi et al. 2018b). Moreover, additional effects specific to TIs such as asymmetric magnon scattering complicate the interpretation of 2ω-Hall measurements (Yasuda et al. 2017).

5.5.4.4. *Current-induced magnetization switching*

SOT-induced magnetization switching was first demonstrated with heavy-metal/FM bilayers, in which the SOTs derive from the spin Hall effect (Miron et al. 2011, 2010). Several groups reported a strong decrease in the switching current density when using TIs ($\sim 10^{10}$ A/m^2 with TIs vs. $\sim 10^{11}$ A/m^2 with heavy metals), consistently with the large measured DL-SOT efficiency (Wu et al. 2019b). The use of ferrimagnets (Han et al. 2017; Wu et al. 2019a), diluted magnetic semiconductors (Khang et al. 2018) or magnetic TIs (Fan et al. 2014; Yasuda et al. 2017) instead of classical 3*d* FMs allowed a further decrease of the critical current by one or two orders of magnitude (10^8–10^9 A/m^2).

5.5.5. *Conclusion and outlook*

There has been mounting experimental evidence of efficient charge-spin interconversion and large SOTs with TIs. However, the reported conversion efficiency shows a large sample-to-sample variability and is presumably still below the maximum achievable. In practical devices, the exact contribution of the TSSs is hardly known due to ill-defined metallic interfaces, charge transfer and orbital hybridization. Employing TIs in real spintronics applications will require better-controlled interfaces in order to further improve the SOT efficiencies. Significant progress has already been achieved to reduce the bulk conduction in TIs by counter-doping. However, interfacial band bending and the tendency of TIs to form alloys with most metals remain problematic. TI/FM heterostructures that incorporate layered FMs (e.g. Fe_3GeTe_2, Cr_2GeTe_6) are expected to display weaker charge transfer and hybridization thanks to their weakly interacting van der Waals interfaces. Shunting in the FM layer can also be suppressed by using magnetic insulators (e.g. $Y_3Fe_5O_{12}$, $Tm_3Fe_5O_{12}$). As the Edelstein length is directly proportional to the relaxation time at the TI surface, suppressing relaxation channels towards bulk states (with FM insulators) and seeking for high mobility TIs should be the best ways to maximize the interconversion efficiency. Here, engineered van der Waals interfaces could play an important role. For instance, enhanced interface mobility and very large SOTs were predicted in TI/graphene/FM heterostructures (Rodriguez-Vega et al. 2017). Another important challenge for the development of non-volatile technologies and spin-logic with TIs concerns their integration into CMOS-compatible fabrication processes. For SOT-MRAMs, TIs should be integrated with CoFeB/MgO/CoFeB magnetic tunnel junctions and support high-temperature annealing without degradation. Beyond SOTs, the unique properties of TIs offer a wealth of opportunities for fundamental research in spintronics. Novel physical effects are expected to take place in van der Waals heterostructures where TIs imprint their strong SOI in adjacent layers by proximity. Other exciting research directions concern TI/antiferromagnet interfaces, at which magnetic proximity effects were evidenced (He et al. 2017; Yang et al. 2020), and opto-spintronics, following reports of ultra-fast (~0.1 ps) spin-to-charge conversion in TIs (Wang et al. 2018).

5.6. References

Ando, K., Takahashi, S., Ieda, J., Kajiwara, Y., Nakayama, H., Yoshino, T., Harii, K., Fujikawa, Y., Matsuo, M., Maekawa, S. et al. (2011). Inverse spin-Hall effect induced by spin pumping in metallic system. *J. Appl. Phys.*, 109, 103913. doi: 10.1063/1.3587173.

Ando, Y., Hamasaki, T., Kurokawa, T., Ichiba, K., Yang, F., Novak, M., Sasaki, S., Segawa, K., Ando, Y., Shiraishi, M. (2014). Electrical detection of the spin polarization due to charge flow in the surface state of the topological insulator Bi1.5Sb0.5Te1.7Se1.3. *Nano Lett.*, 14, 6226–6230. doi: 10.1021/nl502546c.

Asshoff, P.U., Sambricio, J.L., Rooney, A.P., Slizovskiy, S., Mishchenko, A., Rakowski, A.M., Hill, E.W., Geim, A.K., Haigh, S.J., Fal'ko, V.I. et al. (2017). Magnetoresistance of vertical Co-graphene-NiFe junctions controlled by charge transfer and proximity-induced spin splitting in graphene. *2D Materials*, 4(3), 031004. doi: 10.1088/2053-1583/aa7452.

Asshoff, P.U., Sambricio, J.L., Slizovskiy, S., Rooney, A.P., Taniguchi, T., Watanabe, K., Haigh, S.J., Fal'ko, V.I., Grigorieva, I.V., Vera-Marun, I.J. (2018). Magnetoresistance in Co-hBN-NiFe tunnel junctions enhanced by resonant tunneling through single defects in ultrathin hBN barriers. *Nano Lett.*, 18(11), 6954–6960. doi: 10.1021/acs.nanolett.8b02866.

Avci, C.O., Garello, K., Ghosh, A., Gabureac, M., Alvarado, S.F., Gambardella, P. (2015). Unidirectional spin Hall magnetoresistance in ferromagnet/normal metal bilayers. *Nat. Phys.*, 11, 570–575. doi: 10.1038/nphys3356.

Avsar, A., Unuchek, D., Liu, J., Sanchez, O.L., Watanabe, K., Taniguchi, T., Özyilmaz, B., Kis, A. (2017a). Optospintronics in graphene via proximity coupling. *ACS Nano*, 11, 11678.

Avsar, A., Tan, J.Y., Kurpas, M., Gmitra, M., Watanabe, K., Taniguchi, T., Fabian, J., Özyilmaz, B. (2017b). Gate-tunable black phosphorus spin valve with nanosecond spin lifetimes. *Nat. Phys.*, 13, 888.

Avsar, A., Ochoa, H., Guinea, F., Özyilmaz, B., van Wees, J., Vera-Marun, I.J. (2020). Colloquium: Spintronics in graphene and other two-dimensional materials. *Rev. Mod. Phys.*, 92, 021003.

Baibich, M.N., Broto, J.M., Fert, A., Nguyen Van Dau, F., Petroff, F., Etienne, P., Creuzet, G., Friederich, A., Chazelas, J. (1988). Giant magnetoresistance of (001)Fe/(001)Cr magnetic superlattices. *Phys. Rev. Lett.*, 61(21), 2472–2475. doi: 10.1103/PhysRevLett.61.2472.

Banerjee, T., van der Wiel, W.G., Jansen, R. (2010). Spin injection and perpendicular spin transport in graphite nanostructures. *Phys. Rev. B*, 81(21), 214409. doi: 10.1103/physrevb.81.214409.

Basletic, M., Maurice, J.-L., Carrétéro, C., Herranz, G., Copie, O., Bibes, M., Jacquet, É., Bouzehousance, K., Fusil, S., Barthélémy, A. (2008). Mapping the spatial distribution of charge carriers in LaAlO3/SrTiO3 heterostructures. *Nat. Mater.*, 7(8), 621–625. doi: 10.1038/nmat2223.

Behin-Aein, B., Datta, D., Salahuddin, S., Datta, S. (2010). Proposal for an all-spin logic device with built-in memory. *Nat. Nanotechnol.*, 5(4), 266–270. doi: 10.1038/nnano.2010.31.

Beidenkopf, H., Roushan, P., Seo, J., Gorman, L., Drozdov, I., Hor, Y.S., Cava, R.J., Yazdani, A. (2011). Spatial fluctuations of helical Dirac fermions on the surface of topological insulators. *Nat. Phys.*, 7, 939–943. doi: 10.1038/nphys2108.

Benítez, L.A., Sierra, J.F., Savero Torres, W., Arrighi, A., Bonell, F., Costache, M.V., Valenzuela, S.O. (2018). Strongly anisotropic spin relaxation in graphene–transition metal dichalcogenide heterostructures at room temperature. *Nat. Phys.*, 14, 303.

Benítez, L.A., Sierra, J.F., Savero Torres, W., Timmermans, M., Costache, M.V., Valenzuela, S.O. (2019). Investigating the spin-orbit interaction in van der Waals heterostructures by means of the spin relaxation anisotropy. *APL Mater.*, 7, 120701.

Benítez, L.A., Savero Torres, W., Sierra, J.F., Timmermans, M., Garcia, J.H., Roche, S., Costache, M.V., Valenzuela, S.O. (2020). Tunable room-temperature spin galvanic and spin Hall effects in van der Waals heterostructures. *Nat. Mater.*, 19, 170–175. doi: 10.1038/s41563-019-0575-1.

Binda, F., Avci, C.O., Alvarado, S.F., Noël, P., Lambert, C.-H., Gambardella, P. (2021). Spin-orbit torques and magnetotransport properties of α-Sn and β-Sn heterostructures. *Phys. Rev. B.*, 103, 224428. doi: 10.1103/PhysRevB.103.224428.

Blake, P., Hill, E.W., Neto, A.H.C., Novoselov, K.S., Jiang, D., Yang, R., Booth, T.J., Geim, A.K. (2007). Making graphene visible. *Appl. Phys. Lett.*, 91(6), 063124. doi: 10.1063/1.2768624.

Bonell, F., Cuxart, M.G., Song, K., Robles, R., Ordejón, P., Roche, S., Mugarza, A., Valenzuela, S.O. (2017). Growth of twin-free and low-doped topological insulators on BaF_2 (111). *Cryst. Growth Des.*, 17, 4655–4660. doi: 10.1021/acs.cgd.7b00525.

Bonell, F., Goto, M., Sauthier, G., Sierra, J.F., Figueroa, A.I., Costache, M.V., Miwa, S., Suzuki, Y., Valenzuela, S.O. (2020). Control of spin–orbit torques by interface engineering in topological insulator heterostructures. *Nano Lett.* doi: 10.1021/acs.nanolett.0c01850.

Borders, W.A., Pervaiz, A.Z., Fukami, S., Camsari, K.Y., Ohno, H., Datta, S. (2019). Integer factorization using stochastic magnetic tunnel junctions. *Nature*, 573(7774), 390–393. doi: 10.1038/s41586-019-1557-9.

Bréhin, J., Trier, F., Vicente-Arche, L.M., Hemme, P., Noël, P., Cosset-Chéneau, M., Attané, J.-P., Vila, L., Sander, A., Gallais, Y. et al. (2020). Switchable two-dimensional electron gas based on ferroelectric Ca:SrTiO3. *Phys. Rev. Mater.*, 4(041002). doi: 10.1103/PhysRevMaterials.4.041002.

Bruno, F.Y., Walker, S.M., Riccò, S., de la Torre, A., Wang, Z., Tamai, A., Kim, T.K., Hoesch, M., Bahramy, M.S., Baumberger, F. (2019). Band structure and spin–orbital texture of the (111)-KTaO3 2D electron gas. *Adv. Electro. Mater.*, 5(1800860). doi: 10.1002/aelm.201800860.

Bychkov, Y.A. and Rashba, E.I. (1984). Properties of a 2D electron gas with lifted spectral degeneracy. *JETP Lett.*, 39, 78–81.

Caviglia, A.D., Gariglio, S., Reyren, N., Jaccard, D., Schneider, T., Gabay, M., Thiel, S., Hammerl, G., Mannhart, J., Triscone, J.-M. (2008). Electric field control of the LaAlO3/SrTiO3 interface ground state. *Nature*, 456(7222), 624–627. doi: 10.1038/nature07576.

Caviglia, A.D., Gabay, M., Gariglio, S., Reyren, N., Cancellieri, C., Triscone, J.-M. (2010). Tunable rashba spin-orbit interaction at oxide interfaces. *Phys. Rev. Lett.*, 104(126803). doi: 10.1103/PhysRevLett.104.126803.

Cen, C., Thiel, S., Hammerl, G., Schneider, C.W., Andersen, K.E., Hellberg, C.S., Mannhart, J., Levy, J. (2008). Nanoscale control of an interfacial metal-insulator transition at room temperature. *Nat. Mater.*, 7(4), 298–302. doi: 10.1038/nmat2136.

Chauleau, J.Y., Boselli, M., Gariglio, S., Weil, R., de Loubens, G., Triscone, J.-M., Viret, M. (2016). Efficient spin-to-charge conversion in the 2D electron liquid at the LAO/STO interface. *EPL*, 116, 17006. doi: 10.1209/0295-5075/116/17006.

Chaves, A., Azadani, J.G., Alsalman, H., da Costa, D.R., Frisenda, R., Chaves, A.J., Song, S.H., Kim, Y.D., He, D., Zhou, J. et al. (2020). Bandgap engineering of two-dimensional semiconductor materials. *Npj 2D Mater. Appl.*, 4(1). doi: 10.1038/s41699-020-00162-4.

Chen, J.-J., Meng, J., Zhou, Y.-B., Wu, H.-C., Bie, Y.-Q., Liao, Z.-M., Yu, D.-P. (2013). Layer-by-layer assembly of vertically conducting graphene devices. *Nat. Commun.*, 4(1). doi: 10.1038/ncomms2935.

Cobas, E., Friedman, A.L., van't Erve, O.M.J., Robinson, J.T., Jonker, B.T. (2012). Graphene as a tunnel barrier: Graphene-based magnetic tunnel junctions. *Nano Lett.*, 12(6), 3000–3004. doi: 10.1021/nl3007616.

Cobas, E., van' t Erve, O.M.J., Cheng, S.-F., Culbertson, J.C., Jernigan, G.G., Bussman, K., Jonker, B.T. (2016). Room-temperature spin filtering in metallic ferromagnet–multilayer graphene–ferromagnet junctions. *ACS Nano*, 10(11), 10357–10365. doi: 10.1021/acsnano.6b06092.

Cummings, A.W., Garcia, J.H., Fabian, J., Roche, S. (2017). Giant spin lifetime anisotropy in graphene induced by proximity effects. *Phys. Rev. Lett.*, 119, 206601.

Dahal, A., and Batzill, M. (2014). Graphene–nickel interfaces: A review. *Nanoscale*, 6(5), 2548. doi: 10.1039/c3nr05279f.

Dankert, A., Kamalakar, M.V., Wajid, A., Patel, R.S., Dash, S.P. (2014). Tunnel magnetoresistance with atomically thin two-dimensional hexagonal boron nitride barriers. *Nano Res.*, 8(4), 1357–1364. doi: 10.1007/s12274-014-0627-4.

Dankert, A., Geurs, J., Kamalakar, M.V., Charpentier, S., Dash, S.P. (2015). Room temperature electrical detection of spin polarized currents in topological insulators. *Nano Lett.*, 15, 7976–7981. doi: 10.1021/acs.nanolett.5b03080.

Dankert, A., Pashaei, P., Kamalakar, M.V., Gaur, A.P.S., Sahoo, S., Rungger, I., Narayan, A., Dolui, K., Hoque, M.A., Patel, R.S. et al. (2017). Spin-polarized tunneling through chemical vapor deposited multilayer molybdenum disulfide. *ACS Nano*, 11(6), 6389–6395. doi: 10.1021/acsnano.7b02819.

Datta, S. and Das, B. (1990). Electronic analog of the electro-optic modulator. *Appl. Phys. Lett.*, 56, 665.

Dedkov, Y.S., Fonin, M., Laubschat, C. (2008). A possible source of spin-polarized electrons: The inert graphene/Ni(111) system. *Appl. Phys. Lett.*, 92(5), 052506. doi: 10.1063/1.2841809.

Dlubak, B., Martin, M.-B., Deranlot, C., Servet, B., Xavier, S., Mattana, R., Sprinkle, M., Berger, C., Heer, W.A.D., Petroff, F. et al. (2012a). Highly efficient spin transport in epitaxial graphene on SiC. *Nat. Phy.*, 8(7), 557–561. doi: 10.1038/nphys2331.

Dlubak, B., Martin, M.-B., Weatherup, R.S., Yang, H., Deranlot, C., Blume, R., Schloegl, R., Fert, A., Anane, A., Hofmann, S. et al. (2012b). Graphene-passivated nickel as an oxidation-resistant electrode for spintronics. *ACS Nano*, 6(12), 10930–10934. doi: 10.1021/nn304424x.

Dresselhaus, G. (1954). Spin-orbit coupling effects in zinc blende structures. *Phys. Rev.*, 96(556), 280–248.

Dyrdał, A., Barnaś, J., Fert, A. (2020). Spin-momentum-locking inhomogeneities as a source of bilinear magnetoresistance in topological insulators. *Phys. Rev. Lett.*, 124, 046802. doi: 10.1103/PhysRevLett.124.046802.

Edelstein, V.M. (1990). Spin polarization of conduction electrons induced by electric current in two-dimensional asymmetric electron systems. *Solid State Commu.*, 73(3), 233–235. doi: 10.1016/0038-1098(90)90963-C.

Entani, S., Seki, T., Sakuraba, Y., Yamamoto, T., Takahashi, S., Naramoto, H., Takanashi, K., Sakai, S. (2016). Magnetoresistance effect in $Fe_{20}Ni_{80}$/graphene/$Fe_{20}Ni_{80}$ vertical spin valves. *Appl. Phys. Lett.*, 109(8), 082406. doi: 10.1063/1.4961669.

Ertler, C., Konschuh, S., Gmitra, M., Fabian, J. (2009). Electron spin relaxation in graphene: The role of the substrate. *Phys. Rev. B*, 80, 041405.

Fan, Y., Upadhyaya, P., Kou, X., Lang, M., Takei, S., Wang, Z., Tang, J., He, L., Chang, L.-T., Montazeri, M. et al. (2014). Magnetization switching through giant spin–orbit torque in a magnetically doped topological insulator heterostructure. *Nat. Mater.*, 13, 699–704. doi: 10.1038/nmat3973.

Fu, L. and Kane, C.L. (2007). Topological insulators with inversion symmetry. *Phys. Rev. B*, 76, 045302. doi: 10.1103/PhysRevB.76.045302.

Gajek, M., Bibes, M., Barthelemy, A., Bouzehouane, K., Fusil, S., Varela, M., Fontcuberta, J., Fert, A. (2005). Spin filtering through ferromagnetic BiMnO3 tunnel barriers. *Phys. Rev. B*, 72(020406(R)). doi: 10.1103/PhysRevB.72.020406.

Galbiati, M., Tatay, S., Barraud, C., Dediu, A.V., Petroff, F., Mattana, R., Seneor, P. (2014). Spinterface: Crafting spintronics at the molecular scale. *MRS Bull.*, 39(7), 602–607. doi: 10.1557/mrs.2014.131.

Galbiati, M., Tatay, S., Dubois, S.M.-M., Godel, F., Galceran, R., Mañas-Valero, S., Piquemal-Banci, M., Vecchiola, A., Charlier, J.-C., Forment-Aliaga, A. et al. (2019). Path to overcome material and fundamental obstacles in spin valves based on MoS_2 and other transition-metal dichalcogenides. *Phys. Rev. Appl.*, 12(4), 044022. doi: 10.1103/physrevapplied.12.044022.

Ganichev, S.D., Ivchenko, E.L., Bel'kov, V.V., Tarasenko, S.A., Sollinger, M., Weiss, D., Wegscheider, W., Prettl, W. (2002b). Spin-galvanic effect. *Nature*, 417, 153–156. doi: 10.1038/417153a.

Garello, K., Miron, I.M., Avci, C.O., Freimuth, F., Mokrousov, Y., Blügel, S., Auffret, S., Boulle, O., Gaudin, G., Gambardella, P. (2013). Symmetry and magnitude of spin–orbit torques in ferromagnetic heterostructures. *Nat. Nanotechnol.*, 8, 587–593. doi: 10.1038/nnano.2013.145.

Ghiasi, T.S., Ingla-Aynés, J., Kaverzin, A.A., van Wees, B.J. (2017). Large proximity-induced spin lifetime anisotropy in transition-metal dichalcogenide/graphene heterostructures. *Nano Lett.*, 17, 7528.

Ghiasi, T.S., Kaverzin, A.A., Blah, P.J., van Wees, B.J. (2019). Charge-to-spin conversion by the Rashba–Edelstein effect in two-dimensional van der Waals heterostructures up to room temperature. *Nano Lett.*, 19, 5959.

Ghosh, S. and Manchon, A. (2018). Spin-orbit torque in a three-dimensional topological insulator–ferromagnet heterostructure: Crossover between bulk and surface transport. *Phys. Rev. B*, 97, 134402. doi: 10.1103/PhysRevB.97.134402.

Gmitra, M. and Fabian, J. (2015). Graphene on transition-metal dichalcogenides: A platform for proximity spin-orbit physics and optospintronics. *Phys. Rev. B*, 92, 155403.

Godel, F., Kamalakar, M.V., Doudin, B., Henry, Y., Halley, D., Dayen, J.-F. (2014). Voltage-controlled inversion of tunnel magnetoresistance in epitaxial nickel/graphene/MgO/cobalt junctions. *Appl. Phys. Lett.*, 105(15), 152407. doi: 10.1063/1.4898587.

Godel, F., Zatko, V., Carrétéro, C., Sander, A., Galbiati, M., Vecchiola, A., Brus, P., Bezencenet, O., Servet, B., Martin, M.-B. et al. (2020). WS_2 2D semiconductor down to monolayers by pulsed-laser deposition for large-scale integration in electronics and spintronics circuits. *ACS Appl. Nano Mater.*, 3(8), 7908–7916. doi: 10.1021/acsanm.0c01408.

Grollier, J., Querlioz, D., Camsari, K.Y., Everschor-Sitte, K., Fukami, S., Stiles, M.D. (2020). Neuromorphic spintronics. *Nat. Electron.*, 3(7), 360–370. doi: 10.1038/s41928-019-0360-9.

Guillet, T., Marty, A., Beigné, C., Vergnaud, C., Dau, M.-T., Noël, P., Frigerio, J., Isella, G., Jamet, M. (2018). Magnetotransport in Bi_2Se_3 thin films epitaxially grown on Ge(111). *AIP Adv.*, 8, 115125. doi: 10.1063/1.5048547.

Guillet, T., Zucchetti, C., Barbedienne, Q., Marty, A., Isella, G., Cagnon, L., Vergnaud, C., Jaffrès, H., Reyren, N., George, J.-M. et al. (2020a). Observation of large unidirectional Rashba magnetoresistance in Ge(111). *Phys. Rev. Lett.*, 124, 027201. doi: 10.1103/PhysRevLett.124.027201.

Guillet, T., Zucchetti, C., Marchionni, A., Hallal, A., Biagioni, P., Vergnaud, C., Marty, A., Okuno, H., Masseboeuf, A., Finazzi, M. et al. (2020b). Spin orbitronics at a topological insulator-semiconductor interface. *Phys. Rev. B*, 101, 184406. doi: 10.1103/PhysRevB.101.184406.

Guillet, T., Marty, A., Vergnaud, C., Jamet, M., Zucchetti, C., Isella, G., Barbedienne, Q., Jaffrès, H., Reyren, N., George, J.-M. et al. (2021). Large Rashba unidirectional magnetoresistance in the Fe/Ge(111) interface states. *Phys. Rev. B*, 103, 064411. doi: 10.1103/PhysRevB.103.064411.

Han, J. and Liu, L. (2021). Topological insulators for efficient spin–orbit torques. *APL Mater.*, 9, 060901. doi: 10.1063/5.0048619.

Han, W., Kawakami, R.K., Gmitra, M., Fabian, J. (2014). Graphene spintronics. *Nat. Nanotechnol.*, 9, 794.

Han, J., Richardella, A., Siddiqui, S.A., Finley, J., Samarth, N., Liu, L. (2017). Room-temperature spin-orbit torque switching induced by a topological insulator. *Phys. Rev. Lett.*, 119, 077702. doi: 10.1103/PhysRevLett.119.077702.

Hasan, M.Z. and Kane, C.L. (2010). Colloquium: Topological insulators. *Rev. Mod. Phys.*, 82, 3045–3067. doi: 10.1103/RevModPhys.82.3045.

He, Q.L., Kou, X., Grutter, A.J., Yin, G., Pan, L., Che, X., Liu, Y., Nie, T., Zhang, B., Disseler, S.M. et al. (2017). Tailoring exchange couplings in magnetic topological-insulator/antiferromagnet heterostructures. *Nat. Mater.*, 16, 94–100. doi: 10.1038/nmat4783.

He, P., Walker, S.M., Zhang, S.S.-L., Bruno, F.Y., Bahramy, M.S., Lee, J.M., Ramaswamy, R., Cai, K., Heinonen, O., Vignale, G. et al. (2018a). Observation of out-of-plane spin texture in a $SrTiO_3$ two-dimensional electron gas. *Phys. Rev. Lett.*, 120, 266802. doi: 10.1103/PhysRevLett.120.266802.

He, P., Zhang, S.S.-L., Zhu, D., Liu, Y., Wang, Y., Yu, J., Vignale, G., Yang, H. (2018b). Bilinear magnetoelectric resistance as a probe of three-dimensional spin texture in topological surface states. *Nat. Phys.*, 14, 495–499. doi: 10.1038/s41567-017-0039-y.

He, H., Tai, L., Wu, D., Wu, H., Razavi, A., Wong, K., Liu, Y., Wang, K.L. (2021). Enhancement of spin-to-charge conversion efficiency in topological insulators by interface engineering. *APL Mater.*, 9, 071104. doi: 10.1063/5.0049044.

Herling, F., Safeer, C.K., Ingla-Aynés, J., Ontoso, N., Hueso, L.E., Casanova, F. (2020). Gate tunability of highly efficient spin-to-charge conversion by spin Hall effect in graphene proximitized with WSe_2. *APL Mater.*, 8, 071103.

Iqbal, M.Z., Iqbal, M.W., Lee, J.H., Kim, Y.S., Chun, S.-H., Eom, J. (2013). Spin valve effect of NiFe/graphene/NiFe junctions. *Nano Res.*, 6(5), 373–380. doi: 10.1007/s12274-013-0314-x.

Iqbal, M.Z., Iqbal, M.W., Jin, X., Hwang, C., Eom, J. (2015). Interlayer dependent polarity of magnetoresistance in graphene spin valves. *J. Mater. Chem. C*, 3(2), 298–302. doi: 10.1039/c4tc02389g.

Iqbal, M.Z., Siddique, S., Hussain, G., Iqbal, M.W. (2016). Room temperature spin valve effect in the NiFe/Gr–hBN/Co magnetic tunnel junction. *J. Mater. Chem. C*, 4(37), 8711–8715. doi: 10.1039/c6tc03425j.

Iqbal, M.Z., Hussain, G., Siddique, S., Iqbal, M.W. (2017). Graphene spin valve: An angle sensor. *J. Magn. Magn. Mater.*, 432, 135–139. doi: 10.1016/j.jmmm.2017.02.004.

Jamali, M., Lee, J.S., Jeong, J.S., Mahfouzi, F., Lv, Y., Zhao, Z., Nikolic, B.K., Mkhoyan, K.A., Samarth, N., Wang, J.-P. (2015). Giant spin pumping and inverse spin Hall effect in the presence of surface and bulk spin–orbit coupling of topological insulator Bi_2 Se_3. *Nano Lett.*, 15, 7126–7132. doi: 10.1021/acs.nanolett.5b03274.

Jin, M.J., Moon, S.Y., Park, J., Modepalli, V., Jo, J., Kim, S.-I., Koo, H.C., Min, B.-C., Lee, H.-W., Baek, S.-H., Yoo, J.-W. (2017). Nonlocal spin diffusion driven by giant spin Hall effect at oxide heterointerfaces. *Nano Lett.*, 17, 36–43. doi: 10.1021/acs.nanolett.6b03050.

Johansson, A., Göbel, B., Henk, J., Bibes, M., Mertig, I. (2020). Spin and orbital Edelstein effects in a two-dimensional electron gas: Theory and application to AlOx/SrTiO3. *Phys. Rev. Res.*, 3(013275). doi: 10.1103/PhysRevResearch.3.013275.

Johnson, M. and Silsbee, R.H. (1985). Interfacial charge-spin coupling: Injection and detection of spin magnetization in metals. *Phys. Rev. Lett.*, 55, 1790.

Julliere, M. (1975). Tunneling between ferromagnetic films. *Phys. Lett. A*, 54(3), 225–226. doi: 10.1016/0375-9601(75)90174-7.

Kampmeier, J., Borisova, S., Plucinski, L., Luysberg, M., Mussler, G., Grützmacher, D. (2015). Suppressing twin domains in molecular beam epitaxy grown Bi2Te3 topological insulator thin films. *Cryst. Growth Des.*, 15, 390–394. doi: 10.1021/cg501471z.

Kane, C.L. and Mele, E.J. (2005). Z_2 topological order and the quantum spin Hall effect. *Phys. Rev. Lett.*, 95, 146802. doi: 10.1103/PhysRevLett.95.146802.

Karpan, V.M., Giovannetti, G., Khomyakov, P.A., Talanana, M., Starikov, A.A., Zwierzycki, M., van den Brink, J., Brocks, G., Kelly, P.J. (2007). Graphite and graphene as perfect spin filters. *Phys. Rev. Lett.*, 99(17), 176602. doi: 10.1103/physrevlett.99.176602.

Karpan, V.M., Khomyakov, P.A., Starikov, A.A., Giovannetti, G., Zwierzycki, M., Talanana, M., Brocks, G., van den Brink, J., Kelly, P.J. (2008). Theoretical prediction of perfect spin filtering at interfaces between close-packed surfaces of Ni or Co and graphite or graphene. *Phys. Rev. B*, 78(19), 195419. doi: 10.1103/physrevb.78.195419.

Khan, M.F., Kim, H., Nazir, G., Jung, S., Eom, J. (2018). Layer dependent magnetoresistance of vertical MoS_2 magnetic tunnel junctions. *Nanoscale*, 10(35), 16703–16710. doi: 10.1039/c8nr04518f.

Khang, N.H.D., Ueda, Y., Hai, P.N. (2018). A conductive topological insulator with large spin Hall effect for ultralow power spin–orbit torque switching. *Nat. Mater.*, 17, 808–813. doi: 10.1038/s41563-018-0137-y.

Khokhriakov, D., Hoque, A.M., Karpiak, B., Dash, S.P. (2020). Gate-tunable spin-galvanic effect in graphene-topological insulator van der Waals heterostructures at room temperature. *Nat. Commun.*, 11, 3657. doi: 10.1038/s41467-020-17481-1.

Kim, T.H. and Moodera, J.S. (2004). Large spin polarization in epitaxial and polycrystalline Ni films. *Phys. Rev. B*, 69(2), 020403. doi: 10.1103/physrevb.69.020403.

Kim, K.-W., Lee, K.-J., Sinova, J., Lee, H.-W., Stiles, M.D. (2017). Spin-orbit torques from interfacial spin-orbit coupling for various interfaces. *Phys. Rev. B*, 96, 104438. doi: 10.1103/PhysRevB.96.104438.

Kim, Y., Kim, Y.M., Shin, J., Char, K. (2018). LaInO3/BaSnO3 polar interface on MgO substrates. *APL Mater.*, 6(096104). doi: 10.1063/1.5046368.

King, P.D.C., Hatch, R.C., Bianchi, M., Ovsyannikov, R., Lupulescu, C., Landolt, G., Slomski, B., Dil, J.H., Guan, D., Mi, J.L. et al. (2011). Large tunable Rashba spin splitting of a two-dimensional electron gas in Bi_2Se_3. *Phys. Rev. Lett.*, 107, 096802., 10.1103/PhysRevLett.107.096802.

Kochan, D., Gmitra, M., Fabian, J. (2014). Spin relaxation mechanism in graphene: Resonant scattering by magnetic impurities. *Phys. Rev. Lett.*, 112, 116602.

Kondou, K., Yoshimi, R., Tsukazaki, A., Fukuma, Y., Matsuno, J., Takahashi, K.S., Kawasaki, M., Tokura, Y., Otani, Y. (2016). Fermi-level-dependent charge-to-spin current conversion by Dirac surface states of topological insulators. *Nat. Phys.*, 12, 1027–1031. doi: 10.1038/nphys3833.

König, M., Wiedmann, S., Brüne, C., Roth, A., Buhmann, H., Molenkamp, L.W., Qi, X.-L., Zhang, S.-C. (2007). Quantum spin Hall insulator state in HgTe quantum wells. *Science*, 318, 766–770. doi: 10.1126/science.1148047.

Kornblum, L. (2019). Conductive oxide interfaces for field effect devices. *Adv. Mater. Interf.*, 6(15). doi: 10.1002/admi.201900480.

LeClair, P., Swagten, H.J.M., Kohlhepp, J.T., van de Veerdonk, R.J.M., de Jonge, W.J.M. (2000). Apparent spin polarization decay in cu-dusted $Co/Al_2O_3/Co$ tunnel junctions. *Phys. Rev. Lett.*, 84(13), 2933–2936. doi: 10.1103/physrevlett.84.2933.

Lee, J.S., Richardella, A., Hickey, D.R., Mkhoyan, K.A., Samarth, N. (2015). Mapping the chemical potential dependence of current-induced spin polarization in a topological insulator. *Phys. Rev. B*, 92, 155312. doi: 10.1103/PhysRevB.92.155312.

Lesne, E., Fu, Y., Oyarzun, S., Rojas-Sánchez, J.C., Vaz, D.C., Naganuma, H., Sicoli, G., Attané, J.-P., Jamert, M., Jacquet, E. et al. (2016). Highly efficient and tunable spin-to-charge conversion through rashba coupling at oxide interfaces. *Nat. Mater.*, 15(12), 1261–1266. doi: 10.1038/nmat4726.

Li, J., Wang, Z.Y., Tan, A., Glans, P.-A., Arenholz, E., Hwang, C., Shi, J., Qiu, Z.Q. (2012). Magnetic dead layer at the interface between a Co film and the topological insulator Bi_2Se_3. *Phys. Rev. B*, 86, 054430. doi: 10.1103/PhysRevB.86.054430.

Li, C.H., van't Erve, O.M.J., Robinson, J.T., Liu, Y., Li, L., Jonker, B.T. (2014a). Electrical detection of charge-current-induced spin polarization due to spin-momentum locking in Bi_2Se_3. *Nat. Nanotechnol.*, 9, 218–224. doi: 10.1038/nnano.2014.16.

Li, W., Xue, L., Abruña, H.D., Ralph, D.C. (2014b). Magnetic tunnel junctions with single-layer-graphene tunnel barriers. *Phys. Rev. B*, 89(18), 184418. doi: 10.1103/physrevb.89.184418.

Li, C.H., van't Erve, O.M.J., Li, Y.Y., Li, L., Jonker, B.T. (2016). Electrical detection of the helical spin texture in a p-type topological insulator Sb 2 Te 3. *Sci. Rep.*, 6, 29533. doi: 10.1038/srep29533.

Liu, L., Moriyama, T., Ralph, D.C., Buhrman, R.A. (2011). Spin-torque ferromagnetic resonance induced by the spin Hall effect. *Phys. Rev. Lett.*, 106, 036601. doi: 10.1103/PhysRevLett.106.036601.

Liu, L., Pai, C.-F., Li, Y., Tseng, H.W., Ralph, D.C., Buhrman, R.A. (2012). Spin-torque switching with the giant spin Hall effect of tantalum. *Science*, 336, 555–558. doi: 10.1126/science.1218197.

Liu, L., Richardella, A., Garate, I., Zhu, Y., Samarth, N., Chen, C.-T. (2015). Spin-polarized tunneling study of spin-momentum locking in topological insulators. *Phys. Rev. B*, 91, 235437. doi: 10.1103/PhysRevB.91.235437.

Lu, Y., Li, X.W., Gong, G.Q., Xiao, G., Gupta, A., Lecoeur, P., Sun, J.Z., Wang, Y.Y., Dravid, V.P. (1996). Large magnetotunneling effect at low magnetic fields in micrometer-scale epitaxial $La_{0.67}Sr_{0.33}MnO_3$ tunnel junctions. *Phys. Rev. B*, 54(12), R8357–R8360. doi: 10.1103/PhysRevB.54.R8357.

Luo, Y.K., Xu, J., Zhu, T., Wu, G., McCormick, E.J., Zhan, W., Neupane, M.R., Kawakami, R.K. (2017). Opto-valleytronic spin injection in monolayer MoS_2/few-layer graphene. *Nano Lett.*, 17, 3877.

Lv, Y., Kally, J., Zhang, D., Lee, J.S., Jamali, M., Samarth, N., Wang, J.-P. (2018). Unidirectional spin-Hall and Rashba–Edelstein magnetoresistance in topological insulator-ferromagnet layer heterostructures. *Nat. Commun.*, 9, 1–7. doi: 10.1038/s41467-017-02491-3.

Manchon, A., Železný, J., Miron, I.M., Jungwirth, T., Sinova, J., Thiaville, A., Garello, K., Gambardella, P. (2019). Current-induced spin-orbit torques in ferromagnetic and antiferromagnetic systems. *Rev. Mod. Phys.*, 91, 035004. doi: 10.1103/RevModPhys.91.035004.

Manipatruni, S., Nikonov, D.E., Lin, C.-C., Gosavi, T.A., Liu, H., Prasad, B., Huang, Y.-L., Bonturim, E., Ramesh, R., Young, I.A. (2019). Scalable energy-efficient magnetoelectric spin–orbit logic. *Nature*, 565(7737), 35–42. doi: 10.1038/s41586-018-0770-2.

Martin, M.-B., Dlubak, B., Weatherup, R.S., Yang, H., Deranlot, C., Bouzehouane, K., Petroff, F., Anane, A., Hofmann, S., Robertson, J. et al. (2014). Sub-nanometer atomic layer deposition for spintronics in magnetic tunnel junctions based on graphene spin-filtering membranes. *ACS Nano*, 8(8), 7890–7895. doi: 10.1021/nn5017549.

Martin, M.-B., Dlubak, B., Weatherup, R.S., Piquemal-Banci, M., Yang, H., Blume, R., Schloegl, R., Collin, S., Petroff, F., Hofmann, S. et al. (2015). Protecting nickel with graphene spin-filtering membranes: A single layer is enough. *Appl. Phys. Lett.*, 107(1), 012408. doi: 10.1063/1.4923401.

Mendes, J.B.S., Alves Santos, O., Holanda, J., Loreto, R.P., de Araujo, C.I.L., Chang, C.-Z., Moodera, J.S., Azevedo, A., Rezende, S.M. (2017). Dirac-surface-state-dominated spin to charge current conversion in the topological insulator $(Bi_{0.22}Sb_{0.78})_2Te_3$ films at room temperature. *Phys. Rev. B*, 96, 180415. doi: 10.1103/PhysRevB.96.180415.

Meng, J., Chen, J.-J., Yan, Y., Yu, D.-P., Liao, Z.-M. (2013). Vertical graphene spin valve with Ohmic contacts. *Nanoscale*, 5(19), 8894. doi: 10.1039/c3nr03168c.

Miron, I.M., Gaudin, G., Auffret, S., Rodmacq, B., Schuhl, A., Pizzini, S., Vogel, J., Gambardella, P. (2010). Current-driven spin torque induced by the Rashba effect in a ferromagnetic metal layer. *Nat. Mater.*, 9, 230–234. doi: 10.1038/nmat2613.

Miron, I.M., Garello, K., Gaudin, G., Zermatten, P.-J., Costache, M.V., Auffret, S., Bandiera, S., Rodmacq, B., Schuhl, A., Gambardella, P. (2011). Perpendicular switching of a single ferromagnetic layer induced by in-plane current injection. *Nature*, 476, 189–193. doi: 10.1038/nature10309.

Mogi, M., Nakajima, T., Ukleev, V., Tsukazaki, A., Yoshimi, R., Kawamura, M., Takahashi, K.S., Hanashima, T., Kakurai, K., Arima, T. et al. (2019). Large anomalous Hall effect in topological insulators with proximitized ferromagnetic insulators. *Phys. Rev. Lett.*, 123, 016804. doi: 10.1103/PhysRevLett.123.016804.

Mohiuddin, T.M.G., Hill, E., Elias, D., Zhukov, A., Novoselov, K., Geim, A. (2008). Graphene in multilayered CPP spin valves. *IEEE Trans. Magn.*, 44(11), 2624–2627. doi: 10.1109/tmag.2008.2003065.

Myers, E.B., Ralph, D.C., Katine, J.A., Louie, R.N., Buhrman, R.A. (1999). Current-induced switching of domains in magnetic multilayer devices. *Science*, 285(5429), 867–870. doi: 10.1126/science.285.5429.867.

Naganuma, H., Zatko, V., Galbiati, M., Godel, F., Sander, A., Carrétéro, C., Bezencenet, O., Reyren, N., Martin, M.-B., Dlubak, B. et al. (2020). A perpendicular graphene/ ferromagnet electrode for spintronics. *Appl. Phys. Lett.*, 116(17), 173101. doi: 10.1063/ 1.5143567.

Nakagawa, N., Hwang, H.Y., Muller, D.A. (2006). Why some interfaces cannot be sharp. *Nat. Mater.*, 5(3), 204–209. doi: 10.1038/nmat1569.

Noël, P., Thomas, C., Fu, Y., Vila, L., Haas, B., Jouneau, P.-H., Gambarelli, S., Meunier, T., Ballet, P., Attané, J.P. (2018). Highly efficient spin-to-charge current conversion in strained HgTe surface states protected by a HgCdTe layer. *Phys. Rev. Lett.*, 120, 167201. doi: 10.1103/PhysRevLett.120.167201.

Noël, P., Trier, F., Vicente Arche, L.M., Bréhin, J., Vaz, D.C., Garcia, V., Fusil, S., Barthélémy, A., Vila, L., Bibes, M., Attané, J.-P. (2020). Non-volatile electric control of spin–charge conversion in a SrTiO3 Rashba system. *Nature*, 580(7804), 483–486. doi: 10.1038/s41586-020-2197-9.

Novoselov, K.S., Geim, A.K., Morozov, S.V., Jiang, D., Zhang, Y., Dubonos, S.V., Grigorieva, I.V., Firsov, A.A. (2004). Electric field effect in atomically thin carbon films. *Science*, 306(5696), 666–669. doi: 10.1126/science.1102896.

Ohtomo, A. and Hwang, H.Y. (2004). A high-mobility electron gas at the LaAlO3/SrTiO3 heterointerface. *Nature*, 427(January), 423–427.

Ohtomo, A., Muller, D.A., Grazul, J.L., Hwang, H.Y. (2002a). Artificial charge-modulation in atomic-scale perovskite titanate superlattices. *Nature*, 419(September), 378.

Ohtomo, A., Muller, D.A., Grazul, J.L., Hwang, H.Y. (2002b). Epitaxial growth and electronic structure of LaTiOx films epitaxial growth and electronic structure of LaTiOx films. *Appl. Phys. Lett.*, 80(21), 392. doi: 10.1063/1.1481767.

Park, J.-H. and Lee, H.-J. (2014). Out-of-plane magnetoresistance in ferromagnet/graphene/ ferromagnet spin-valve junctions. *Phys. Rev. B*, 89(16). doi: 10.1103/physrevb.89.165417.

Piquemal-Banci, M., Galceran, R., Caneva, S., Martin, M.-B., Weatherup, R.S., Kidambi, P.R., Bouzehouane, K., Xavier, S., Anane, A., Petroff, F. et al. (2016). Magnetic tunnel junctions with monolayer hexagonal boron nitride tunnel barriers. *Appl. Phys. Lett.*, 108(10), 102404. doi: 10.1063/1.4943516.

Piquemal-Banci, M., Galceran, R., Martin, M.-B., Godel, F., Anane, A., Petroff, F., Dlubak, B., Seneor, P. (2017). 2D-MTJs: Introducing 2D materials in magnetic tunnel junctions. *J. Phys. D: Appl. Phys.*, 50(20), 203002. doi: 10.1088/1361-6463/aa650f.

Piquemal-Banci, M., Galceran, R., Godel, F., Caneva, S., Martin, M.-B., Weatherup, R.S., Kidambi, P.R., Bouzehouane, K., Xavier, S., Anane, A. et al. (2018). Insulator-to-metallic spin-filtering in 2D-magnetic tunnel junctions based on hexagonal boron nitride. *ACS Nano*, 12(5), 4712–4718. doi: 10.1021/acsnano.8b01354.

Piquemal-Banci, M., Galceran, R., Dubois, S.M.-M., Zatko, V., Galbiati, M., Godel, F., Martin, M.-B., Weatherup, R.S., Petroff, F., Fert, A. et al. (2020). Spin filtering by proximity effects at hybridized interfaces in spin-valves with 2D graphene barriers. *Nat. Commun.*, 11(1). doi: 10.1038/s41467-020-19420-6.

Prenat, G., Jabeur, K., Vanhauwaert, P., Pendina, G.D., Oboril, F., Bishnoi, R., Ebrahimi, M., Lamard, N., Boulle, O., Garello, K. et al. (2016). Ultra-fast and high-reliability SOT-MRAM: From cache replacement to normally-off computing. *IEEE Trans. Multi-Scale Comput. Syst.*, 2, 49–60. doi: 10.1109/TMSCS.2015.2509963.

Raes, B., Scheerder, J.E., Costache, M.V., Bonell, F., Sierra, J.F., Cuppens, J., Van de Vondel, J., Valenzuela, S.O. (2016). Determination of the spin-lifetime anisotropy in graphene using oblique spin precession. *Nat. Commun.*, 7, 11444.

Raes, B., Cummings, A.W., Bonell, F., Coastache, M.V., Sierra, J.F., Roche, S., Valenzuela, S.O. (2017). Spin precession in anisotropic media. *Phys. Rev. B*, 95, 085403.

Ramesh, R. and Spaldin, N.A. (2007). Multiferroics: Progress and prospects in thin films. *Nat. Mater.*, 3(January), 21–29.

Reifsnyder Hickey, D., Azadani, J.G., Richardella, A.R., Kally, J.C., Lee, J.S., Chang, H., Liu, T., Wu, M., Samarth, N., Low, T. et al. (2019). Structure and basal twinning of topological insulator Bi_2Se_3 grown by MBE onto crystalline $Y_3Fe_5O_{12}$. *Phys. Rev. Mater.*, 3, 061201. doi: 10.1103/PhysRevMaterials.3.061201.

Reina, A., Jia, X., Ho, J., Nezich, D., Son, H., Bulovic, V., Dresselhaus, M.S., Kong, J. (2009). Large area, few-layer graphene films on arbitrary substrates by chemical vapor deposition. *Nano Lett.*, 9(1), 30–35. doi: 10.1021/nl801827v.

Reinle-Schmitt, M.L., Cancellieri, C., Li, D., Fontaine, D., Medarde, M., Pomjakushina, E., Schneider, C.W., Gariglio, S., Ghosez, P., Triscone, J.-M., Willmott, P.R. (2012). Tunable conductivity threshold at polar oxide interfaces. *Nat. Commun.*, 3(932). doi: 10.1038/ncomms1936.

Richardella, A., Kandala, A., Lee, J.S., Samarth, N. (2015). Characterizing the structure of topological insulator thin films. *APL Mater.*, 3, 083303. doi: 10.1063/1.4926455.

Rödel, T.C., Fortuna, F., Sengupta, S., Frantzeskakis, E., Le Fèvre, P., Bertran, F., Mercey, B., Matzen, S., Agnus, G., Maroutian, T., Lecoeur, P., Santander-Syro, A.F. (2016). Universal fabrication of 2D electron systems in functional oxides. *Adv. Mater.*, 28(10), 1976–1980. doi: 10.1002/adma.201505021.

Rodriguez-Vega, M., Schwiete, G., Sinova, J., Rossi, E. (2017). Giant Edelstein effect in topological-insulator–graphene heterostructures. *Phys. Rev. B*, 96, 235419. doi: 10.1103/PhysRevB.96.235419.

Rojas-Sánchez, J.-C., Oyarzún, S., Fu, Y., Marty, A., Vergnaud, C., Gambarelli, S., Vila, L., Jamet, M., Ohtsubo, Y., Taleb-Ibrahimi, A. et al. (2016). Spin to charge conversion at room temperature by spin pumping into a new type of topological insulator: α-Sn films. *Phys. Rev. Lett.*, 116, 096602. doi: 10.1103/PhysRevLett.116.096602.

Rougemaille, N., N'Diaye, A.T., Coraux, J., Vo-Van, C., Fruchart, O., Schmid, A.K. (2012). Perpendicular magnetic anisotropy of cobalt films intercalated under graphene. *Appl. Phys. Lett.*, 101(14), 142403. doi: 10.1063/1.4749818.

Safeer, C.K., Ingla-Aynés, J., Herling, F., Garcia, J.H., Vila, M., Ontoso, N., Calvo, M.R., Roche, S., Hueso, L.E., Casanova, F. (2019a). Room-temperature spin Hall effect in graphene/MoS_2 van der Waals heterostructures. *Nano Lett.*, 19, 1074.

Safeer, C.K., Ontoso, N., Ingla-Aynés, J., Herling, F., Pham, V.T., Kurzmann, A., Ensslin, K., Chuvilin, A., Robredo, I., Vergniory, M.G. et al. (2019b). Large multidirectional spin-to-charge conversion in low-symmetry semimetal $MoTe_2$ at room temperature. *Nano Lett.*, 19, 8758.

Sánchez, J.C.R., Vila, L., Desfonds, G., Gambarelli, S., Attané, J.P., De Teresa, J.M., Magén, C., Fert, A. (2013). Spin-to-charge conversion using Rashba coupling at the interface between non-magnetic materials. *Nat. Commun.*, 4, 2944. doi: 10.1038/ncomms3944.

Scanlon, D.O., King, P.D.C., Singh, R.P., de la Torre, A., Walker, S.M., Balakrishnan, G., Baumberger, F., Catlow, C.R.A. (2012). Controlling bulk conductivity in topological insulators: Key role of anti-site defects. *Adv. Mater.*, 24, 2154–2158. doi: 10.1002/adma.201200187.

Scheiderer, P., Pfaff, F., Gabel, J., Kamp, M., Sing, M., Claessen, R. (2015). Surface-interface coupling in an oxide heterostructure: Impact of adsorbates on LaAlO3/SrTiO3. *Phys. Rev. B*, 92(195422). doi: 10.1103/PhysRevB.92.195422.

Seneor, P., Dlubak, B., Martin, M.-B., Anane, A., Jaffres, H., Fert, A. (2012). Spintronics with graphene. *MRS Bull.*, 37(12), 1245–1254. doi: 10.1557/mrs.2012.277.

Sharma, P.R., Gautam, P., Nam, J., Kim, K.S., Noh, H. (2020). Magnetoresistance effect in a vertical spin valve fabricated with a dry-transferred CVD graphene and a resist-free process. *Mater. Res. Express*, 7(8), 085603. doi: 10.1088/2053-1591/abad03.

Shen, K., Vignale, G., Raimondi, R. (2014). Microscopic theory of the inverse Edelstein effect. *Phys. Rev. Lett.*, 112, 096601. doi: 10.1103/PhysRevLett.112.096601.

Shi, S., Wang, A., Wang, Y., Ramaswamy, R., Shen, L., Moon, J., Zhu, D., Yu, J., Oh, S., Feng, Y., Yang, H. (2018). Efficient charge-spin conversion and magnetization switching through the Rashba effect at topological-insulator/Ag interfaces. *Phys. Rev. B*, 97, 041115. doi: 10.1103/PhysRevB.97.041115.

Sinova, J., Valenzuela, S.O., Wunderlich, J., Back, C.H., Jungwirth, T. (2015). Spin Hall effects. *Rev. Mod. Phys.*, 87, 1213.

Tang, J., Chang, L.-T., Kou, X., Murata, K., Choi, E.S., Lang, M., Fan, Y., Jiang, Y., Montazeri, M., Jiang, W. et al. (2014). Electrical detection of spin-polarized surface states conduction in (Bi0.53Sb0.47)2Te3 topological insulator. *Nano Lett.*, 14, 5423–5429. doi: 10.1021/nl5026198.

Tang, C., Chang, C.-Z., Zhao, G., Liu, Y., Jiang, Z., Liu, C.-X., McCartney, M.R., Smith, D.J., Chen, T., Moodera, J.S. et al. (2017). Above 400-K robust perpendicular ferromagnetic phase in a topological insulator. *Sci. Adv.*, 3, e1700307. doi: 10.1126/sciadv.1700307.

Tarakina, N.V., Schreyeck, S., Luysberg, M., Grauer, S., Schumacher, C., Karczewski, G., Brunner, K., Gould, C., Buhmann, H., Dunin-Borkowski, R.E. et al. (2014). Suppressing twin formation in Bi2Se3 thin films. *Adv. Mater. Inter.*, 1, 1400134. doi: 10.1002/admi.201400134.

Tombros, N., Jozsa, C., Popinciuc, M., Jonkman, H.T., van Wees, B.J. (2007). Electronic spin transport and spin precession in single graphene layers at room temperature. *Nature*, 448(7153), 571–574. doi: 10.1038/nature06037.

Trier, F., Vaz, D.C., Bruneel, P., Noël, P., Fert, A., Vila, L., Attané, J.-P., Barthélémy, A., Gabay, M., Jaffrès, H., Bibes, M. (2020). Electric-field control of spin current generation and detection in ferromagnet-free SrTiO3-Based nanodevices. *Nano Lett.*, 20(1), 395–401. doi: 10.1021/acs.nanolett.9b04079.

Tserkovnyak, Y., Brataas, A., Bauer, G.E.W. (2002). Enhanced gilbert damping in thin ferromagnetic films. *Phys. Rev. Lett.*, 88, 117601. doi: 10.1103/PhysRevLett.88.117601.

Tsymbal, E.Y. and Kohlstedt, H. (2006). Tunneling across a ferroelectric. *Science*, 313(5784), 181–183. doi: 10.1126/science.1126230.

Tsymbal, E.Y. and Zutic, I. (eds) (2011). *Handbook of Spin Transport and Magnetism*. CRC Press, New York.

Tsymbal, E.Y., Mryasov, O.N., LeClair, P.R. (2003). Spin-dependent tunnelling in magnetic tunnel junctions. *Journal of Physics: Condensed Matter*, 15(4), R109–R142. doi: 10.1088/0953-8984/15/4/201.

Tuan, D.V., Soriano, D., Ortmann, F., Valenzuela, S.O., Roche, S. (2014). Pseudospin-driven spin relaxation mechanism in graphene. *Nat. Phys.*, 10, 857.

Vaklinova, K., Hoyer, A., Burghard, M., Kern, K. (2016). Current-induced spin polarization in topological insulator–graphene heterostructures. *Nano Lett.*, 16, 2595–2602. doi: 10.1021/acs.nanolett.6b00167.

Valenzuela, S.O. (2009). Non local electronic spin detection, spin accumulation and the spin Hall effect. *Int. J. Mod. Phys.*, B 23, 2413.

Vas'ko, F.T. (1979). Spin splitting in the spectrum of two-dimensional electrons due to the surface potential. *JETP Lett.*, 541 [Online]. Available at: http://www.jetpletters.ac.ru/ps/1368/article_20698.pdf.

Vaz, D.C. (2018). Spin-to-charge current conversion in SrTiO3-based two-dimensional electron gases. PhD thesis, Sorbonne Université, Paris.

Vaz, D.C., Lesne, E., Sander, A., Naganuma, H., Jacquet, E., Santamaria, J., Barthélémy, A., Bibes, M. (2017). Tuning up or down the critical thickness in LaAlO3/SrTiO3 through in situ deposition of metal overlayers. *Adv. Mater.*, 29(28), 1700486. doi: 10.1002/adma.201700486.

Vaz, D.C., Barthélémy, A., Bibes, M. (2018). Oxide spin-orbitronics: New routes towards low-power electrical control of magnetization in oxide heterostructures. *Jpn. J. Appl. Phys*, 57(9). doi: 10.7567/JJAP.57.0902A4.

Vaz, D.C., Noël, P., Johansson, A., Göbel, B., Bruno, F.Y., Singh, G., McKeown-Walker, S., Trier, F., Vicente-Arche, L.M., Sander, A. et al. (2019). Mapping spin–charge conversion to the band structure in a topological oxide two-dimensional electron gas. *Nat. Mater.*, 18(11), 1187–1193. doi: 10.1038/s41563-019-0467-4.

Vaz, D.C., Trier, F., Dyrdał, A., Johansson, A., Garcia, K., Barthélémy, A., Mertig, I., Barnaś, J., Fert, A., Bibes, M. (2020). Determining the Rashba parameter from the bilinear magnetoresistance response in a two-dimensional electron gas. *Phys. Rev. Mater.*, 4, 071001. doi: 10.1103/PhysRevMaterials.4.071001.

ViolBarbosa, C.E., Shekhar, C., Yan, B., Ouardi, S., Ikenaga, E., Fecher, G.H., Felser, C. (2013). Direct observation of band bending in the topological insulator Bi_2Se_3. *Phys. Rev. B*, 88, 195128. doi: 10.1103/PhysRevB.88.195128.

Voerman, J.A., Li, C., Huang, Y., Brinkman, A. (2019). Spin-momentum locking in the gate tunable topological insulator BiSbTeSe2 in non-local transport measurements. *Adv. Electron. Mater.*, 5, 1900334. doi: 10.1002/aelm.201900334.

de Vries, E.K., Kamerbeek, A.M., Koirala, N., Brahlek, M., Salehi, M., Oh, S., van Wees, B.J., Banerjee, T. (2015). Towards the understanding of the origin of charge-current-induced spin voltage signals in the topological insulator Bi_2Se_3. *Phys. Rev. B*, 92, 201102. doi: 10.1103/PhysRevB.92.201102.

Walsh, L.A., Smyth, C.M., Barton, A.T., Wang, Q., Che, Z., Yue, R., Kim, J., Kim, M.J., Wallace, R.M., Hinkle, C.L. (2017). Interface chemistry of contact metals and ferromagnets on the topological insulator Bi_2Se_3. *J. Phys. Chem. C*, 121, 23551–23563. doi: 10.1021/acs.jpcc.7b08480.

Wang, W., Narayan, A., Tang, L., Dolui, K., Liu, Y., Yuan, X., Jin, Y., Wu, Y., Rungger, I., Sanvito, S., Xiu, F. (2015a). Spin-valve effect in NiFe/MoS_2 /NiFe junctions. *Nano Lett.*, 15(8), 5261–5267. doi: 10.1021/acs.nanolett.5b01553.

Wang, Y., Deorani, P., Banerjee, K., Koirala, N., Brahlek, M., Oh, S., Yang, H. (2015b). Topological surface states originated spin-orbit torques in Bi_2Se_3. *Phys. Rev. Lett.*, 114, 257202. doi: 10.1103/PhysRevLett.114.257202.

Wang, H., Kally, J., Lee, J.S., Liu, T., Chang, H., Hickey, D.R., Mkhoyan, K.A., Wu, M., Richardella, A., Samarth, N. (2016). Surface-state-dominated spin-charge current conversion in topological-insulator--ferromagnetic-insulator heterostructures. *Phys. Rev. Lett.*, 117, 076601. doi: 10.1103/PhysRevLett.117.076601.

Wang, Y., Ramaswamy, R., Motapothula, M., Narayanapillai, K., Zhu, D., Yu, J., Venkatesan, T., Yang, H. (2017). Room-temperature giant charge-to-spin conversion at the SrTiO3-LaAlO3 oxide interface. *Nano Lett.*, 17(12), 7659–7664. doi: 10.1021/acs.nanolett.7b03714.

Wang, X., Cheng, L., Zhu, D., Wu, Y., Chen, M., Wang, Y., Zhao, D., Boothroyd, C.B., Lam, Y.M., Zhu, J.-X. et al. (2018). Ultrafast spin-to-charge conversion at the surface of topological insulator thin films. *Adv. Mater.*, 30, 1802356. doi: 10.1002/adma.201802356.

Wang, H., Liu, Y., Wu, P., Hou, W., Jiang, Y., Li, X., Pandey, C., Chen, D., Yang, Q., Wang, H. et al. (2020). Above room-temperature ferromagnetism in wafer-scale two-dimensional van der Waals Fe3GeTe2 tailored by a topological insulator. *ACS Nano*, 14, 10045–10053. doi: 10.1021/acsnano.0c03152.

Weatherup, R.S., Dlubak, B., Hofmann, S. (2012). Kinetic control of catalytic CVD for high-quality graphene at low temperatures. *ACS Nano*, 6(11), 9996–10003. doi: 10.1021/nn303674g.

Weatherup, R.S., Amara, H., Blume, R., Dlubak, B., Bayer, B.C., Diarra, M., Bahri, M., Cabrero-Vilatela, A., Caneva, S., Kidambi, P.R. et al. (2014). Interdependency of subsurface carbon distribution and graphene–catalyst interaction. *J. Am. Chem. Soc.*, 136(39), 13698–13708. doi: 10.1021/ja505454v.

Willmott, P.R., Pauli, S.A., Herger, R., Schlepütz, C.M., Martoccia, D., Patterson, B.D., Delley, B., Clarke, R., Kumah, D., Cionca, C., Yacoby, Y. (2007). Structural basis for the conducting interface between LaAlO3 and SrTiO3. *Phys. Rev. Lett.*, 99(155502). doi: 10.1103/PhysRevLett.99.155502.

Wu, H., Xu, Y., Deng, P., Pan, Q., Razavi, S.A., Wong, K., Huang, L., Dai, B., Shao, Q., Yu, G. et al. (2019a). Spin-orbit torque switching of a nearly compensated ferrimagnet by topological surface states. *Adv. Mater.*, 31, 1901681. doi: 10.1002/adma.201901681.

Wu, H., Zhang, P., Deng, P., Lan, Q., Pan, Q., Razavi, S.A., Che, X., Huang, L., Dai, B., Wong, K. et al. (2019b). Room-temperature spin-orbit torque from topological surface states. *Phys. Rev. Lett.*, 123, 207205. 10.1103/PhysRevLett.123.207205.

Yan, W., Txoperena, O., Llopis, R., Dery, H., Hueso, L.E., Casanova, F. (2016). A two-dimensional spin field-effect switch. *Nat. Commun.*, 7, 13372.

Yang, H., Vu, A.D., Hallal, A., Rougemaille, N., Coraux, J., Chen, G., Schmid, A.K., Chshiev, M. (2015). Anatomy and giant enhancement of the perpendicular magnetic anisotropy of cobalt–graphene heterostructures. *Nano Lett.*, 16(1), 145–151. doi: 10.1021/acs.nanolett.5b03392.

Yang, C.-Y., Pan, L., Grutter, A.J., Wang, H., Che, X., He, Q.L., Wu, Y., Gilbert, D.A., Shafer, P., Arenholz, E. et al. (2020). Termination switching of antiferromagnetic proximity effect in topological insulator. *Sci. Adv.*, 6, eaaz8463. doi: 10.1126/sciadv.aaz8463.

Yasuda, K., Tsukazaki, A., Yoshimi, R., Takahashi, K.S., Kawasaki, M., Tokura, Y. (2016). Large unidirectional magnetoresistance in a magnetic topological insulator. *Phys. Rev. Lett.*, 117, 127202. doi: 10.1103/PhysRevLett.117.127202.

Yasuda, K., Tsukazaki, A., Yoshimi, R., Kondou, K., Takahashi, K.S., Otani, Y., Kawasaki, M., Tokura, Y. (2017). Current-nonlinear Hall effect and spin-orbit torque magnetization switching in a magnetic topological insulator. *Phys. Rev. Lett.*, 119, 137204. doi: 10.1103/PhysRevLett.119.137204.

Yoshimi, R., Tsukazaki, A., Kikutake, K., Checkelsky, J.G., Takahashi, K.S., Kawasaki, M., Tokura, Y. (2014). Dirac electron states formed at the heterointerface between a topological insulator and a conventional semiconductor. *Nat. Mater.*, 13, 253–257. doi: 10.1038/nmat3885.

Yu, L. and Zunger, A. (2014). A polarity-induced defect mechanism for conductivity and magnetism at polar-nonpolar oxide interfaces. *Nat. Commun.*, 5(5118). doi: 10.1038/ncomms6118.

Zatko, V., Galbiati, M., Dubois, S.M.-M., Och, M., Palczynski, P., Mattevi, C., Brus, P., Bezencenet, O., Martin, M.-B., Servet, B. et al. (2019). Band-structure spin-filtering in vertical spin valves based on chemical vapor deposited WS\lesssub\greater2\less/sub\greater. *ACS Nano*, 13(12), 14468–14476. doi: 10.1021/acsnano.9b08178.

Zatko, V., Dubois, S.M.-M., Godel, F., Carrétéro, C., Sander, A., Collin, S., Galbiati, M., Peiro, J., Panciera, F., Patriarche, G. et al. (2021). Band-gap landscape engineering in large-scale 2D semiconductor van der Waals heterostructures. *ACS Nano*, 15(4), 7279–7289. doi: 10.1021/acsnano.1c00544.

Zhang, S. and Fert, A. (2016). Conversion between spin and charge currents with topological insulators. *Phys. Rev. B*, 94, 184423. doi: 10.1103/PhysRevB.94.184423.

Zhang, Y., He, K., Chang, C.-Z., Song, C.-L., Wang, L.-L., Chen, X., Jia, J.-F., Fang, Z., Dai, X., Shan, W.-Y. et al. (2010). Crossover of the three-dimensional topological insulator Bi 2 Se 3 to the two-dimensional limit. *Nat. Phys.*, 6, 584–588. doi: 10.1038/nphys1689.

Zhang, G., Qin, H., Chen, J., He, X., Lu, L., Li, Y., Wu, K. (2011). Growth of topological insulator Bi_2Se_3 thin films on SrTiO3 with large tunability in chemical potential. *Adv. Funct. Mater.*, 21, 2351–2355. doi: 10.1002/adfm.201002667.

Zhao, K., Xing, Y., Han, J., Feng, J., Shi, W., Zhang, B., Zeng, Z. (2017). Magnetic transport property of NiFe/WSe2/NiFe spin valve structure. *J. Magn. Magn. Mater.*, 432, 10–13. doi: 10.1016/j.jmmm.2017.01.066.

Zhao, B., Karpiak, B., Khokhriakov, D., Johansson, A., Hoque, A., Xu, X., Jiang, Y., Mertig, I., Dash, S.P. (2020). Unconventional charge–spin conversion in Weyl-Semimetal WTe_2. *Adv. Mater.*, 32, 2000818.

Zhou, G., Tang, G., Li, T., Pan, G., Deng, Z., Zhang, F. (2017). Graphene-passivated cobalt as a spin-polarized electrode: Growth and application to organic spintronics. *J. Phys. D: Appl. Phys.*, 50(9), 095001. doi: 10.1088/1361-6463/aa5445.

Zhu, Z., Cheng, Y., Schwingenschlögl, U. (2012). Band inversion mechanism in topological insulators: A guideline for materials design. *Phys. Rev. B*, 85, 235401. doi: 10.1103/PhysRevB.85.235401.

Zou, K., Ismail-Beigi, S., Kisslinger, K., Shen, X., Su, D., Walker, F.J., Ahn, C.H. (2015). LaTiO3/KTaO3 interfaces: A new two-dimensional electron gas system. *APL Mater.*, 3(036104). doi: 10.1063/1.4914310.

Žutić, I., Fabian, J., Sarma, S.D. (2004). Spintronics: Fundamentals and applications. *Rev. Mod. Phys.*, 76(2), 323–410. doi: 10.1103/revmodphys.76.323.

6

Valleytronics in 2D Materials

Steven A. VITALE
MIT Lincoln Laboratory, Lexington, Massachusetts, USA

6.1. Introduction

The isolation of high-quality monolayer two-dimensional materials has provided an unprecedented opportunity to explore the unique physics of 2D solid-state systems. The seminal work on graphene (Novoselov et al. 2004, 2005; Zhang et al. 2005) inspired several thousand publications over the following five years and over 100,000 in the subsequent decade. The impact of graphene on the physics of solid-state materials cannot be overstated, but this chapter would not be included in a book of potential Beyond CMOS technologies if graphene were the only known 2D material. In 2010, isolation and optical characterization of monolayer molybdenum disulfide changed the landscape of 2D materials (Mak et al. 2010; Splendiani et al. 2010). Figure 6.1 shows the atomic structure of a typical MX_2 van der Waals material where M is a transition metal atom (e.g. Mo, W) and X is a chalcogen (S, Se or Te). In a van der Waals material, the crystal forms sheets with no covalent or ionic bonding between layers. The layers are held together only by van der Waals forces. Since the in-plane bonding is strong and the out-of-plane bonding is weak, individual layers may be peeled off, as with the famous "Scotch tape" exfoliation technique used to isolate monolayer graphene from a crystal of graphite (Novoselov et al. 2004).

This new class of materials, the monolayer Transition Metal Dichalcogenides (TMDs), opened the possibility for novel electronic devices based on 2D materials.

For a color version of all figures in this chapter, see www.iste.co.uk/cresti/beyondcmos.zip.

Beyond-CMOS, coordinated by Alessandro CRESTI.

Importantly, monolayer TMDs, unlike their bulk counterparts, are a direct gap material with the bandgap conveniently located within the visible range. But what truly differentiates 2D TMDs is the ability to address and manipulate the *valley pseudospin* of real and quasi-particles within the material. This chapter focuses on this aspect of 2D TMDs and their potential to transform microelectronic architectures, and we will also consider how graphene may play a role.

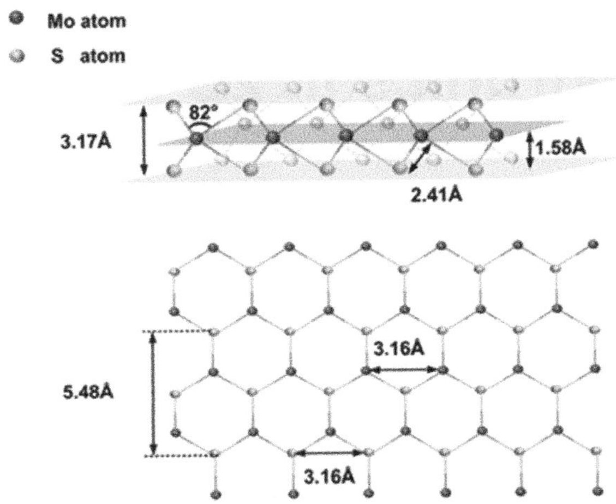

Figure 6.1. *Crystal structure of monolayer MoS_2. Reproduced with permission (Li et al. 2016). Copyright 2016, Taylor & Francis Group*

The chapter is organized as follows. Section 6.2 provides an introduction to excitons and valley physics. Section 6.3 discusses valley lifetime, information transport, and operations, and introduces key concepts such as valley polarization. Section 6.4 surveys the state of valleytronic materials and device demonstrations. Section 6.5 draws upon all of the earlier material to assess the state of valleytronic computing and the remaining challenges to be overcome to establish valleytronics as a practical information processing paradigm. This chapter is not a complete review of the literature, but instead presents the background necessary to understand valleytronic phenomena and the outlook for valleytronic computing.

6.2. Exciton and valley physics

The first use of the word "valleytronics" of which the author is aware is in Rycerz et al. (2007) who used it to describe a theoretical graphene-based device. Just as "electronics" implies manipulation of electrons for some useful purpose, valleytronics implies manipulation of valley. But what does that mean? First, we will introduce the concept of a valley and then turn to a discussion of excitons, which provides the background for a deeper dive into valley physics.

6.2.1. *Introduction to valleys and excitons*

The band structure of a canonical 2D material, MoS_2, is shown in Figure 6.2a. The horizontal axis of the band structure diagram represents a certain path through the Brillouin zone of the material. The full Brillouin zone representation of 2D material requires a 3D plot as shown in Figure 6.2b. Such a representation is visually helpful but usually not easy to interpret quantitatively so we will focus on the "slice" of the full Brillouin zone as shown in Figure 6.2a. A *valley* is a point of extrema in the bands. The absolute minimum energy points in the conduction band and the valence band are the most important ones and typically when we refer to a valley, it is only those two points that are relevant. Note that in the valence band, because the carriers are positively charged holes, the minimum energy is higher on the band diagram.

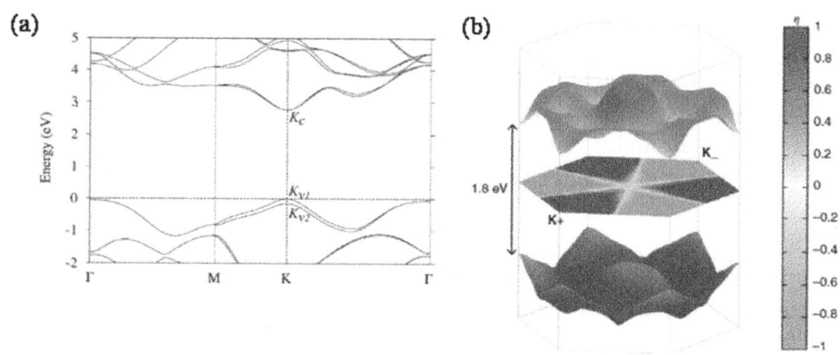

Figure 6.2. *Band structure of monolayer MoS_2 (a) along the high symmetry points. Reproduced with permission (Cheiwchanchamnangij and Lambrecht 2012). Copyright 2012, American Physical Society. (b) Across the entire Brillouin zone. Reproduced with permission (Cao et al. 2012). Copyright 2012, Springer Nature*

Figure 6.3 shows the band structure of MoS_2 in bulk, four-layer, two-layer and monolayer form. Valleys exist in all of these structures, but the band structure evolves as the material is thinned down. The dramatic turn of events that initiated the field of valleytronics was the discovery (Mak et al. 2010; Splendiani et al. 2010) that monolayer MoS_2 is a direct gap material, meaning the minimum energy points of the valence and conduction bands occur at the same location in momentum space, labeled as the K point. This implies that we can efficiently create charge carriers at that specific point in momentum space (valley) through photoexcitation, and that efficient recombination of those carriers will also occur only at that valley.

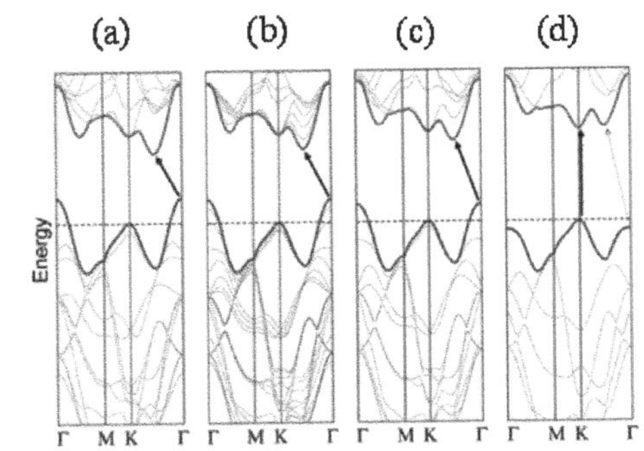

Figure 6.3. *Calculated band diagram of MoS_2 in (a) bulk, (b) four-layer, (c) bilayer, (d) monolayer films. Reproduced with permission (Splendiani et al. 2010). Copyright 2010, American Chemical Society*

Though carrier recombination and photoemission is possible in so-called indirect gap materials, where the band extrema occur at different points in the Brillouin zone (silicon is a good example), the emission is very inefficient, because momentum conservation requires a coincident phonon of the correct momentum; an unlikely event.

This brings us to a discussion of excitons. Excitons are bound electron hole pairs, as illustrated schematically in Figure 6.4a. When a photon with energy greater than or equal to the bandgap is absorbed by a semiconductor material, an electron is promoted from the valence band to the conduction band. In some materials, such as silicon, the electron and hole tend to drift away from each other and become free carriers; thus, silicon is a good photodetector. In other materials, loosely termed excitonic materials, the electron and hole tend to co-exist in a bound state called an exciton. The exciton is often treated as a single quasi-particle though it is a hydrogenic state with overlap between the electron and hole wavefunctions. The exciton quasi-particle is a boson, and is typically in a singlet ($s_z = 0$) or triplet ($s_z = -1, 0, +1$) state. For the $s_z = \pm 1$ cases, the exciton spin is $\mathbf{S} = \pm\hat{\mathbf{z}}$, where the z-axis is perpendicular to the plane of the 2D material.

An exciton can be visualized as an electron and hole co-rotating around their common center of mass, not unlike a hydrogen atom, though in the case of an exciton the electron and hole effective masses are roughly equal. Excitons exhibit a Rydberg series of hydrogenic energy states: 1s, 2s, etc. as illustrated in Figure 6.4d. There are several researchers currently studying valleytronic effects in the higher energy states (Goryca et al. 2019), but for our discussion, we need only consider the 1s state. If the excitation photon has energy greater than the bandgap, the excess energy will either impart kinetic energy to the exciton (a hot exciton) or it will dissipate into the crystal lattice.

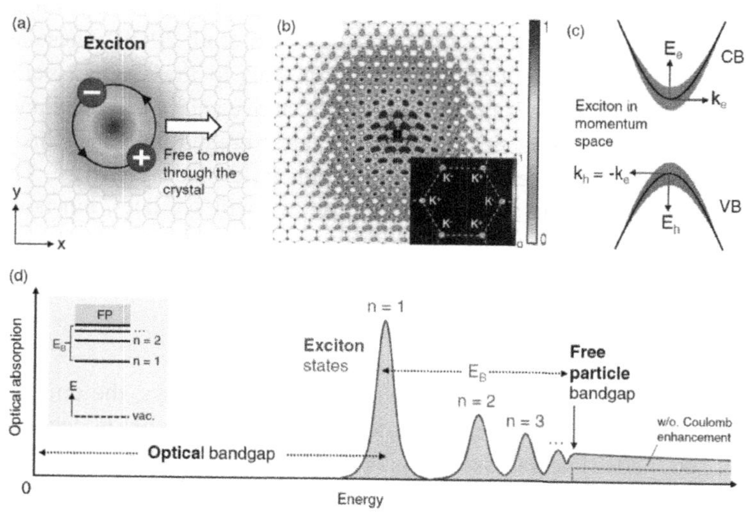

Figure 6.4. *Excitons in 2D TMDs. (a) Illustration of co-rotation of electron and hole about center of mass. (b) Exciton wave function for MoS_2 showing extended size in real space due to strong confinement in momentum space. (c) Width of bands indicating modulus squared of electron and hole wavefunctions near K points. (d) Illustration of relationship between optical bandgap (E_{exc}), free particle bandgap (E_g) and exciton binding energy (E_B) along with excited exciton electronic states. Reproduced with permission (Wang et al. 2018a). Copyright 2018, American Physical Society*

As we might expect, an electron and hole in close proximity is a quasi-stable state. The two will recombine at some point. When we speak of a long-lived exciton in solid-state systems, we are typically discussing lifetimes of only 1 ns. Exciton lifetimes measured in femtoseconds or picoseconds are typical, and only in special cases are the exciton lifetimes microseconds or longer. The exciton lifetime is related to its binding energy. Strongly bound excitons with significant electron–hole wavefunction overlap have short lifetimes, and weakly bound excitons have longer lifetimes.

Finite exciton binding energy implies that when in a bound state, the electron and hole energy levels are not precisely at the band edges. The relationship is given by:

$$E_g = E_{exc} + E_B \qquad [6.1]$$

and is illustrated in Figure 6.4d. The exciton binding energy is challenging to determine precisely, because it is difficult to measure the true bandgap energy, but high magnetic field measurements have determined 2D TMD exciton binding energies to range from 167 meV to 231 meV (Goryca et al. 2019). These are very large compared to typical bulk materials, due in part to their monolayer nature. Unlike in bulk crystals, there is no high-k material above and below the plane of the exciton to screen the fringing field between the electron and hole, and this lack of dielectric screening greatly enhances the electron–hole binding energy. This will have significant implications for valleytronics as we will see later, but now let us turn to valley physics.

6.2.2. *Valley physics*

Though essentially all semiconductor materials have valleys, the unique nature of valleys in hexagonal monolayer 2D materials was first explored in detail by Xiao (Xiao et al. 2007; 2012), and this section is indebted to that work.

First, a valleytronic material must have at least two degenerate valleys in the band structure. If there is a single valley, there is no contrast possible, i.e. there is no "0" versus "1" valley state that can be used for computation or other functions. In principle, there can be more than two distinct degenerate valley states, but in all 2D TMD materials, there are exactly two distinct valleys. Degenerate means that in the unperturbed state carriers in the two valleys have equal energy. To be useful, the valleys should be degenerate, or nearly so, else all the carriers will fall in to the lower energy valley, and it will be very expensive (in terms of energy) to populate the opposite valley. Though many materials possess energy-degenerate valleys, usually it is not possible to address each valley independently as there is no external field that couples to one valley and not the other. What distinguishes 2D TMDs is that they possess two intrinsic properties, which couple to electrical and magnetic fields with contrasting response in the two valleys.

Consider the simplified band structure of a generic 2D TMD in Figure 6.5a. These materials have a hexagonal crystal structure in real space, which inverts to a hexagonal Brillouin zone (rotated by 30 degrees) in momentum space. The valleys in this material are at the high-symmetry K points at the corners of the Brillouin zone. The six K points are located at distance 4π/3a, where a is the lattice constant, from the center of the Brillouin zone (the Γ point where k = 0). In MoS_2, a = 3.166 Å and the K-point represents a fairly large crystal momentum. Since the K points are well separated in momentum space, they should be stable, relatively long-lived valleys that require a short wavelength phonon (again, an unlikely event) for a carrier to jump from one to another.

There are two inequivalent K-points: K and K'. Note that some authors use the notation -K instead of K', these are equivalent notations, and in this chapter, we will try to consistently use K'. The K and K' valleys are points of equal but opposite crystal momentum. Crystal momentum is a difficult thing to visualize as it does not imply the exciton, electron or hole is undergoing translational motion in real space. Figure 6.5b shows one possible representation of this situation, where the wavefunction of an electron in the K valley of the conduction band rotates in a counter-clockwise fashion around the hexagonal lattice. An electron in the K' valley rotates in the clockwise direction. It is said that the K and K' valleys are time-reversed images of one another, as the electron rotation illustration should make clear. This mental picture will be useful below.

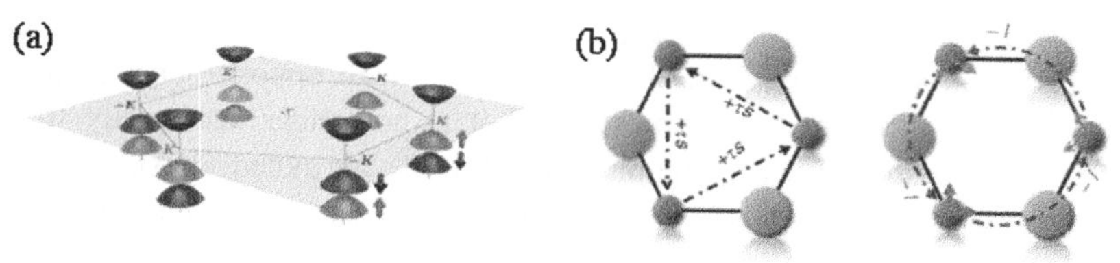

Figure 6.5. *(a) Inequivalent K and K' valleys in Brillouin zone. Reproduced with permission (Xiao et al. 2012). Copyright 2012, American Physical Society. (b) Two components of electron wavefunction self-rotation about hexagonal unit cell. Reproduced with permission (Cao et al. 2012). Copyright 2012, Springer Nature*

All hexagonal crystals have K points, but the crystal structure must possess certain symmetries to exhibit valley contrasting behavior. The lattice must possess time-reversal symmetry, but lack inversion symmetry. Time reversal symmetry implies an electron will traverse the same path through the material if time runs forwards or backwards (though its direction of motion along that path will reverse). Time reversal symmetry is not present in magnetic materials. Maxwell's equations are not time symmetric when B is not equal to 0, so the presence of magnetic atoms in the lattice will cause an electron's path to differ if time runs forwards versus backwards. Generally, we only consider non-magnetic materials to be valleytronic materials. The absence of inversion symmetry means there exists no point in the unit cell through which we can reflect all of the atoms back on to a position occupied by the same kind of atom.

The canonical 2D TMDs (MoS_2, WS_2, $MoSe_2$, WSe_2) are time-symmetric and lack inversion symmetry and so are candidate valleytronic materials. Graphene possesses time-reversal symmetry, but it is also inversion symmetric as there is a point of inversion symmetry in the center of the hexagonal unit cell. Although it is possible to distort bilayer graphene to break the inversion symmetry using strain or applied fields, that is generally a weak effect and we will not focus on bilayer graphene here. Valley contrasting effects can also arise through quantum confinement, when graphene is pattern into very fine nanoribbons; we consider this briefly at the end of the chapter.

These symmetries manifest two important phenomena: contrasting Berry curvature $\mathbf{\Omega}$ and orbital magnetic moment $\mathbf{m}$. These in turn give rise to two physically observable phenomena: the Valley Hall Effect and optical circular dichroism, respectively. Both $\mathbf{\Omega}$ and $\mathbf{m}$ are the pseudovectors; thus, in a symmetric homogenous medium (say, air), they reverse sign under time-reversal symmetry, $\boldsymbol{\Omega}(\boldsymbol{k}) = -\boldsymbol{\Omega}(-\boldsymbol{k})$, but are invariant under spatial inversion, $\boldsymbol{\Omega}(\boldsymbol{k}) = \boldsymbol{\Omega}(-\boldsymbol{k})$. In this case, the only allowed solution is $\mathbf{\Omega}$ (and $\mathbf{m}$) equal to zero. However, if the medium is not inversion symmetric, then electrons moving with opposite crystal momenta will experience different environments, and it is no longer necessarily true that $\boldsymbol{\Omega}(\boldsymbol{k}) = \boldsymbol{\Omega}(-\boldsymbol{k})$, so non-zero solutions are allowed. We noted above that the 2D TMDs are non-magnetic, observe time-reversal symmetry, and thus, the counter-rotating wavepackets at K and –K are time reversed images of one another. It is therefore still required that at the K points $\boldsymbol{\Omega}(\boldsymbol{K}) = -\boldsymbol{\Omega}(-\boldsymbol{K})$. Thus, we see qualitatively that contrasting $\mathbf{\Omega}$ (and $\mathbf{m}$) requires only certain symmetry conditions, but the magnitude of these quantities, and by implication how useful they are, depend on further details of the crystal structure.

Turning now to a more formal description, the Bloch electrons have an orbital magnetic moment given by:

$$\boldsymbol{m}(\boldsymbol{k}) = \frac{-ie}{2\hbar} \times \langle \nabla_k u(\boldsymbol{k})| \times \left(H(\boldsymbol{k}) - E(\boldsymbol{k})\right)|\nabla_k u(\boldsymbol{k})\rangle \qquad [6.2]$$

where $u(\boldsymbol{k})$ is the periodic part of the electron Bloch function, $H(\boldsymbol{k})$ is the Bloch Hamiltonian, and $E(\boldsymbol{k})$ is the energy dispersion in the lowest conduction band. The Bloch function is a wavefunction of a quantum particle, for example, an electron or a hole, moving in a periodic crystal lattice. The crystal lattice, composed of various types of atoms and their arrangements, produces a distinct potential energy distribution. The allowed Bloch functions are solutions of the Schrödinger equation with the specified periodic potential induced by the lattice. The Bloch theorem shows that the solutions must be a plane wave modulated by a (spatially) periodic function,

$$\psi(\boldsymbol{r}) = e^{i\boldsymbol{k}\cdot\boldsymbol{r}}u(\boldsymbol{r}) \tag{6.3}$$

The function $u(\boldsymbol{k})$ in equation [6.2] is the periodic part after Fourier transform of [6.3] into k-space. Calculating the orbital magnetic moment in general requires a numerical approach with an example shown in Figure 6.6a. The normalized orbital magnetic moment is +1 at the K point and -1 at the K' point, which suggests if we could measure the orbital magnetic moment of a particle, we could identify its valley state.

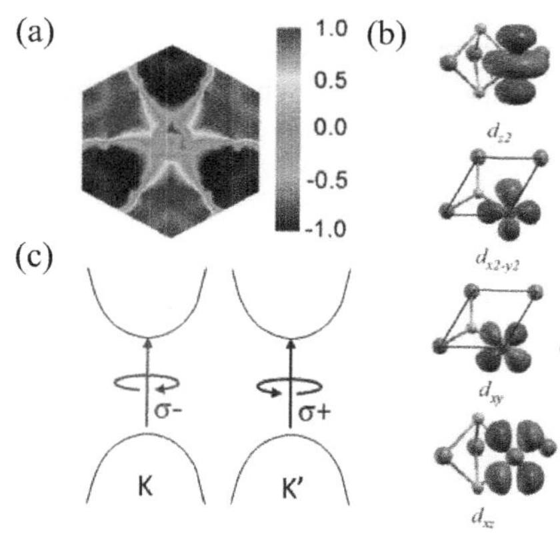

Figure 6.6. *Optical circular dichroism in TMD monolayers. (a) Calculated* $\mathbf{m}\ (\mathbf{k})\cdot\mathbf{z}$ *normalized by* $\mu_B^*(\mathbf{k})$ *across Brillouin zone. Reproduced with permission (Song et al. 2015). Copyright 2015, Springer Nature. (b) Orbitals associated with the TMD metal atoms. Reproduced with permission (Feng et al. 2012). Copyright 2012, American Physical Society. (c) Schematic optical selection rule at K and K'*

An instructive analytic solution for **m** is possible in limited cases. The relevant $\boldsymbol{k}\cdot\boldsymbol{p}$ Hamiltonian at the band edges in the vicinity of K and K' is given by a simple expression:

$$\hat{H} = at(\tau_z k_x\sigma_x + k_y\sigma_y) + \frac{\Delta}{2}\sigma_z \tag{6.4}$$

where a is the lattice spacing, t is the nearest neighbor hopping integral, $\tau_z = \pm 1$ is the valley index, $k_{x,y}$ is the distance (in k-space) from the K point, σ is the Pauli matrix element, and Δ is the bandgap. Values for these parameters can be found in

(Xiao et al. 2012). The x and y coordinates are aligned with the d-orbitals of the M atom, and z points out-of-plane (Figure 6.6b). This Hamiltonian is sometimes described as the "massive Dirac fermion model" which simply means it applies to electrons with non-zero effective mass (unlike in graphene, for example).

Since we are discussing 2D materials with only $k_x\hat{\mathbf{x}}$ and $k_y\hat{\mathbf{y}}$ components, **m** always points out-of-plane, i.e. $\boldsymbol{m} = m_z\hat{\mathbf{z}}$. Using equations [6.2] and [6.4] in the limit of parabolic bands in close proximity to the K points,

$$\boldsymbol{m}(k) = \frac{-2a^2t^2\Delta}{4a^2t^2k^2+\Delta^2}\frac{e}{2\hbar}\tau_z\hat{\mathbf{z}} \qquad [6.5]$$

It is immediately recognizable that for $\tau_z = \pm 1$ corresponding to the K and K' points, **m** is equal in magnitude and opposite in sign; thus, we say that valleytronic materials have *contrasting* orbital magnetic moment at the K points, which is a key feature. In the present coordinate system where k=0 at the valley extrema (K points), we can rewrite equation [6.5] as:

$$\boldsymbol{m}(K) = \mu_B^*\tau_z\hat{\mathbf{z}} \qquad [6.6]$$

$$\mu_B^* = \frac{e\hbar}{2m_e^*} \qquad [6.7]$$

$$m_e^* = \frac{2\hbar^2\Delta}{3a^2t^2} \qquad [6.8]$$

Equations [6.6]–[6.8] show that in the vicinity of the K points, the orbital magnetic moment has the same form as the traditional electron spin magnetic moment, where the electron rest mass in this case is replaced by an effective mass derived from band structure. Analogous to the electron real spin s = ±½, we now have a *valley pseudospin* $\tau_z = \pm 1$. Just as we can couple to real spin using magnetic fields, we will see that we can couple to valley pseudospin using optical fields.

As long as the valley pseudospin is a good quantum number (in the vicinity of the K points in this case), we can construct new optical selection rules with knowledge of the topology of the atomic orbitals as illustrated in Figure 6.6b. Near the K points, the valance band is composed primarily of the metal atom $\left(d_{x^2-y^2} + i\tau d_{xy}\right)/\sqrt{2}$ orbital with m = $2\tau_z$. The conduction band is primarily the metal atom d_{z^2} orbital with m = 0. Therefore, promoting an electron from the valence band to the conduction band requires $\Delta m = -2\tau_z$. (Mak et al. 2018) We write the optical selection rules in terms of conservation of angular momentum as:

$$l_p\hbar = \Delta\tau_z\hbar - \Delta m\hbar + \Delta m_{exc}\hbar + 3N\hbar \qquad [6.9]$$

The left-hand side represents the angular momentum of the incoming photon, $\Delta\tau_z\hbar$ is the change in valley pseudospin angular momentum, Δm is the change in electron angular momentum, $\Delta m_{exc}\hbar$ is the angular momentum acquired by the resulting exciton (singlet = 0, triplet = 1), and the final term represents dissipation of excess momentum to the threefold rotationally symmetric lattice (Xiao et al. 2015). N can be zero or a positive integer. From equation [6.9] for single photon excitation, transitions at K are allowed by σ- photons and transitions at K' are allowed by σ^+ photons, in both cases generating a singlet exciton as illustrated in Figure 6.6c. Creating a triplet state requires a two-photon process and requires an additional spin-selection rule. We note that there is a discrepancy in the literature about which polarity (σ^- or σ^+) couples to which valley (K or K') (see Xu et al. (2014) and Schaibley et al. (2016) vs. Cao et al. (2012) and Mak et al. (2018), for example). This appears to arise from different conventions of sign for σ and/or K and is not fundamentally important.

The same optical selection rule requires that when the excited carrier relaxes back to the valence band, it will emit a σ^+ photon from the K' valley and a σ- photon from the K valley. These 2D TMDs thus are predicted to exhibit strong *optical circular dichroism*, first experimentally observed in the seminal work by Cao et al. (2012), Mak et al. (2012), Sallen et al. (2012) and Zeng et al. (2012). The ability to initialize and readout specific valley states optically is very powerful and allows access to a new degree of freedom, τ_z, forming the essence of valleytronics.

Now, we turn to the discussion of Berry curvature to explore how valleytronics allows us to manipulate individual valley states spatially. The Berry curvature describes the geometric properties of the electronic bands and is key to understanding band-related topology. The equations of motion for an electron in a periodic solid (i.e. a crystal) are as follows:

$$\dot{\boldsymbol{r}} = \frac{1}{\hbar}\frac{\partial E_n(\boldsymbol{k})}{\partial \boldsymbol{k}} - \dot{\boldsymbol{k}} \times \Omega_n(\boldsymbol{k}) \quad [6.10]$$

$$\hbar\dot{\boldsymbol{k}} = -e\boldsymbol{E} - e\dot{\boldsymbol{r}} \times \boldsymbol{B} \quad [6.11]$$

where E_n is energy of the nth energy band (e.g. the lowest conduction band) and Ω_n is the Berry curvature in that band. Note that if the Berry curvature is zero, as is the case in free space or a disordered medium, these reduce to the familiar equations of motion for an electron in electric and magnetic fields.

The Berry curvature is calculated from the Bloch functions of the charge carriers,

$$\boldsymbol{\Omega_n}(\boldsymbol{k}) = \nabla_{\boldsymbol{k}} \times \langle u_n(\boldsymbol{k})|i\nabla_{\boldsymbol{k}}|u_n(\boldsymbol{k})\rangle \quad [6.12]$$

see Figure 6.7 where this has been calculated for MoS_2. The Berry curvature is 19 Å^2 at the K points and -19 Å^2 at the K' points. Using the Hamiltonian [6.4] in the vicinity of the K points, we can obtain an analytic expression for the Berry curvature in the lowest conduction band:

$$\boldsymbol{\Omega}_c(k) = \frac{-2a^2t^2\Delta}{(4a^2t^2k^2+\Delta^2)^{3/2}}\tau_z\hat{\mathbf{z}} \qquad [6.13]$$

Very near the K points (k = 0),

$$\boldsymbol{\Omega}_c(k) = \frac{-2a^2t^2}{\Delta^2}\tau_z\hat{\mathbf{z}} \qquad [6.14]$$

Again we see that the Berry curvature is equal in magnitude and opposite in sign for $\pm K$ corresponding to $\tau_z = \pm 1$. Equation [6.10] reveals that finite Berry curvature gives rise to an anomalous electron velocity transverse to the electric field. This is the Valley Hall Effect, and unlike the traditional Hall Effect, it occurs in the absence of a magnetic field. Contrasting the Berry curvature implies that electrons in the K valley will move in one transverse direction, whereas electrons in the K' valley will move in the opposite transverse direction. Thus, in valleytronic materials, we can spatially separate carriers based on their valley pseudospin.

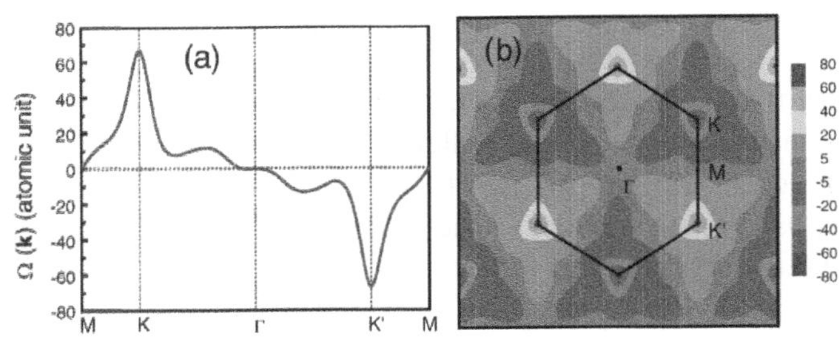

Figure 6.7. *Berry curvature calculated (a) at high symmetry points, (b) across entire Brillouin zone. Units are (Bohr radius)^2. Reproduced with permission (Feng et al. 2012). Copyright 2012, American Physical Society*

6.2.3. *Spin orbit coupling and exotic excitons*

The above section derived the basic measurable quantities of valleytronics without any need to invoke spin orbit coupling (SOC). The relationship between spin and valley is sometimes intertwined, so we should emphasize here that valley contrasting $\boldsymbol{\Omega}$ and **m** is independent of real spin or spin-orbit coupling in 2D TMDs.

The interplay between spin and valley greatly enriches the physics and possible applications of valleytronic materials, however.

Spin orbit coupling derives from the orbital motion of the electron around the nucleus. Sitting in the rest frame of the electron, we will observe the positively charged nucleus orbiting the electron. This current produces a magnetic field that interacts with the spin magnetic moment **μ** of the electron, causing a Zeeman energy shift:

$$\Delta E = -\boldsymbol{\mu} \cdot \boldsymbol{B} = \frac{g_e \mu_B}{\hbar} \boldsymbol{S} \cdot \boldsymbol{B} \qquad [6.15]$$

where **S** is the spin angular momentum vector of the electron. This is the effect responsible for fine structure in atomic spectra. If we make the following substitution for **B** in terms of the momentum of the electron **p**:

$$\boldsymbol{B} = \frac{-1}{c} (\boldsymbol{v} \times \boldsymbol{E}) = \frac{-1}{m_e c} (\boldsymbol{p} \times \boldsymbol{E}) \qquad [6.16]$$

and include a relativistic correction, we have:

$$\Delta E = \frac{(g_e - 1)\mu_B}{\hbar m_e c} \boldsymbol{S} \cdot (\mathbf{p} \times \boldsymbol{\nabla V}) \qquad [6.17]$$

For a hydrogen atom, the spin-orbit splitting between S=±½ can be calculated analytically, but in a solid, it must be solved computationally across **k** (recall **p** = ℏ**k**). An example showing the pronounced spin-split bands near K is illustrated in Figure 6.8. The Hamiltonian near the K points now gains an additional term,

$$\hat{H} = at(\tau_z k_x \sigma_x + k_y \sigma_y) + \frac{\Delta}{2} \sigma_z - \frac{\lambda}{2} \tau_z (\sigma_z - 1) s_z \qquad [6.18]$$

where λ is the SOC energy. By equation [6.17], the spin-orbit coupling is proportional to the gradient of the nuclear electrostatic potential suggesting high Z atoms will have larger band splitting. The valence band SOC was calculated to be 153 meV for MoS_2, 173 meV for $MoSe_2$ and 439 meV for both WS_2 and WSe_2 (Roldan and Guinea 2017). In all four cases, the CB splitting was calculated to be < 40 meV, the difference being related to the character of the atomic orbitals in the bands.

SOC strongly affects the exciton physics in 2D TMDs. A schematic band diagram at the K points is shown in Figure 6.8. Due to the strong spin splitting in the valence band, we see that two different excitons, labeled A and B, are possible. The A exciton is the lower energy, arising from the upper spin orbit split valence band,

and the B exciton is higher energy arising from the lower band. Many experiments have verified the coexistence of A and B exciton populations through photoluminescence (Mak et al. 2010).

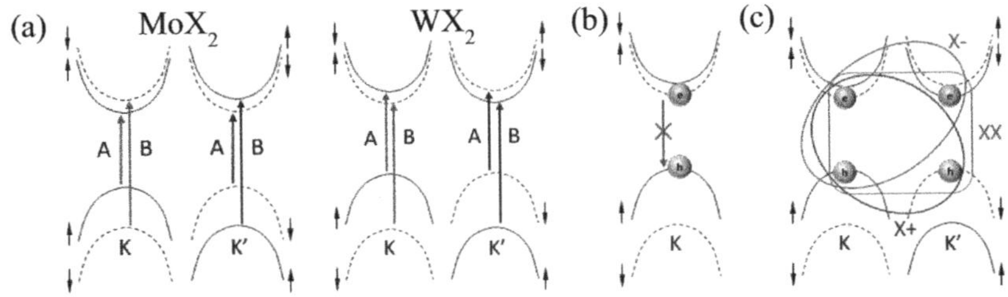

Figure 6.8. *Schematic band structure near K points illustrating, (a) bright excitons in MoX_2 and WX_2, (b) dark exciton and (c) trions and biexcitons*

The details of the spin texture between the A and B excitons and the K and K' valleys changes depending on the specific material. Among the usual MX_2 TMDs, the chalcogen (S or Se) does not change the band structure qualitatively so we need consider only the MoX_2 and WX_2 cases in Figure 6.8a. Note the larger SOC in WX_2 and that the lowest spin orbit split conduction band and the highest valence band are of the same spin in MoX_2, but of different spin in WX_2.

The small but non-negligible conduction band SOC can favor dark excitons. If an electron is promoted to the upper spin orbit split conduction band, it is possible for it to relax to the lower conduction band (Figure 6.8b). To do so requires a spin flip induced by a magnetic impurity or other scattering mechanisms. Optical selection rules forbid radiative recombination in this case; thus, they are termed "dark" (Chongyun et al. 2018). Dark excitons are probably long-lived compared to bright excitons (~100 ps vs. ~1 ps), because of the lack of a radiative recombination pathway.

Figure 6.8c illustrates a few more exotic exciton states. Excitons can capture a free electron or hole to form a charged three-body quasi-particle called a trion. It is generally believed that the electron or hole occupies the opposite valley from that of the exciton, thus creating a mixed-valley state. Finally, we note the existence of biexcitions, which may be slightly favorable energetically. Biexciton states can exist with both excitons in the same valley or in opposite valleys.

6.3. Valley lifetime, transport and operations

To process information using valleytronics, we must be able to transport and perform operations on the valley index. To do that with high fidelity, the valley information must persist for long enough to complete the required series of operations and store the resulting data to some register. Valley lifetime is the characteristic timescale of the exponential equilibration rate of valley information, and is defined as the time for 1/e of the information carriers to scatter into the opposite valley. We use the term valley lifetime to describe a population in a single valley; later, we will consider valley coherence, a term we reserve to describe the phase relationship between particles in two opposite valleys or a particle in a superposition state of both valleys. We will first discuss valley lifetime before considering how valley information can be transported and manipulated.

6.3.1. *Valley lifetime*

The early work in the field generally involved measuring the exciton valley lifetime through circularly polarized photoluminescence (PL) experiments. A circularly polarized laser with photon energy greater than that of the bandgap is used to excite the 2D TMD material and a photoluminescence spectrum is recorded. If σ^+ light is used for excitation, then per the previous calculations we would expect only σ^+ photoluminescence to be emitted. The measured degree of *valley polarization* is given by:

$$P_+ = \frac{I_+ - I_-}{I_+ + I_-} \qquad [6.19a]$$

$$P_- = \frac{I_- - I_+}{I_+ + I_-} \qquad [6.19b]$$

where $P_\pm$ is the polarization under σ^+ or σ^- light, and $I_\pm$ is the intensity of emitted σ^+ or σ^- light. Even with perfect valley initialization, the measured polarization may be less than 100%, because an exciton may scatter into the opposite valley before it recombines. Valley polarization is related to the valley lifetime t_v and the exciton recombination lifetime t_{exc}:

$$P = \frac{P_0}{1 + \frac{t_{exc}}{t_v}} \qquad [6.20]$$

Intuitively, if the exciton recombination lifetime is long, there is a greater probability it will scatter to the opposite valley before it emits a photon and valley polarization will decrease. We can achieve very high valley polarization by decreasing t_{exc}. For example, by changing the substrate from quartz to graphene, the exciton lifetime decreases significantly due to rapid charge transfer from the TMD to graphene (Miyauchi et al. 2018). This technique has been used to increase the valley polarization of WS_2 at room temperature (Lorchat et al. 2018). Similarly by using GaN as a substrate, valley polarization of MoS_2 at room temperature was increased (Wan et al. 2018). But it is important to note that for practical valleytronic computing, reducing exciton lifetime is not a promising approach as we must allow time to perform operations on the information carrier before the information is emitted via a photon.

The Kerr rotation is a complementary technique to measure valley polarization. A population of valley-polarized excitons is created using circularly polarized light as before. But instead of measuring PL, a linearly polarized probe beam is reflected off the sample surface. An exciton and free carrier valley population imbalance causes anisotropy in refractive index and absorption with respect to the two helicities of circularly polarized light. This anisotropy will rotate the linearly polarized beam upon reflection from the TMD surface. An excess population in the K valley will rotate the probe beam in one direction, and an excess in the other valley will rotate the beam in the opposite direction. Typical rotation angles are small, about 15 μrad, but measurable to high accuracy (McCormick et al. 2017). The Kerr rotation is a qualitative measurement of valley polarization, but importantly, it is sensitive to dark excitons and free carriers, which are not visible in PL studies.

In equation [6.20], P_o is the initial polarization limited, for example, by TMD material defects and quality of laser polarization. Other experimental factors influencing polarization measurements include the laser wavelength, sample temperature and substrate material. If the excitation laser energy is well above the exciton resonance, the resultant hot excitons can more readily scatter to the opposite valley. Similarly, valley scattering increases with sample temperature; a high degree of polarization measured at 4 K can decrease to zero at room temperature. Interaction between the 2D TMD and the substrate can have a strong effect on P_o, and the recent trend is to use a 2D material sandwich where the monolayer TMD is encapsulated between layers of hBN to isolate it from the substrate and from the environment.

A model for valley lifetime was proposed by Miyauchi et al. (2018):

$$t_v = \frac{C\hbar[1-exp(-E_F/kT)]^2}{\Gamma_h} \qquad [6.21]$$

where C is a material-dependent constant, E_F is the Fermi level, and Γ_h is the PL linewidth. Γ_h increases with temperature and with exciton density, i.e. the PL emission is subject to temperature and pressure broadening. Equations [6.20] and [6.21] predict the two important features: valley polarization decreases with both temperature and exciton density.

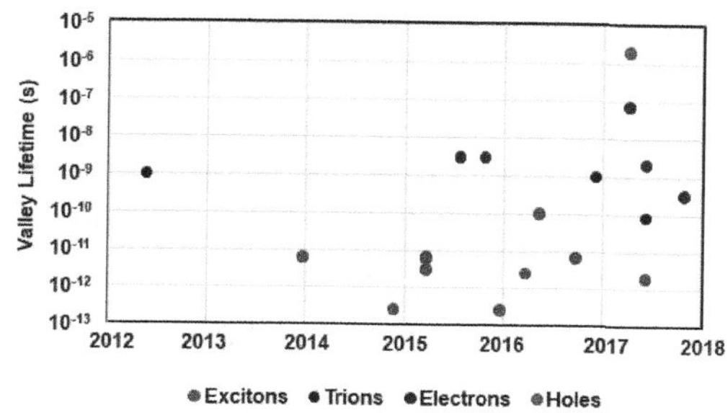

Figure 6.9. *Valley lifetime measured by TRPL or TRKR from various authors. Reproduced with permission (Vitale et al. 2018). Copyright 2018, Wiley*

Experimentally, a simple steady-state PL polarization measurement cannot resolve P_o, T_{exc} and t_v independently. To do so requires time resolved experiments using a pulsed excitation source and time-gated detection. The decay curves of total PL and circularly polarized PL allow calculation of these quantities. There is a growing database of valley polarization measurements in the literature; some results are plotted in Figure 6.9. Valley lifetime for excitons is in the range of 0.1–100 ps.

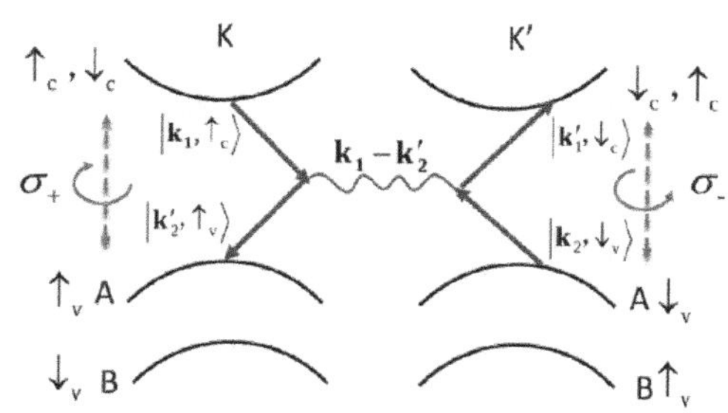

Figure 6.10. *Electron–hole exchange interaction. Reproduced with permission (Yu Wu 2014). Copyright 2014, American Physical Society*

The short valley lifetime is surprising as we emphasized that the valley state is protected from scattering by (1) the large momentum mismatch between K and K' and (2) strong spin-orbit coupling that requires a spin flip or a very energetically unfavorable transition in the valence band. The latter point is referred to as "spin-valley locking" or "spin protection of valley". Based on these considerations, we should expect the valley lifetime to be quite long.

Several spin–relaxation mechanisms including D'yakonov-Perel' and Elliott-Yafet have been proposed but not convincingly demonstrated. However, there is support for the electron–hole exchange interaction as a primary mechanism for exciton valley depolarization (Yu and Wu 2014). A Feynman diagram for this process is shown in Figure 6.10. The exchange interaction is a long-range quantum mechanical coupling between electron and hole wavefunctions, which causes the exciton to process between the K and K' states. The precession rate was calculated to be proportional to the exciton center of mass momentum, p:

$$\omega(P) \approx \sqrt{5}C\alpha_{\tau}p/\hbar \qquad [6.22]$$

where C is proportional to the distance between the exciton and hole in the exciton, and α_{τ} is a constant determined form the crystal lattice structure. At large p, the time for a π precession was calculated to be only 13 fs. That the precession rate increases with the center-of-mass momentum of the exciton is consistent with experimental results showing valley polarization decreases as the laser photon energy increases farther above resonance (Hanbicki et al. 2017).

The exchange interaction is quite fundamental, and it could impose a strict limit on exciton valley lifetime even as sample quality and experimental techniques improve. However, Feng has shown that valley lifetime increases with electrostatic doping, an effect they attribute to screening of the exchange interaction (Feng et al. 2019). Returning to Figure 6.9, we see that charged trions exhibit longer valley lifetime, from 100 to 1,000 ps, which may be consistent with this theory, and suggests there could be pathways to suppress the exchange interaction.

Long recombination lifetime (100 ns – 1 μs) and ultra-long valley lifetimes (up to 40 μs) are observed in interlayer excitons (Kim et al. 2017; Nagler et al. 2017; Miller et al. 2017; Zhang et al. 2019). An interlayer exciton is split between two stacked layers of 2D TMD, with the hole resident in one layer and the electron resident in the other layer. The formation process and band diagram of a heterogeneous interlayer exciton is shown in Figure 6.11. The band alignment should be such that it is energetically advantageous for the electron to be in one layer and the hole to be in the opposite layer, else the bilayer exciton will not form. Typically, we will pump the A exciton resonance of one of the two layers, but once excited, the electron or hole will quickly tunnel into the other layer. The stack must consist of two different TMD materials, as a homogeneous bilayer of TMD restores inversion symmetry and is not a valleytronic material.

The interlayer spacing (~0.6 nm) still allows sufficient electron–hole wavefunction overlap to create an exciton state, but the forced separation weakens exciton binding energy and the electron–hole exchange interaction. These are both favorable from the perspective of valleytronic information processing: a weakly bound exciton will have a longer recombination lifetime and a weaker exchange interaction implies a longer valley lifetime. We note that interlayer dark excitons have also been experimentally inferred with a valley lifetime of about 1 μs (Chongyun et al. 2018).

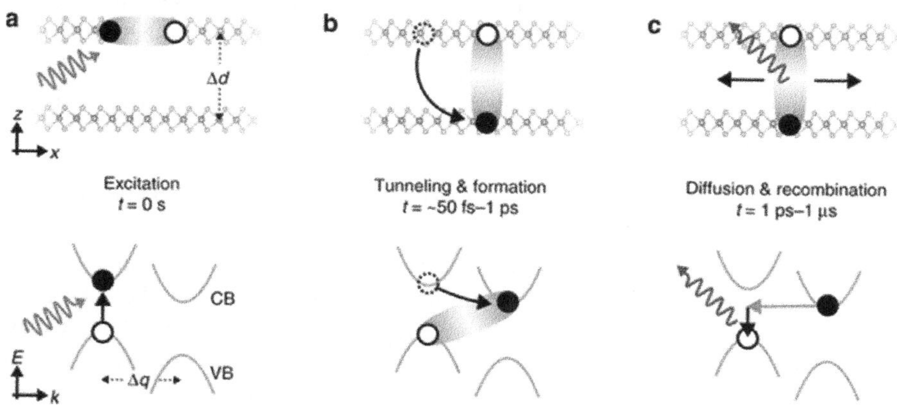

Figure 6.11. *Formation process of an interlayer exciton. Reproduced with permission (Lin 2021). Copyright 2021, Springer Nature*

Alternatively, excitons can be localized to defect sites intentionally generated by electron beam (Moody et al. 2018) or He ion beam (Klein et al. 2018) irradiation. The defect bound excitons can be distinguished from regular excitons as they appear at a significantly lower energy indicating association with states in the bandgap. They show long recombination and valley lifetimes, of the order of 100 ns and 1 μs, respectively (Moody et al. 2018). Engineered defect sites suggest the possibility of patterning the TMD material with defects in a deterministic way, which could be a powerful technique to build valleytronic devices.

We will not discuss the valley lifetime of other excitonic species in detail, and it will suffice here to say that there is currently no compelling reason to expect that B excitons, biexcitons (Shao-Yu et al. 2018) or more exotic species will have favorable properties for valleytronic information processing. Instead, we move on to the valley lifetime of free carriers.

Valley polarized excitons may transfer polarization to resident carriers by scattering processes or by non-radiative recombination of the hole with a resident carrier with the opposite valley. The exchange interaction is weak for unbound free carriers as there is less wavefunction overlap with nearby holes. Also, free carriers very near the band edges should have a long lifetime, because the phase space for electron–phonon scattering vanishes at the band edge where the density of states goes to zero (Ciccarino et al. 2018). Figure 6.9 shows that the valley lifetime of free electrons is in the range of 1–70 ns as measured by time resolved Kerr rotation (TRKR) (Luyi et al. 2015; Yang et al. 2015; Dey et al. 2017; McCormick et al. 2017; Yan et al. 2017), where the Kerr rotation angle is monitored as a function of time after a fast pulse of circularly polarized light. Since the excitons recombine in < 0.1 ns, in n-type (p-type) TMD material, the long-lived signal is attributed to an excess of valley-polarized free electrons (holes).

The free electron valley lifetime decreases with temperature, in one case, from 3 ns at 4 K to 0.2 ns at 40 K (Luyi et al. 2015). The record valley lifetime for electrons was measured to be 130 ns at 5 K (Dey et al. 2017). In this case, the electron doping was electrostatically controlled by a back gate. When the polarity of the back gate was reversed inducing hole-doping, an extraordinary hole valley polarization lifetime of 2 μs was observed. The longer hole valley lifetime is due to the strong spin-protection of the valley state in the valence band.

6.3.2. *Valley transport*

To perform useful computation, we must be able to transport valley information from one operation to another. If the valley information carrier is an electron or hole, electrostatic transport of the particle itself is straightforward, but the valley state may not be preserved. We define a valley mean free path λ_v based on the residual valley polarization P_v as the carrier moves through the crystal:

$$P_v = P_o e^{-\lambda_v x} \qquad [6.23]$$

For traditional CMOS, information mean free path is not an important quantity as the information is encoded in the charge. If a carrier, say an electron, is somehow lost, an equivalent electron will be promoted from the valence band or drawn from the source to restore the population to the Fermi level. Valleytronics, by contrast, is based on a non-equilibrium population of valley states. If the valley state is lost by scattering or recombination, it will not spontaneously be restored.

There have not yet been many studies of valley mean free path. Electron and hole λ_v can be determined by measuring valley-specific non-local resistance, often through the Valley Hall Effect using a Hall bar structure. In 2D TMDs, λ_v is approximately 1 μm for electrons (Wu et al. 2019), which is longer than the electron mean free path estimated from electron mobility by about a factor of 50. In gated bilayer graphene (recall that valley-specific states are present in monolayer graphene but cannot be measured as there is no contrast between them), valley mean free path was also measured to be 1–1.6 μm (Shimazaki et al. 2015; Sui et al. 2015). We might expect λ_v to be much longer in monolayer graphene due to its long electron mean free path, ~0.5 μm at room temperature. Although we cannot selectivity address either valley uniquely in monolayer graphene, it may be possible to inject valley information from another material. This opens the possibility of using 2D TMDs for initialization, manipulation and readout of valley information, while using graphene for valley information transport. The challenge then becomes transferring the valley pseudospin from the TMD to graphene with minimal loss. The first experiments demonstrating this heterogeneous transfer of valley information have recently been reported with promising results (Avsar et al. 2017; Luo et al. 2017).

Exciton λ_v is longer, of the order of a few microns (Unuchek et al. 2019), presumably because of the weaker electrostatic scattering associated with an uncharged exciton. Unfortunately, exciton transport is primarily by isotropic diffusion, which makes it challenging to move the valley information efficiently in the desired direction. Interestingly, the same work reports using an electrostatic field to act upon the persistent dipole moment of an interlayer exciton to define a preferred drift direction. The initial results were of modest efficiency, but an approach like this will almost certainly be necessary for exciton valley information to be passed through multiple physical logic gates.

We close this section on valley transport by noting that in valleytronic materials, dissipationless topological currents can propagate along edge modes, where the K and K' currents are counter propagating (Gao et al. 2018). This could be exploited to perform extremely low-loss information transport although more study is required.

6.3.3. *Valley operations*

At this time, there is no new logic paradigm comprising some set of valley-specific Boolean operations, so we default to trying to reproduce the conventional set of operations (AND, NOT, NOR, XOR, etc.) but using valleytronic particles. CMOS logic operates through pull-up and pull-down networks that instantiate a voltage (basis set of $[V_{cc}, 0]$) on the output pin to define a binary [1, 0] state. But what will be on the output pin of a valleytronic logic gate? We consider this question in further detail in section 6.5, but for now, it will suffice that the output will be some ensemble of particles in a particular valley state. The basis set may be

[K, K'], for example. Or the presence or absence of a particle in a valley could constitute a basis set [K or K', 0] where "0" could mean either no carriers or an equal mixture of K and K' states representing zero valley polarization. Alternatively, it may be convenient to use singlet and triplet valley states:

$$0: \frac{1}{\sqrt{2}}(|K\rangle - |K'\rangle) \quad [6.24a]$$

$$1: \frac{1}{\sqrt{2}}(|K\rangle + |K'\rangle) \quad [6.24b]$$

The best basis set for valleytronic computing has not been determined, but let us say we will use [K, K']. Then, we need to figure out how to perform the simplest logic gate K → K', a NOT gate or inverter. Figure 6.12 shows these states mapped on to the Bloch sphere. The Bloch sphere is commonly used to describe single qubit operations for quantum computing (a topic we will discuss at the end of this chapter); for readers unfamiliar with the Bloch sphere representation, we now describe it briefly.

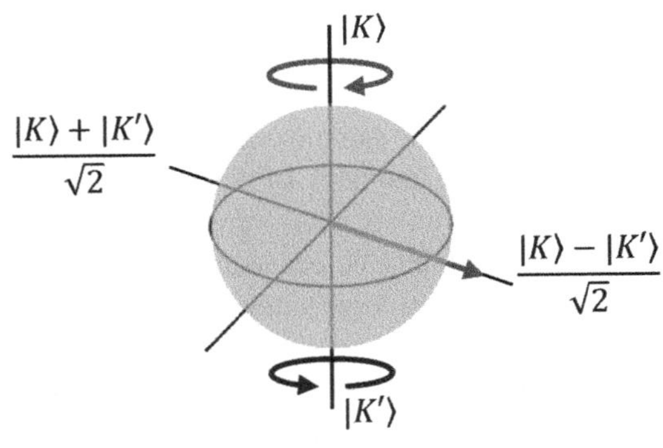

Figure 6.12. *Valley states on the Bloch sphere*

Here, the K state is at the North Pole of the sphere, and it can be initialized using σ^- light. The K' state is at the South Pole and is initialized by σ^+ light. Linearly polarized light will initialize a superposition state (equation [6.24]) on the equator of the sphere. Any point on the equator can be selected by rotating the angle of the linearly polarized light in real space in reference to the crystal axes. Points between the pole and the equator are selected using the appropriate mix of circular and linear polarization, i.e. elliptically polarized light.

How to rotate a state initialized at K or K' to some other point on the Bloch sphere is an active area of research. Unlike real spin which will process in real space about an applied magnetic field, a deterministic precession of valley pseudospin through the poles has not yet been shown. Possible methods to achieve a controlled rotation between the poles include modulation of the valley exciton exchange field or flipping the spin of the particle to encourage an energetically favorable valley hopping.

By contrast, rotation of a superposition state along the equator has been experimentally demonstrated (Ye et al. 2017), providing a path to a valid NOT operation, i.e. a π rotation from the 0 to 1 states in equation [6.24]. Suppose we create an energy splitting ΔE between the normally degenerate K and K' states. The wavefunctions in these states will evolve at slightly different frequencies, the frequency difference given by:

$$\Delta\omega = \frac{\Delta E}{\hbar} \qquad [6.25]$$

A coherent superposition state initialized at t=0 with equal amplitude in both states (a linearly polarized state on the equator) will then process along the equator according to the equation:

$$|\psi(t)\rangle = \frac{1}{\sqrt{2}}(|K\rangle + e^{-i\Delta\omega t}|-K\rangle) \qquad [6.26]$$

Given $\Delta\omega$, by using a pulse of duration τ, we can induce the state to rotate by π, $\pi/2$ or any other amount limited only by the coherence time of the state. To make an analogy with familiar terminology in quantum systems, we can equate T1 (population lifetime) with t_v (valley lifetime), T2 (coherence dephasing time) with t_{vc} (valley coherence time) and T2* (dephasing by processes other than population loss) with t_{vc}*, giving:

$$\frac{1}{t_{vc}} = \frac{1}{2t_v} + \frac{1}{t_{vc*}} \qquad [6.27]$$

The upper limit on valley coherence time is $2t_v$, but it could be much shorter. Initial results suggest that exciton valley coherence time is only about 0.5 ps (Wang et al. 2016; Ye et al. 2017) and trion valley coherence time is about the same (Titze et al. 2018). But with a sufficiently large ΔE, we could implement a very fast NOT gate or inverter using a π-pulse, whose duration is given by,

$$t = \hbar\frac{\Delta\emptyset}{\Delta E} \qquad [6.28]$$

where $\Delta\emptyset$ is the desired phase gate. With $\Delta E = 18$ meV (Sie et al. 2015) and $\Delta\emptyset = \pi$, $t = 115$ fs so even a 0.5-ps valley coherence time could support a simple logic operation, although the fidelity of such a fast gate needs to be determined.

The authors are not aware of any measurements of electron or hole valley coherence time, but given the long electron and hole valley lifetimes, this topic is of great importance.

Many papers have explored how to break the valley degeneracy and create a large energy splitting, ΔE. Currently, there are two ways to accomplish this for excitons: Zeeman splitting and the optical Stark effect. In the first case, a strong magnetic field couples directly to the exciton valley pseudospin and induces a Zeeman effect, very similar to the magnetic field induced energy level splitting in atoms arising from coupling to real spin. This technique is straightforward, but has the disadvantage that the splitting is quite weak: only 0.11 meV/Tesla in WSe_2 (Aivazian et al. 2015). This implies tens of Tesla would be required to execute a π-rotation within the valley coherence time, which is probably impractical and certainly would be energy inefficient. Nevertheless, a $\pi/6$ rotation in WSe_2 has been demonstrated using a 9-T field (Wang et al. 2016).

To avoid a large external magnetic field, some researchers have been exploring co-integrating 2D TMDs with magnetic materials, such as EuS (Moody et al. 2016) or CrI_3 (Zhong et al. 2017) as illustrated in Figure 6.13. Through the magnetic proximity effect, impressive Zeeman splitting representing effective fields greater than 100 T is shown. This promises to be a powerful technique to achieve large valley energy splitting, with the caveat that it cannot be modulated after fabrication.

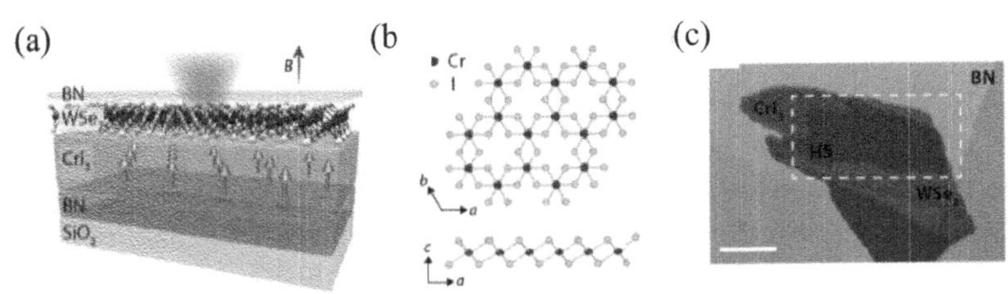

Figure 6.13. *Magnetic proximity effect in CrI_3/WSe_2 heterostructure. (a) Schematic, (b) CrI_3 crystal structure, (c) photomicrograph of heterostructure. Reproduced with permission (Zhong et al. 2017). Copyright 2017, AAAS*

The second way to break valley degeneracy is through the optical (or AC) Stark effect. This is related to the conventional Stark effect: in the presence of an electric field, a dipole will have one energy if it is aligned parallel with the electric field and a higher energy if it is antiparallel to the field. Whereas a DC electric field is sufficient to split energy levels of an atom (due to the electric field-induced dipole), there is no equivalent stationary dipole associated with an in-plane exciton in a 2D TMD. However, since an exciton at the K points can be visualized as a wavefunction rotating around the hexagonal lattice, a circularly polarized laser field resonant with this motion will appear as a DC electric field in the exciton reference frame. The optical Stark effect coupling to valley helicity was first shown (Sie et al. 2015) in WS_2 and was subsequently used in WSe_2 to demonstrate (Ye et al. 2017) the first coherent rotation of a valley state along the equator of the Bloch sphere. This technique is more complicated than the Zeeman splitting, but has produced an energy splitting equivalent to a 60-T magnetic field, and could be enhanced further if a creative way to increase the effective dipole moment of the exciton was developed.

Achieving a rotation around the Block sphere provides us with a powerful single input logic gate. But to perform computing, we also need two input valleytronic logic gates, and although development is ongoing at this time, there is no published work.

6.4. Valleytronic devices and materials

With a solid understanding of valley polarized excitons and free carriers, as well as the tasks to be performed with them, we can now turn to the recent literature to sample the experimental demonstration of primitive valleytronic devices. "Primitive" here means that the devices should be understood as the building blocks which will come together to form valleytronic logic gates, and does not mean the effort required to execute these devices was less than a *tour de force*.

The discussion of valleytronics to this point assumes the valleytronic material is a perfect monolayer (or heterogeneous bilayer) of single-crystal 2D TMD. An amorphous TMD film will probably not exhibit significant valleytronic effects. Real materials will be somewhere in-between, and we now introduce the concept of "valleytronic-quality" material. Valleytronic quality means the material exhibits sufficient valley lifetime, valley coherence time, and valley mean free path to perform efficient logic gate operations, and further that the material is uniform in these qualities at useful length scales. The latter might be 10x10 μm for individual devices, 10x10 mm for circuits and 100x100 mm for wafer-scale manufacturing. Exfoliated material derived from natural or synthetic bulk crystals is often of high quality, but the crystal size after layer transfer rarely exceeds 10–20 μm.

There have been several reports of wafer-scale growth of 2D TMDs though the material characterization typically does not include measurements of the essential valleytronic properties. In fact, forming a single crystal of 2D TMD at the chip or wafer scale is a formidable challenge, although progress is being made. It is not possible to grow a bulk crystal from melt as is done with Si or Ge. Homoepitaxial growth suffers from lack of a starting substrate, so the usual approach is heteroepitaxial growth on sapphire or quartz in a tube furnace or custom epitaxy chamber. Grown material is generally considered to be of lesser quality than exfoliated material, but this does not appear to be fundamental.

The challenges of growing a crystalline monolayer TMD are quite unlike those of growing bulk films such as GaN or InP. TMD growth nucleates at many points on the substrate and each nuclei grows laterally, usually in a triangular shape. Isolated triangles can grow to about 40 μm on a side, at which point: (1) the triangles merge forming a monolayer with significant disorder where the randomly oriented triangles meet and (2) second and third monolayers begin to grow on top before the first layer has completely merged (see Figure 6.14).

Figure 6.14. *TEM image of WS_2 grown by vapor phase epitaxy on a sapphire substrate, showing merged monolayer with second and third layers (triangles) on top. Reproduced with permission (Vitale et al. 2018). Copyright 2018, Wiley*

Preventing adlayer growth may be very difficult, so approaches have been developed to remove the undesirable adlayers. One method employs a finely tuned electrochemical polishing process (Chakraborty et al. 2019) to remove the upper layers, which are weakly attached by van der Waals forces. It does not remove the bottom monolayer; however, as it is more strongly attached by interaction with the substrate. Another approach is a wafer-scale exfoliation technique, where a stress-tuned nickel film is used in place of Scotch tape to delicately peel off individual layers of 2D TMD (Shim et al. 2018). This approach has the advantage of being able to select and transfer any layer of the stack, not just the bottom one.

After addressing wafer-scale single-crystal growth, we will still need to contend with defects. Defect density in the small triangular grains is $10^{12} - 10^{13}$ cm^{-2} or ~1,000 times higher than in microelectronic-grade silicon (Rhodes et al. 2019). Quantitative analysis of how defect density affects valleytronic properties is not yet available, but such measurements will be required to understand the defect level that is "good enough" to enable high performance valleytronic devices. One proposed method employs room-temperature photoluminescence mapping in conjunction with a machine learning algorithm to predict valleytronic quality based on the heterogeneity within the sample (Tanaka et al. 2019). Another proposed method is to measure the ratio of A/B exciton PL (McCreary et al. 2018). An exciton can decay by radiative and non-radiative channels:

$$\frac{1}{t_{exc}} = \frac{1}{t_R} + \frac{1}{t_{NR}} + \frac{1}{t_{BA}} \qquad [6.29]$$

where t_{exc} is the exciton recombination lifetime, t_R is the radiative recombination lifetime, and t_{NR} is the non-radiative recombination lifetime. The final term is available only to the B exciton and is the time for energy relaxation from the B to A exciton state. Since PL intensity is proportional to t_{exc}/t_R, and t_{NR} is assumed proportional to the defect density, we can show using equation [6.29] for the A and B excitons that a low B/A ratio implies lower defect density.

Related work has demonstrated that the non-radiative recombination channel is enhanced by the presence of magnetic dopants (Duan et al. 2019) and by sulfur vacancies (Amani et al. 2015). By eliminating these defects impressive PL yield increases of 35x and 100x have been achieved. This suggests there is significant opportunity to improve the valleytronic quality of 2D TMDs by reducing impurities and vacancies.

Moving beyond the material, we now consider device processing. Substrate interactions including charge puddles can affect device performance, as can environmental adsorbates. These interactions may unintentionally dope the TMD, creating an uncontrolled free carrier or trion population resulting in significant differences in valleytronic properties from device-to-device and from lab-to-lab. Substrate and environmental effects can be reduced by encapsulating the TMD layer between layers of hBN (Figure 6.15), thus resulting in higher mobility and longer valley lifetime although at the cost of increased fabrication complexity.

Patterning of monolayer TMDs can be done with high fidelity using conventional photolithography and plasma etching; an example is provided in Figure 6.16. Since the TMD layer is so thin, a very short Ar plasma sputtering step will etch the film.

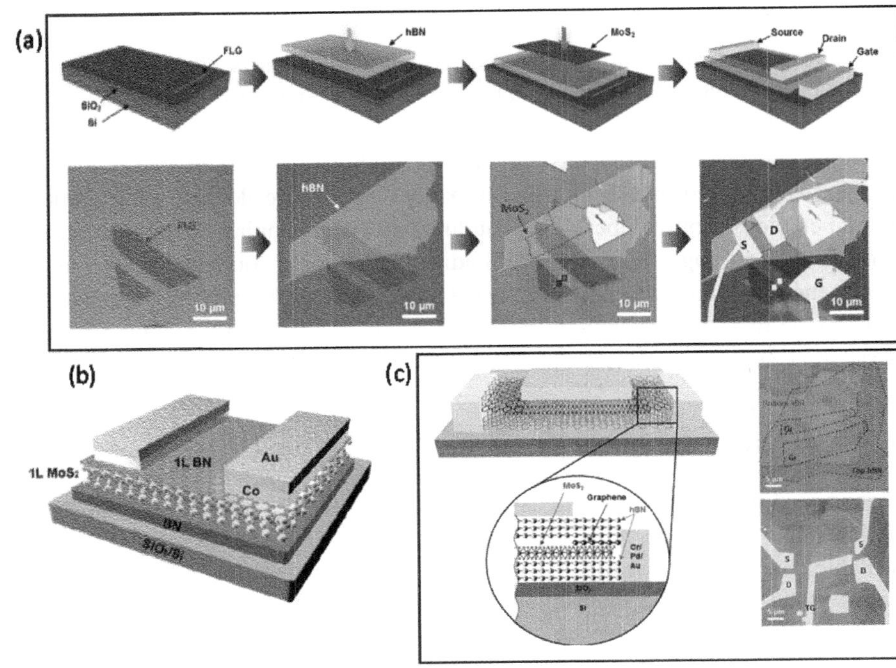

Figure 6.15. *MoS_2 mobility and transistor enhancement by hBN isolation. (a) Flexible and transparent FET. Reproduced with permission (Lee et al. 2013), Copyright 2013, American Chemical Society. (b) MoS_2 electroptical device isolated by hBN. Reproduced with permission (Cui et al. 2017). Copyright 2017, American Chemical Society. (c) Graphene-contacted MoS_2 FET with hBN top and bottom isolation. Reproduced with permission (Lee et al. 2015). Copyright 2015, American Chemical Society*

Low resistance electrical contact to the TMD is essential to build high performance devices. In general, good ohmic contact to 2D materials is challenging as the thickness of the material does not allow for the formation of some equivalent of a silicide, and their van der Waals nature often results in poor adhesion. Surface contamination strongly affects contact resistance as well; it has been shown that UHV metal deposition produces much lower resistance contact than conventional deposition systems (English et al. 2016). Being a zero bandgap semimetal, graphene can form ohmic contacts to a range of materials although the challenge then becomes contacting the graphene to external interconnects. That notwithstanding, a high-quality MoS_2 device with graphene contacts has shown excellent results (Wang et al. 2018b).

Figure 6.16. *Raman image of the in-plane E_{2g} mode of WS_2 grown by vapor phase epitaxy and patterned at a large-scale (figure is 400 microns on a side) using optical lithography and plasma etching. Reproduced with permission (Vitale et al. 2018). Copyright 2018, Wiley*

In spite of these challenges, several valleytronic devices have been demonstrated. Helicity-resolved PL from MoS_2 was enhanced by coupling with a chiral metasurface, which is a step towards non-local routing of valley information (Li et al. 2018). Gated electrostatic control of WSe_2 PL and polarization was demonstrated using a graphene-hBN sandwich structure as shown in Figure 6.17 (Chakraborty et al. 2019). Although the degree of polarization control was modest it represents: (1) an important step towards all-electrical manipulation of a valleytronic device and (2) a possible demonstration that the electron–hole exchange interaction can be suppressed by using an electric field to separate the electron and hole wavefunctions.

Another device consisting of an hBN-encapsulated $MoSe_2/WSe_2$ heterostructure takes advantage of the out-of-plane persistent dipole characteristic of interlayer excitons (Ciarrocchi et al. 2019). A key feature of this device is the dual gates that provide independent control of electrostatic doping and electric field. This allowed the researchers to show another form of a NOT gate operation demonstrating the promise of interlayer excitons for valleytronic information processing.

A valleytronic transistor concept employing graphene is shown in Figure 6.18. Although graphene is not natively a valleytronic material, confinement effects arising when the material is patterned into a nanoribbon or quantum dot theoretically permit observable valley contrasting behavior. Here, the Rashba effect induces a phase difference between K and K' valleys in the device channel; through appropriate electrostatic control of phase, we can turn the valley current on and off like a transistor. Using 2D TMDs instead of graphene may avoid the need for precisely patterned nanoribbons and may also provide stronger valley contrast, i.e. a higher on/off ratio.

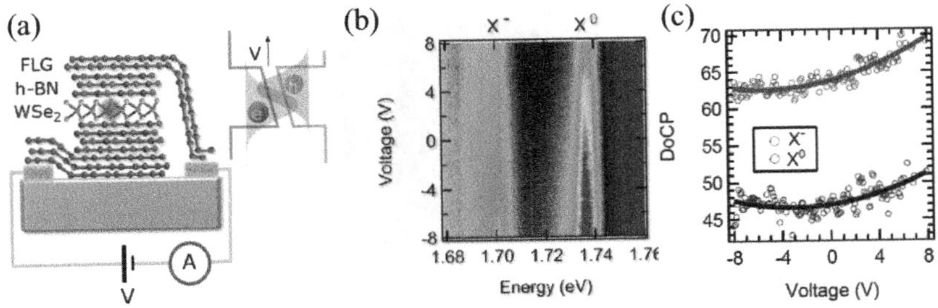

Figure 6.17. *(a) Schematic of gated WSe_2 device, (b) electrostatic control of neutral exciton and trion emission and (c) valley polarization. Reproduced with permission (Chakraborty et al. 2019). Copyright 2019, OSA*

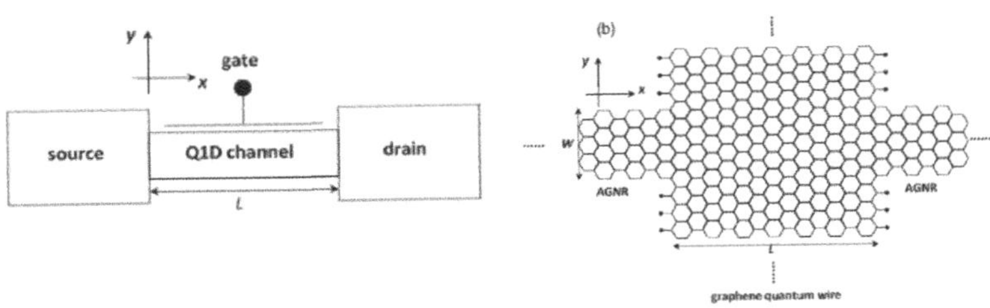

Figure 6.18. *Concept for a valleytronic transistor based on the evolution of the phase of the K and K' states between the source and drain. Reproduced with permission (Lee et al. 2012). Copyright 2012, American Physical Society*

6.5. Valleytronic computing

This section takes us into the future as we will consider how to move forward from rudimentary devices towards practical computing platforms. First, we consider valleytronic classical computing from a performance/power perspective. Next, we consider possible valleytronic information processing circuit architectures. Finally, we look at how valleytronics may provide an alternative route to quantum computing.

6.5.1. *Classical computing – power and performance*

General purpose computing based on CMOS logic faces several challenges. First, the energy efficiency of computation, often measured in GOPS/Watt, where GOPS stands for Giga Operations Per Second, must continue to improve as the supply of power (particularly in mobile and IoT devices) is limited. Second, the devices should continue to scale to smaller size as the computational capability is proportional to the number of resources that can be paced on a chip. Finally, for many data-intensive problems such as image recognition, the energy cost of moving data from logic to cache and from cache to DRAM is extremely high, much larger than the energy consumed by the logic itself.

The first two issues are usually addressed by reducing operating voltage (while maintaining current through transistor improvements) and by physical shrinking of transistors and interconnects. The third requires bringing logic and memory physically closer or changing to a non von Neumann architecture.

Valleytronics will offer an orthogonal approach to these issues if it can employ the valley degree of freedom in a revolutionary way. We could speculate that the operating voltage of a valleytronic device will be of the order of the valley splitting energy. For robust room temperature operation, we may need a splitting of 100 meV, which would offer a 25x improvement in energy cost over the likely end-of-roadmap CMOS operating voltage of 0.5 V, as the device switching energy is equal to ½ CV^2.

This 25x improvement is compelling, but where are we now? Consider a notional CNOT gate, which is logically equivalent to a reversible XOR with two inputs and two outputs. Here, the inputs and outputs are optical. The information carrier is an exciton, where the valley pseudospin represents the 0 or 1 state as defined in equation [6.24]. CNOT input Y is an exciton initialized as 0 or 1. The other input (X) is an optical control pulse: if present (1), the control laser pulse rotates the exciton input from the 0 state to the 1 state and if the control laser pulse is not present (0), the exciton is unchanged. Readout occurs by measuring the chirality of the PL emitted when the exciton recombines. We will calculate the energy cost of an inverting gate operation, X=1. Initializing the input Y will requires energy E_i, given by,

$$E_i = \frac{n_{exc}E_g}{f_i} = \frac{n_{exc}E_g}{\alpha P} \qquad [6.30]$$

where n_{exc} is the number of excitons required to generate an output well above the noise level, f_i is the initialization efficiency, α is the absolute absorption of the TMD, and P is the polarization fidelity. State evolution is performed by applying a control pulse, which rotates state Y through a phase evolution of $-i\omega t$. For a phase shift $\Delta\varphi=\pi/2$ within a time τ, we need to induce an energy splitting between the 0 and 1 states,

$$\Delta E = \hbar\omega = \hbar\frac{\Delta\varphi}{\tau} \qquad [6.31]$$

The electric field of the laser control pulse can be related to the energy splitting by

$$\Delta E = \hat{d} \cdot \vec{E} = d|E| \qquad [6.32]$$

where $\hat{d}$ is the effective electric dipole of the exciton, $\vec{E}$ is the laser field, and for the right-hand expression, the laser polarization has been aligned with the dipole. The laser pulse energy is given by:

$$P_L\tau = I_L A\tau = \left(\frac{c\varepsilon_o}{2}|E|^2\right)A\tau \qquad [6.33]$$

where P_L is the laser power, I_L is the laser intensity, and A is the illuminated area. Substituting for $|E|$, we find an expression for the control pulse energy,

$$E_c = A\tau\frac{c\varepsilon_o}{2}\left(\frac{\Delta E}{d}\right)^2 = A\frac{c\varepsilon_o}{2\tau}\left(\frac{\pi\hbar}{2d}\right)^2 \qquad [6.34]$$

The control pulse energy decreases with valley lifetime and is quadradically sensitive to the effective exciton dipole moment.

We will assume that the output exciton converts to a photon with unity yield, which is then routed to a photodiode and a charge storage register. Readout efficiency f_r is a function of the waveguide attenuation α_W and waveguide length L,

$$f_r = 10^{-\alpha_w L} \qquad [6.35]$$

The total energy cost of the inverting gate operation is now given by,

$$E_{tot} = \frac{1}{f_r}(E_i + E_c) = \frac{1}{f_r}\left(\frac{n_{exc}E_g}{\alpha P} + A\frac{c\varepsilon_o}{2\tau}\left(\frac{\pi\hbar}{2d}\right)^2\right) \qquad [6.36]$$

To evaluate this expression, we use some plausible values: α_W = 1 dB/cm, L = 10 μm, n_{exc} = 10, E_g = 1.8 eV, α = 0.01, P = 10%, A = 1 μm^2, τ = 0.4 ps and d = 50 Debye. The energy per CNOT operation is 6.112 fJ, and is nearly equally split between the initialization and control pulse energy demands. This is ~ 100x greater than the energy cost of a 14 nm CMOS inverter. Therefore, valleytronic logic employing optical exciton initialization and state rotation is not currently an energy-efficient form of computing, but we will return to this at the end of the chapter.

6.5.2. *Classical computing – architecture*

To realize a revolutionary valleytronic computer, first we must select a valleytronic particle. We have presented the valleytronic qualities associated with multiple species of excitons, electrons and holes, and we acknowledge there is no consensus on which particle is the best one to use. Excitons are easy to initialize and to readout optically. But deterministic exciton transport is very difficult. Among the various exciton species, interlayer excitons seem the most promising as they are bright, have the longest lifetime, and the persistent dipole oriented in the out-of-plane direction allows for more control. Conversely, electrons and holes are easy to transport and have long valley lifetime, but valley polarized free carriers are costly to generate in an excitonic material, because we first create an exciton and then some fraction of those excitons dissociate into free carriers or transfer the valley information to resident free carriers. Free electrons and holes benefit from long recombination and valley lifetime. Deciding between interlayer excitons, electrons or holes will probably be done in conjunction with the next topic.

The second challenge is to construct a plausible set of valley logic gates. Logic gates using free-carriers are well established, but if we simply use the same gates with valley polarized electron and hole current, there will be no practical benefit. The logic function must exploit the valley degree of freedom in a novel way to demonstrate a performance/power advantage over conventional transistor-based logic. There have been a few such proposals. First, a complete set of Boolean logic gates was designed based on the valley spin-valve concept (Tao et al. 2020). The authors claim the device could operate below 60 mV/decade subthreshold swing (the thermodynamic CMOS limit) if the dimensions of the device were sufficiently small. Another complete set of gates was proposed using valley-assisted tunneling as the information transport mechanism (Ang et al. 2017). The devices were all electrostatic, had a 100:1 on:off ratio (in theory) and provided for logical reversibility. Finally, theoretical all excitonic NOT and AND gates have been proposed (Espinosa-Ortega Liew 2013). Although this last work makes no explicit use of the valley index, it could be a step towards valleytronic exciton logic.

Although they represent interesting constructs, these logic gate proposals do not yet convincingly demonstrate benefit over conventional silicon CMOS with respect to power, speed or scalability. Therefore, there remains an opportunity to develop valleytronic logic gates that explicitly use the valley pseudospin of free carriers to some practical advantage.

The third challenge is that we must be able to cascade these logic gate functions to create complex circuits. To be competitive, these logic gates must employ the same particle on the inputs, outputs and control gates. A device that uses valley-specific charge on a control gate and outputs valley-specific photons is certainly possible to construct, but it is unlikely we could build compact and energy-efficient complex circuits using that device. Cascading gates also implies there must be some mechanism for valley gain to overcome inevitable losses and environmental noise, and to allow for fan-out. There has been no device yet reported that demonstrates a mechanism for valley gain.

An all-valleytronic computer is a compelling concept, but it is also useful to consider how valleytronic devices may co-exist with other platforms. For example, proposed non-volatile spin-logic devices require magnetic fields or high current to switch state, which adversely affects energy efficiency. If spin-polarized valley currents can be generated in the valleytronic material and the spin information transferred to the spintronic device, the costly magnetic fields may be eliminated. Furthermore, valleytronic devices offer rapid generation of valley/spin polarized carriers which may improve the speed of typically slow all-spintronic systems. Polarized valley/spin currents arising from the Valley Hall Effect have been experimentally demonstrated (Shimazaki et al. 2015), but they have not yet been combined with a spintronic device.

Another potential synergy would be to use spintronic devices for memory and valleytronic devices for logic. Non-volatile spin-based memory elements are well established, but they have not been able to outperform CMOS for logic. Valleytronics may eventually provide fast and efficient logical gates, but since the valley information is held in fragile, non-equilibrium states, it is difficult to envision an effective valleytronic memory. Since valley information can also be encoded in spin through spin orbit coupling, transferring that information to spintronic registers may offer the best of both technologies.

6.5.3. *Quantum computing*

Quantum computing is perhaps the ultimate “Beyond CMOS” technology as it promises to solve NP-hard problems that rapidly scale to become intractable by classical computing. Valley-based qubits may be attractive compared to other modalities such as superconducting artificial atoms, trapped ions or Majorana fermions. For example, it is conceptually easier to integrate millions of TMD-based qubits on a microchip than it is to address millions of individual trapped ions. Gate speeds should be much faster as well. Superconducting qubits require temperatures as low as 15 mK for noise suppression, whereas valleytronic effects are robust at 4 K and persist up to room temperature. It should be possible to integrate a much larger number of qubits onto a single 4 K platform than a single 15 mK platform, thus allowing a more practical path to scaling to a useful number of qubits. Since valley qubits are likely to be addressed optically through micron-scale integrated photonic waveguides, there is a packing density advantage over superconducting qubits, which are limited by the size of microwave transmission lines. Valleytronics also offers a wider basis set of $[|\uparrow_K\rangle, |\uparrow_{K\prime}\rangle, |\downarrow_K\rangle, |\downarrow_{K\prime}\rangle]$, and the combination of spin and valley degrees of freedom may allow for new quantum operations.

There are disadvantages to valley qubits as well. For example, the large spin-orbit coupling may increase the sensitivity of valley qubits to perturbations from the external environment, resulting in limited coherence time. Also, entanglement between two valley qubits has not yet been demonstrated. An entanglement experiment could be performed by measuring time correlated valley-polarized PL from two excitons in close proximity. A next step would be to demonstrate a controllable coupling to entangle specific pairs of valley states.

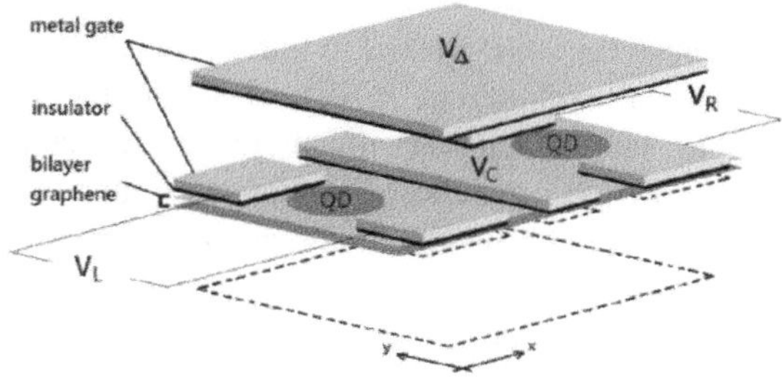

Figure 6.19. *Proposed valley qubit device. Reproduced with permission (Wu et al. 2013). Copyright 2013, American Physical Society*

In the previous section, we discussed how to perform single valley qubit rotations on the Bloch sphere. There have been a few proposals for how to perform two-valley-qubit gate operations although curiously they are based on graphene. Figure 6.19 illustrates a device composed of two graphene quantum dots. The orientation of the valley pseudospin places the pair in either a singlet or triplet state. According to the authors, the pseudospin can be rotated by an in-plane electric field between two parallel gates and the coupling between the qubits can be modulated by another control gate. If experimentally demonstrated, this device would represent a significant advancement in valley-based quantum computing.

6.5.4. *Outlook*

The valley pseudospin offers a robust new degree of freedom for information processing. Unfortunately, the state of the art for simple valleytronic operations falls short of commercial CMOS technology by a large factor. However, it is still early in the development of valleytronic logic devices and state manipulation techniques. Key parameters in the device performance model include the effective dipole moment, polarization fidelity and the valley lifetime. Orders of magnitude improvement are possible in the dipole moment and valley lifetime, and as much as 10x in valley polarization. There are several active areas of research to improve these parameters, including heterostructures of TMD layers to allow robust interlayer excitons, improved synthesis techniques to reduce valley scattering centers and techniques to provide isolation from the substrate and the environment. As research continues, these parameters will improve and it may not be long before the energy efficiency of valleytronic logic operations exceeds that of conventional CMOS technology.

In a way, the challenges for valleytronics to impact quantum computing are less daunting than for classical computing, if only because the "competition" for quantum computing has set the bar very much lower. Extreme device density, scalability, speed and energy efficiency are not very important for quantum computing but they are absolutely essential for general purpose computing. Few publications discuss these issues as they may appear to be "showstoppers" for valleytronic computing. But in order to move forward, we must acknowledge these challenges and work to resolve them. One exception is the publication by Cha that clearly presents the challenges to outperform CMOS (Cha et al. 2018). The work then demonstrates a WSe_2-graphene-Bi_2Se_3 heterodevice that combines the appropriate aspects of valleytronics, spintronics and topological transport in an intelligent way. Fortunately, the last 10 years of fundamental research on valley physics and manipulation of valleytronic properties have provided a solid base from which to address the architectural and engineering challenges to realize valleytronic computing.

6.6. References

Aivazian, G., Gong, Z., Jones, A.M., Chu, R.-L., Yan, J., Mandrus, D.G., Zhang, C., Cobden, D., Yao, W., Xu, X. (2015). Magnetic control of valley pseudospin in monolayer WSe_2. *Nature Physics*, 11(2), 148–152.

Amani, M., Lien, D.-H., Kiriya, D., Xiao, J., Azcatl, A., Noh, J., Madhvapathy, S.R., Addou, R., Kc, S., Dubey, M. et al. (2015). Near-unity photoluminescence quantum yield in MoS_2. *Science,* 350(6264), 1065–1068.

Ang, Y.S., Yang, S.A., Zhang, C., Ma, Z., Ang, L.K. (2017). Valleytronics in merging Dirac cones: All-electric-controlled valley filter, valve, and universal reversible logic gate. *Physical Review B*, 96(24), 245410.

Avsar, A., Unuchek, D., Liu, J., Sanchez, O.L., Watanabe, K., Taniguchi, T., Ozyilmaz, B., Kis, A. (2017). Optospintronics in graphene via proximity coupling. *ACS Nano*, 11(11), 11678–11686.

Cao, T., Wang, G., Han, W., Ye, H., Zhu, C., Shi, J., Niu, Q., Tan, P., Wang, E., Liu, B. et al. (2012). Valley-selective circular dichroism of monolayer molybdenum disulphide. *Nature Communications*, 3(6), 887.

Cha, S., Noh, M., Kim, J., Son, J., Bae, H., Lee, D., Kim, H., Lee, J., Shin, H.-S., Sim, S. et al. (2018). Generation, transport and detection of valley-locked spin photocurrent in WSe_2/graphene/Bi_2Se_3 heterostructures. *Nature Nanotechnology*, 13(10), 910–914.

Chakraborty, C., Mukherjee, A., Qiu, L., Vamivakas, A.N. (2019). Electrically tunable valley polarization and valley coherence in monolayer WSe_2 embedded in a van der Waals heterostructure. *Optical Materials Express*, 9(3), 1479–1487.

Cheiwchanchamnangij, T. and Lambrecht, W.R.L. (2012). Quasiparticle band structure calculation of monolayer, bilayer, and bulk MoS_2. *Physical Review B (Condensed Matter and Materials Physics)*, 85(20), 205302.

Chongyun, J., Weigao, X., Rasmita, A., Zumeng, H., Ke, L., Qihua, X., Wei-Bo, G. (2018). Microsecond dark-exciton valley polarization memory in two-dimensional heterostructures. *Nature Communications*, 9(1), 753.

Ciarrocchi, A., Unuchek, D., Avsar, A., Watanabe, K., Taniguchi, T., Kis, A. (2019). Polarization switching and electrical control of interlayer excitons in two-dimensional van der Waals heterostructures. *Nature Photonics*, 13(2), 131–136.

Ciccarino, C.J., Christensen, T., Sundararaman, R., Narang, P. (2018). Dynamics and spin-valley locking effects in monolayer transition metal dichalcogenides. *Nano Letters*, 18(9), 5709–5715.

Cui, X., Shih, E.-M., Jauregui, L.A., Chae, S.H., Kim, Y.D., Li, B., Seo, D., Pistunova, K., Yin, J., Park, J.-H. et al. (2017). Low temperature ohmic contact to monolayer MoS_2 by van der Waals bonded Co/h-BN electrodes. *Nano Letters*, 17(8), 4781–4786.

Dey, P., Yang, L., Robert, C., Wang, G., Urbaszek, B., Marie, X., Crooker, S.A. (2017). Gate-controlled spin-valley locking of resident carriers in WSe_2 monolayers. *Physical Review Letters*, 119(13), 137401.

Duan, H., Guo, P., Wang, C., Tan, H., Hu, W., Yan, W., Ma, C., Cai, L., Song, L., Zhang, W. et al. (2019). Beating the exclusion rule against the coexistence of robust luminescence and ferromagnetism in chalcogenide monolayers. *Nature Communications*, 10(1), 1584.

English, C.D., Shine, G., Dorgan, V.E., Saraswat, K.C., Pop, E. (2016). Improved contacts to MoS_2 transistors by ultra-high vacuum metal deposition. *Nano Letters*, 16(6), 3824–3830.

Espinosa-Ortega, T. and Liew, T.C.H. (2013). Complete architecture of integrated photonic circuits based on AND and NOT logic gates of exciton polaritons in semiconductor microcavities. *Physical Review B (Condensed Matter and Materials Physics)*, 87(19), 195305 (195305).

Feng, W., Yao, Y., Zhu, W., Zhou, J., Yao, W., Xiao, D. (2012). Intrinsic spin Hall effect in monolayers of group-VI dichalcogenides: A first-principles study. *Physical Review B (Condensed Matter and Materials Physics)*, 86(16), 165108.

Feng, S., Cong, C., Konabe, S., Zhang, J., Shang, J., Chen, Y., Zou, C., Cao, B., Wu, L., Peimyoo, N. et al. (2019). Engineering valley polarization of monolayer WS_2: A physical doping approach. *Small*, 15(12), 1805503.

Gao, F., Xue, H., Yang, Z., Lai, K., Yu, Y., Lin, X., Chong, Y., Shvets, G., Zhang, B. (2018). Topologically protected refraction of robust kink states in valley photonic crystals. *Nature Physics*, 14(2), 140–144.

Goryca, M., Li, J., Stier, A.V., Taniguchi, T., Watanabe, K., Courtade, E., Shree, S., Robert, C., Urbaszek, B., Marie, X. et al. (2019). Revealing exciton masses and dielectric properties of monolayer semiconductors with high magnetic fields. *Nature Communications*, 10(1), 4172.

Hanbicki, A.T., Currie, M., Kioseoglou, G., Hellberg, C.S., Friedman, A.L., Jonker, B.T. (2017). Optical polarization of excitons and trions under continuous and pulsed excitation in single layers of WSe_2. *Nanoscale*, 9(44), 17422–17428.

Kim, J., Jin, C., Chen, B., Cai, H., Zhao, T., Lee, P., Kahn, S., Watanabe, K., Taniguchi, T., Tongay, S. et al. (2017). Observation of ultralong valley lifetime in WSe_2/MoS_2 heterostructures. *Science Advances*, 3(7), 1700518.

Klein, J., Kuc, A., Nolinder, A., Altzschner, M., Wierzbowski, J., Sigger, F., Kreupl, F., Finley, J.J., Wurstbauer, U., Holleitner, A.W. et al. (2018). Robust valley polarization of helium ion modified atomically thin MoS_2. *2D Materials*, 5(1), 011007.

Lee, M.K., Lue, N.Y., Wen, C.K., Wu, G.Y. (2012). Valley-based field-effect transistors in graphene. *Physical Review B (Condensed Matter and Materials Physics)*, 86(16), 165411.

Lee, G.-H., Yu, Y.-J., Cui, X., Petrone, N., Lee, C.-H., Choi, M.S., Lee, D.-Y., Lee, C., Yoo, W.J., Watanabe, K. et al. (2013). Flexible and transparent MoS2 field-effect transistors on hexagonal boron nitride-graphene heterostructures. *ACS Nano*, 7(9), 7931–7936.

Lee, G.-H., Cui, X., Kim, Y.D., Arefe, G., Zhang, X., Lee, C.-H., Ye, F., Watanabe, K., Taniguchi, T., Kim, P. et al. (2015). Highly stable, dual-gated MoS2 transistors encapsulated by hexagonal boron nitride with gate-controllable contact, resistance, and threshold voltage. *ACS Nano*, 9(7), 7019–7026.

Li, M., Shi, J., Liu, L., Yu, P., Xi, N., Wang, Y. (2016). Experimental study and modeling of atomic-scale friction in zigzag and armchair lattice orientations of MoS2. *Science and Technology of Advanced Materials*, 17(1), 189–199.

Li, Z., Liu, C., Rong, X., Luo, Y., Cheng, H., Zheng, L., Lin, F., Shen, B., Gong, Y., Zhang, S. et al. (2018). Tailoring MoS_2 valley-polarized photoluminescence with super chiral near-field. *Advanced Materials*, 30(34), 1801908.

Lin, K.-Q. (2021). A roadmap for interlayer excitons. *Light: Science & Applications*, 10(1), 99(93).

Lorchat, E., Azzini, S., Chervy, T., Taniguchi, T., Watanabe, K., Ebbesen, T.W., Genet, C., Berciaud, S. (2018). Room-temperature valley polarization and coherence in transition metal dichalcogenide-graphene van der Waals heterostructures. *ACS Photonics*, 5(12), 5047–5054.

Luo, Y.K., Xu, J., Zhu, T., Wu, G., McCormick, E.J., Zhan, W., Neupane, M.R., Kawakami, R.K. (2017). Opto-valleytronic spin injection in monolayer MoS_2/few-layer graphene hybrid spin valves. *Nano Letters*, 17(6), 3877–3883.

Luyi, Y., Sinitsyn, N.A., Chen, W., Yuan, J., Zhang, J., Lou, J., Crooker, S.A. (2015). Long-lived nanosecond spin relaxation and spin coherence of electrons in monolayer MoS_2 and WS_2. *Nature Physics*, 11(10), 830–834.

Mak, K.F., Lee, C., Hone, J., Shan, J., Heinz, T.F. (2010). Atomically thin MoS_2: A new direct-gap semiconductor. *Physical Review Letters*, 105(13), 136805.

Mak, K.F., He, K., Shan, J., Heinz, T.F. (2012). Control of valley polarization in monolayer MoS_2 by optical helicity. *Nature Nanotechnology*, 7(8), 494–498.

Mak, K.F., Xiao, D., Shan, J. (2018). Light-valley interactions in 2D semiconductors. *Nature Photonics*, 12(8), 451–460.

McCormick, E.J., Newburge, M.J., Luo, Y., McCreary, K.M., Singh, S., Martin, I.B., Cichewicz Jr., E.J., Jonker, B.T., Kawakami, R.K. (2017). Imaging spin dynamics in monolayer WS_2 by time-resolved kerr rotation microscopy. *2D Materials*, 5(1), 011010.

McCreary, K.M., Hanbicki, A.T., Sivaram, S.V., Jonker, B.T. (2018). A- and B-exciton photoluminescence intensity ratio as a measure of sample quality for transition metal dichalcogenide monolayers. *APL Materials*, 6(11), 111106.

Miller, B., Steinhoff, A., Pano, B., Klein, J., Jahnke, F., Holleitner, A., Wurstbauer, U. (2017). Long-lived direct and indirect interlayer excitons in van der Waals heterostructures. *Nano Letters*, 17(9), 5229–5237.

Miyauchi, Y., Konabe, S., Feijiu, W., Wenjin, Z., Hwang, A., Hasegawa, Y., Lizhong, Z., Mouri, S., Minglin, T., Eda, G. et al. (2018). Evidence for line width and carrier screening effects on excitonic valley relaxation in 2D semiconductors. *Nature Communications*, 9(1), 2598.

Moody, G., Hao, K., Dass, C.K., Singh, A., Xu, L., Tran, K., Chen, C.-H., Li, M.-Y., Li, L.-J., Clark, G. et al. (eds) (2016). Coherent quantum dynamics of excitons in monolayer transition metal dichalcogenides. *Ultrafast Phenomena and Nanophotonics XX*, February 15–18, San Francisco, CA, SPIE, 97461T.

Moody, G., Tran, K., Lu, X., Autry, T., Fraser, J.M., Mirin, R.P., Yang, L., Li, X., Silverman, K.L. (2018). Microsecond valley lifetime of defect-bound excitons in monolayer WSe_2. *Physical Review Letters*, 121(5), 057403.

Nagler, P., Plechinger, G., Ballottin, M.V., Mitioglu, A., Meier, S., Paradiso, N., Strunk, C., Chernikov, A., Christianen, P.C.M., Schuller, C. et al. (2017). Interlayer exciton dynamics in a dichalcogenide monolayer heterostructure. *2D Materials*, 4(2), 025112.

Novoselov, K.S., Geim, A.K., Morozov, S.V., Jiang, D., Zhang, Y., Dubonos, S.V., Grigorieva, I.V., Firsov, A.A. (2004). Electric field effect in atomically thin carbon films. *Science*, 306(5696), 666–669.

Novoselov, K.S., Geim, A.K., Morozov, S.V., Jiang, D., Katsnelson, M.I., Grigorieva, I.V., Dubonos, S.V., Firsov, A.A. (2005). Two-dimensional gas of massless Dirac fermions in graphene. *Nature*, 438(7065), 197–200.

Rhodes, D., Chae, S.H., Ribeiro-Palau, R., Hone, J. (2019). Disorder in van der Waals heterostructures of 2D materials. *Nature Materials*, 18(6), 541–549.

Roldan, R. and Guinea, F. (2017). Electronic properties and strain engineering in semiconducting transition metal dichalcogenides. In *2D Materials Properties and Devices*, Avouris, P., Heinz, T.F., Low, T. (eds). Cambridge University Press, Cambridge.

Rycerz, A., Tworzydlo, J., Bennakker, C.W.J. (2007). Valley filter and valley valve in graphene. *Nature Physics*, 3(March), 172–175.

Sallen, G., Bouet, L., Marie, X., Wang, G., Zhu, C.R., Han, W.P., Lu, Y., Tan, P.H., Amand, T., Liu, B.L. et al. (2012). Robust optical emission polarization in MoS_2 monolayers through selective valley excitation. *Physical Review B (Condensed Matter and Materials Physics)*, 86(8), 081301.

Schaibley, J.R., Yu, H., Clark, G., Rivera, P., Ross, J.S., Seyler, K.L., Yao, W., Xu, X. (2016). Valleytronics in 2D materials. *Nature Reviews Materials*, 1(11), 16055.

Shao-Yu, C., Goldstein, T., Taniguchi, T., Watanabe, K., Jun, Y. (2018). Coulomb-bound four- and five-particle intervalley states in an atomically-thin semiconductor. *Nature Communications*, 9(1), 317.

Shim, J., Bae, S.-H., Kong, W., Lee, D., Qiao, K., Nezich, D., Park, Y.J., Zhao, R., Sundaram, S., Li, X. et al. (2018). Controlled crack propagation for atomic precision handling of wafer-scale two-dimensional materials. *Science*, 362(6415), 665.

Shimazaki, Y., Yamamoto, M., Borzenets, I.V., Watanabe, K., Taniguchi, T., Tarucha, S. (2015). Generation and detection of pure valley current by electrically induced berry curvature in bilayer graphene. *Nature Physics*, 11(12), 1032–1036.

Sie, E.J., McIver, J.W., Lee, Y.-H., Fu, L., Kong, J., Gedik, N. (2015). Valley-selective optical Stark effect in monolayer WS_2. *Nature Materials*, 14(3), 290–294.

Song, Z., Quhe, R., Liu, S., Li, Y., Feng, J., Yang, Y., Lu, J., Yang, J. (2015). Tunable valley polarization and valley orbital magnetic moment Hall effect in honeycomb systems with broken inversion symmetry. *Scientific Reports*, 5(1), 13906.

Splendiani, A., Sun, L., Zhang, Y., Li, T., Kim, J., Chim, C.-Y., Galli, G., Wang, F. (2010). Emerging photoluminescence in monolayer MoS_2. *Nano Letters*, 10(4), 1271–1275.

Sui, M., Chen, G., Ma, L., Shan, W.-Y., Tian, D., Watanabe, K., Taniguchi, T., Jin, X., Yao, W., Xiao, D. et al. (2015). Gate-tunable topological valley transport in bilayer graphene. *Nature Physics*, 11(12), 1027–1031.

Tanaka, K., Hachiya, K., Zhang, W., Matsuda, K., Miyauchi, Y. (2019). Machine-learning analysis to predict the exciton valley polarization landscape of 2D semiconductors. *ACS Nano*, 13(11), 12687–12693.

Tao, L.L., Naeemi, A., Tsymbal, E.Y. (2020). Valley-spin logic gates. *Physical Review Applied*, 13(5), 054043.

Titze, M., Bo, L., Xiang, Z., Ajayan, P.M., Hebin, L. (2018). Intrinsic coherence time of trions in monolayer $MoSe_2$ measured via two-dimensional coherent spectroscopy. *Physical Review Materials*, 2(5), 054001.

Unuchek, D., Ciarrocchi, A., Avsar, A., Sun, Z., Watanabe, K., Taniguchi, T., Kis, A. (2019). Valley-polarized exciton currents in a van der Waals heterostructure. *Nature Nanotechnology*, 14(12), 1104–1109.

Vitale, S.A., Nezich, D., Varghese, J.O., Kim, P., Gedik, N., Jarillo-Herrero, P., Xiao, D., Rothschild, M. (2018). Valleytronics: Opportunities, challenges, and paths forward. *Small*, 14(38), 1801483.

Wan, Y., Xiao, J., Li, J., Fang, X., Zhang, K., Fu, L., Li, P., Song, Z., Zhang, H., Wang, Y. et al. (2018). Epitaxial single-layer MoS_2 on GaN with enhanced valley helicity. *Advanced Materials*, 30(5), 1703888.

Wang, G., Marie, X., Liu, B.L., Amand, T., Robert, C., Cadiz, F., Renucci, P., Urbaszek, B. (2016). Control of exciton valley coherence in transition metal dichalcogenide monolayers. *Physical Review Letters*, 117(18), 187401.

Wang, G., Chernikov, A., Glazov, M.M., Heinz, T.F., Marie, X., Amand, T., Urbaszek, B. (2018a). Colloquium: Excitons in atomically thin transition metal dichalcogenides. *Reviews of Modern Physics*, 90(2), 021001.

Wang, K., De Greve, K., Jauregui, L.A., Sushko, A., High, A., Zhou, Y., Scuri, G., Taniguchi, T., Watanabe, K., Lukin, M.D. et al. (2018b). Electrical control of charged carriers and excitons in atomically thin materials. *Nature Nanotechnology*, 13(2), 128–132.

Wu, G.Y., Lue, N.Y., Chen, Y.C. (2013). Quantum manipulation of valleys in bilayer graphene. *Physical Review B (Condensed Matter and Materials Physics)*, 88(12), 125422.

Wu, Z., Zhou, B.T., Cai, X., Cheung, P., Liu, G.-B., Huang, M., Lin, J., Han, T., An, L., Wang, Y. et al. (2019). Intrinsic valley Hall transport in atomically thin MoS_2. *Nature Communications*, 10(1), 611.

Xiao, D., Yao, W., Niu, Q. (2007). Valley-contrasting physics in graphene: Magnetic moment and topological transport. *Physical Review Letters*, 99(23), 236809.

Xiao, D., Liu, G.-B., Feng, W., Xu, X., Yao, W. (2012). Coupled spin and valley physics in monolayers of MoS_2 and other group-VI dichalcogenides. *Physical Review Letters*, 108(19), 196802.

Xiao, J., Ye, Z., Wang, Y., Zhu, H., Wang, Y., Zhang, X. (2015). Nonlinear optical selection rule based on valley-exciton locking in monolayer WS_2. *Light: Science & Applications*, 4(12), e366–e366.

Xu, X., Yao, W., Xiao, D., Heinz, T.F. (2014). Spin and pseudospins in layered transition metal dichalcogenides. *Nature Physics*, 10(5), 343–350.

Yan, T., Yang, S., Li, D., Cui, X. (2017). Long valley relaxation time of free carriers in monolayer WSe_2. *Physical Review B*, 95(24), 241406.

Yang, L., Chen, W., McCreary, K.M., Jonker, B.T., Lou, J., Crooker, S.A. (2015). Spin coherence and dephasing of localized electrons in monolayer MoS_2. *Nano Letters*, 15(12), 8250–8254.

Ye, Z., Sun, D., Heinz, T.F. (2017). Optical manipulation of valley pseudospin. *Nature Physics*, 13(1), 26–29.

Yu, T. and Wu, M.W. (2014). Valley depolarization due to intervalley and intravalley electron-hole exchange interactions in monolayer MoS_2. *Physical Review B (Condensed Matter and Materials Physics)*, 89(20), 205303.

Zeng, H., Dai, J., Yao, W., Xiao, D., Cui, X. (2012). Valley polarization in MoS_2 monolayers by optical pumping. *Nature Nanotechnology*, 7(8), 490–493.

Zhang, Y., Tan, Y.-W., Stormer, H.L., Kim, P. (2005). Experimental observation of the quantum Hall effect and Berry's phase in graphene. *Nature*, 438(7065), 201–204.

Zhang, L., Gogna, R., Burg, G.W., Horng, J., Paik, E., Chou, Y.-H., Kim, K., Tutuc, E., Deng, H. (2019). Highly valley-polarized singlet and triplet interlayer excitons in van der Waals heterostructure. *Physical Review B*, 100(4), 041402.

Zhong, D., Seyler, K.L., Linpeng, X., Cheng, R., Sivadas, N., Huang, B., Schmidgall, E., Taniguchi, T., Watanabe, K., Mcguire, M.A. et al. (2017). Van der Waals engineering of ferromagnetic semiconductor heterostructures for spin and valleytronics. *Science Advances*, 3(5), 1603113.

7

Molecular Electronics: Electron, Spin and Thermal Transport through Molecules

Dominique VUILLAUME
Institute of Electronic Microelectronics and Nanotechnology (IEMN), CNRS, Villeneuve d'Ascq, France

7.1. Introduction

It is widely accepted that molecular electronics (ME) started in 1971 when Mann and Kuhn were the first to measure electron transport and demonstrate the tunneling effect through a monolayer of organic molecules (alkyl chains) connected between two electrodes (Mann and Kuhn 1971). In 1974, Aviram and Ratner went a step further when they theoretically proposed the concept of a molecular rectifier (molecular diode) (Aviram and Ratner 1974). They introduced the idea that a donor–acceptor molecule can rectify the current as in the p–n semiconductor junctions. We nevertheless note that the foundations of these ideas are the former works by R.S. Mulliken, who developed the concept of molecular orbitals (in the 1940s) and later made seminal works on charge transfers in donor–acceptor complexes (Mulliken 1949). Another seminal contribution, namely that protein molecules can conduct electricity, was proposed by A. Szent-Györgyi in the same period (Szent-Györgyi 1941). The reader interested in a historical review on the foundations of ME is referred to Hush (2003).

For a color version of all figures in this chapter, see www.iste.co.uk/cresti/beyondcmos.zip.

Beyond-CMOS, coordinated by Alessandro CRESTI.

The Aviram-Ratner molecular rectifier was experimentally demonstrated in 1990 by Ashwell et al. (1990; Martin et al. 1993) and then extensively studied by many groups, especially by Metzger et al. (1997; Vuillaume et al. 1999; Krzeminski et al. 2001), the reader is referred to an extended review published recently (Metzger 2015). Nowadays, due to molecular engineering and optimization of the molecule/electrode interactions, the performances of the molecular rectifiers, i.e. the rectification ratios (ratio of the current thought the device at two opposite voltage polarities) are in excess of 10^5 on a par with their semiconductor counterparts (Chen et al. 2017).

ME aims at "information processing using photo-, electro-, iono-, magneto-, thermo-, mechanico or chemio-active effects at the scale of structurally and functionally organized molecular architectures" (adapted from Lehn (1988)). ME is of great interest for basic science at the nanoscale, and because molecules are quantum objects by nature, chemists can tailor their properties at the synthesis level opening many perspectives for new experiments.

This chapter reviews recent results on the physics of electron transport in molecular devices. The chapter is organized as follows. A brief description of molecular junction (MJ) technology is first given followed by an introduction to the basic physics of electron transport through MJs from DC to ~20 GHz. Then, several sections review selected results on spin-dependent transport, plasmonics, quantum interferences, thermal transport and electronic noise in ME devices.

7.2. How to make a molecular junction

The term molecular junction (MJ) refers to a simple device where molecules (from a monolayer to a single molecule) are connected between two electrodes (metal or semiconductor), Figure 7.1. The number of molecules in the MJ depends on the size of these electrodes, and the MJs are usually classified as "large-area MJ" and "single (or a few) MJ". Albeit the Langmuir–Blodgett method was used in the early times of ME to deposit molecules on the electrode surfaces (only large-area MJs), chemisorption is nowadays widely used because it is also suitable at the nanoscale to attach few molecules between electrodes of a nanometer dimension. In that case, the molecules are equipped with anchor groups at both (or only one) ends. The chemical nature of the anchor group is chosen depending on the nature of the electrodes to permit a chemical reaction with the electrode by forming a chemical bond between the molecule and the electrode. Archetypes of anchor groups are thiols (–SH) on metals (Au, Ag, etc.) and alkenes (–C=C–) on hydrogenated-silicon surfaces, and a detailed review is given in Schreiber (2000), Love et al. (2005), and Vilan and Cahen (2017). This approach is widely versatile to adapt the molecules on the electrodes of interest. However, this chemical link also has a pronounced effect on the global electronic properties of the MJs, mainly governing the electronic coupling between molecules and electrodes (see below in this chapter), and this point has been extensively studied (see a review Jia and Guo (2013)). For large-area MJs (left part of Figure 7.1), in a vertical configuration, the top electrodes (typically with a lateral size > few μm, roughly speaking $\gtrsim 10^8$ molecules in the MJs) can be fabricated by metal evaporation. However, due to the extreme thickness of the molecular layer (monolayer), the metal atoms can easily diffuse into the monolayers resulting in nanoscale metallic shorts and a low yield of fabrication. This drawback can be avoided by intercalating a thin layer of a highly conductive polymer ("organic metal"), acting as a protective buffer between the monolayer and the metal electrode (Akkerman et al. 2006) and soft metal transfer techniques have been developed (Vilan and Cahen 2002; Loo et al. 2003; Jeong et al. 2017). The other solution, most widely used for laboratory experiments, is the use of soft metal contacts such as mercury drops (Mann and Kuhn 1971; Holmlin et al. 2001) or eutectic GaIn drop contacts (Chiechi et al. 2008); the latter being combined with microfluidic systems for a better control and stability (Karuppannan et al. 2019).

At the nanoscale, scanning tunneling microscope (STM) and conductive probe atomic force microscope (C-AFM) are the tools of choice to study a single or a few (typically $\lesssim$ 100) molecules (right part of Figure 7.1). For single-molecule experiments, STM break-junctions (STM-BJ) (Xu and Tao 2003) and mechanically controlled break junctions (MCBJ) (Reed et al. 1997) have been developed (see a review in Xiang et al. (2013) and Schwarz and Lörtscher (2014)). In these BJ approaches, the two electrodes are repeatedly (up to thousands or more times) moved close and apart and a molecule bridges the electrode gap when its size matches the electrode gap. The signature of the molecule conductance appears as plateaus in the current vs. gap electrode distance traces. C-AFM is used to gently contact (weak loading force) the monolayer (Wold and Frisbie 2000; Cui et al. 2001; Morita and Lindsay 2007) on large surfaces and to measure the current voltage at a few hundred places on the monolayer. Another approach is the use of tiny nanodot electrodes (nanodot–molecule junction, NMJ) (Clement et al. 2011; Smaali et al. 2012). In this latter case, hundreds to thousands of nanodots (typically 5–40 nm in diameter) are fabricated by e-beam lithography and can be measured in a "one shot" single C-AFM image. Note that all these techniques require a large number of measurements and a solid statistical analysis to get reliable conductance values of the MJs. Finally, another approach is to use a 2D network of metal nanoparticles (NP)-molecule network (NMN) (van der Molen et al. 2009; Liao et al. 2015; Viero et al. 2015, 2018). The basic element is an NP–molecule–NP junction with few molecules bridging the gap between two adjacent NPs (typically around 10 nm in diameter), its conductance being extracted from the global conductance of the NMN, if the NP–molecule–NP elements are reasonably ordered and the network structure is known (e.g. via scanning electron microscopy).

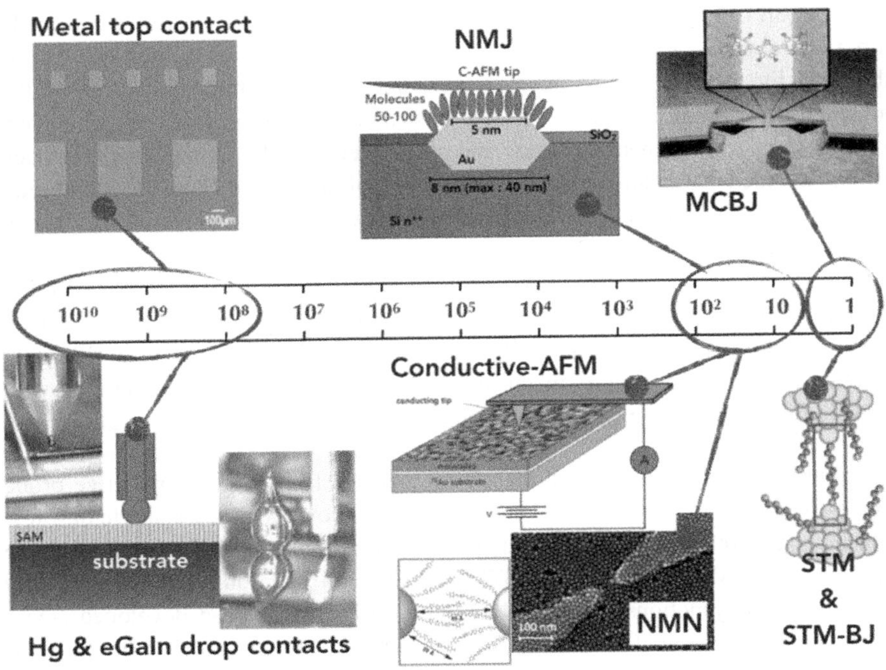

Figure 7.1. *An overview of several types of molecular junctions (MJs) along a scale of the approximate number of molecules in the MJ: from "large-area MJ" on the left to few molecules and single-molecule junction on the right*

More details on these technological issues can be found in Haick and Cahen (2008) and Vuillaume (2008, 2010).

7.3. Electron transport in molecular devices: back to basics

We briefly expose how electrical charges (electrons and holes) are transported through a molecule (or an ensemble of molecules, e.g. a molecular monolayer) connected between two electrodes. For more details, the reader is referred to textbooks Bourgoin et al. (2007) and Cuevas and Scheer (2010).

Figure 7.2 shows the energy diagram of a molecular junction. In this simplified scheme, the electrodes are modeled by their Fermi energy, ε_F, and the molecule by their quantized energy levels (molecular orbital, MO) whose highest occupied molecular orbital (HOMO, ε_H) and lowest unoccupied molecular orbital (LUMO, ε_L) play the biggest role in charge transport. All energies are referred to the vacuum energy level. Two other parameters are required, Γ_L and Γ_R, to model the interactions of the molecules with the left and right electrodes (Figure 7.2a). These energies correspond to the interaction between the electrons in the molecules and the electron cloud in the electrodes, i.e. the coupling between the molecular orbitals and the density of states in the electrodes. As a result of this hybridization of the MOs with the delocalized states of electrons in the electrodes, the MOs are broadened by a quantity $\Gamma = \Gamma_L + \Gamma_R$. This coupling energy, typically ~ 0.1 to 10 meV, mainly dictates the physical nature of the electron transport. In the weak coupling regime (~ 0.1 meV), the molecule can be considered as a quantum dot with an ultra-thin tunnel barrier at the molecule/electrode interface. More precisely, this regime corresponds to $\Gamma << U$, with U the additional energy, i.e. the energy required to add an electron to the LUMO (U_{N+1}-U_N) or to remove one electron from the HOMO (U_{N-1}-U_N) and U_N being the total energy of the system with N electrons. In this regime, Coulomb blockade (at low enough temperature) and incoherent transport (the electron wave function can be perturbed during the transport) are usually observed. This regime (Coulomb blockade) is not discussed in this chapter; for more details, see Cuevas and Scheer (2010) and Thijssen and Van Der Zant (2008). In the medium–strong coupling regime (considered here), the transport is usually coherent and can be classified as off-resonant transport (Figure 7.2b) or resonant transport (Figure 7.2c) (see next section). The coupling energy is determined by the physical or chemical nature of the molecule/electrode interface. The weak coupling case generally corresponds to physisorbed molecules or molecules chemisorbed on electrodes covered by a very thin insulating layer (e.g. native oxide on silicon or on Al, metal covered by few atomic layers of insulating materials like NaCl or KBr on Ag, for instance). For anchoring molecules on metal (e.g. Au), molecules equipped with thiol-anchoring groups at their ends are usually used giving a larger coupling energy, but other anchoring groups have also been studied (e.g. amine, pyridine, cyano, carboxyl), in each case, the coupling being strongly dependent on the atomic detail configuration between the anchoring group and the electrode surface (see reviews in Jia and Guo (2013), Su et al. (2016), Hybertsen and Venkataraman (2016); Evers et al. (2020)). In both coupling regimes, and whether the transport is coherent or incoherent, we also distinguish elastic (at constant energy) and inelastic electron transport (Figure 7.2d), where the energy of the electron is modified by various interactions (e.g. with the vibration modes of the chemical bonds in the molecule).

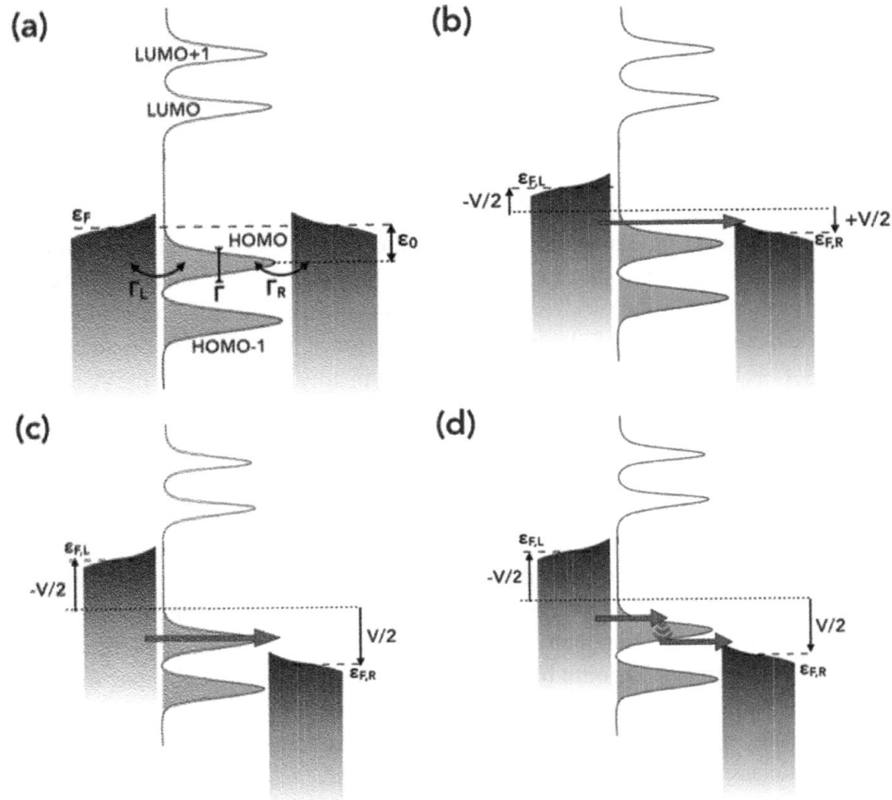

Figure 7.2. *(a) Simplified energy diagram of a molecular junction. (b) Coherent, off-resonant, electron transport, eV < 2ε₀. (c) Coherent, resonant, electron transport, eV ≥ 2ε₀. (d) Inelastic electron transport*

7.4. Electron transport: DC and low frequency

At low voltages (Figure 7.2b), the quantum tunneling effect is the dominant electron transport mechanism. At higher voltages (Figure 7.2c), resonant transport through one of the molecular orbitals (here HOMO for illustration) is allowed as soon as $eV = \pm 2\varepsilon_0$, with ε_0 the energy position of the involved MO with respect to the electrode Fermi energy. The key characteristic of such a molecular junction is its current versus voltage, I(V), curve. The I(V) characteristic is described by the Büttiker–Imry–Landauer formalism as (Büttiker et al. 1985):

$$I(V)=\frac{2e}{h}\int T(E)\left[f(E,\varepsilon_{F,L})-f(E,\varepsilon_{F,R})\right]dE \quad [7.1]$$

where T(E) is the transmission coefficient (electron transmission probability) through the molecule, f is the Fermi–Dirac statistics $f(E,\varepsilon)=\left[1+\exp\frac{E-\varepsilon}{k_BT}\right]^{-1}$, e is the electron charge, k_B is the Boltzmann constant, T is the temperature, h is the Planck constant and $\left|\varepsilon_{F,L}-\varepsilon_{F,R}\right|=eV$ and V is the applied voltage. T(E) is determined using first-principles calculations, such as non-equilibrium Green's functions (NEGF) combined with density functional theory (DFT); see Cuevas Scheer (2010). Figure 7.3 shows such a calculated T(E) and the corresponding I(V) curves for an azobenzene derivative molecule contacted between Co and Au electrodes and for several conformation of the molecule in the molecular junction (Thomas et al. 2021). If we consider that there is no peak of T(E) near ε_F (i.e. off-resonance, ε_F far enough from the LUMO and HOMO levels) and a low temperature, the conductance of the molecular junction can be simplified (from equation [7.1]) to:

$$G=\frac{2e^2}{h}T(\varepsilon_F)=G_0T(\varepsilon_F) \quad [7.2]$$

with G_0 the quantum of conductance (77.5 µS). A simple analytical model can be derived if we consider that i) a single MO (either LUMO or HOMO) dominates the charge transport, ii) the voltage mainly drops at the molecule/electrode interface (i.e. no shift of the molecular orbital with the applied voltage) and iii) the MO broadening is described by a Lorentzian or Breit–Wigner distribution (Datta 1995; Cuevas and Scheer 2010). Then, the transmission coefficient can be simplified and written as:

$$T(E)=\frac{4\Gamma_L\Gamma_R}{(E-\varepsilon_0)^2+(\Gamma_L+\Gamma_R)^2} \quad [7.3]$$

In this single energy-level model, at a low temperature, we get the following analytical expression (Datta 1995):

$$I(V)=\frac{8e}{h}\frac{\Gamma_L\Gamma_R}{\Gamma_L+\Gamma_R}\left(\arctan\left(\frac{\varepsilon_0+\frac{\Gamma_L}{\Gamma_L+\Gamma_R}eV}{\Gamma_L+\Gamma_R}\right)-\arctan\left(\frac{\varepsilon_0-\frac{\Gamma_R}{\Gamma_L+\Gamma_R}eV}{\Gamma_L+\Gamma_R}\right)\right) \quad [7.4]$$

with $\varepsilon_0 = \varepsilon_{L/H}-\varepsilon_F$ depending on the MO involved in the transport. This equation is used to fit the experimental data and to extract the parameters of the models ε_0, Γ_L and Γ_R. At a low bias (off-resonant transport), this equation gives an S-like shape I(V) curve with a step-like increase of the current (in absolute value) when approaching the resonant transport condition and a plateau for voltages larger than $V = 2\varepsilon_0/e$ (in absolute value). Figure 7.4 shows simulated I(V) curves with various values of the parameters ε_0, Γ_L and Γ_R. The parameter ε_0 mainly controls the position of the resonant transport step and the curvature of the S-like shape at the low bias ($V < 2\varepsilon_0/e$) (Figure 7.4a), while the coupling energy Γ_L and Γ_R dictate the current amplitude (Figure 7.4b). Asymmetric I(V) curves (rectification diode behavior) are obtained with $\Gamma_L \neq \Gamma_R$, a situation experimentally encountered when the MOs involved in the electron transport are geometrically located in an asymmetric position in the junction, for example, when the molecule is closer to an electrode than the other (Lenfant et al. 2003; Nijhuis et al. 2009; Chen et al. 2017), or when the molecule is coupled to the electrodes via two chemically different anchoring groups (Figure 7.5) (Brunner 2013). Note that this single energy-level model has to be used with caution. Equation [7.4] is a low-temperature approximation, and it can be used at room temperature for voltages below the resonant transport conditions (Grüter et al. 2005; Brunner 2013), since the temperature broadening of the Fermi function is not taken into account.

Similarly, due to its simplicity, the same I(V) curve can account for two different situations: an electron transport through the HOMO strongly coupled to the right electrode (red curve in Figure 7.4c) and an electron transport through the LUMO strongly coupled to the left electrode (green squares in Figure 7.4c). Finally, this model does not capture some important effects like the fact that the MO energy level in the junctions and the coupling parameters are dependent on the applied voltage. Recently, it has been demonstrated that fixing the low-bias conductance value and the Seebeck coefficient in the fit protocol improve the determination of the MO energy level (Delmas et al. 2020). Another approach to measure ε_0 is to determine the voltage at which it is aligned with the Fermi energy of one of the electrodes. In this technique, known as transition voltage spectroscopy (TVS), plotting abs(V^2/I) versus V gives two peaks (positive and negative threshold voltages: V_{T+} and V_{T-})

when the levels are aligned (Figure 7.4d) (Beebe et al. 2006, 2008; Huisman et al. 2009; Mirjani et al. 2011; Bâldea 2012). The MO energy level is given by Bâldea (2012):

$$\left|\varepsilon_0\right| = 2\frac{e\left|V_{T+}V_{T-}\right|}{\sqrt{V_{T+}^2 + 10\left|V_{T+}V_{T-}\right|/3 + V_{T-}^2}} \qquad [7.5]$$

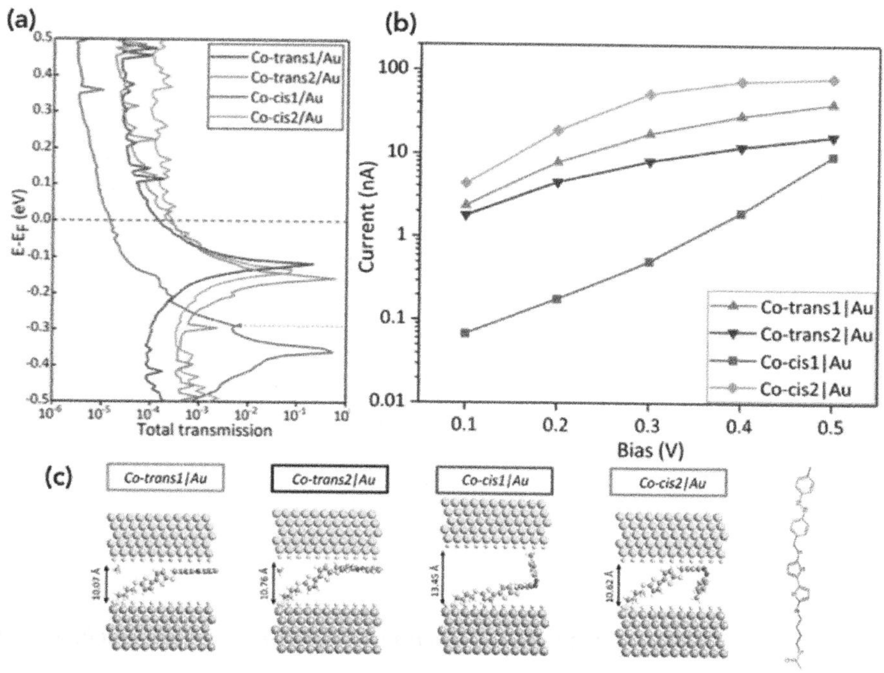

Figure 7.3. *(a) Calculated (DFT and NEGF) electron transmission coefficient T(E) and (b) the corresponding I(V) curves using equation [7.1] for a Co–molecule–Au junction. (c) Calculated optimized geometry for four conformations of the azobenzene–bithiophene–alkylthiol molecule (chemical structure shown at the right side). From Thomas et al. (2021) Royal Society of Chemistry*

However, this approach is subject to the same cautions as the one energy-level model since equations [7.5] and [7.4] are based on the same approximations. In addition, energy levels associated with defects or impurities at the molecule/electrode interfaces (e.g. silicon and metal electrodes covered by ultra-thin native oxides) can be detected and wrongly attributed to the molecule (Ricoeur et al. 2012). In conclusion, these analyses of the experimental I(V) curves have to be considered as qualitative and must be carefully checked versus ab initio calculations.

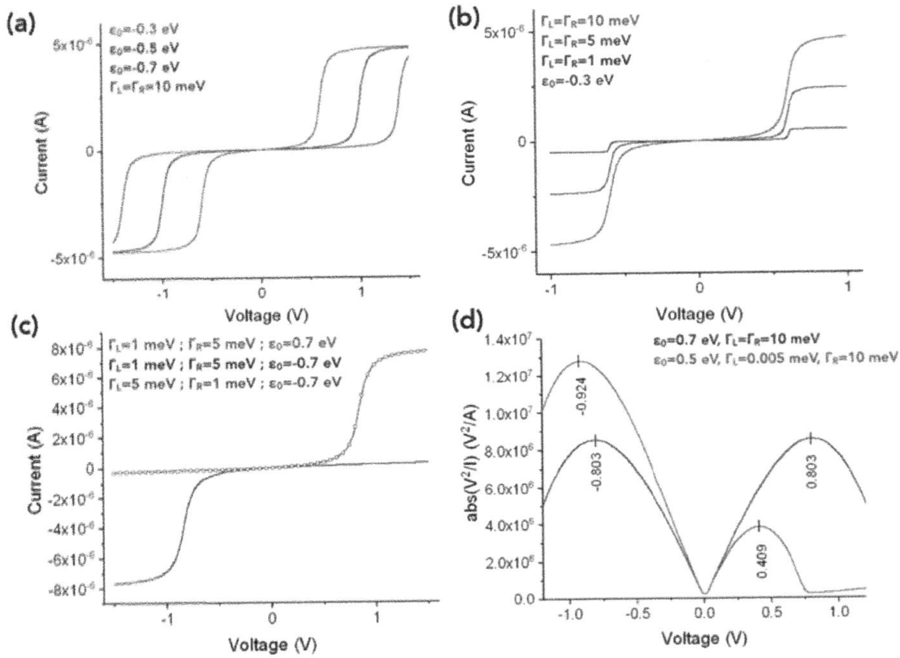

Figure 7.4. *Analytical modelling of electron transport in molecular junctions*

NOTES ON FIGURE 7.4. *(a) Simulated I(V) curves for different values of the energy level ε_0 ($\Gamma_L = \Gamma_R = 10$ meV). (b) Simulated I(V) curves for different values of coupling energy $\Gamma_L = \Gamma_R$ ($\varepsilon_0 = 0.3$ eV). (c) Simulated I(V) curves with $\Gamma_L \neq \Gamma_R$. (d) Transition voltage spectroscopy (TVS): plot of abs(V^2/I) vs. V and determination of the negative and positive threshold voltages (V_{T-} and V_{T+}) for two sets of parameters. Red curve: V_{T-} = -0924 V, V_{T+} = 0.409 V, given ε_0 = 0.5 V with equation [7.5]. Blue curve: $V_{T-} = V_{T+}$ = 0.803 V, given ε_0 = 0.7 V with equation [7.5].*

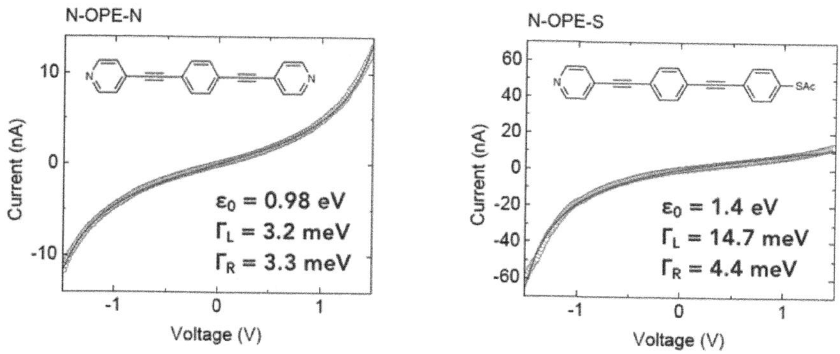

Figure 7.5. *Typical examples of the fits of the one energy-level model on molecular junctions with symmetric anchoring groups N-OPE-N and asymmetric anchoring groups N-OPE-S (OPE: olygo(phenylene ethynylene); N and S refer to pyridine- and thiol-anchoring groups, respectively). The single-molecule I(V) curves are measured by MCBJ (mechanically controlled break junction). After Brunner (2013)*

Inside molecular junctions with more than a single molecule, the classical electrical circuitry laws are no longer valid, i.e. the conductance of N molecules associated in parallel is no longer the sum of the conductance of each individual molecule. This feature is due to molecule–molecule interactions, for example, π–π molecular interactions (Magoga Joachim 1999; Reuter et al. 2011). Reuter et al. (Reuter et al. 2012) theoretically predicted that these molecule–molecule interactions should induce an asymmetry of the conductance histogram distribution (no longer Gaussian). This prediction was recently verified by varying the density of interacting molecules in nanodot/molecule junctions (NMJs) connected by conductive AFM (Trasobares et al. 2017). In these experiments (Figure 7.6), a classical log-normal distribution was observed for diluted molecules (weak interaction) – Figure 7.6d. On the contrary, the conductance histograms for densely packed molecules (strong interactions) clearly exhibited an asymmetric distribution with a tail towards the low conductance values (red arrow, Figure 7.6c) as a fingerprint of these interactions (Reuter et al. 2012). From this distribution, we determined the π–π interaction energy (30–35 meV) in good agreement with first-principles calculations (Trasobares et al. 2017).

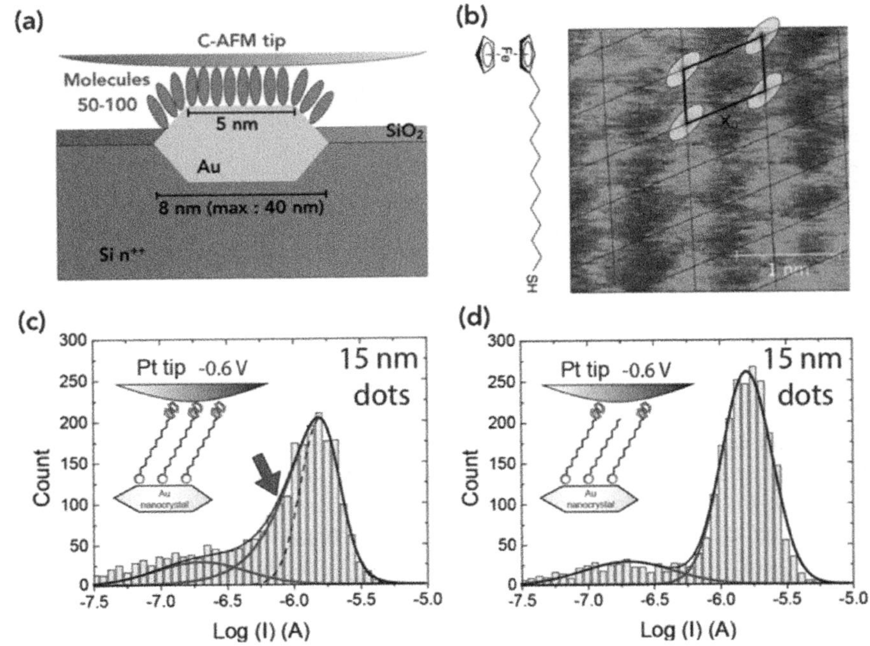

Figure 7.6. *Nanodot-molecule-junction (NMJ)*

NOTES ON FIGURE 7.6.– *(a) Scheme (cross section) of the nanodot–molecule–junction (NMJ). Tiny gold nanodots (diameter 5–40 nm) are fabricated (e-beam lithography) on highly doped Si substrate and covered by a monolayer of chemisorbed ferrocenyl-undecanethiol molecules. (b) Chemical structure of the molecule (Fc-C_{11}-SH) and high-resolution image of the ferrocenyl moieties (orange spot) organized on the top flat surface of the Au nanodot measured by STM in UHV. (c) Current histogram acquired by C-AFM on ~ 3,000 NMJs for a densely packed monolayer (to favor the intermolecular π–π interactions) and (d) for a diluted monolayer (~ 10 alkylthiol chains per ferrocenyl-undecanethiol). From Trasobares et al. (2017) American Chemical Society.*

7.5. Electron transport at high frequencies

The typical electron transit time through a molecular junction considered as a tunnel barrier (off-resonant transport) is given by:

$$\tau = \sqrt{\frac{m}{2\Delta}}d \qquad [7.6]$$

with m the electron mass, d the molecule length and Δ the tunnel barrier (i.e. ε_L-ε_F or ε_F-ε_H), predicting transit times of few femtoseconds considering d = 1 nm and Δ = 1 eV and in the ps regime for resonant transport (Nitzan 2001; Ke et al. 2010). Thus, operations of ME devices in the THz regime seems theoretically possible. However, up to now, molecular electronics devices were characterized in the DC regime, or at frequencies below MHz. We have recently demonstrated a molecular diode working at a frequency of 18 GHz with an estimation of the cut-off frequency of 520 GHz (Trasobares et al. 2016). Such a value is on a par with the performance of RF-silicon Schottky diodes. These results were obtained by combining a tiny molecular diode made of about few tens of ferrocenyl-alkylthiol molecules chemically grafted on a nanodot Au electrode (5–20 nm in diameter) and connected with the tip of a homemade modified interferometer scanning microwave microscope (iSMM) simultaneously measuring the DC current and the microwave reflection signal S_{11} (Dargent et al. 2013; Wang et al. 2014). Figure 7.7 shows the diode rectification behavior at DC, ~4 and 18 GHz. We clearly observe a rectification behavior of ~12 dB of the measured microwave reflection signal S_{11} at 4 GHz (~4 dB at 18 GHz). The dynamic conductance vs. voltage curves at 18 GHz (deduced from the S_{11} measurements) and the DC conductance (measured simultaneously) are similar (Figure 7.6d), demonstrating that the molecular rectification behavior is preserved up to 18 GHz. These results demonstrate that molecular electronics is prone to high-frequency operation, and open perspectives to explore theoretically predicted exotic effects such as conductance amplification due to dynamic resonance of electron transport (Guyon et al. 2005; Wu et al. 2005) and THz molecular switches (Orellana and Claro 2003).

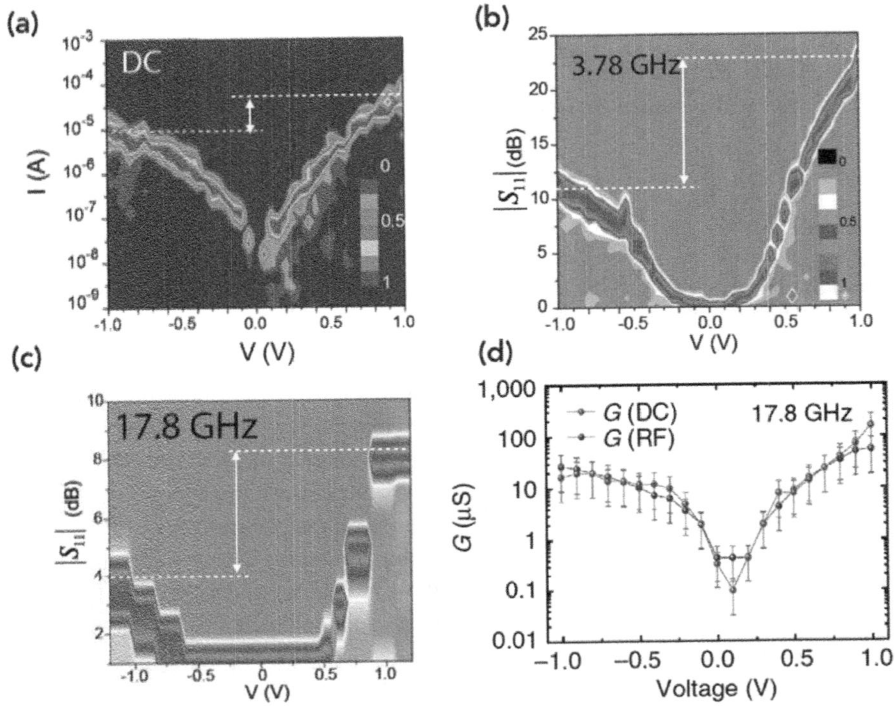

Figure 7.7. *Histograms ("heat maps")*

NOTES ON FIGURE 7.7.– *(a) DC current versus DC voltage acquired on a network of 100 ferrocenyl-undecanethiol molecular junctions (NMJs, Figure 7.5a) and the simultaneously measured amplitude of the microwave S_{11} parameter at (b) 3.8 GHz and (c) 18 GHz. (d) Comparison of the DC conductance and the microwave conductance at 18 GHz (extracted from the S_{11}-V data). From Trasobares et al. (2016), creative commons CC BY license.*

7.6. Spin-dependent electron transport in molecular junctions

Using the spin magnetic moment of the electron instead of its charge to encode and process information (spin electronics or spintronics) is a field of research increasingly studied. This spin can only have two quantum states (+ ½, we speak of "spin up" or - ½ "spin down"). This binary system thus allows encoding a logical "bit" "0 or 1". For organic materials and molecules, the weak spin–orbit coupling of carbon and the weak hyperfine interaction of the electron spins with the nucleus therefore suggests longer spin lifetimes (time during which the spin can remain polarized "up" or "down" before reversing due to various interactions with its environment). Tunnel magnetoresistance (TMR) through a ferromagnetic metal (FM)–molecular monolayer–FM junction was first reported in 2004 (Petta et al. 2004). The principle is identical to that of inorganic spin–valve junctions, and the tunnel insulator is here played by a monolayer of alkyl chains, sandwiched between two ferromagnetic (Ni) electrodes. A TMR of 16% (at 4K) has been observed, but with a strong dispersion from one sample to another (e.g. positive and negative TMR have been observed for the same system). The presence of localized defects in the molecular tunnel barrier could be responsible for this dispersion. A correlation between the TMR decay and the diffraction of spins at the metal–molecule interface by molecular vibration modes was suggested on Ni–octanethiol–Co junctions (Wang and Richter 2006). These results were rationalized and the TMR performances improved by a detailed study of the spin-dependent hybridization at FM electrode/molecule interface, demonstrating that the spin polarization of the molecule-functionalized FM electrodes can be inverted or enhanced depending on the spin-dependent coupling strength between the molecules and the FM electrodes (Barraud et al. 2010), Figure 7.8a. This has led to the birth of a new research field on molecular interfaces, dubbed "spinterface" (Sanvito 2011; Galbiati 2014; Delprat et al. 2018) with many confirmations spanning from STM (Figure 7.8b) to photoemission experiments (Iacovita et al. 2008; Atodiresei et al. 2010; Javaid et al. 2010; Schmaus et al. 2011; Kawahara et al. 2012; Lach et al. 2012; Djeghloul et al. 2013). Progress was obtained recently in the fabrication of molecular junctions on FM electrodes, for example, by the chemical grafting of alkylphosphonic derivative monolayers on $La_{0.7}Sr_{0.3}MnO_3$ (Tatay et al. 2012; Galbiati et al. 2015; Poggini et al. 2016) with a reported TMR up to 10^4 % (Figure 7.8c) and with a significant stability up to high voltages (few volts on nm-thick monolayers) (Galbiati et al. 2012; Tatay et al. 2016). Simple molecules (alkyl chains) were used in these works and more functional (or stimuli responsive) molecules are mandatory for more elaborated molecular devices (Vuillaume 2010; Xiang et al. 2016), for example, redox molecules for memory and photochromes for electro-optical molecular devices.

Beyond this first step, we should expect that the possibility of tuning the resistance as well as the magnetoresistance thanks to the spin-dependent hybridization at the FM electrode/molecule interface could lead to a new class of devices combining analogical and digital properties. When photoswitch molecules are used with FM electrodes, the spin-polarized electron transport through the FM–molecule–FM junctions will depend on the conformation of the molecules and the molecule/electrode atomic contact geometry as evaluated from theoretical studies (Wang et al. 2013; Ulman et al. 2014; Zeng et al. 2016). However, the chemical grafting of monolayers of functional molecules (more complex structures than simple alkyl chains) on FM electrodes remain challenging. We have recently reported the optically induced conductance switching at the nanoscale (conductive-AFM) of diarylethene derivative self-assembled monolayers (SAMs) on $La_{0.7}Sr_{0.3}MnO_3$ electrodes (Thomas et al. 2020) and observed a weak conductance switching of the diarylethene molecular junctions (*closed* isomer/*open* isomer conductance ratios $R_{c/o} < 10$), partly hidden under some conditions by the optically induced conductance switching of the $La_{0.7}Sr_{0.3}MnO_3$ substrate, and conductance ratio (*cis*/*trans* isomers) of about 20 for azobenzene derivatives on Co (Thomas et al. 2021) (Figure 7.8d). These performances are lower than those of molecular junctions of the same azobenzene derivatives on gold electrodes for which conductance ratios up to $\approx 7x10^3$ were measured (Smaali et al. 2010) and $R_{c/o} \approx 100$ were calculated and measured for diarylethene derivatives (Van Dyck et al. 2013; Jia et al. 2016). These results call more experimental and theoretical works to design stimuli-responsive molecular spin valves with higher performances.

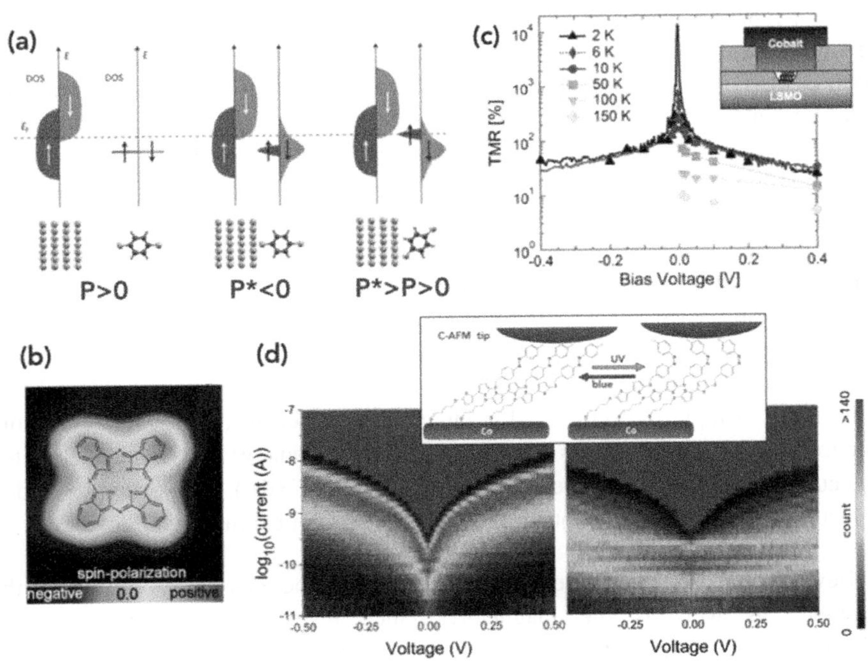

Figure 7.8. *Examples of molecular spintronics*

NOTES ON FIGURE 7.8.– *(a) Scheme of the spinterface concept. A ferromagnetic electrode with a positive spin polarization (higher density of spin-up states) is* *functionalized with a molecule. Depending on the shift and broadening of the molecular orbitals, the spin polarization P* of the functionalized electrode can be inverted (P*<0) or enhanced (P*>P>0). From Sanvito (2010) Nature. (b) An experimental example of the spinterface effect. Spin-polarized STM image of a single metal-free phthalocyanine molecule (H_2Pc) on a Fe surface showing the spin polarization inversion with respect to the Fe surface. From Atodiresei et al. (2010) American Physical Society. (c) TMR up 10^4 (at 2K) measured in a LSMO/tetradecane phosphonic acid monolayer/Co spin valve. From Tatay et al. (2016), with permission from the Institute of Physics. (d) Current–voltage 2D histograms of Co–azobenzene–bithiophene–alkylthiol monolayer/C-AFM tip for the "trans" and "cis" conformation of the molecules. The conformational switch is triggered by light (UV light for trans-to-cis, and visible light for cis-to-trans). From Thomas et al. (2021) Royal Society of Chemistry.*

Another approach to molecular spintronics consists of using one (or more) magnetic molecule in a molecular junction. The first experiments were performed with paramagnetic molecules (terpyridinyl lingands) complexing a magnetic ion (Co) (Park et al. 2002). In a regime of strong coupling of the molecule with the electrodes and at low temperatures (<25–30 K), a Kondo effect, characteristic of the presence of an unpaired spin electron, has been observed. In a transistor configuration (with a bottom gate electrode), and with a molecule containing two magnetic centers (two vanadium atoms), it has been shown that this effect can be inhibited or activated by playing on the charge state of the molecule by applying a gate voltage (switching between the Coulomb blockade regime and the Kondo regime) (Liang et al. 2002). These first demonstrations of a molecular spin transistor opened the door to other possibilities such as the magnetic molecular spin valve (a magnetic molecule between two ferromagnetic electrodes) – see a review in Bogani Wernsdorfer (2008) or the implementation of spin qubits in a magnetic molecule electrically controlled by an STM tip, for example (Lehmann et al. 2007). It is also possible to use the spin state of a molecule attached to a non-magnetic one-dimensional conductor (e.g. a carbon nanotube) to modulate the electronic transport in this conductor. This modulation could be detected directly by measuring the current or by using an ultra-sensitive carbon nanotube nano-squid that allows measuring small variations of magnetic flux (Cleuziou et al. 2006). The topics of Kondo physics and molecular magnet electronics have been observed and studied in a wide variety of molecules. The reader is referred to recent review papers for more details (Scott and Natelson 2010; Gehring et al. 2019; Coronado 2020).

7.7. Molecular electronic plasmonics

In nanophotonics, a promising approach is devices based on plasmons – oscillations of electrons at metal–dielectric interfaces – that can operate at optical frequencies. Unlike photons, plasmons can be manipulated with metal nanostructures of a few tens of nanometers, i.e. far below the diffraction limit. The ability to capture and confine light in nanoscale structures below 100 nm in the form of so-called surface plasmons (SPs) may transform current technologies. SPs confine and enhance local electromagnetic fields near surfaces of metallic nanostructures at optical frequencies and have the ability to propagate along sub-diffractive metallic waveguide opening up new perspectives for integrated opto-electronic circuits at the nanoscale (Barnes et al. 2003; Lal et al. 2007; Gramotnev and Bozhevolnyi 2010; Schuller et al. 2010; Du et al. 2017). On the other hand, molecular-scale electronics operates at the length scale of few nanometers; thus, combining molecular devices and plasmonics is an opportunity to study electron/plasmon interaction in the quantum regime.

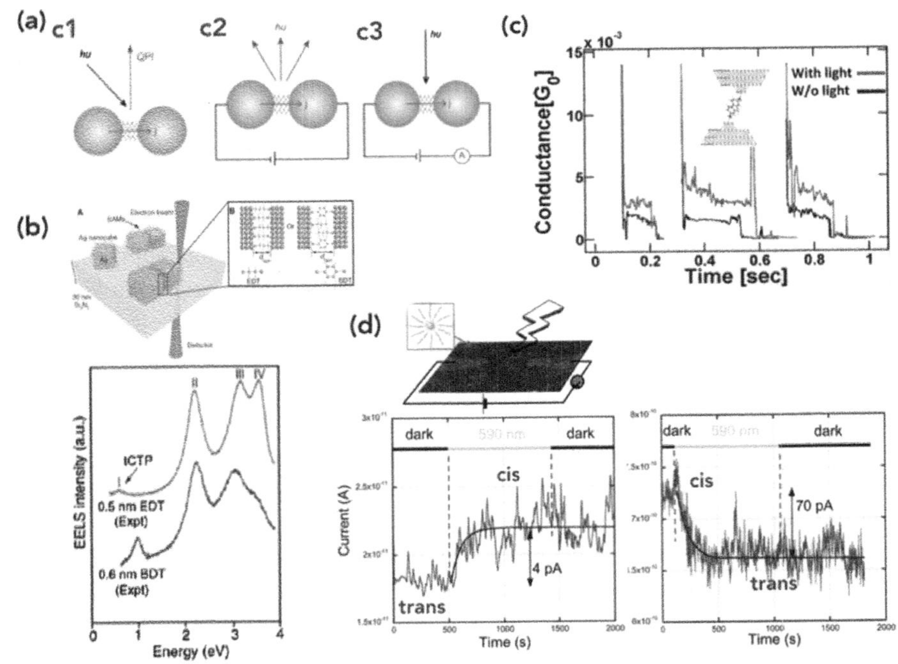

Figure 7.9. *Molecular plasmonics*

NOTES ON FIGURE 7.9.– *(a) Schemes of operation principles for molecular electronic plasmonics. From Wang Nijhuis (2016) Elsevier. (b) Direct observation of the tunnel charge transfer plasmon (tCTP) in Ag nanocube dimers connected by two types of molecules. From Tan et al. (2014), Science. (c) Plasmon-assisted tunneling through a single diaminofluorene molecule junction. From Vadai et al. (2013) American Chemical Society. (d) Plasmon-induced isomerization of azobenzene derivatives and the corresponding variation of electron transport properties of a 2D network of Au nanoparticles (10 nm diameter) capped with azobenzene bithiophene molecules. From Stievenard et al. (2018) Royal Society of Chemistry.*

Figure 7.9a shows several possible ways how molecular tunneling junctions based on electrode–molecule–electrode configurations can be combined with plasmonics. For simplicity, the electrodes are drawn as spherical plasmonic resonators, but other forms (rods, planar surface, etc.) are possible. The molecular component can be present in the form of a single molecule or a self-assembled monolayer (SAM). In the configuration c1 (Figure 7.9a), molecular junctions are used to study quantum plasmonics. Two closely spaced plasmonic resonators are bridged by a SAM onto which plasmons are excited. Usually this is done by incident light or by an electron beam inside a transmission electron microscope. These plasmons induce an electric field across the gap resulting in quantum mechanical tunneling across the molecules leading to quantum plasmon resonances (QPR) such as the so-called charge transfer plasmon (CTP) modes (Zuloaga et al. 2009; Esteban et al. 2012). In this dimer configuration, if the gap is below $< \sim 1$ nm, a new quantum plasmon, the tunneling charge transfer plasmon mode (tCTP), has been observed in junctions consisting of two silver nanocubes separated by SAM of 1,2-ethanedithiolates (EDT) or 1,4-benzenedithiolates (BDT) (Figure 7.9b). The emission energy of the tCTP can be tuned by changing the molecular structures (and by doing so the tunneling barrier height) bridging two plasmonic resonators.

In the configuration c2 (Figure 7.9a), a voltage is applied across the molecular gap. The tunneling charge carriers then excite plasmon modes in the electrode material, either directly by the tunneling charge carriers or via electroluminescence from the molecules inside the junction (Reecht et al. 2014). In this molecular electronic plasmon excitation devices, the properties of the generated plasmons can be controlled by the molecules inside the devices: i) the polarization of the plasmon depends on the tilt angle of the SAM; ii) the frequency of the plasmon depends on the applied bias and iii) the bias-selective plasmon excitation in only one direction of the bias using a molecular diode (Du et al. 2017).

The reverse process, i.e. the coupling of a plasmon to tunneling charge carriers, can also happen (Arielly et al. 2011). In this configuration (c3, Figure 7.9a), the plasmons are excited in the junction via an external light source and they couple to the tunneling charge carriers and increase the tunneling current across the junction (so-called optical rectification). In this case, molecular electronics is applied to detect plasmons. A first mechanism is plasmon-assisted tunneling (PAT), because the plasmon field modulates the tunnel barrier in the junction. A second, indirect, mechanism is possible if the molecules absorb the plasmon energy and generate electron–hole pairs. A typical example of PAT is shown in Figure 7.9c for a single molecule (diaminofluorene) junction (Vadai et al. 2013). The plasmon field enhancement in the molecular junction is about 10^3. Finally, if it is known that the redox and/or conformational states of molecules linked to plasmonic nanostructures can shift the SP frequency (Zheng et al. 2009; Joshi et al. 2014; Schaming et al. 2014). It has been recently demonstrated that SP can also be used to induce the isomerization of molecular switches (i.e. azobenzene derivatives) and then a significant change of the current in the devices (Figure 7.9d) (Stievenard et al. 2018). A plasmon-induced resonance energy transfer (PIRET) mechanism (Li et al. 2015) is likely responsible for this effect. This plasmon-induced isomerization (PII) is faster (about a factor 10) than the usual isomerization triggered by UV–visible light, which may be helpful for light-driven molecular memory and light-reconfigurable molecular circuits.

A detailed and comprehensive review on molecular electronic plasmonic has been recently published in Wang and Nijhuis (2016).

7.8. Quantum interference and thermal transport

Thermoelectricity at the nanoscale is based on the seminal work of Hicks and Dresselhaus, who have established theoretically that one-dimensional quantum wires (formed by adjacent metal atoms) should lead to highly efficient thermoelectric systems compared to 2D and 3D systems (Hicks and Dresselhaus 1993). They demonstrated that lowering the dimensionality of the systems improves the electronic quantities that govern thermoelectricity. In molecular junction, the first theoretical analysis of the thermoelectric properties was reported by Paulsson Datta (2003) for a benzenedithiol molecule. Since that, many works have theoretically and experimentally shown that molecular junctions indeed exhibit very interesting thermoelectric characteristics; they are described in a series of review papers (Lambert 2015; Rincon-Garcia et al. 2016; Cui et al. 2017; Gehring et al. 2019).

The thermoelectric properties are described by three simple equations:

$$S = -\frac{\Delta V}{\Delta T} \quad [7.7]; \quad ZT = \frac{\sigma S^2}{\kappa} = \frac{G_{el} S^2}{G_{th}} \quad [7.8]; \quad PF = \sigma S^2 \text{ or } G_{el} S^2 \quad [7.9]$$

where S is the Seebeck coefficient or thermopower (ΔV being the difference of voltage generated by applying a difference of temperature ΔT between the electrodes – Figure 7.10). The efficiency of energy conversion is evaluated by the thermoelectric figure of merit ZT. This metric is used regardless of the materials and systems used to fabricate the thermoelectric devices and can be used for comparison: the higher is ZT, the best is the material. In these equations, σ and κ are the electrical and thermal conductivities, respectively, or equivalently at the molecular level (single and few molecules devices) where dimensions are ill-defined, the electronic and thermal conductances are G_{el} and G_{th}, respectively. PF is the power factor. The thermal conductivity/conductance is the sum of a vibrational contribution and an electronic contribution, $\kappa = \kappa_v + \kappa_e$ ($G_{th} = G_{th,v}+G_{th,e}$). The vibrational contribution is negligible at low temperatures, but it increases with temperature and can eventually dominate the electronic contribution (this latter being significant if the molecule/material is enough conducting). However, in molecular junction, the electronic contribution is usually negligible with $G_{th,v}$ of the order of 1–70 $pW.K^{-1}$ (Rincon-Garcia et al. 2016; Cui et al. 2017).

The Seebeck coefficient can be described within the Landauer formalism:

$$S = -\frac{\pi^2 k_B^2 T}{3e} \frac{\partial \ln T(E)}{\partial E}\bigg|_{E=\varepsilon_F} = -\frac{\pi^2 k_B^2 T}{3e} \frac{1}{T(\varepsilon_F)} \frac{\partial T(E)}{\partial E}\bigg|_{E=\varepsilon_F} \quad [7.10]$$

with T(E) the transmission probability of electrons through the molecular junction. Since the sign of S is related to the slope of T(E), it is a fingerprint whether the charge transport through the molecular junction occurs via the LUMO or the HOMO, i.e. it depends on the direction of the electron flow (Figure 7.10a). It is positive if going to the hot side (thus promoted by HOMO, Figure 7.10a) and negative the other way around, i.e. electron transport through the LUMO (Figure 7.10b). Experimental and theoretical values of S range from 0.5 to 30 $\mu V.K^{-1}$ (Rincon-Garcia et al. 2016; Cui et al. 2017).

From the analysis of the known values of S, G_{el} and G_{th} of a large variety of molecular junction, it may be inferred that molecular junctions are prone to a high thermoelectric capacity with expected ZT higher than in bulk systems (values of 3 to 4 have been theoretically predicted) (Noori et al. 2017). However, this requires that high values of S and G_{el} and a low value of G_{th} are obtained for the same molecule, which remains a challenge in molecular electronics. For instance, the electronic contribution to thermal conductance, $G_{th,el}$, is a function of electronic conductance G_{el}; thus, these parameters cannot be tuned independently.

Several factors can strongly influence T(E) and its derivative, and thus the thermoelectric properties (Figure 7.10). Firstly, the position of the HOMO/LUMO levels with respect of the Fermi energy. Moving these levels close to ε_F increases the electronic conductance G_{el} and the slope $\partial \ln(T(E))/\partial E$ (thus the Seebeck coefficient S) – Figure 7.10c. Secondly, a decrease of the coupling energies (Γ_L and Γ_R) increases the slope of ln(T(E)) at ε_F, but it decreases the conductance (see equation [7.4]), Figure 7.10d, thus at the risk of decreasing the power factor PF (equation [7.9]). Thirdly, quantum interferences (QI) have been theoretically proposed (Finch et al. 2009) to improve S and ZT by introducing narrower and asymmetry peaks in T(E) around the Fermi level (Figure 7.10e).

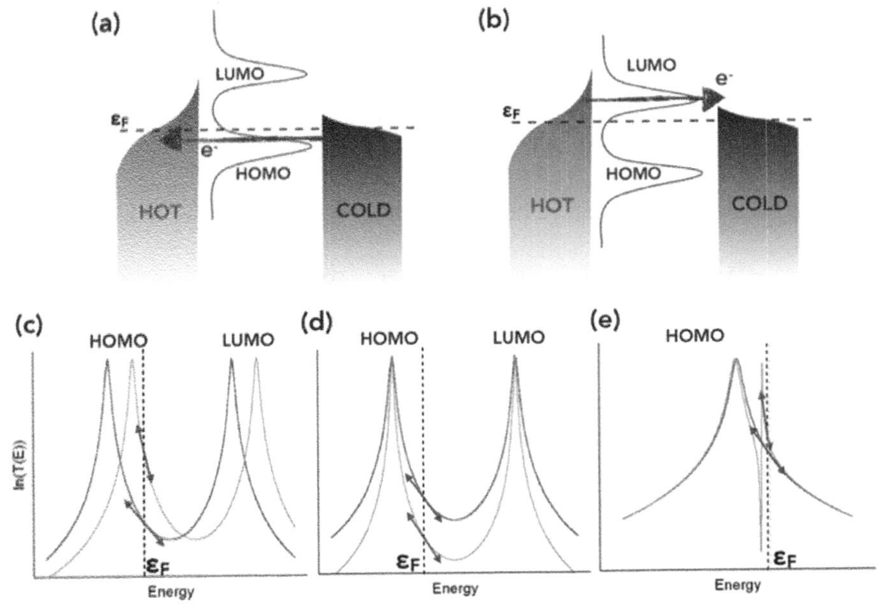

Figure 7.10. *Energy scheme of a molecular junction submitted to a temperature gradient (shown by the red and blue broadened Fermi–Dirac functions in the electrodes)*

NOTES ON FIGURE 7.10.– *When the HOMO is close to the Fermi energy (a), the net flow of electrons goes to the hot electrodes, while it goes to the cold electrode if the LUMO is close to the Fermi energy (b). Typical transmission coefficient T(E) and slope* $\partial ln(T(E))/\partial E$ *(red arrows) illustrating three ways to improve the Seebeck coefficient: (c) shift of the molecular orbitals closer to* ε_F*, (d) reducing the electrode coupling energy* Γ *and (c) introducing a sharp resonance (here a Fano resonance) near the Fermi energy. In (c) and (d) the blue lines are simulations from equation [7.3] with* ε_0 *= -0.5/0.5 eV (HOMO/LUMO),* $\Gamma_L = \Gamma_R$ *= 10 meV; the orange line in (c) with* ε_0 *= -0.3/0.7 eV (HOMO/LUMO),* $\Gamma_L = \Gamma_R$ *= 10 meV; the orange in (d) with* ε_0 *= -0.5/0.5 eV (HOMO/LUMO),* $\Gamma_L = \Gamma_R$ *= 5 meV. In (e), Fano resonance equation [7.11] with* ε_0 *= -0.5 eV,* Γ *= 10 meV,* ε_1 *= -0.4 and u = 5 meV (orange curve) and without (blue curve) Fano resonance (*ε_1 *and u = 0).*

Quantum interferences in molecular junctions (Sautet and Joachim 1988; Magoga and Joachim 1999) can arise in several situations. The first one occurs if the electron transmission is mediated by a combination of different molecular orbitals (Figure 7.11a). The quantum mechanics wavefunctions (described by amplitude and phase) of the electrons passing through one or another orbital may have their phases similar or in opposition (phase difference of π) resulting in constructive or destructive interferences, respectively, that increase or decrease the conductance (a dip is created in T(E) near ε_F). The typical example is the benzene-based molecular junctions with connections to the electrodes in para or meta positions (Figure 7.11b), and several experimental results were reported for several π-conjugated short molecules (Mayor et al. 2003; Aradhya et al. 2012; Ballmann et al. 2012; Arroyo et al. 2013a, 2013b, Li et al. 2016; Garner et al. 2018).

Another possible QI is the destructive Fano resonance (Fano 1961), which occurs if, near the Fermi energy, the molecular junction has a delocalized molecular orbital between the two electrodes and a discrete energy level localized on another part of the molecule (Figure 7.11c), for example, due to a chemical moiety on a side of the molecule backbone (i.e. a pendant group). If these two energy levels are close in energy and the localized state is coupled with the delocalized one but not directly to the electrodes, charge transport resonances will occur because of quantum confinement of electrons and T(E) will display an asymmetric Fano resonance/ anti-resonance, thus resulting in a pronounced variation of its derivative and then an enhancement of the thermopower (Figure 7.10e). In this case, T(E) is no longer described by equation [7.3], but it can be written as:

$$T(E)=\frac{4\Gamma_L\Gamma_R}{\left(E-\varepsilon_0-\frac{u^2}{E-\varepsilon_1}\right)^2+\left(\Gamma_L+\Gamma_R\right)^2} \qquad [7.11]$$

with ε_0 and ε_1 the energy levels of the delocalized and localized states, respectively, Γ_L and Γ_R the coupling energies of the delocalized states to the electrodes and u the coupling energy between the two states (Figure 7.11c).

Experimental and theoretical results have been published evocating Fano resonance to explain the electron transport of specifically designed molecular junctions (Papadopoulos et al. 2006; Solomon et al. 2008a, 2008b; Fracasso et al. 2011; Guédon et al. 2012; Kim et al. 2012; García-Suárez et al. 2013; Rabache et al. 2013; Darwish et al. 2014; Valkenier et al. 2014; Carlotti et al. 2016). Among them, a typical example concerns anthraquinone derivatives that are cross-conjugated molecules showing a lower conductance compared to their linearly conjugated counterparts (Figure 7.11d). The cross-conjugated anthraquinone displays a pronounced dip of around 0 bias that is the fingerprint of a destructive Fano anti-resonance (Fracasso et al. 2011; Darwish et al. 2012; Guédon et al. 2012; Rabache et al. 2013).

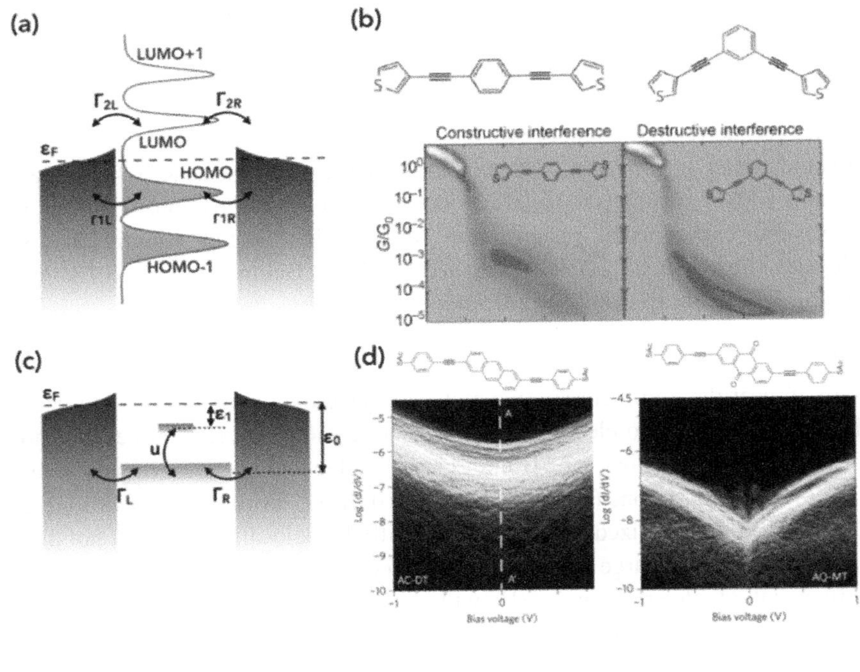

Figure 7.11. *Quantum interferences in molecular junctions*

NOTES ON FIGURE 7.11.– *(a) Quantum interference energy scheme when two energy levels (here HOMO and LUMO) are combined in the electron transmission. (b) Typical examples of constructive and destructive QIs in a benzene derivative molecule connected to the electrodes in the para and meta positions, respectively. The MCBJ measurements show a plateau at a higher conductance for the molecule in the para position. From Arroyo et al. (2013b) Wiley. (c) Energy scheme for a molecular junction with a Fano resonance between a delocalized state (purple) and a localized state (red) coupled with the delocalized state but not to the electrodes. (d) Comparison of the conductance–voltage curves (conductive-AFM measurements) of a cross-conjugated anthraquinone and a linearly conjugated analog. The anthraquinone molecular junction shows a lower current with a marked dip in the conductance at 0 bias. From Guédon et al. (2012), Nature.*

Albeit ZT values as high as theoretically predicted (ZT $\approx$ 4 at room temperature (Noori et al. 2017) are not yet experimentally demonstrated using this QI approach, these findings pave the way towards important improvements and higher level of sophistication of the thermoelectric molecular devices by adapting its intrinsic components (nature of electrodes, molecular backbone and interfacial coupling) to obtain the required efficiency at working conditions.

7.9. Noise in molecular junctions

Noise (stochastic fluctuations of a signal or quantity) is a ubiquitous phenomenon in nature and the low-frequency noise (LFN) or the so-called 1/f noise (also known as flicker noise, discovered by Johnson (1925) in vacuum tubes) is observed in many fields (physics, astrophysics, biology, economics, technology, etc.), not only in electronic devices (Ward and Greenwood 2007). In molecular electronics, noise spectroscopy has been studied this last decade (or so) to reveal subtle electron transport phenomena not observed using conventional DC current measurements.

Noise in electronic devices is characterized by its frequency dependence according to its physical origin. Basically, the time-dependent current in any electronic device is written as $I(t) = \langle I \rangle + \delta I(t)$, with $\langle I \rangle$ the average current and $\delta I(t)$ the time-dependent fluctuations. Due to the stochastic nature of noise, the noise is more conveniently characterized in the frequency domain, via Fourier transformation, using its power spectral density (PSD), $S_I(f) = \langle \delta I(t)^2 \rangle / \Delta f$, where $\langle \delta I(t)^2 \rangle$ is the variance measured at a frequency f over a bandwidth Δf (Briaire and Vandamme 1998). The most basic noise is the thermal noise (or Johnson–Nyquist noise), which is due to the random thermal motion of electrons in the devices or materials and exists in any of them. This noise is independent of the frequency (white noise) and given by a current PSD $S_I = 4k_BT/R$, with k_B the Boltzmann constant, T the temperature and R the resistance. It is usually the noise floor of the device. In the following, I briefly describe some recent results on 1/f noise, RTS (random telegraph signal) noise and shot noise in molecular junctions. For more details, see a recently published review (Kim and Song 2021).

The 1/f noise refers to any fluctuations with a PSD proportional to $1/f^n$ with n usually close to 1, but more generally $0.5 \lesssim n \lesssim 2$, the value of n depending on the physical origin of the noise. In bulk materials and devices, 1/f noise is related to the number fluctuation or mobility fluctuation phenomena; see a review in Weissman (1988). The first one corresponds to the fluctuations of the number of charge carriers in the devices, which can arise due to trapping/detrapping of charges by traps (i.e. defects) in the materials and/or at interfaces (McWhorter 1957). In the second one (Hooge 1969), the mobility of charges fluctuates due to scattering by various events (impurities, defects, electron-phonon interactions, etc.). Note that both mechanisms can be active simultaneously. The current noise PSD is then strictly proportional to 1/f and scales as I^2 with I the DC current (Weissman 1988). Such a $1/f^n$ noise was observed (Figure 7.12a) in molecular junctions, silicon-alkyl chain (C_8H_{37})-Al, with $1 < n < 1.2$ (n slightly increasing with the applied voltage) (Clément et al. 2007). The normalized current PSD (S_I/I^2) reveals a bump at voltages between 0.4 and 1 V (Figure 7.12b), which was attributed to localized states (localized energy levels associated to defects/impurities) into the alkyl tunnel barrier. This behavior was consistently modeled by an energy-dependent trap-induced tunnel current and considering the effect of trapping/detrapping of electrons by a two-level system (Figure 7.12b) (Rogers and Buhrman 1984; Kogan 1996). This model also predicts the current PSD, S_I, scales as $(\partial I/\partial V)^2$ as experimentally observed (Clément et al. 2007). The bump in the S_I/I^2 versus V is sample dependent (Figure 7.12b) and it is likely due to an energy-peaked (modeled by a Gaussian) distribution of energy states in resonance with the Fermi energy windows for $V > 0.4$ V. The same result (bump in the normalized PSD) was later observed in Au–hexanedithiol–Au junctions and similarly ascribed to localized states (defects) in the molecular junctions (Kim et al. 2010). It is likely that these localized states are induced by the top metal (large area) deposition on the molecular monolayer in these cases. At the nanoscale, noise measurements on SAMs of alkylthiols and short π-conjugated molecules (4-mercaptopyridine, benzenedithiol) on Au by C-AFM showed that the current PSD is proportional to $1/f^2$ (Cho et al. 2017, 2018). This behavior was attributed to molecule–electrode interface bond (i.e. S–Au bonds) fluctuations that induce conductance fluctuations in the molecular junctions.

In tiny devices or for a small number of noise sources, RTS noise is observed in the time domain. RTS noise manifests as abrupt, discrete, current switching between two distinct levels, Figure 7.12(c) and (d) (albeit more levels are also observed), and it has been widely observed and characterized in nanoelectronic devices (Kirton Uren 1989). Figure 7.12c shows the first observation of RTS noise in a single (or a few) molecular junctions (Donhauser et al. 2001). In this experiment, a single or a few oligo-phenylene-ethynylene (OPE) derivatives were inserted in a monolayer of insulating molecules (alkylthiols) chemisorbed on Au surface and current images at a fixed bias were acquired by STM over time. The STM images showed a stochastic two-level switching of conductance of the OPE derivatives (Figure 7.12c). This behavior was explained by the fluctuations of the S–Au bonds at the molecule/metal interface, the S–Au bond being randomly broken/reformed or the degree of hybridization between the thiol group and the metal can change, thus modulating the molecular junction conductance (Moore et al. 2006). More recently, RTS noise was also observed in single-molecule experiments (STM-BJ and MCBJ) for various molecules: di(methylthio)-stilbene, dithiadecane at low temperatures (4–70 K) by STM-BJ (Adak et al. 2015) and oligophenyl at room temperature by MCBJ (Xiang et al. 2014) and MCBJ with alkanedithiol and oligo phenylene ethynylene (Brunner et al. 2014). Flicker noise was also observed at room temperature in the STM-BJ experiments with exponent n = 1.7 and the PSD was found to scale with $(\partial I/\partial V)^2$ if the molecules are weakly coupled to the two electrodes and to about $(\partial I/\partial V)$ for a molecule more strongly coupled at the two sides (Adak et al. 2015). For the oligophenyl MCBJ experiments, the PSD exhibits $1/f^2$ dependence as expected for RTS noise (Lorentzian PSD, $S_I \propto \tau/(1+(2\pi f\tau)^2)$, with τ the effective time constant of the RTS noise). In both experiments, the authors concluded that the RTS noise is due to the dynamic rearrangement of the molecule–metal bonds.

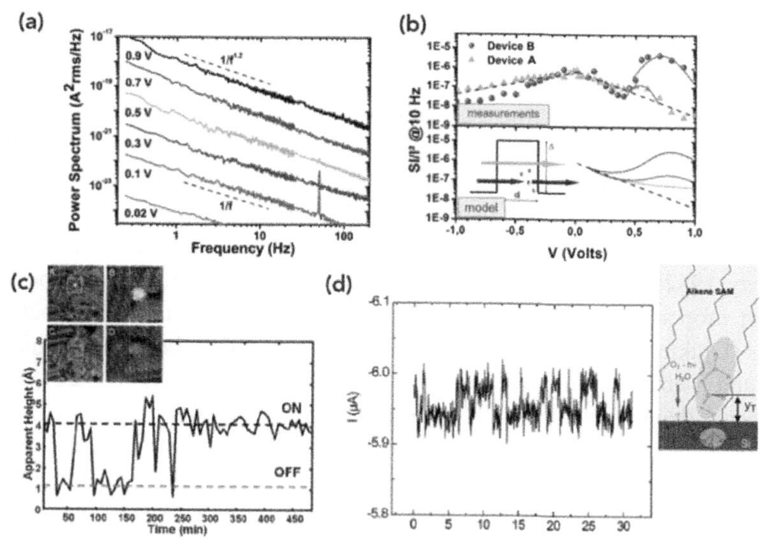

Figure 7.12. *Noise and fluctations in molecular junctions*

NOTES ON FIGURE 7.12.– *(a) Current PSD, S_I, in Si–alkyl–Al junctions at several voltages. (b) Normalized S_I/I^2 for two Si–alkyl–Al junctions versus the applied voltage and scheme of the trap-assisted tunneling model explaining the bump at V > 0.4V in the experiments, the bump amplitude being related to the density of defects in the molecular junctions. From Clément et al. (2007) American Physical Society. (c) STM images showing a typical RTS noise (two-level fluctuations) observed for a single OPE molecule in a matrix of alkylthiol monolayer on Au. From Donhauser et al. (2001) Science. (d) RTS noise due to an electrochemical reaction (redox) of water molecules trapped at the interface between silicon and alkyl molecules. From Clément et al. (2012) American Chemical Society.*

RTS noise was also used to study electrochemical reactions in the molecular junctions, such as redox reactions that can modify the transport mechanisms in the devices. In molecular junctions containing a low density of redox molecules (ferrocenyl-hexanethiol) diluted in a monolayer of non-redox molecules (alkylthiols), RTS noise was observed, while no RTS was detected in a pure non-redox molecular junction (Arielly et al. 2014). It was also demonstrated that a low density of water molecules trapped at the silicon–alkyl interface can induce RTS noise in the molecular junctions (Figure 7.12d). The underlying mechanism was suggested to come from electrochemical charge transfer reaction between the silicon substrate and H^+/H_2 redox couple with a characteristic energy level close to the Fermi energy of the silicon electrode (Clément et al. 2012).

Finally, shot noise (discovered by Schottky (1918) in vacuum electronic tubes) is also an interesting phenomenon in molecular junctions. Shot noise is due to the discrete and quantized nature of electrons transmitted through the device. It is related to the transmission coefficient $T_m(E)$, where m denotes the number of transmission channels in the junction. At low temperatures ($k_BT \ll eV$) and low transmission limit ($T_m \ll 1$), the shot noise PSD is simply $S_I = 2eVG_0\sum_m T_m = 2eI$, with G_0 the conductance quantum, V the applied voltage and I the DC current (Ambrozy 1982; Blanter Büttiker 2000). Shot noise was used to study the electron transport properties in single-molecule junctions with deuterium (D_2) between Pt contacts (Djukic Van Ruitenbeek 2006; Tewari et al. 2019), benzene (Kiguchi et al. 2008) and benzenedithiol (Karimi et al. 2016; OuYang et al. 2018). These experiments demonstrated that the shot noise in the Pt–D_2–Pt junction is strongly reduced and that a single channel, with a high transmission probability, dominates the electron transport. Thus, in such a nearly fully transparent channel, the electrons are transmitted coherently in the molecular junctions and the shot noise vanishes (Djukic and Van Ruitenbeek 2006). In addition, it was found that, at higher voltages, shot noise spectroscopy could be used to detect molecular vibrations (Tewari et al. 2019). In a Pt–benzene–Pt molecular junction, where the benzene is connected to the electrodes by direct Pt–C bonds, the DC conductance is high (0.1–1G_0). The shot noise measurements revealed that the electron transport occurs via several channels, the precise number depending on the molecule/electrode conformation (Kiguchi et al. 2008). For Au–benzenedithiol–Au junctions with a large range of conductance (0.01–0.24G_0), shot noise measurements demonstrated that electrons are always transmitted through a single conduction channel. The authors concluded that the Au–S bond at the interface is the limiting factor and that a direct tunneling from the Au contact to the benzene ring is hindered in that case (Karimi et al. 2016).

7.10. Conclusion and further reading

In this chapter, selected results are presented and discussed. More results and discussions are available in recent review papers. In addition to the review papers already cited in this chapter, several other review papers are of interest on ME in general (Aradhya and Venkataraman 2013; Xiang et al. 2016; Vilan et al. 2017; Evers et al. 2020) and on more specific topics:

– ME on silicon and other semiconductor platforms (Vilan and Cahen 2017);

– molecular spintronics (Cinchetti et al. 2017; Coronado 2020; Kumar and Ruben 2021);

– thermal transport in MJs (Cui et al. 2017; Wang et al. 2019);

– focalized on theory and computational methods (Nitzan 2001; Xue and Ratner 2006; Thijssen and Van Der Zant 2008).

From the perspective of applications, and although no ME devices are commercially available so far (and probably never will be), basic research in ME has significantly improved our fundamental understanding of physics at the nanoscale. Nevertheless, ME devices are envisioned to complement semiconductor devices by providing new functions or already existing functions at a simpler process level and at a lower cost by virtue of their self-organization capabilities. In addition to the general review papers mentioned above, significant results to implement electronic devices at the molecular scale are reviewed on:

– molecular diodes (Metzger 2015);

– molecular switches towards memory devices (Irie 2000; Kuhr et al. 2004; Fuentes et al. 2011; Fahrenbach et al. 2013);

– more exploratory research on ME for unconventional computing such as neuromorphic computing and quantum computing can be found in Lehmann et al. (2007), Avizienis et al. (2012), Beiu et al. (2012), Vincent et al. (2012), Bose et al.

(2015), Demis et al. (2015), Godfrin et al. (2017), Tanaka et al. (2018), Viero et al. (2018) and Gaita-Ariño et al. (2019).

7.11. References

Adak, O., Rosenthal, E., Meisner, J., Andrade, E.F., Pasupathy, A.N., Nuckolls, C., Hybertsen, M.S., Venkataraman, L. (2015). Flicker noise as a probe of electronic interaction at metal–single molecule interfaces. *Nano Letters*, 15(6), 4143–4149.

Akkerman, H.B., Blom, P.W.M., De Leeuw, D.M., De Boer, B. (2006). Towards molecular electronics with large-area molecular junctions. *Nature*, 441(4), 69–72.

Ambrozy, A. (1982). *Electronic Noise*. McGraw-Hill, New York.

Aradhya, S.V. and Venkataraman, L. (2013). Single-molecule junctions beyond electronic transport. *Nature Nanotechnology*, 8(6), 399–410.

Aradhya, S.V., Meisner, J.S., Krikorian, M., Ahn, S., Parameswaran, R., Steigerwald, M.L., Nuckolls, C., Venkataraman, L. (2012). Dissecting contact mechanics from quantum interference in single-molecule junctions of stilbene derivatives. *Nano Letters*, 12(3), 1643–1647.

Arielly, R., Ofarim, A., Noy, G., Selzer, Y. (2011). Accurate determination of plasmonic fields in molecular junctions by current rectification at optical frequencies. *Nano Letters*, 11(7), 2968–2972.

Arielly, R., Vadai, M., Kardash, D., Noy, G., Selzer, Y. (2014). Real-time detection of redox events in molecular junctions. *Journal of the American Chemical Society*, 136(6), 2674–2680.

Arroyo, C.R., Frisenda, R., Moth-Poulsen, K., Seldenthuis, J.S., Bjørnholm, T., Van Der Zant, H.S. (2013a). Quantum interference effects at room temperature in OPV-based single-molecule junctions. *Nanoscale Research Letters*, 8(1), 234.

Arroyo, C.R., Tarkuc, S., Frisenda, R., Seldenthuis, J.S., Woerde, C.H.M., Eelkema, R., Grozema, F.C., Van Der Zant, H.S.J. (2013b). Signatures of quantum interference effects on charge transport through a single benzene ring. *Angewandte Chemie International Edition*, 52, 3152–3155.

Ashwell, G.J., Sambles, J.R., Martin, A.S., Parker, W.G., Szablewski, M. (1990). Rectifying characteristics of Mg/(C16H33-Q3CNQ LB film)/Pt structures. *J. Chem. Soc., Chem. Commun.*, 19, 1374–1376.

Atodiresei, N., Brede, J., Lazić, P., Caciuc, V., Hoffmann, G., Wiesendanger, R., Blügel, S. (2010). Design of the local spin polarization at the organic-ferromagnetic interface. *Physical Review Letters*, 105(6), 066601.

Aviram, A. and Ratner, M.A. (1974). Molecular rectifiers. *Chem. Phys. Lett.*, 29(2), 277–283.

Avizienis, A.V., Sillin, H.O., Martin-Olmos, C., Shieh, H.-H., Aono, M., Stieg, A.Z., Gimzewski, J.K. (2012). Neuromorphic atomic switch networks. *PloS One*, 7(8), e42772.

Bâldea, I. (2012). Ambipolar transition voltage spectroscopy: Analytical results and experimental agreement. *Physical Review B*, 85(3), 035442.

Ballmann, S., Härtle, R., Coto, P.B., Elbing, M., Mayor, M., Bryce, M.R., Thoss, M., Weber, H.B. (2012). Experimental evidence for quantum interference and vibrationally induced decoherence in single-molecule junctions. *Physical Review Letters*, 109(5), 056801.

Barnes, W.L., Dereux, A., Ebbesen, T.W. (2003). Surface plasmon subwavelength optics. *Nature*, 424(6950), 824–830.

Barraud, C., Seneor, P., Mattana, R., Fusil, S., Bouzehouane, K., Deranlot, C., Graziosi, P., Hueso, L., Bergenti, I., Dediu, V. et al. (2010). Unravelling the role of the interface for spin injection into organic semiconductors. *Nature Physics*, 6(8), 615–620.

Beebe, J.M., Kim, B., Gadzuk, J.W., Frisbie, C.D., Kushmerick, J.G. (2006). Transition from direct tunneling to field emission in metal-molecule-metal junctions. *Physical Review Letters*, 97(2), 026801.

Beebe, J.M., Kim, B., Frisbie, C.D., Kushmerick, J.G. (2008). Measuring relative barrier heights in molecular electronic junctions with transition voltage spectroscopy. *ACS Nano*, 2(5), 827–832.

Beiu, V., Calame, M., Cuniberti, G., Gamrat, C., Konkoli, Z., Vuillaume, D., Wendin, G., Yitzchaik, S. (2012). Aspects of computing with locally connected networks. *AIP Conference Proceedings*, 1479, 1875–1879.

Blanter, Y.M. and Büttiker, M. (2000). Shot noise in mesoscopic conductors. *Physics Reports*, 336(1), 1–166.

Bogani, L. and Wernsdorfer, W. (2008). Molecular spintronics using single-molecule magnets. *Nature Materials*, 7(3), 179–186.

Bose, S.K., Lawrence, C.P., Liu, Z., Makarenko, K.S., Van Damme, R.M.J., Broersma, H.J., Van Der Wiel, W.G. (2015). Evolution of a designless nanoparticle network into reconfigurable Boolean logic. *Nature Nanotechnology*, 10, 1048–1053.

Bourgoin, J.P., Vuillaume, D., Goffman, M., Filoramo, A. (2007). Molecular electronics. In *Nanoscience, Nanotechnologies and Nanophysics*, Dupas, C. and Lahmani, M. (eds). Springer, Berlin.

Briaire, J. and Vandamme, L.K.J. (1998). Uncertainty in Gaussian noise generalized for cross-correlation spectra. *Journal of Applied Physics*, 84(8), 4370–4374.

Brunner, J. (2013). Gaining microscopic insight into molecular junctions by transport experiments. PhD thesis, Basel University.

Brunner, J., González, M.T., Schönenberger, C., Calame, M. (2014). Random telegraph signals in molecular junctions. *Journal of Physics: Condensed Matter*, 26(47), 474202.

Büttiker, M., Imry, Y., Landauer, R., Pinhas, S. (1985). Generalized many-channel conductance formula with application to small rings. *Physical Review B*, 31(10), 6207–6215.

Carlotti, M., Kovalchuk, A., Wächter, T., Qiu, X., Zharnikov, M., Chiechi, R.C. (2016). Conformation-driven quantum interference effects mediated by through-space conjugation in self-assembled monolayers. *Nature Communications*, 7(1), 13904.

Chen, X., Roemer, M., Yuan, L., Du, W., Thompson, D., Del Barco, E., Nijhuis, C.A. (2017). Molecular diodes with rectification ratios exceeding 105 driven by electrostatic interactions. *Nature Nanotechnology*, 12(8), 797–803.

Chiechi, R.C., Weiss, E.A., Dickey, M.D., Whitesides, G.M. (2008). Eutectic gallium–indium (EGaIn): A moldable liquid metal for electrical characterization of self-assembled monolayers. *Angewandte Chemie International Edition*, 47(1), 142–144.

Cho, D., Lee, H., Shekhar, S., Yang, M., Park, J.Y., Hong, S. (2017). Direct mapping of electrical noise sources in molecular wire-based devices. *Scientific Reports*, 7(1), 43411.

Cho, D., Shekhar, S., Lee, H., Hong, S. (2018). Nanoscale mapping of molecular vibrational modes via vibrational noise spectroscopy. *Nano Letters*, 18(2), 1001–1009.

Cinchetti, M., Dediu, V.A., Hueso, L.E. (2017). Activating the molecular spinterface. *Nature Materials*, 16(5), 507–515.

Clément, N., Pleutin, S., Seitz, O., Lenfant, S., Vuillaume, D. (2007). 1/fγ tunnel current noise through Si-bound alkyl monolayers. *Physical Review B*, 76, 205407.

Clément, N., Patriarche, G., Smaali, K., Vaurette, F., Nishiguchi, K., Troadec, D., Fujiwara, A., Vuillaume, D. (2011). Large array of sub-10-nm single-grain au nanodots for use in nanotechnology. *Small*, 7(18), 2607–2613.

Clément, N., Guerin, D., Pleutin, S., Godey, S., Vuillaume, D. (2012). Role of hydration on the electronic transport through molecular junctions on silicon. *The Journal of Physical Chemistry C*, 116(33), 17753–17763.

Cleuziou, J.-P., Wernsdorfer, W., Bouchiat, V., Ondarçuhu, T., Monthioux, M. (2006). Carbon nanotube superconducting quantum interference device. *Nature Nanotechnology*, 1(1), 53–59.

Coronado, E. (2020). Molecular magnetism: From chemical design to spin control in molecules, materials and devices. *Nature Reviews Materials*, 5, 87.

Cuevas, J.C. and Scheer, E. (2010). *Molecular Electronics: An Introduction to Theory and Experiment*, World Scientific, New Jersey.

Cui, X.D., Primak, A., Zarate, X., Tomfohr, J., Sankey, O.F., Moore, A.L., Moore, T.A., Gust, D., Harris, G., Lindsay, S.M. (2001). Reproductible measurement of single-molecule conductivity. *Science*, 294, 571–574.

Cui, L., Miao, R., Jiang, C., Meyhofer, E., Reddy, P. (2017). Perspective: Thermal and thermoelectric transport in molecular junctions. *The Journal of Chemical Physics*, 146(9), 092201.

Dargent, T., Haddadi, K., Lasri, T., Clément, N., Ducatteau, D., Legrand, B., Tanbakuchi, H., Theron, D. (2013). An interferometric scanning microwave microscope and calibration method for sub-fF microwave measurements. *Review of Scientific Instruments*, 84(12), 123705.

Darwish, N., Díez-Pérez, I., Da Silva, P., Tao, N., Gooding, J.J., Paddon-Row, M.N. (2012). Observation of electrochemically controlled quantum interference in a single anthraquinone-based norbornylogous bridge molecule. *Angewandte Chemie International Edition*, 51, 3203–3206.

Darwish, N., Aragonès, A.C., Darwish, T., Ciampi, S., Díez-Pérez, I. (2014). Multi-responsive photo- and chemo-electrical single-molecule switches. *Nano Letters*, 14(12), 7064–7070.

Datta, S. (1995). *Electronic Transport in Mesoscopic Systems*. Cambridge University Press.

Delmas, V., Diez-Cabanes, V., Van Dyck, C., Scheer, E., Costuas, K., Cornil, J. (2020). On the reliability of acquiring molecular junction parameters by Lorentzian fitting of I/V curves. *Physical Chemistry Chemical Physics*, 22(46), 26702–26706.

Delprat, S., Galbiati, M., Tatay, S., Quinard, B., Barraud, C., Petroff, F., Seneor, P., Mattana, R. (2018). Molecular spintronics: The role of spin-dependent hybridization. *J. Phys. D: Appl. Phys.*, 51(47), 473001.

Demis, E.C., Aguilera, R., Sillin, H.O., Scharnhorst, K., Sandouk, E.J., Aono, M., Stieg, A.Z., Gimzewski, J.K. (2015). Atomic switch networks–nanoarchitectonic design of a complex system for natural computing. *Nanotechnology*, 26(20), 204003.

Djeghloul, F., Ibrahim, F., Cantoni, M., Bowen, M., Joly, L., Boukari, S., Ohresser, P., Bertran, F., Le Fèvre, P., Thakur, P. et al. (2013). Direct observation of a highly spin-polarized organic spinterface at room temperature. *Scientific Reports*, 3, 1272.

Djukic, D. and Van Ruitenbeek, J.M. (2006). Shot noise measurements on a single molecule. *Nano Letters*, 6(4), 789–793.

Donhauser, Z.J., Mantooth, B.A., Kelly, K.F., Bumm, L.A., Monnell, J.D., Stapleton, J.J., Price, D.W., Rawlett, A.M., Allara, D.L., Tour, J.M. et al. (2001). Conductance switching in single molecules through conformational changes. *Science*, 292, 2303–2307.

Du, W., Wang, T., Chu, H.-S., Nijhuis, C.A. (2017). Highly efficient on-chip direct electronic–plasmonic transducers. *Nature Photon*, 11(10), 623–627.

Esteban, R., Borisov, A.G., Nordlander, P., Aizpurua, J. (2012). Bridging quantum and classical plasmonics with a quantum-corrected model. *Nature Communications*, 3, 825.

Evers, F., Korytár, R., Tewari, S., Van Ruitenbeek, J.M. (2020). Advances and challenges in single-molecule electron transport. *Reviews of Modern Physics*, 92(3), 035001.

Fahrenbach, A.C., Warren, S.C., Incorvati, J.T., Avestro, A.-J., Barnes, J.C., Stoddart, J.F., Grzybowski, B.A. (2013). Organic switches for surfaces and devices. *Advanced Materials*, 25, 331–348.

Fano, U. (1961). Effects of configuration interaction on intensities and phase shifts. *Physical Review*, 124(6), 1866–1878.

Finch, C.M., García-Suárez, V.M., Lambert, C.J. (2009). Giant thermopower and figure of merit in single-molecule devices. *Physical Review B*, 79(3), 033405.

Fracasso, D., Valkenier, H., Hummelen, J.C., Solomon, G.C., Chiechi, R.C. (2011). Evidence for quantum interference in SAMs of arylethynylene thiolates in tunneling junctions with eutectic Ga–In (EGaIn) top-contacts. *Journal of the American Chemical Society*, 133(24), 9556–9563.

Fuentes, N., Martin-Lasanta, A., Alvarez De Cienfuegos, L., Ribagorda, M., Parra, A., Cuerva, J. (2011). Organic-based molecular switches for molecular electronics. *Nanoscale*, 3(10), 4003–4014.

Gaita-Ariño, A., Luis, F., Hill, S., Coronado, E. (2019). Molecular spins for quantum computation. *Nature Chemistry*, 11, 301–309.

Galbiati, M. (2014). Spinterface: Crafting spintronics at the molecular scale. *MRS Bulletin*, 39, 602.

Galbiati, M., Barraud, C., Tatay, S., Bouzehouane, K., Deranlot, C., Jacquet, E., Fert, A., Seneor, P., Mattana, R., Petroff, F. (2012). Unveiling self-assembled monolayers' potential for molecular spintronics: Spin transport at high voltage. *Advanced Materials*, 24(48), 6429–6432.

Galbiati, M., Tatay, S., Delprat, S., Barraud, C., Cros, V., Jacquet, E., Coloma, F., Choueikani, F., Otero, E., Ohresser, P. et al. (2015). Influence of alkylphosphonic acid grafting on the electronic and magnetic properties of La2/3Sr1/3MnO3 surfaces. *Applied Surface Science*, 353, 24–28.

García-Suárez, V.M., Ferradás, R., Ferrer, J. (2013). Impact of Fano and Breit-Wigner resonances in the thermoelectric properties of nanoscale junctions. *Physical Review B*, 88(23), 235417.

Garner, M.H., Li, H., Chen, Y., Su, T.A., Shangguan, Z., Paley, D.W., Liu, T., Ng, F., Li, H., Xiao, S. et al. (2018). Comprehensive suppression of single-molecule conductance using destructive σ-interference. *Nature*, 558(7710), 415–419.

Gehring, P., Thijssen, J.M., Van Der Zant, H.S.J. (2019). Single-molecule quantum-transport phenomena in break junctions. *Nature Reviews Physics*, 1(6), 381–396.

Godfrin, C., Ferhat, A., Ballou, R., Klyatskaya, S., Ruben, M., Wernsdorfer, W., Balestro, F. (2017). Operating quantum states in single magnetic molecules: Implementation of Grover's quantum algorithm. *Physical Review Letters*, 119(18), 187702.

Gramotnev, D.K. and Bozhevolnyi, S.I. (2010). Plasmonics beyond the diffraction limit. *Nature Photonics*, 4(2), 83–91.

Grüter, L., Cheng, F., Heikkilä, T.T., González, M.T., Diederich, F., Schönenberger, C., Calame, M. (2005). Resonant tunnelling through a C60 molecular junction in a liquid environment. *Nanotechnology*, 16(10), 2143–2148.

Guédon, C.M., Valkenier, H., Markussen, T., Thygesen, K.S., Hummelen, J.C., Van Der Molen, S.J. (2012). Observation of quantum interference in molecular charge transport. *Nature Nanotechnology*, 7, 305.

Guyon, R., Jonckheere, T., Mujica, V., CréPieux, A., Martin, T. (2005). Current and noise in a model of an alternating current scanning tunneling microscope molecule-metal junction. *The Journal of Chemical Physics*, 122(14), 144703.

Haick, H. and Cahen, D. (2008). Making contact: Connecting molecules electrically to the macroscopic world. *Progress in Surface Science*, 83, 217–261.

Hicks, L.D. and Dresselhaus, M.S. (1993). Thermoelectric figure of merit of a one-dimensional conductor. *Physical Review B*, 47(24), 16631–16634.

Holmlin, R.E., Haag, R., Chabinyc, M.L., Ismagilov, R.F., Cohen, A.E., Terfort, A., Rampi, M.A., Whitesides, G.M. (2001). Electron transport through thin organic films in metal-insulator-metal junctions based on self-assembled monolayers. *Journal of the American Chemical Society*, 123(21), 5075–5085.

Hooge, F.N. (1969). $1/f$ noise is no surface effect. *Physics Letters A*, 29, 139–140.

Huisman, E.H., Guédon, C.M., Van Wees, B.J., Van Der Molen, S.J. (2009). Interpretation of transition voltage spectroscopy. *Nano Letters*, 9(11), 3909–3913.

Hush, N.S. (2003). An overview of the first half-century of molecular electronics. *Annals of the New York Academy of Sciences*, 1006, 1–20.

Hybertsen, M.S. and Venkataraman, L. (2016). Structure–property relationships in atomic-scale junctions: Histograms and beyond. *Accounts of Chemical Research*, 49(3), 452–460.

Iacovita, C., Rastei, M.V., Heinrich, B.W., Brumme, T., Kortus, J., Limot, L., Bucher, J.P. (2008). Visualizing the spin of individual cobalt-phthalocyanine molecules. *Physical Review Letters*, 101(11), 116602.

Irie, M. (2000). Diarylethenes for memories and switches. *Chemical Reviews*, 100(5), 1685–1716.

Javaid, S., Bowen, M., Boukari, S., Joly, L., Beaufrand, J.-B., Chen, X., Dappe, Y., Scheurer, F., Kappler, J.-P., Arabski, J. et al. (2010). Impact on interface spin polarization of molecular bonding to metallic surfaces. *Physical Review Letters*, 105(7), 077201.

Jeong, H., Kim, D., Xiang, D., Lee, T. (2017). High-yield functional molecular electronic devices. *ACS Nano*, 11(7), 6511–6548.

Jia, C. and Guo, X. (2013). Molecule–electrode interfaces in molecular electronic devices. *Chemical Society Reviews*, 42(13), 5642.

Jia, C., Migliore, A., Xin, N., Huang, S., Wang, J., Yang, Q., Wang, S., Chen, H., Wang, D., Feng, B. et al. (2016). Covalently bonded single-molecule junctions with stable and reversible photoswitched conductivity. *Science (New York, NY)*, 352(6292), 1443–1445.

Johnson, J.B. (1925). The Schottky effect in low frequency circuits. *Physical Review*, 26(1), 71–85.

Joshi, G.K., Blodgett, K.N., Muhoberac, B.B., Johnson, M.A., Smith, K.A., Sardar, R. (2014). Ultrasensitive photoreversible molecular sensors of azobenzene-functionalized plasmonic nanoantennas. *Nano Letters*, 14, 532–540.

Karimi, M.A., Bahoosh, S.G., Herz, M., Hayakawa, R., Pauly, F., Scheer, E. (2016). Shot noise of 1,4-benzenedithiol single-molecule junctions. *Nano Letters*, 16(3), 1803–1807.

Karuppannan, S.K., Hongting, H., Troadec, C., Vilan, A., Nijhuis, C.A. (2019). Ultrasmooth and photoresist-free micropore-based egain molecular junctions: Fabrication and how roughness determines voltage response. *Advanced Functional Materials*, 29(38), 1904452.

Kawahara, S.L., Lagoute, J., Repain, V., Chacon, C., Girard, Y., Rousset, S., Smogunov, A., Barreteau, C. (2012). Large magnetoresistance through a single molecule due to a spin-split hybridized orbital. *Nano Letters*, 12(9), 4558–4563.

Ke, S.-H., Liu, R., Yang, W., Baranger, H. (2010). Time-dependent transport through molecular junctions. *The Journal of Chemical Physics*, 132(23), 234105–234106.

Kiguchi, M., Tal, O., Wohlthat, S., Pauly, F., Krieger, M., Djukic, D., Cuevas, J.C., Ruitenbeek, J.M.V. (2008). Highly conductive molecular junctions based on direct binding of benzene to platinum electrodes. *Physical Review Letters*, 101(4), 046801.

Kim, Y. and Song, H. (2021). Noise spectroscopy of molecular electronic junctions. *Applied Physics Reviews*, 8, 011303.

Kim, Y., Song, H., Kim, D., Lee, T., Jeong, H. (2010). Noise characteristics of charge tunneling via localized states in metal-molecule-metal junctions. *ACS Nano*, 4(8), 4426–4430.

Kim, Y., Garcia-Lekue, A., Sysoiev, D., Frederiksen, T., Groth, U., Scheer, E. (2012). Charge transport in azobenzene-based single-molecule junctions. *Physical Review Letters*, 109(22), 226801.

Kirton, M.J. and Uren, M.J. (1989). Noise in solid-state microstructures: A new perspective on individual defects, interface states and low frequency (1/f) noise. *Advances in Physics*, 38(4), 367–468.

Kogan, S.M. (1996). *Electronic Noise and Fluctuations in Solids*. Cambridge University Press.

Krzeminski, C., Delerue, C., Allan, G., Vuillaume, D., Metzger, R. (2001). Theory of electrical rectification in a molecular monolayer. *Physical Review B.*, 64(8), 085405.

Kuhr, W.G., Gallo, A.R., Manning, R.W., Rhodine, C.W. (2004). Molecular memories based on CMOS platform. *MRS Bulletin*, 29(11), 838–842.

Kumar, K.S. and Ruben, M. (2021). Sublimable spin-crossover complexes: From spin-state switching to molecular devices. *Angewandte Chemie International Edition*, 60(14), 7502–7521.

Lach, S., Altenhof, A., Tarafder, K., Schmitt, F., Ali, M.E., Vogel, M., Sauther, J., Oppeneer, P.M., Ziegler, C. (2012). Metal–organic hybrid interface states of a ferromagnet/organic semiconductor hybrid junction as basis for engineering spin injection in organic spintronics. *Advanced Functional Materials*, 22(5), 989–997.

Lal, S., Link, S., Halas, N.J. (2007). Nano-optics from sensing to waveguiding. *Nature Photonics*, 1(11), 641–648.

Lambert, C.J. (2015). Basic concepts of quantum interference and electron transport in single-molecule electronics. *Chemical Society Reviews*, 44(4), 875–888.

Lehmann, J., Gaita-Arino, A., Coronado, E., Loss, D. (2007). Spin qubits with electrically gated polyoxometalate molecules. *Nature Nanotechnology*, 2(5), 312–317.

Lehn, J.-M. (1988). Supramolecular chemistry-scope and perpspectives. Molecules, supermolecules and molecular devices (Nobel lecture). *Angewandte Chemie International Edition in English*, 27, 89–112.

Lenfant, S., Krzeminski, C., Delerue, C., Allan, G., Vuillaume, D. (2003). Molecular rectifying diodes from self-assembly on silicon. *Nano Letters*, 3(6), 741–746.

Li, J., Cushing, S.K., Meng, F., Senty, T.R., Bristow, A.D., Wu, N. (2015). Plasmon-induced resonance energy transfer for solar energy conversion. *Nature Photon*, 9, 601–608.

Li, Y., Mol, J.A., Benjamin, S.C., Briggs, G.A.D. (2016). Interference-based molecular transistors. *Scientific Reports*, 6(1), 33686.

Liang, W., Shores, M.P., Bockrath, M., Long, J.R., Park, H. (2002). Kondo resonance in a single-molecule transistor. *Nature*, 417(725–729).

Liao, J., Blok, S., Van Der Molen, S.J., Diefenbach, S., Holleitner, A.W., Schönenberger, C., Vladyka, A., Calame, M. (2015). Ordered nanoparticle arrays interconnected by molecular linkers: Electronic and optoelectronic properties. *Chem. Soc. Rev.*, 44, 999–1014.

Loo, Y.-L., Lang, D.V., Rogers, J.A., Hsu, J.W.P. (2003). Electrical contacts to molecular layers by nanotransfer printing. *Nano Letters*, 3(7), 913–917.

Love, J.C., Estroff, L.A., Kriebel, J.K., Nuzzo, R.G., Whitesides, G.M. (2005). Self-assembled monolayers of thiolates on metals as a form of nanotechnology. *Chemical Reviews*, 105(4), 1103–1169.

Magoga, M. and Joachim, C. (1999). Conductance of molecular wires connected or bonded in parallel. *Physical Review B*, 59, 16011–16021.

Mann, B. and Kuhn, H. (1971). Tunneling through fatty acid salt monolayers. *J. Appl. Phys.*, 42(11), 4398–4405.

Martin, A.S., Sambles, J.R., Ashwell, G.J. (1993). Molecular rectifier. *Physical Review Letters*, 70(2), 218–221.

Mayor, M., Weber, H.B., Reichert, J., Elbing, M., Von Hänisch, C., Beckmann, D., Fischer, M. (2003). Electric current through a molecular rod–relevance of the position of the anchor groups. *Angewandte Chemie International Edition*, 42(47), 5834–5838.

McWhorter, A.L. (1957). 1/f noise and germanium surface properties. In *Semiconductor Surface Physics*, Kingdton, R.H. (ed.). University of Pennsylvania Press.

Metzger, R.M. (2015). Unimolecular Electronics. *Chemical Reviews*, 115, 5056–5115.

Metzger, R.M., Chen, B., Hopfner, U., Lakshmikantham, M.V., Vuillaume, D., Kawai, T., Wu, X.L., Tachibana, H., Hughes, T.V., Sakurai, H. et al. (1997). Unimolecular electrical rectification in hexadecylquinolinium tricyanoquinodimethanide. *Journal of the American Chemical Society*, 119(43), 10455–10466.

Mirjani, F., Thijssen, J.M., Van Der Molen, S.J. (2011). Advantages and limitations of transition voltage spectroscopy: A theoretical analysis. *Physical Review B*, 84(11), 115402.

Moore, A.M., Dameron, A.A., Mantooth, B.A., Smith, R.K., Fuchs, D.J., Ciszek, J.W., Maya, F., Yao, Y., Tour, J.M., Weiss, P.S. (2006). Molecular engineering and measurements to test hypothesized mechanisms in single molecule conductance switching. *J. Am. Chem. Soc.*, 128(6), 1959–1967.

Morita, T. and Lindsay, S. (2007). Determination of single molecule conductances of alkanedithiols by conducting-atomic force microscopy with large gold nanoparticles. *Journal of the American Chemical Society*, 129(23), 7262–7263.

Mulliken, R.S. (1949). Quelques aspects de la théorie des orbitales moléculaires. *J. Chim. Phys.*, 46, 497–542.

Nijhuis, C., Reus, W., Whitesides, G. (2009). Molecular rectification in metal–SAM–metal oxide–metal junctions. *Journal of the American Chemical Society*, 131(49), 17814–17827.

Nitzan, A. (2001). Electron transmission through molecules and molecular interfaces. *Annu. Rev. Phys. Chem.*, 52, 681.

Noori, M., Sadeghi, H., Lambert, C.J. (2017). High-performance thermoelectricity in edge-over-edge zinc-porphyrin molecular wires. *Nanoscale*, 9(16), 5299–5304.

Orellana, P. and Claro, F. (2003). A terahertz molecular switch. *Physical Review Letters*, 90(17), 178302.

Ouyang, B., Hsu, B.C., Chen, Y.-C. (2018). Quantum noise of current in a 1,4-benzenedithiol single-molecule junction: First-principle calculations. *The Journal of Physical Chemistry C*, 122(33), 19266–19272.

Papadopoulos, T., Grace, I., Lambert, C. (2006). Control of electron transport through Fano resonances in molecular wires. *Physical Review B*, 74(19), 4.

Park, J., Pasupathy, A.N., Goldsmith, J.I., Chang, C., Yaish, Y., Petta, J.R., Rinkoski, M., Sethna, J.P., Abrunas, H.D., McEuen, P.L. et al. (2002). Coulomb blockade and the Kondo effect in single-atom transistors. *Nature*, 417, 722–725.

Paulsson, M. and Datta, S. (2003). Thermoelectric effect in molecular electronics. *Physical Review B*, 67, 241403(R).

Petta, J.R., Slater, S.K., Ralph, D.C. (2004). Spin-dependent transport in molecular tunnel junctions. *Physical Review Letters*, 93(13), 136601.

Poggini, L., Cucinotta, G., Pradipto, A.-M., Scarrozza, M., Barone, P., Caneschi, A., Graziosi, P., Calbucci, M., Cecchini, R., Dediu, V.A. et al. (2016). An organic spin valve embedding a self-assembled monolayer of organic radicals. *Advanced Materials Interfaces*, 3(14), 1500855.

Rabache, V., Chaste, J., Petit, P., Della Rocca, M.L., Martin, P., Lacroix, J.-C., McCreery, R.L., Lafarge, P. (2013). Direct observation of large quantum interference effect in anthraquinone solid-state junctions. *Journal of the American Chemical Society*, 135(28), 10218–10221.

Reecht, G., Scheurer, F., Speisser, V., Dappe, Y.J., Mathevet, F., Schull, G. (2014). Electroluminescence of a polythiophene molecular wire suspended between a metallic surface and the tip of a scanning tunneling microscope. *Physical Review Letters*, 112(4), 047403.

Reed, M.A., Zhou, C., Muller, C.J., Burgin, T.P., Tour, J.M. (1997). Conductance of a molecular junction. *Science*, 278, 252–254.

Reuter, M.G., Seideman, T., Ratner, M.A. (2011). Molecular conduction through adlayers: Cooperative effects can help or hamper electron transport. *Nano Letters*, 11(11), 4693–4696.

Reuter, M.G., Hersam, M.C., Seideman, T., Ratner, M.A. (2012). Signatures of cooperative effects and transport mechanisms in conductance histograms. *Nano Letters*, 12(5), 2243–2248.

Ricoeur, G., Lenfant, S., Guerin, D., Vuillaume, D. (2012). Molecule/electrode interface energetics in molecular junction: A “transition voltage spectroscopy” study. *Journal of Physical Chemistry C*, 116(39), 20722–20730.

Rincon-Garcia, L., Evangeli, C., Rubio-Bollinger, G., Agraït, N. (2016). Thermopower measurements in molecular junctions. *Chemical Society Reviews*, 45, 4285–4306.

Rogers, C.T. and Buhrman, R.A. (1984). Composition of 1/f noise in metal-insulator-metal tunnel junctions. *Physical Review*, 53(1272–1275).

Sanvito, S. (2010). Molecular spintronics: The rise of spinterface science. *Nature Physics*, 6(8), 562–564.

Sanvito, S. (2011). Molecular spintronics. *Chemical Society Reviews*, 40(6), 3336.

Sautet, P. and Joachim, C. (1988). Electronic interference produced by a benzene embedded in a polyacetylene chain. *Chemical Physics Letters*, 153(6), 511–516.

Schaming, D., Nguyen, V.-Q., Martin, P., Lacroix, J.-C. (2014). Tunable plasmon resonance of gold nanoparticles functionalized by electroactive bisthienylbenzene oligomers or polythiophene. *The Journal of Physical Chemistry C*, 118(43), 25158–25166.

Schmaus, S., Bagrets, A., Nahas, Y., Yamada, T.K., Bork, A., Bowen, M., Beaurepaire, E., Evers, F., Wulfhekel, W. (2011). Giant magnetoresistance through a single molecule. *Nature Nanotechnology*, 6, 185–189.

Schottky, W. (1918). Über spontane Stromschwankungen in verschiedenen Elektrizitätsleitern. *Annalen der Physik*, 362(23), 541–567.

Schreiber, F. (2000). Structure and growth of self-assembling monolayers. *Progress in Surface Science*, 65(5–8), 151–256.

Schuller, J.A., Barnard, E.S., Cai, W., Jun, Y.C., White, J.S., Brongersma, M.L. (2010). Plasmonics for extreme light concentration and manipulation. *Nature Materials*, 9(3), 193–204.

Schwarz, F. and Lörtscher, E. (2014). Break-junctions for investigating transport at the molecular scale. *Journal of Physics Condensed Matter: An Institute of Physics Journal*, 26(47), 474201.

Scott, G. and Natelson, D. (2010). Kondo resonances in molecular devices. *ACS Nano*, 4(7), 3560–3579.

Smaali, K., Lenfant, S., Karpe, S., Ocafrain, M., Blanchard, P., Deresmes, D., Godey, S., Rochefort, A., Roncali, J., Vuillaume, D. (2010). High on-off conductance switching ratio in optically-driven self-assembled conjugated molecular systems. *ACS Nano*, 4(4), 2411–2421.

Smaali, K., Clement, N., Patriarche, G., Vuillaume, D. (2012). Conductance statistics from a large array of sub-10 nm molecular junctions. *ACS Nano*, 6(6), 4639–4647.

Solomon, G., Andrews, D., Van Duyne, R., Ratner, M. (2008a). When things are not as they seem: Quantum interference turns molecular electron transfer "rules" upside down. *Journal of the American Chemical Society*, 130(25), 7788–7789.

Solomon, G.C., Andrews, D.Q., Goldsmith, R.H., Hansen, T., Wasielewski, M.R., Van Duyne, R.P., Ratner, M.A. (2008b). Quantum interference in acyclic systems: Conductance of cross-conjugated molecules. *Journal of the American Chemical Society*, 130(51), 17301–17308.

Stievenard, D., Guerin, D., Lenfant, S., Leveque, G., Nijhuis, C.A., Vuillaume, D. (2018). Electrical detection of plasmon-induced isomerization in molecule-nanoparticle network devices. *Nanoscale*, 10, 23122–23130.

Su, T.A., Neupane, M., Steigerwald, M.L., Venkataraman, L., Nuckolls, C. (2016). Chemical principles of single-molecule electronics. *Nature Reviews Materials*, 1(3), 16002.

Szent-Györgyi, A. (1941). Towards a new biochemistry? *Science*, 93(2426), 609–611.

Tan, S.F., Wu, L., Yang, J.K.W., Bai, P., Bosman, M., Nijhuis, C.A. (2014). Quantum plasmon resonances controlled by molecular tunnel junctions. *Science (New York)*, 343(6178), 1496–1499.

Tanaka, H., Akai-Kasaya, M., Termehyousefi, A., Hong, L., Fu, L., Tamukoh, H., Tanaka, D., Asai, T., Ogawa, T. (2018). A molecular neuromorphic network device consisting of single-walled carbon nanotubes complexed with polyoxometalate. *Nature Communications*, 9(1), 2693.

Tatay, S., Barraud, C., Galbiati, M., Seneor, P., Mattana, R., Bouzehouane, K., Deranlot, C., Jacquet, E., Forment-Aliaga, A., Jegou, P. et al. (2012). Self-assembled monolayer-functionalized half-metallic manganite for molecular spintronics. *ACS Nano*, 6(10), 8753–8757.

Tatay, S., Galbiati, M., Delprat, S., Barraud, C., Bouzehouane, K., Collin, S., Deranlot, C., Jacquet, E., Seneor, P., Mattana, R. et al. (2016). Self-assembled monolayers based spintronics: From ferromagnetic surface functionalization to spin-dependent transport. *Journal of Physics Condensed Matter: An Institute of Physics Journal*, 28(9), 094010.

Tewari, S., Sabater, C., Van Ruitenbeek, J. (2019). Identification of vibration modes in single-molecule junctions by strong inelastic signals in noise. *Nanoscale*, 11(41), 19462–19467.

Thijssen, J.M. and Van Der Zant, H.S.J. (2008). Charge transport and single-electron effects in nanoscale systems. *Physica Status Solidi (b)*, 245, 1455–1470.

Thomas, L., Guerin, D., Quinard, B., Jacquet, E., Mattana, R., Seneor, P., Vuillaume, D., Mélin, T., Lenfant, S. (2020). Conductance switching at the nanoscale of diarylethene derivative self-assembled monolayers on La 0.7Sr 0.3MnO 3. *Nanoscale*, 12, 8268–8276.

Thomas, L., Arbouch, I., Guerin, D., Wallart, X., Van Dyck, C., Mélin, T., Cornil, J., Vuillaume, D., Lenfant, S. (2021). Conductance switching of azobenzene-based self-assembled monolayers on cobalt probed by UHV conductive-AFM. *Nanoscale*, 13, 6977–6990.

Trasobares, J., Vuillaume, D., Theron, D., Clément, N. (2016). A 17 GHz molecular rectifier. *Nature Communications*, 7, 12850.

Trasobares, J., Rech, J., Jonckheere, T., Martin, T., Aleveque, O., Levillain, E., Diez-Cabanes, V., Olivier, Y., Cornil, J., Nys, J.P. et al. (2017). Estimation of π-π electronic couplings from current measurements. *Nano Letters*, 17(5), 3215–3224.

Ulman, K., Narasimhan, S., Delin, A. (2014). Tuning spin transport properties and molecular magnetoresistance through contact geometry. *The Journal of Chemical Physics*, 140(4), 044716.

Vadai, M., Nachman, N., Ben-Zion, M., BüRkle, M., Pauly, F., Cuevas, J.-C., Selzer, Y. (2013). Plasmon-induced conductance enhancement in single-molecule junctions. *The Journal of Physical Chemistry Letters*, 4(17), 2811–2816.

Valkenier, H., Guédon, C.M., Markussen, T., Thygesen, K.S., Van Der Molen, S.J., Hummelen, J.C. (2014). Cross-conjugation and quantum interference: A general correlation? *Physical Chemistry Chemical Physics*, 16(2), 653–662.

Van Der Molen, S.J., Liao, J., Kudernac, T., Agustsson, J.S., Bernard, L., Calame, M., Van Wees, B.J., Feringa, B.L., Schönenberger, C. (2009). Light-controlled conductance switching of ordered metal-molecule-metal devices. *Nano Letters*, 9(1), 76–80.

Van Dyck, C., Geskin, V., Kronemeijer, A.J., De Leeuw, D.M., Cornil, J. (2013). Impact of derivatization on electron transmission through dithienylethene-based photoswitches in molecular junctions. *Physical Chemistry Chemical Physics*, 15(12), 4392–4404.

Viero, Y., Copie, G., Guerin, D., Krzeminski, C., Vuillaume, D., Lenfant, S., Cleri, F. (2015). High conductance ratio in molecular optical switching of functionalized nanoparticle self-assembled nanodevices. *The Journal of Physical Chemistry C*, 119, 21173–21183.

Viero, Y., Guerin, D., Alibart, F., Lenfant, S., Vuillaume, D. (2018). Light-stimulable molecules/nanoparticles multi-terminal networks for switchable logical function and reservoir computing. *Advanced Functional Materials*, 28, 1801506.

Vilan, A. and Cahen, D. (2002). Soft contact deposition onto molecularly modified GaAs. Thin metal film flotation: Principles and electrical effects. *Advanced Functional Materials*, 12(11–12), 795–807.

Vilan, A. and Cahen, D. (2017). Chemical modification of semiconductor surfaces for molecular electronics. *Chemical Reviews*, 117(5), 4624–4666.

Vilan, A., Aswal, D., Cahen, D. (2017). Large-area, ensemble molecular electronics: Motivation and challenges. *Chemical Reviews*, 117(5), 4248–4286.

Vincent, R., Klyatskaya, S., Ruben, M., Wernsdorfer, W., Balestro, F. (2012). Electronic read-out of a single nuclear spin using a molecular spin transistor. *Nature*, 488(7411), 357–360.

Vuillaume, D. (2008). Molecular-scale electronics. *Comptes rendus physique*, 9(1), 78–94.

Vuillaume, D. (2010). Molecular nanoelectronics. *Proceedings of the IEEE*, 98(12), 2111–2123.

Vuillaume, D., Chen, B., Metzger, R.M. (1999). Electron transfer through a monolayer of hexadecylquinolinium tricyanoquinodimethanide. *Langmuir*, 15(11), 4011–4017.

Wang, T. and Nijhuis, C.A. (2016). Molecular electronic plasmonics. *Applied Materials Today*, 3, 73–86.

Wang, W. and Richter, C.A. (2006). Spin-polarized inelastic electron tunneling spectroscopy of a molecular magnetic tunnel junction. *Applied Physics Letters*, 89(15), 153105.

Wang, Y., Che, J.G., Fry, J.N., Cheng, H.-P. (2013). Reversible spin polarization at hybrid organic–ferromagnetic interfaces. *The Journal of Physical Chemistry Letters*, 4(20), 3508–3512.

Wang, F., Clément, N., Ducatteau, D., Troadec, D., Tanbakuchi, H., Legrand, B., Dambrine, G., Theron, D. (2014). Quantitative impedance characterization of sub-10 nm scale capacitors and tunnel junctions with an interferometric scanning microwave microscope. *Nanotechnology*, 25(40), 405703.

Wang, K., Meyhofer, E., Reddy, P. (2019). Thermal and thermoelectric properties of molecular junctions. *Advanced Functional Materials*, 30, 1904534.

Ward, L.M. and Greenwood, P.E. (2007). 1/f noise. *Scholarpedia*, 2(1537).

Weissman, M.B. (1988). 1/f noise and other slow, nonexponential kinetics in condensed matter. *Reviews of Modern Physics*, 60(2), 537–571.

Wold, D.J. and Frisbie, C.D. (2000). Formation of metal-molecule-metal tunnel junctions: Microcontacts to alkanethiol monolayers with a conducting AFM tip. *Journal of the American Chemical Society*, 122(12), 2970–2971.

Wu, J., Wang, B., Wang, J., Guo, H. (2005). Giant enhancement of dynamic conductance in molecular devices. *Physical Review B*, 72(19), 195324.

Xiang, D., Jeong, H., Lee, T., Mayer, D. (2013). Mechanically controllable break junctions for molecular electronics. *Advanced Materials*, 25(35), 4845–4867.

Xiang, D., Lee, T., Kim, Y., Mei, T., Wang, Q. (2014). The origin of discrete current fluctuations in single molecule junction. *Nanoscale*, 6(13396–13401).

Xiang, D., Wang, X., Jia, C., Lee, T., Guo, X. (2016). Molecular-scale electronics: From concept to function. *Chemical Reviews*, 116(7), 43418–44440.

Xu, B. and Tao, N.J. (2003). Measurement of single-molecule resistance by repeated formation of molecular junctions. *Science*, 301, 1221–1223.

Xue, Y. and Ratner, M.A. (2006). Molecular electronics: From physics to computing. In *Nanotechnology: Science and Computation*, Chen, J., Jonoska, N., Rozenberg, G. (eds). Springer Berlin Heidelberg, Berlin, Heidelberg.

Zeng, J., Chen, K.-Q., Deng, X., Long, M. (2016). Light-driven strong spin valve effects in an azobenzene-based spin optoelectronic device. *Journal of Physics D: Applied Physics*, 49(41), 415104.

Zheng, Y.B., Yang, Y.-W., Jensen, L., Fang, L., Juluri, B.K., Flood, A.H., Weiss, P.S., Stoddart, J.F., Huang, T.J. (2009). Active molecular plasmonics: Controlling plasmon resonances with molecular switches. *Nano Letters*, 9(2), 819–825.

Zuloaga, J., Prodan, E., Nordlander, P. (2009). Quantum description of the plasmon resonances of a nanoparticle dimer. *Nano Letters*, 9(2), 887–891.

8

Superconducting Quantum Electronics

Sasan RAZMKHAH[1] **and Pascal FEBVRE**[2]

[1] Department of Electrical and Electronics Engineering,
TOBB University of Economics and Technology, Ankara, Turkey
[2] IMEP-LAHC, Université Savoie Mont Blanc, Le Bourget du Lac, France

8.1. Introduction

8.1.1. *A little bit of history*

Superconducting quantum electronics is formally born in 1911, three years after Heike Kamerlingh Onnes was able to liquefy helium at the University of Leiden in the Netherlands. With his student Gilles Holst, they studied the electrical resistivity at very low temperatures of pure metals like mercury, tin or lead. At this time, it was not clear whether the resistance of such metals would decrease at low temperatures because of less lattice vibrations or increase because the electrons would be frozen. What they observed in April 1911 was that instead, the resistance of mercury, while decreasing continuously with lower temperatures, suddenly drops to zero under a certain temperature, called critical temperature Tc, of about 4 K for mercury (Onnes 1991). Onnes realized that he was in front of a new state of the matter that he called superconductivity. This state was not understood but clearly observed: it became apparent after a while that this behavior does not abide by classic physical laws. It took time, also in the context of hectic progress due to World War I, to realize that this new state of matter could only be explained with quantum physics. As such, the mercury circuit measured in the laboratory of Kamerlingh Onnes in 1911 was a pure, though very simple, superconducting quantum circuit.

Beyond-CMOS, coordinated by Alessandro CRESTI.

Figure 8.1. *Heike Kamerlingh Onnes discovered in 1911 the zero resistance of mercury below a critical temperature of about 4.2 K*

Two other properties were subsequently found: namely the expulsion of magnetic fields by superconductors, which is responsible for the always impressive experiments of magnetic levitation, called the Meissner or Meissner-Ochsenfeld effect (Meissner and Ochsenfeld 1933) from the names of the two German physicists who discovered it in 1933; and finally the specific heat transition (Bray 1974), associated with the phase transition between a normal metal and a superconducting metal at the critical temperature.

It took two decades until the London brothers (London et al. 1935) could set some phenomenological and macroscopic approaches based on Maxwell equations to explain the absence of resistance. The two other properties were also explained with phenomenological considerations: after the existence of an energy gap in superconductors was established by Daunt and Mendelssohn (Daunt et al. 1946), we had to wait until 1958 for John Bardeen, Leon Cooper and John Schrieffer to propose a microscopic theory, called now the BCS theory, based on the existence of Cooper pairs in superconductors, that are pairs of electrons of opposite spins that find themselves in a more stable state in certain conditions, due to attractive interactions through the superconductor lattice. Their theory gave them the Nobel Prize in Physics in 1972, after the one of Kamerlingh Onnes in 1913.

Figure 8.2. *From left to right, Walter Meissner, Robert Ochsenfeld, Fritz London and Heinz London were four physicists who contributed to elaborate a macroscopic phenomenological approach to understand superconductivity*

Since Cooper pairs are composed of two electrons, they are bosons and possess as such all the properties of the Bose-Einstein statistical law, like interference and diffraction, since they follow a collective behavior associated with a common macroscopic quantum state, like photons. Their behavior can be modeled by using a macroscopic complex wavefunction whose phase has no equivalent in classical physics, with the advantage of Cooper pairs being part of a solid-state circuit, namely an electrical circuit easier to handle and more compact than a free-air photonic system.

Figure 8.3. *From left to right, Bardeen, Cooper and Schrieffer proposed in 1956 a microscopic theory of superconductivity, for which they were awarded the Nobel Prize in Physics in 1972*

8.1.2. *The Josephson junction*

It was during his PhD thesis at Cambridge University, UK, in the 1960s that Brian Josephson came with an elegant approach (Josephson 1962) to derive the fact that, since superconductors can be understood through a macroscopic wavefunction, two superconductors placed at a very short distance from each other, of the order of the atomic distance, can interact in a quantum way. Indeed, since the probability of presence of Cooper pairs from one superconductor becomes non-zero in the volume of the second superconductor at the vicinity of the first superconductor, currents of Cooper pairs can quantum-mechanically cross the gap, even in the absence of the voltage difference between the two superconductors. Resolving Schrödinger equation of such a simply coupled system, Josephson showed that a DC current can cross the barrier in the absence of voltage difference, while an ac current appears when a DC voltage is imposed, with the very remarkable property that the frequency of the ac current, now called now "ac Josephson current", is proportional to the DC voltage. The ratio of voltage to frequency is given by a fundamental constant, $h/2e$ (h is the Planck constant and e is the charge of the electron), and is the basis for what is called Josephson metrology, used to develop the Josephson voltage standard and calibrate instruments. For this theory, Brian Josephson won the Nobel Prize in Physics in 1973. Josephson junctions, the devices made of two superconductors separated by a thin barrier, have become the main active elements of superconductor electronics for quantum detectors, analogue, digital or pure quantum electronics. It is in some way the equivalent of transistors for semiconductor technologies, though the physical principles of operation are fundamentally different.

Figure 8.4. *Brian Josephson predicted in 1962 that a quantum macroscopic electrical current can cross a thin barrier separating two superconductors. He was awarded the Nobel Prize in Physics in 1973. The Josephson junction derived from this prediction is the main active element of superconductor electronics*

Josephson junctions (JJs) are simple devices made of two superconductors coupled through a thin barrier, which can be made of one of several layers of insulating, normal metal, or ferromagnetic material. All the physics of the device is modeled through the "cross" symbol as shown in Figure 8.5. The wavefunctions of each superconductor, associated with the probability of presence of Cooper pairs in the volume, are represented to illustrate the coupling. Intrinsically, the two superconductors also form a capacitor CJ that is modeled in parallel with the ideal junction. The different properties of the barrier, in particular its resistance, are taken into account in the JJ model through the "cross" symbol. The model takes into account the physical properties of both charge carriers, namely Cooper pairs and quasiparticles. Nevertheless, up to now at least, the Josephson junction device is used by taking advantage of either Cooper pairs only, for example, in Josephson mixers (Schoelkopf et al. 1993), SQUIDs (Clarke and Braginski 2006a) or SFQ digital electronics (Likharev and Semenov 1991a), or quasiparticles only (by suppressing or reducing Cooper pairs currents with a magnetic field); for instance, in the case of superconductor–insulator–superconductor (SIS) mixers (Febvre et al. 1994). The case of flux-flow oscillators based on long Josephson junctions (Kinev et al. 2019) is more complex to analyze and requires a full approach of the dynamics of all charge carriers in the junction. Design parameters usually allow favoring one mode or the other. In practice, modeling the global behavior of a Josephson junction embedded in a circuit is not trivial and simulations are usually made with strong assumptions to simplify the numerical codes. In this context, one model often used to model junctions is the resistively and capacitively shunted junction (RCSJ) model used in particular for digital electronics. It accounts for the resistance of the barrier due to quasiparticles by adding a resistor in parallel with the basic model of Figure 8.5. The RCSJ model is presented in Figure 8.6. The parallel resistor is the normal resistance R_N (the resistance of the junction just above the critical temperature) in the simplest case; it can be R_N in parallel with a shunt resistor R_{shunt} as in single-quantum-flux (SFQ) digital logic, or a nonlinear *R(V)* curve obtained from the DC *I–V* curve due to quasiparticle currents.

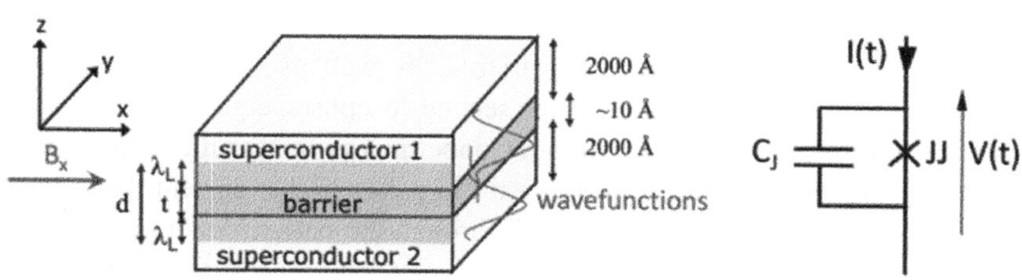

Figure 8.5. *Schematic view of a Josephson junction with its equivalent electrical circuit. When a magnetic field is applied to the JJ, it penetrates at the surface of the superconductors over the λ_L London penetration depth. For a color version of this figure, see www.iste.co.uk/cresti/beyondcmos.zip*

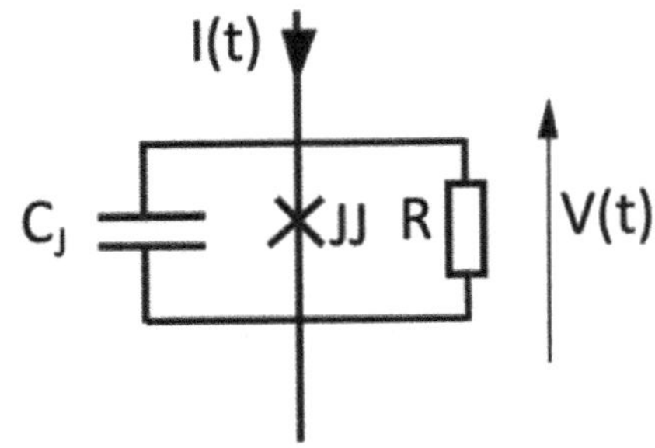

Figure 8.6. *Schematic associated with the RCSJ model. The resistance can be either constant for the simplest model or nonlinear to take into account some first-order physical features of quasiparticles*

8.1.2.1. *Josephson equations*

The respective positions of the two superconductors located at atomic distance from each other allow their wavefunctions to penetrate through the barrier and influence the state of the other superconductor. Under these conditions, a coupling of quantum origin is present and Cooper pairs can tunnel through the barrier without consuming any energy. The term "*tunneling*" is used when a charge can pass through a barrier of potential that it would not normally according to the laws of classical physics. From the coupling of the superconductors through the barrier, Brian Josephson derived the two following equations, simply called Josephson equations.

$$I = I_c \sin(\varphi) \quad [8.1]$$

$$\frac{\partial \varphi}{\partial t} = \frac{2e}{\hbar} V \quad [8.2]$$

The phase φ is the difference of the wavefunction phases of the two superconductors. It is a fundamental variable, as well as the current I flowing through the junction. As can be seen in the second Josephson equation, the voltage is the derivative of the phase. The maximum Cooper pair current that can flow through the junction is the critical current Ic. If we try to pass a higher current than Ic, some quasiparticles will necessarily flow as well through the barrier, leading to a voltage drop caused by the barrier's normal resistance, or the shunt resistor if there is one. Depending on the junction parameters (capacitance, resistance and critical current), the behavior of the device will be different, like in any electrical circuit. An important parameter for Josephson junctions is the dimensionless McCumber parameter (McCumber 1968) defined by:

$$\beta_c = \frac{2\pi}{\Phi_0} R^2 C I_c \qquad [8.3]$$

where $\Phi_0 = \frac{h}{2e}$ is the quantum of magnetic flux, fundamental quantity defined by the Planck constant and the charge of the electron. Depending on this parameter, the current–voltage (*I–V*) characteristic of the Josephson junction can exhibit, or not, a hysteresis, which is a manifestation of its electrodynamics.

Figure 8.7 shows *I–V* curves for different values of β_c. Hysteresis occurs when $\beta_c > 1$ and the return current (current obtained when the voltage goes back to zero) decreases rapidly at the transition from the resistive state to the superconducting state for higher β_c. To eliminate this hysteresis in many electronic applications, a parallel shunt resistor is connected to the Josephson junction so that $\beta_c \approx 1$. The *I–V* characteristics of a junction based on the RCSJ model can be drawn from time-domain simulations with appropriate computer-aided design (CAD) software like JOINUS (Razmkhah and Febvre 2020).

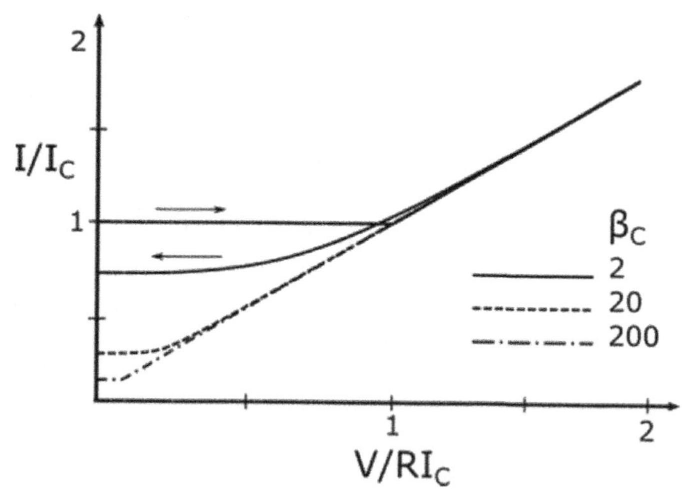

Figure 8.7. *I–V characteristic of a Josephson junction for different values of the β_c McCumber parameter (Chesca et al. 2005)*

8.1.2.2. *Influence of the magnetic field*

The phase of the superconductor wavefunction is intimately connected to the local magnetic flux on the device. It is in fact a very general property of electromagnetic systems that can be analyzed with a quantum approach. If we consider a constant and spatially invariant magnetic field of coordinates (B_x, B_y) in the xy-plane of the Josephson junction's barrier (horizontal plane in the case of Figure 8.5), its local phase across the junction's section can be written as (Barone and Paternò 1982):

$$\varphi(x,y) = \frac{2\pi}{\Phi_0} d\left(B_x y - B_y x\right) \quad [8.4]$$

where d is the thickness where the magnetic flux is present: d is the sum of the barrier thickness t and of the London penetration depths λ_L of each superconductor $d = t + 2\lambda_L$. λ_L is equivalent to the skin depth of normal metals, but it occurs in the DC regime for superconductors.

We see in equation [8.4] that, if the magnetic field is, for example, along the x direction only, then the phase is proportional to the product of d by y by B_x which is the area of the junction's barrier cross-section perpendicular to the magnetic field (extended to the London penetration depth zones) multiplied by the magnetic field, hence the magnetic flux Φ_x crossing the junction at the barrier location, where magnetic fields are allowed to penetrate. As a result, the phase is:

$$\varphi(y) = 2\pi \frac{\Phi_x}{\Phi_0} \quad [8.5]$$

Consequently, the phase can vary along the y direction only and then the Cooper pairs currents, also called Josephson currents, crossing the junction also vary along y like the sinus of the phase (see equation [8.1]). For specific values of the magnetic field, we can obtain phase variations such that the currents are positive on one half of the junction section and negative on the other half. The total resulting current is then zero. For other values of the magnetic field, the total current can vary from I_c to 0. This effect is exactly identical to the well-known diffraction phenomenon in optics. The magnetic field plays the role of the light angle in the case of a slit crossed by monochromatic light. This is not surprising since the magnetic field modulates the phase of the wavefunction the same way as the angle of the light modifies the phase of the electric field in diffraction experiments due to the path difference. The total Cooper pairs current in the junction is analogous to the total electric field crossing the diffractive slit. Both systems deal with bosons that can be in phase or out of phase. As a result, a diffraction law by the magnetic field can also be established for Josephson junctions, given by equation [8.6]. The corresponding diffraction pattern curve is shown in Figure 8.8.

$$\mathrm{I}(B_x) = \mathrm{I}_c \left| \frac{\sin\left(\pi \frac{\Phi_x}{\Phi_0}\right)}{\pi \frac{\Phi_x}{\Phi_0}} \right| \quad [8.6]$$

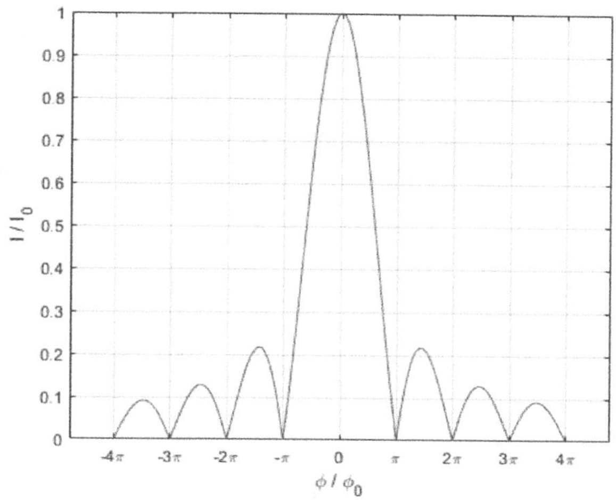

Figure 8.8. *Effect of the magnetic field on the critical current of a Josephson junction. For a color version of this figure, see www.iste.co.uk/cresti/beyondcmos.zip*

This diffraction effect is used, for example, to suppress Josephson currents in SIS receivers for radioastronomy. But, more importantly, it is also one manifestation of the more general interference-diffraction principle used by superconducting quantum interference devices (SQUIDs), which are still today the most sensitive magnetometers.

8.1.3. *Superconducting quantum interference devices (SQUIDs)*

By analogy with the Young double-slit interference experiment in optics, two Josephson junctions are placed in parallel and their total critical current, whose value depends on the magnetic field inside the loop formed by the two junctions, is a measure of the magnetic field. Formally, the SQUID is an interferometer, as its name tells. The magnetic field in the SQUID loop is analogous to the distance between the two slits in the Young experiment. More generally, superconducting quantum interference filters (SQIFs), composed of many Josephson junctions, are also used for sensitive magnetometry experiments at microwave frequencies (Mitchell et al. 2016; Crété et al. 2021) and are equivalent to gratings in optics.

8.1.3.1. *The DC SQUID*

The DC SQUID consists of two identical Josephson junctions, of critical current *Ic*, connected in parallel within the superconducting ring and is used to measure the magnetic flux passing through the ring. In Figure 8.9, the conventional representation of DC SQUID is in part (a) and the schematic representation in part (b). In part (a), the gray regions represent the superconducting material, and the black regions 1 and 2 represent the insulating layer forming the Josephson junction. Part (b) shows the RCSJ models of Josephson junctions 1 and 2. Elements I_{N1} and I_{N2} represent the thermal noise current sources originating from the shunt resistances of the junctions.

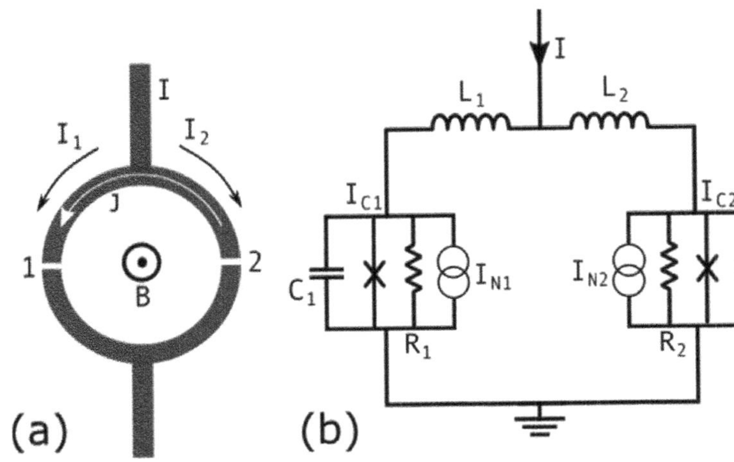

Figure 8.9. *Schematic of DC SQUID magnetometer (Chesca et al. 2005)*

The DC SQUID measures the magnetic field using the fluxoid quantization property of superconducting rings (the fluxoid takes into account the phase of Josephson junctions which is a purely quantum feature and does not correspond to a geometrical flux). The total phase of the DC SQUID ring must be an integer multiple of 2π due to the interference of the superconducting wavefunctions of each arm (otherwise the phase could take several values each time an additional turn is made and that does not correspond to a physical observable state). When the magnetic flux applied to the DC SQUID does not induce an exact multiple of 2π for the total phase of the ring (in this case, the magnetic flux is not an integer of the magnetic flux Φ_0), the difference is brought by a phase difference at the terminals of each Josephson junction. The phases of the two junctions must follow equation [8.1], meaning that an induced current, called screening current, circulates along the ring (current J in Figure 8.9), drawing an induced magnetic flux that brings the total flux in the loop to a multiple of Φ_0 (and the total phase to a multiple of 2π). In practice, this is possible as long as the induced current stays below the critical current of the Josephson junctions.

The induced screening current J, hence the magnetic field, can be sensed by measuring the I–V characteristic. If an external magnetic field is applied to the ring, as shown in Figure 8.10, the DC SQUID induces a current J to cancel the magnetic flux (modulo Φ_0), as said before. This induced current is $\mathrm{J} = I_c$ for a magnetic field corresponding to Φ_0. Additionally, the two junctions share equally the DC SQUID biasing current I since they are identical. Therefore, the currents in each junction are given by equations [8.7] and [8.8].

$$I_1 = \frac{I}{2} + \mathrm{J} \qquad [8.7]$$

$$I_2 = \frac{I}{2} - \mathrm{J} \qquad [8.8]$$

To read the magnetic flux, a biasing current $I = 2I_c$ is applied to the DC SQUID so that $I_1 = I_2 = I_c$ in the absence of magnetic field, for $J = 0$. Each junction is at the verge of switching so that the least applied magnetic field will make either the first or the second junction to switch to a non-zero voltage mode since its current will surpass I_c, changing the shape of the I–V curve as seen in Figure 8.11. As the magnetic flux is increased, the DC SQUID voltage increases until the applied magnetic flux reaches $\Phi_0/2$, which corresponds to a difference of phase of π between the two junctions and consequently to $\mathrm{J} = I_c/2$. Since the critical current of the junction is I_c then the maximum applied DC current that can flow in the DC SQUID to keep it a zero voltage corresponds to $I_1 = I_c$ and to $I = I_c$ from equation [8.7]. Hence, $I_2 = 0$ from equation [8.8]. The DC biasing current only goes to the left junction and the total critical current of the SQUID in the presence of an applied flux $\Phi_0/2$ is reduced to I_c, half of its $2I_c$ value in the absence of magnetic field. When the magnetic field continues to increase above $\Phi_0/2$, the values of currents just mentioned before correspond to a current in the first junction that is above I_c. Since this is not possible, the induced current J flips to the opposite value $-I_c/2$ as can be seen in Figure 8.10. This corresponds to a total change of screening current of $2\mathrm{J} = I_c$, which corresponds to a sudden change of flux of Φ_0 in the SQUID loop, as can also be seen in the bottom part of Figure 8.10. This effect can only be analyzed with a dynamic analysis of the DC SQUID. Just after the switching $\mathrm{J} = -I_c/2, I_1 = 0, I_2 = I_c$: there is no apparent change in the SQUID critical current on the I–V curve, but from this point, it starts to increase again and the SQUID DC voltage starts to decrease when the applied flux is increased till Φ_0. In other words, the highest voltage and the lowest critical current under constant biasing supply, are reached for an external applied field of $(n + 1/2)\ \Phi_0$. The induced current is again zero when the applied magnetic field reaches the next multiple of Φ_0. This cycle continues periodically as shown in Figure 8.11.

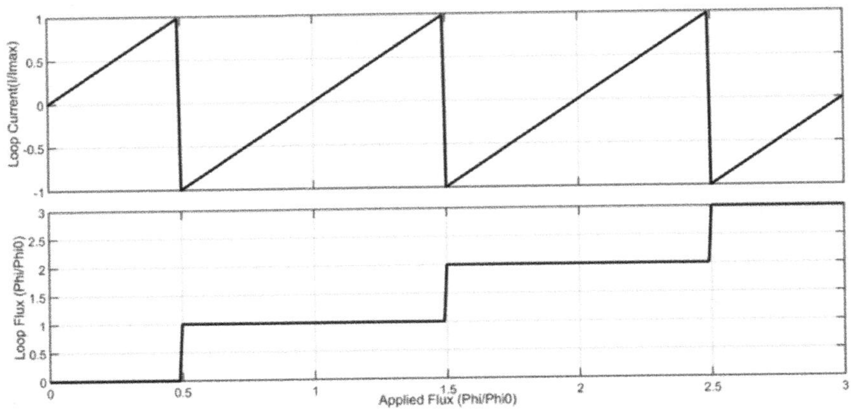

Figure 8.10. *Top: induced current in the biased DC SQUID loop when an external magnetic flux is applied with a linear increase. Bottom: the corresponding variation of the magnetic flux in the ring*

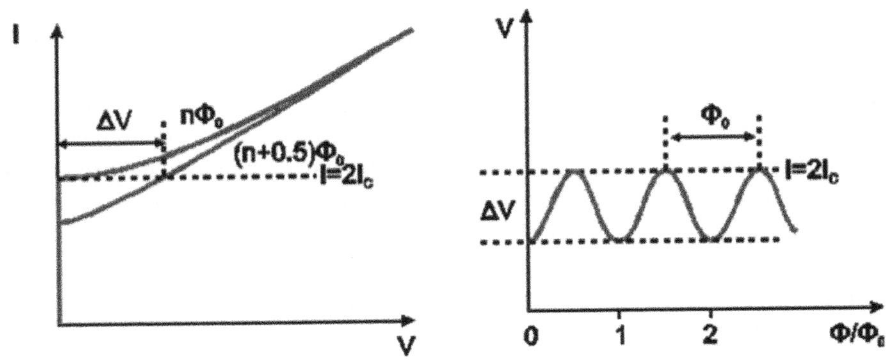

Figure 8.11. *DC SQUID I–V characteristic and DC SQUID voltage in the presence of an external applied magnetic field. For a color version of this figure, see www.iste.co.uk/cresti/beyondcmos.zip*

8.1.3.2. *Superconducting quantum interference filters (SQIFs)*

New devices with new properties can be created by increasing the number of junctions. The multi-junction interferometer is a system of N Josephson junctions connected in parallel with two superconducting electrodes. It is called a SQIF and is an analogue of an optical grating in terms of interferometric properties.

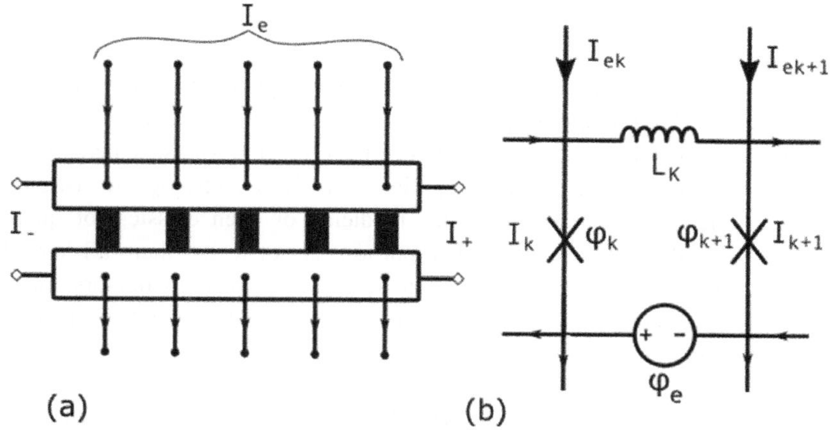

Figure 8.12. *(a) Multijunction interferometer and (b) equivalent circuit (Likharev 1986)*

To analyze this circuit, the flux quantization condition can be written as:

$$\varphi_k - \varphi_{k+1} = \frac{2\pi}{\Phi_0}\Phi_k \quad [8.9]$$

where Φ_k is the flux in the loop comprising junctions k and $k+1$. Depending on the number of junctions, different interferometer patterns can be produced, as seen in Figure 8.13. The impedance on the interferometer can be controlled with the biasing applied to the junctions.

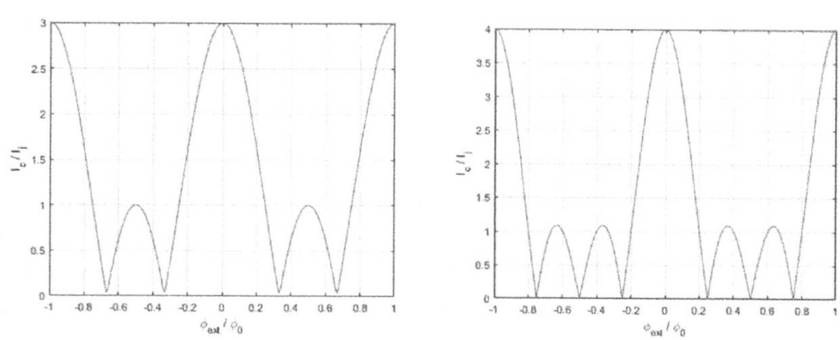

Figure 8.13. *Interference pattern of a SQIF made of: (a) three parallel Josephson junctions and (b) four parallel Josephson junctions. For a color version of this figure, see www.iste.co.uk/cresti/beyondcmos.zip*

8.1.4. *Emergence of superconductor electronics*

Since superconductors are based on the same type of particles as photons, namely bosons, a large range of specific properties can be derived for electrical and "electronic" circuits based on superconductors. Over the years, superconductors have been used in a wide range of applications, from metrology to astronomy, geophysics, telecommunications systems, medicine or even classical or quantum computing, as just a few examples. The purpose is not to list them all here, but, rather, the important features that lead to the use of superconductors must be emphasized here, since they are the main drivers for when we have to choose between technologies.

The first important feature is that superconductors are always associated with very low energies, and then very low power consumption, since their energy gap, of the order of a few milli-electron-volts, is about one thousandth of the one of semiconductors. Currents are usually in the range of a few microamperes to some hundreds of microamperes.

Secondly, for the same reason associated with the low energy gap, characteristic frequencies of superconductors are also about one thousandth of the ones of semiconductors, making them particularly suitable at THz frequencies or at very high microwave frequencies for electronic systems, unlike semiconductors that perform very well at infrared frequencies due to their higher gap. The Josephson voltage to frequency ratio mentioned earlier for metrology of about 484 GHz/mV clearly shows the frequency domain of performance for such materials.

At last, their pure quantum behavior at the macroscopic scale, that is the scale of a full electronic circuit, leads to unique properties that can be used for quantum electronics, be it for quantum sensors, quantum-classical or even pure quantum processing.

On top of that, since the current state of the art requires cooling, at least below minus 200°C, to have good superconducting properties, such systems inherently benefit from the low thermal energy of the environment and make superconducting devices particularly valuable to measure and process very faint signals, in particular when quantum regime manifestations can occur.

We are now going to review a few of the best usages of superconductors related to the micro- and nano-electronics field.

8.2. Passive superconducting electronics

The simplest circuits that can be derived from superconducting materials are simple devices, namely printed circuits made of thin superconductor films usually deposited on a quartz, sapphire or silicon substrate to realize frequency filters (Zhang et al. 1995; Matthaei 2003), resonators (Shen et al. 1992) with very high quality factors (Q-factors), antennas (Hansen 1990; Lancaster et al. 1992), low-loss waveguides (Sage et al. 2011) or low-loss transmission lines (Polonsky et al. 1993). Therefore, designers must consider impedance matching, energy dissipation and nonlinear effects. Due to the non-dissipative nature and penetration depth of superconductors, radio frequency circuits designed with such materials have high Q-factors (Megrant et al. 2012) and good properties at conventional frequencies. To understand this, we discuss below the surface impedance of superconductors and their behavior in the radiofrequency (RF) regime.

8.2.1. *Surface impedance of superconductors*

Even if superconductors exhibit zero DC resistance, in practice not all electrons are assembled in Cooper pairs at temperatures comprised between the absolute and the critical temperatures. Usually, this can be taken into account by what is called the two-fluid model, which considers that charge carriers are a combination of Cooper pairs and electrons. Since Cooper pairs flow without losses they, in some way, short-circuit other lossy channels so that the effective resistance is indeed null. Nevertheless, as soon as variable signals are applied to the superconductor, transient signals appear since Maxwell equations are still valid. Currents and a magnetic field are present at the surface of superconductors with a skin depth depending on its physical properties. Then, electrons start playing a role. To see it from a quantum point of view, an incoming signal at a given frequency can be seen as photons of energy proportional to the signal frequency. If the frequency is high enough, it can, combined with thermal noise associated with the ambient electromagnetic field energy, break some Cooper pairs back into quasiparticles (which are electrons whose physical quantities are not independent). This effect is naturally overwhelming when the incoming photons frequency reaches the gap frequency of the superconductor, typically of the order of 700 GHz for Niobium. Then, most of Cooper pairs can be broken, the superconductor becomes again a normal metal, some energy is absorbed and can be accounted through RF losses from an electromagnetic point of view. The surface impedance of superconductors is the ratio of the difference of voltage created by such process on a superconductor film, divided by the currents present in the skin depth of the film that produces Joule heating due to the residual conductivity of superconductors in the presence of quasiparticles.

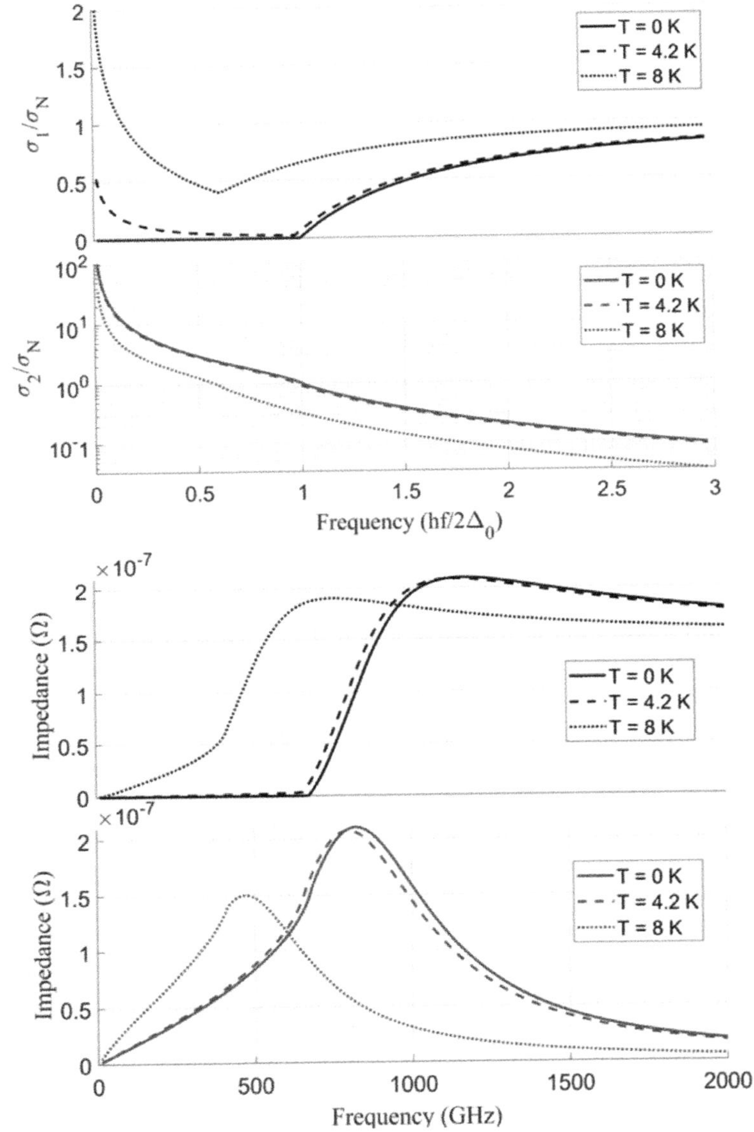

Figure 8.14. *Real part (black) and imaginary part (red) of the complex conductivity normalized to the conductivity of the normal metal just above Tc (2 top curves) and of the surface impedance (2 bottom curves) of thin films for the AIST HSTP superconductor process at three different temperatures. For a color version of this figure, see www.iste.co.uk/cresti/beyondcmos.zip*

The surface impedance of superconductors and the energy propagation in superconductors can be discussed from two points of view, first by using macroscopic and phenomenological approaches such as London's theory to calculate the penetration depth and the current distribution with Beam's theory (Tinkham 1996). The second and more precise approach is based on the microscopic BCS theory. In this second approach, the surface impedance is calculated with Mattis–Bardeen's theory (Mattis and Bardeen 1958; Mazin 2005) and the resulting complex impedance can be used in a finite element solver to calculate the energy dissipation in the circuit (Mattis and Bardeen 1958). While the second approach is more precise, it does not have an analytical solution and needs a numerical approach that can be time-consuming. Figure 8.14 shows a calculation based on the Mattis–Bardeen theory of superconducting films made with the HSTP process (Nagasawa et al. 2014) developed in Japan at AIST.

8.2.2. *Superconductor waveguides and transmission lines*

Any radiation is made of a collection of photons of a certain frequency and therefore energy. This energy travels through a medium, called a waveguide when the energy is guided. The idea of the transmission of energy in a waveguide was first mathematically explained by Lord Rayleigh at the end of the 19th century. There are different types of waveguides. Superconductor-based waveguides and other types of transmission lines for very low-loss transmission of energy are usually based on thin superconductor films and are crucial elements for several applications, in particular to propagate microwave signals and also for readout electronics and readout of quantum bits (qu-bits) states.

Figure 8.15 and Figure 8.17 show two types of waveguides studied with the Sonnet software (Sonnet Software n/d). The respective simulated frequency responses are shown in Figure 8.17 and Figure 8.18. Sonnet does not have a superconductor model itself, but surface impedance values derived from the Mattis–Bardeen theory can be inserted and work well if the temperature is low enough.

Coplanar waveguides are better suited when higher impedances are required, while striplines are more suitable for lower-impedance integrated circuits. Connections made between the elements of integrated circuits should be able to carry RF signals at near light speed up to distances of a few millimeters without any distortion. For this purpose, superconducting microstrip connections are used. Thus, data communication between elements can be realized at about the third of light speed of vacuum with low dissipation (Razmkhah and Bozbey 2016; Razmkhah 2018; Le Roux et al. 2021).

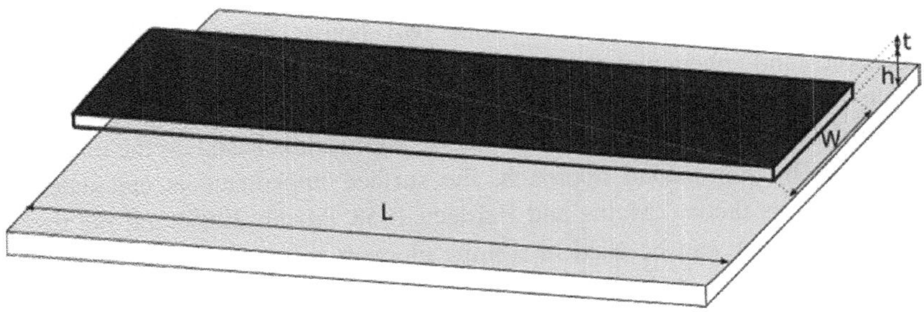

Figure 8.15. *Superconducting stripline studied with Sonnet software. For a color version of this figure, see www.iste.co.uk/cresti/beyondcmos.zip*

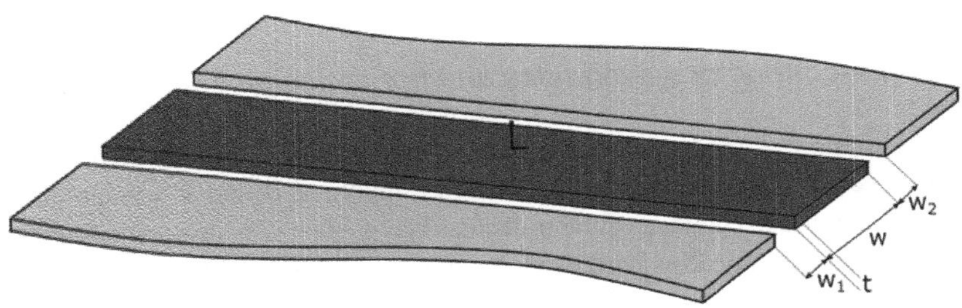

Figure 8.16. *Superconducting coplanar waveguide studied with Sonnet software. For a color version of this figure, see www.iste.co.uk/cresti/beyondcmos.zip*

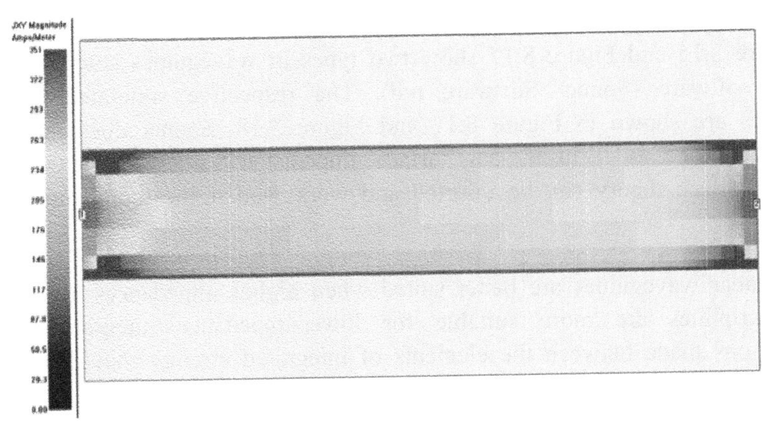

Figure 8.17. *Stripline simulation results obtained with Sonnet software. For a color version of this figure, see www.iste.co.uk/cresti/beyondcmos.zip*

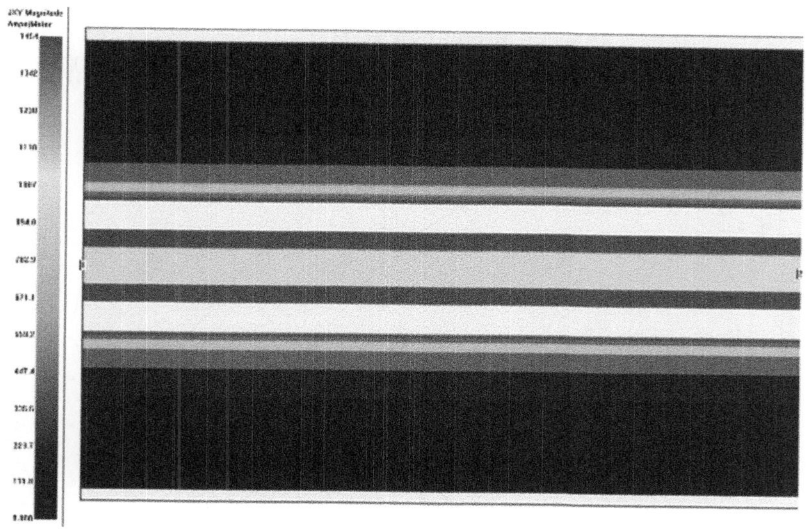

Figure 8.18. *Coplanar waveguide simulation results in Sonnet software. For a color version of this figure, see www.iste.co.uk/cresti/beyondcmos.zip*

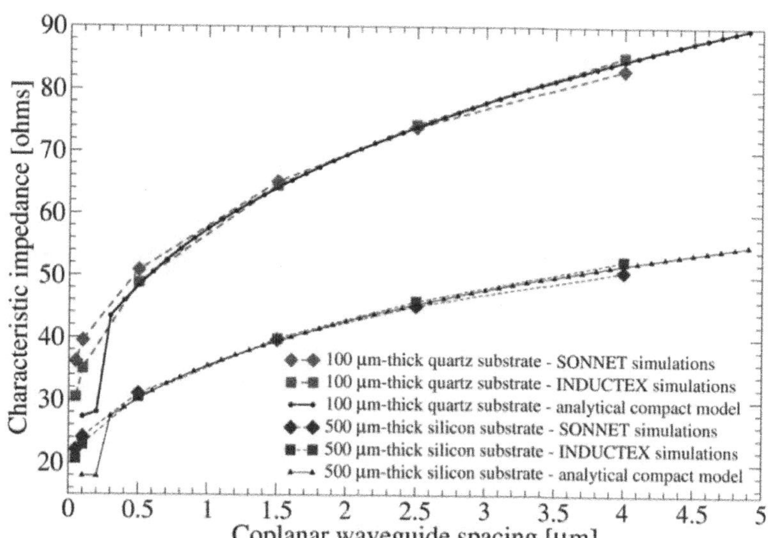

Figure 8.19. *Calculation of the characteristic impedance of superconducting coplanar waveguides based on numerical techniques with Sonnet and Inductex software, as well as with a compact model based on analytical formulae. For a color version of this figure, see www.iste.co.uk/cresti/beyondcmos.zip*

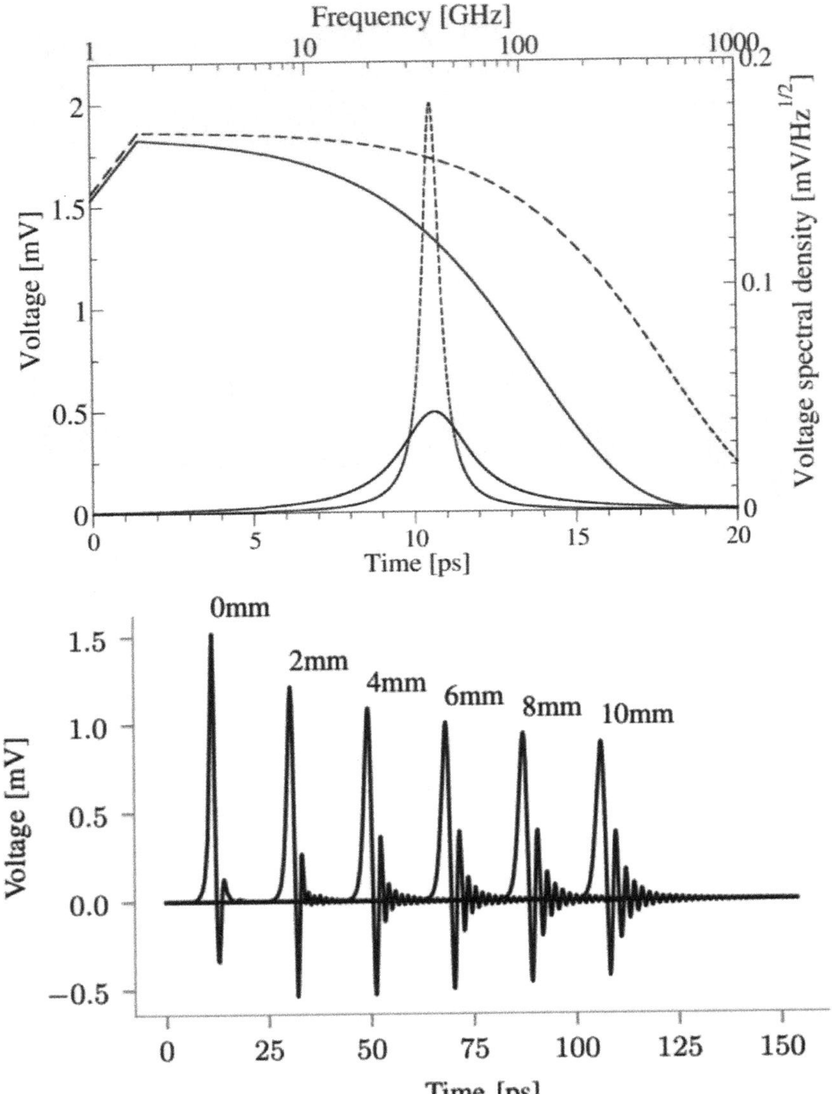

Figure 8.20. *Short single-flux-quantum theoretical pulse generated by a Josephson junction along with its spectral density (top). An example of resulting pulse shape (Le Roux et al. 2021) after propagation over a few millimeters along a superconducting transmission line (bottom). For a color version of this figure, see www.iste.co.uk/cresti/beyondcmos.zip*

It is also possible to derive simplified compact models of such passive elements based on analytical formalism initially developed for normal metal microwave lines (Febvre et al. 1995). This is useful when complex circuits must be simulated since time-domain numerical techniques require computing power and simulation time. In this case, an analytical method based on libraries of compact models is preferable. Figure 8.19 shows that numerical calculations made with the Sonnet and Inductex (Fourie et al. 2011) software exhibit results in agreement with the ones obtained from compact models.

However, passive transmission lines are also used to propagate very short pulse voltages, called SFQ pulses emitted by Josephson junctions working in digital mode (see section 8.4.1). In this case, the frequency spectrum is wide (see top of Figure 8.20), ranging over several hundreds of GHz, and has some energy at frequencies above the gap frequency of the superconductor, which results in absorption and dispersion as can be shown at the bottom of Figure 8.20.

8.2.3. *Superconducting antennas*

Superconducting antennas have been mostly used in integrated receivers (Buttgenbach et al. 1992) for THz applications for radioastronomy and security. They present nearly the same properties as normal metal antennas since the geometrical shape of antennas is the primary feature from which emission or reception properties can be derived. The main difference is that they are usually fabricated on superconducting chips, which are small (less than 1 cm^2 typically) and are consequently usually developed for frequencies above a few hundreds of GHz, with some geometrical features in the μm range (Zmuidzinas and LeDuc 1992). The other advantage is the lower losses due to the use of superconducting materials.

8.2.4. *Superconducting filters*

One main application of superconducting filters concerns the microwave field and is linked to the very low losses of superconducting films that allow having high Q-factors and consequently very narrow bandwidths whenever high-frequency selectivity is required in a microwave system. Also, since filters are made of simple elements, namely inductors and capacitors, they can also be easily fabricated with less mature thin-film technologies based on high–critical-temperature (high-Tc) superconductors (Marcilhac et al. 1999, 2001; Picard et al. 2004). Another application is the sub-THz field, mostly for integrated receivers in radioastronomy, where filters are used to separate incoming frequency signals from the down-converted output microwave signals at the intermediate frequency (Febvre et al. 1993, 1994).

Figure 8.21 shows an example of low-pass superconducting thin-film structure to filter out signals above 100 GHz for radioastronomy applications (Olofsson et al. 1998).

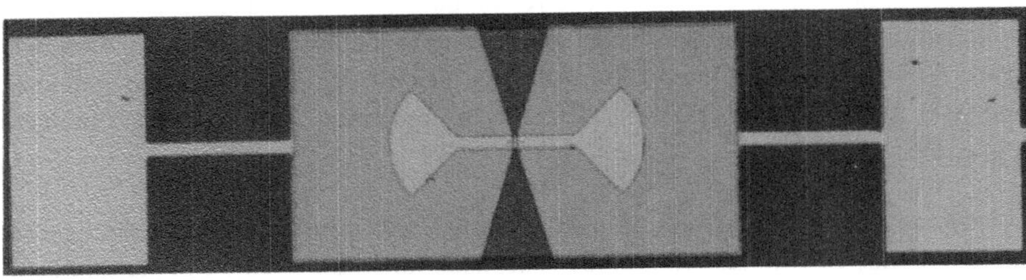

Figure 8.21. *Example of superconducting low-pass filter integrated with a high-frequency pass-band filter (center) for radioastronomy application. For a color version of this figure, see www.iste.co.uk/cresti/beyondcmos.zip*

8.2.5. *Microwave switches*

Microwave switches have a long history, especially associated with the invention of radars during World War II. Cryogenic microwave switches were studied since conventional circuits do not function correctly at low temperatures. They were investigated in particular for detectors and sensors due to the lower-noise background at cryogenic temperatures. In recent years, microwave switches got traction again due to their use in quantum technologies for computing and communication. Cryogenic switches use different technologies from CMOS ones, which are dissipative and do not perform the same as at room temperature. Cryogenic switches are compact and can reach high frequencies. Hence they are used in quantum technologies (Mueller and Herd 1993; Hornibrook et al. 2015; Frasca and Charbon 2019; Pauka et al. 2021). Nevertheless, the semiconductor switches are coupled with superconductor transmission lines to benefit from the absence of dissipation, higher speed (~c/3) and generally better performance at cryogenic temperatures.

Therefore, it makes sense to manufacture superconducting switches for better integration. For quantum technologies, superconductor switches have the advantage of keeping the quantum state as signals propagate along the circuit (Yeh and Anlage 2013; Hornibrook et al. 2015; Pechal et al. 2016a; Das et al. 2018). Historically, Brix et al. (Birx and Scalapino 1979) designed one of the first superconductor switches at cryogenic temperatures with a response time of a few nanoseconds, which encouraged other groups to design a range of other switches based on different architectures. Some of them are based on superconductor micro electronic mechanical systems (MEMs) that can make contact as the current flows through the control port and connect two superconductors together (Noel et al. 2003; Bogozi et al. 2005; Attar et al. 2014). Another architecture that has been used in recent years is based on ferromagnetic superconductors that can be switched by applying a low local magnetic field (Tagirov 1999; Larkin et al. 2012; Nguyen et al. 2020). However, such switches are not compatible with quantum circuits due to their magnetic moment. Thermal switches like cryotrons are also of interest, as the first superconductor transistors were based on such switches (Gray and Kampwirth 1984; Ma and Yang 1990; McCaughan et al. 2019). Nevertheless, there are drawbacks due to the change between normal and superconducting states, which limits the ultimate speed. Nano cryotrons and hot electron switches have recently solved the speed problem by switching up to tens of GHz. They can also keep the quantum state as the wave passes through them (Ye et al. 2014; Baek et al. 2015; Pechal et al. 2016b; Baghdadi et al. 2020; Ritter et al. 2021).

8.3. Superconducting detectors

The development of complex electronics systems requires the need of active and nonlinear devices and transducers to provide a large range of functionalities. The transistor obviously plays this role in semiconductor electronics. On its side, superconducting electronics benefit from the nonlinear properties of films associated with the phase transition between the normal state and the superconducting state, at the vicinity of the critical temperature T_c. For instance, thermal effects are directly used in cryotrons (Buck 1956) to make switches, as mentioned before, for a range of detectors like bolometers (Mazin 2005), transition edge sensors (TES) (Gerrits et al. 2016), hot electron bolometers (HEBs) (Shurakov et al. 2016) or superconductor nanowire single photon detector (SNSPDs) (Engert et al. 2013). They are also indirectly used through the modification of the complex surface impedance of superconducting films through the modification of their kinetic inductance. This last effect is used, for instance, in kinetic inductance detectors (KIDs) (Day et al. 2003; Baselmans 2012) used for astronomy applications and also for multiplexing purposes, in particular for the development of multi-qubits quantum systems.

Other applications are associated with “large currents” systems such as power lines (Larbalestier et al. 2010), fault current limiter switches (Leung 2000), magnetic resonance imaging (MRI) systems (Lvovsky et al. 2013) or energy storage (Buckles and Hassenzahl 2000). They are out of scope of this review.

Since the focus of the chapter is on quantum technologies, we will discuss the use of such detectors in quantum applications.

8.3.1. *Transition edge sensors (TES)*

Transition edge sensors (TES) or transition edge detectors (TED) are a group of sensors that make advantage of the steep resistance transition between the superconductor state with zero resistance and the normal metal resistive state. They are among the most sensitive thermal sensors and single particle detectors (Cabrera et al. 1998; Irwin and Hilton 2005; Smith et al. 2012; Ullom and Bennett 2015; Gerrits et al. 2016). TEDs such as bolometers are biased at the thermal resistance transition point so that any small temperature disturbance induces a large change in resistance.

TEDs consist of a very thin superconductor film deposited on the substrate surface. Their shape can be optimized to cover a large area. Figure 8.22(a) demonstrates a yttrium barium copper oxide (YBCO) high-temperature superconductor TED. Figure 8.22(b) shows the resistance of this sensor versus temperature; the sensor is biased at the point of maximum change. The sensitivity of TEDs is determined by the surface absorption, the sharpness of R–T transition and the substrate thickness and conductance (Kokabi et al. 2009). Even conventional high-temperature superconductor TEDs are two orders of magnitude more sensitive than their semiconductor counterparts and can be used as quantum detectors (Silberhorn 2007).

a)

b)

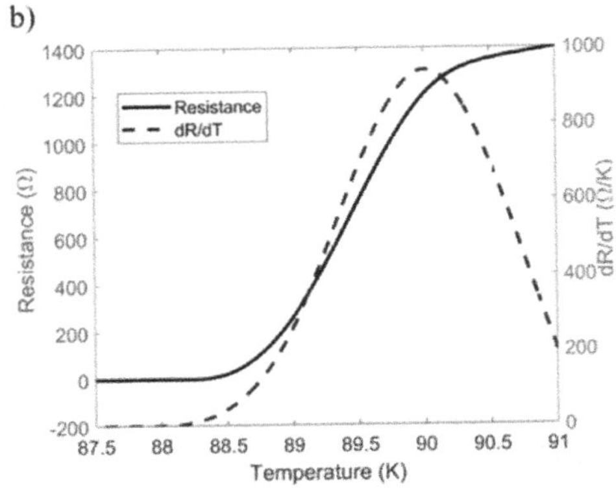

Figure 8.22. *a) High-temperature superconductor transition edge detector; b) resistance versus transition temperature and its derivative (Ansari et al. 2021). For a color version of this figure, see www.iste.co.uk/cresti/beyondcmos.zip*

8.3.2. *Superconductor nanowire single-photon detectors (SNSPDs)*

Superconductor nanowire single photon detectors (SNSPDs) are a group of sensors with a very high sensitivity. SNSPDs are made of a very narrow thin-film superconductor meandered ribbon deposited over the surface of the substrate so that it can cover a large detection area like TEDs. These detectors are also very suitable for being used as arrays (Engert et al. 2013).

SNSPDs are biased with a DC current in their superconductor state near the critical temperature point. When a photon is absorbed at the surface, it breaks superconductivity at the absorption point and the sensor becomes resistive so that a voltage occurs. When the detector recovers and becomes superconducting again, the voltage returns to zero. The voltage pulse at the sensor's output depends on the incident photons on its surface. Figure 8.23 shows the detection process in SNSPDs[1]. The detector's recovery speed depends on the cooling method and on the nanowire geometry. It can work up to GHz frequencies, but at the cost of sensitivity. SNSPDs can even detect microwave photons (Sinclair et al. 2019; Chen et al. 2020).

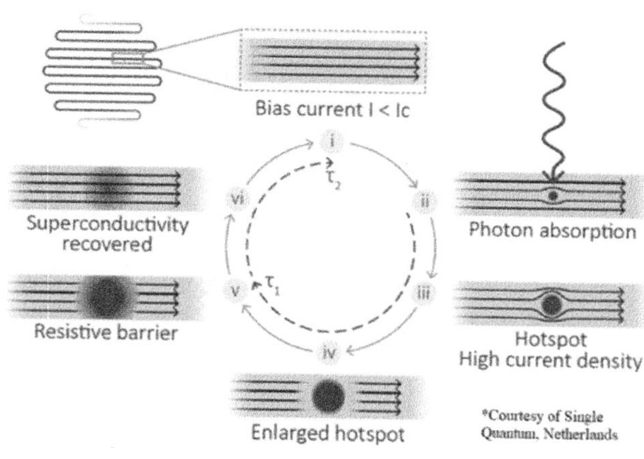

Figure 8.23. *Principle of photon detection by an SNSPD detector. For a color version of this figure, see www.iste.co.uk/cresti/beyondcmos.zip*

8.3.3. *Kinetic inductance detectors (KIDs)*

Kinetic inductance is caused by the existence of electrons in the presence of an electromagnetic solicitation. In superconductors, it originates more precisely from

1. See: https://singlequantum.com/technology/snspd.

quasiparticles whose state variables are not independent since they emanate from broken Cooper pairs. The kinetic inductance can be calculated from Mattis–Bardeen equations (Mattis and Bardeen 1958). In case a DC current is present in the superconductor near the critical current, the more accurate Usadel's equations (Usadel 1970; Ivanov et al. 1978) should be used for kinetic inductance calculations. Such is the case for most KID detectors.

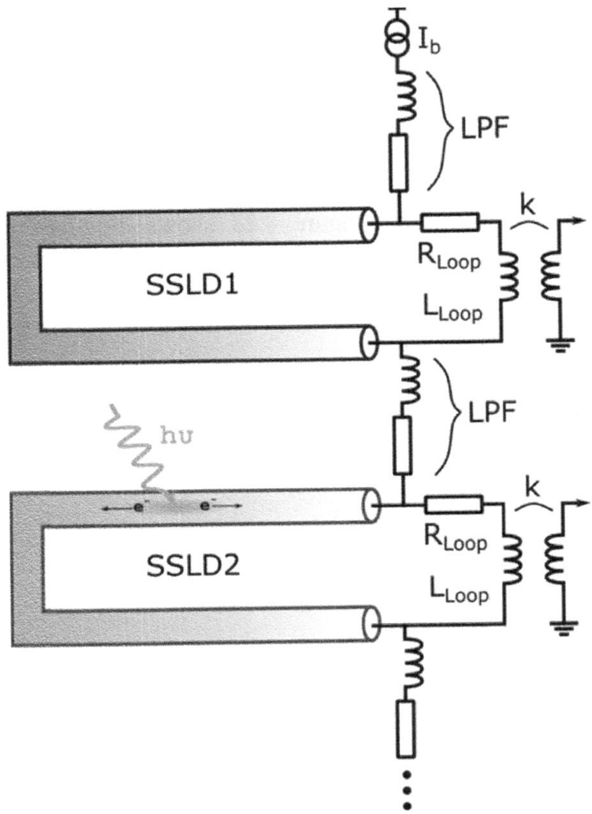

Figure 8.24. *Array of superconductor kinetic inductance detectors connected with low pass filters (Tanaka et al. 2021). SSLD stands for superconductor stripline detector*

Kinetic inductance detectors (KID) were first introduced by Day et al. (2003). KIDs work on a very simple principle and hence they are the easiest sensors to design among photon or particle detectors. A KID is a very thin superconductor stripline that has an absorber layer on top of it. It is usually DC current-biased at I_B near the critical current for more sensitivity. When a particle hits the absorber, the particle energy breaks a few Cooper pairs. This momentarily increases the density of quasiparticles. The change in quasiparticles causes a change in the kinetic inductance value and, since a bias current is present on the stripline, it generates an oscillating voltage pulse varying as $d(L_K I_B)/dt$, where L_K is the kinetic inductance. The recovery time is in order of nanoseconds, hence KIDs can reach near GHz detection speed. Their simple structure, sensitivity as well as their detection and recovery speed make KIDs one of the most promising quantum detectors.

A detector that uses any of the three quantum effects namely, entanglement, superposition or quantum interference or a combination of these effects is classified as a quantum detector. Due to quantum nature of the detectors principles, they are more accurate and exhibit higher sensitivity. Superconductor detectors are among commercially used quantum detectors.

8.4. Superconducting digital electronics

After the Josephson junction discovery, many new applications surfaced and the Josephson junction has become to date the most versatile active element used for superconducting analogue, digital and quantum electronics. These applications include magnetometers and gradiometers based on RF and DC SQUIDS (Clarke, Braginski 2006b), Josephson voltage standards (Benz, Hamilton 1996), analogue-to-digital converters (Mukhanov et al. 2004; Ren et al. 2015), quasi-one-junction SQUIDs (QOS) (Brock et al. 2005), digital logic circuits (Likharev 1986; Likharev and Semenov 1991), quantum computing, quantum communication and quantum sensing. We will discuss some of its applications further.

The biggest impact that semiconductor technology has had in the last few decades is in the field of computing. Josephson junction-based logic circuits can contribute to this field dramatically. Although superconductor logic circuits have some limits due to cooling and fabrication, they have advantages over semiconductors in terms of speed and power consumption. Over the years and after heavy investments, semiconductors have been optimized with advanced fabrication processes. Nevertheless, speed limits have been reached for nearly a decade, while the energy wall and the ultimate densities of integration make progress more difficult and costly. Though other alternatives exist based on different approaches like distributed or in-memory computing, superconductor logic is based on a disruptive technology. It has shown a very reliable operation of circuits at least 10 times faster, with energy consumption orders of magnitude lower than the semiconductor state of the art[2].

2. See: https://irds.ieee.org/editions.

There are different approaches to use superconductor circuits in digital mode. Some include voltage state or flux state logic (Kadin 1999; Rose-Innes 2012). Superconductor circuits have been demonstrated with low- and high-temperature superconductor (HTS) materials. However, the technology based on HTS materials is not yet mature enough to make complex circuits. Low-temperature superconductor (LTS) materials are easier to work with, and complex circuits can be manufactured to date with a complexity surpassing 1 million devices per cm^2 (Tolpygo et al. 2021). Moreover, commercial nanoelectronics processes are available for niobium-based superconductors[3].

For every logic circuit design, a switching process is needed to change the signal path and store the data bit in a memory. In semiconductors, the transistor is the switch and the capacitor stores the data. In superconductor circuits, the switch is a Josephson junction or a SQUID, and the memory is the inductance of the loop in which the JJ is inserted. Before SFQ electronics were invented and even before the invention of the Josephson junction, cryotron superconductor digital circuits were introduced (Buck 1956). These circuits used two different superconductor materials with different critical magnetic fields H_c. The memory was a simple superconductor loop that could store magnetic flux and therefore a permanent DC current. As the magnetic field reaches the critical value, the material becomes resistive and switches to the normal mode. However, this technology was abandoned as the speed of the gates was limited by the L/R time constant where L is the loop inductance and R is the resistance of the device in the normal mode. Semiconductors could compete with this speed and had better margins of operation.

In this chapter, we investigate two different Josephson-based logics, namely the single flux quantum (SFQ) logic and the adiabatic quantum flux parametron (AQFP) logic. Cells and basic circuits fabricated with these logics are shown, advantages and disadvantages are investigated and finally the fabrication processes are described.

8.4.1. *Single flux quantum (SFQ) logic*

SFQ was first introduced as an alternative superconductor logic by K.K. Likharev et al. in the late 1980s (Likharev 1986; Likharev and Semenov 1991). In this technology, the Josephson junction acts as a switch and the data is stored as a magnetic flux in superconducting loops. As mentioned before, SFQ logic has a very low power consumption and therefore is considered as an alternative solution for computing centers (ITRS 2004; NSA 2005). The integration capabilities and the high speed of SFQ circuits are also dependent on the fabrication process. Various

3. See: https://www.fluxonics.eu.

groups have worked on processors based on the SFQ technology (Bunyk et al. 2003; Dorojevets and Bunyk 2003; Fujimaki et al. 2008; Filippov et al. 2011, 2012b; Dorojevets et al. 2013a, 2013b).

8.4.1.1. *Principle of operation*

In SFQ circuits, the data is carried by a voltage pulse propagating at the guided speed of light. Its area (the product of voltage by time over the pulse duration) corresponds to the single magnetic flux quantum $\Phi_0 = h/2e$, from which originated the name single-flux-quantum electronics (SFQ electronics). The pulses have a width of a few picoseconds, and therefore the circuits can function at hundreds of gigahertz in theory, associated with the plasma frequency of Josephson junctions, which can reach 1 THz for niobium-based devices. For instance, a toggle-flip-flop (TFF) cell was reported to work at 770 GHz (Chen et al. 1998, 1999). Since the storage in SFQ circuits is in the form of quanta of magnetic flux associated with DC permanent currents and no voltage, circuits are current-biased, typically from 75 to 90% of the critical currents of JJs. Figure 8.25 shows the typical *I*–*V* characteristic of a Josephson junction used in SFQ logic. The McCumber parameter β_c is about 1.

Figure 8.25 illustrates the mechanism of SFQ pulse generation in SFQ circuits. For this to happen, it is important to point out that the Josephson junction must be placed in parallel with an inductor, which is in practice either a shunt superconducting wire forming a loop with the junction or a closed circuit connected to the junction's terminals. As the junction is biased with I_b close to its critical current value I_c, when a low current excitation comes to the junction (e.g. provided by an SFQ pulse or a DC current), the total current becomes higher than I_c at some point and a voltage appears across the junction terminals. The voltage transition that is shown with path 2 in Figure 8.25 takes place in about 1 ps as the current redistributes over the circuit due to a strong change of the nonlinear junction impedance during the event. With a standard fabrication process, the pulse voltage amplitude ranges from 250 to 500 μV but can reach more than 1 mV for a more advanced process associated with thinner thicknesses of insulating quantum-tunneling barriers of Josephson junctions.

After the pulse has reached its maximum, the junction returns to the starting point through path 3. During the pulse generation, the phase of the junction jumps by 2π, then returns to the initial state with no memory of the voltage event that occurred. A good analogy of the junction can be made with a pendulum whose angle is the junction phase, while its momentum corresponds to the current. The voltage pulse event corresponds to the angular frequency of the pendulum making one full turn. The physics of the Josephson junction tells us that the area of the voltage pulse is quantized such that:

$$\int V(t)dt = \frac{h}{2e} = \Phi_0 = 2.07\ Wb = 2.07\ mV \cdot ps = \ 2.07\ mA \cdot pH \qquad [8.10]$$

where the quantum of magnetic flux Φ_0 is a fundamental quantity. Since the area is quantized, the technological process determines the shape of the pulse, from wider duration with lower pulse height for standard processes, until higher pulse height with shorter duration for the most aggressive processes.

The physical quantity corresponding to a single-flux quantum Φ_0 carries the digital information in SFQ logic. If a pulse is generated between two instants set by a clock, it is considered as a logical "*1*". In the absence of pulse, the signal is coded as a logical "*0*".

Once the voltage pulse has been generated across the junction's terminals, it can either propagate at the speed of light into the circuit connected to the junction if the inductive loop is too small to accommodate a magnetic flux quantum Φ_0 because $LI_c < \Phi_0$, or, in case $LI_c > \Phi_0$, the SFQ pulse can be stored in the junction's loop in the form of a current $I < I_c$ with $LI = \Phi_0$ (see equation [8.10]). The logic information is then stored in the circuit under the form of a permanent current. The loop acts as a local memory.

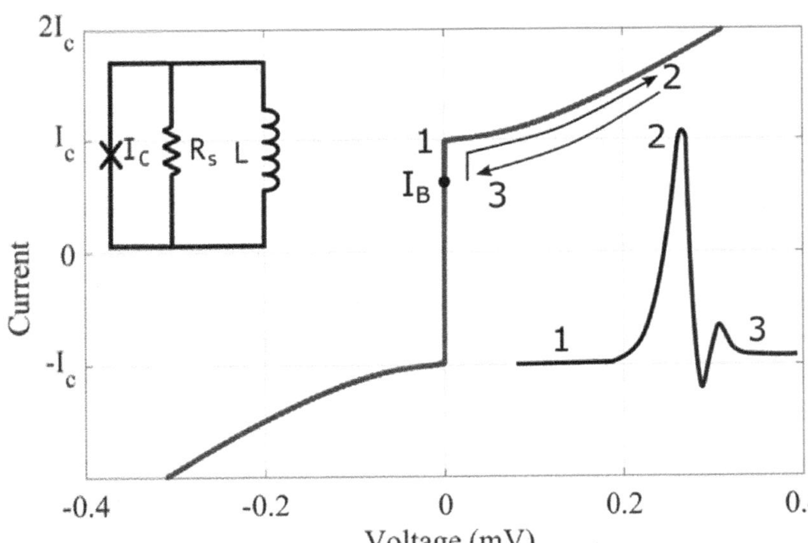

Figure 8.25. *I–V characteristic of a resistively shunted Josephson junction, biased at I_B. The voltage follows the indicated path when an SFQ pulse hits the junction. For a color version of this figure, see www.iste.co.uk/cresti/beyondcmos.zip*

8.4.1.2. *Control and readout interface cells*

The very short and small-amplitude voltage pulses are impossible to observe with traditional equipment. However, it is necessary to generate SFQ pulses and read the output state of circuits to interact with SFQ circuits. Therefore, two interface circuits have been introduced to convert the DC or low-frequency signal to SFQ pulses and vice versa. The DCSFQ converter is a cell with a threshold, like a Schmidt trigger but with a current threshold, which generates an SFQ pulse each time its total bias current passes over that threshold (typically 1 mA). Figure 8.26 illustrates its operation, and the short width, low amplitude of SFQ pulses is obvious.

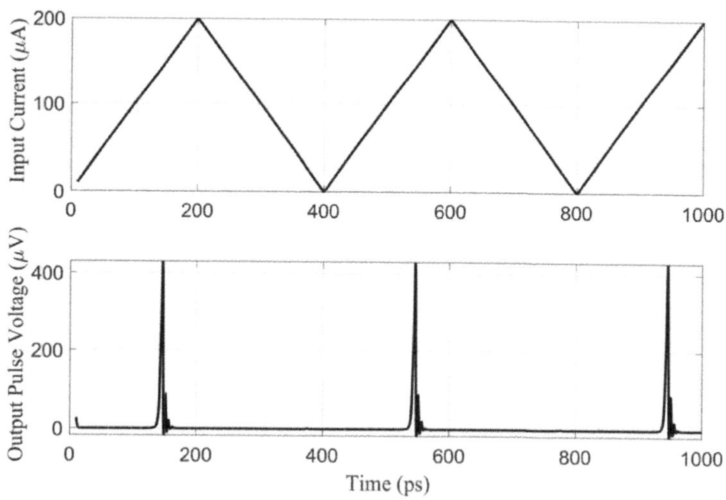

Figure 8.26. *Conversion of DC or low-frequency signal in SFQ pulses by a DCSFQ converter cell*

Since SFQ pulses do not keep their voltage, they can be considered as return-to-zero (RZ) circuits. The interface to read SFQ pulses is called an SFQDC converter cell. It converts incoming SFQ pulses to a readable non-return-to-zero (NRZ) signal. The SFQDC converter circuit acts like a TFF whose state switches whenever it detects an incoming SFQ pulse. Before any incoming SFQ pulse, the state is zero and the average output voltage of the circuit is also zero. An incoming SFQ pulse makes an internal JJ switch and stores a magnetic flux. The new state is such that the output stage is an oscillating voltage at very high frequency (in fact, it is a fast train of SFQ pulses) whose average is consequently non-zero. The measurement equipment cannot follow the fast changes since it low-pass filters the signal and only sees the non-zero average voltage. It should be noted that the voltage state value of the SFQDC converter is not important. What matters is the instant for which the voltage transition occurs, be it from zero to non-zero or the opposite. This is like a transition edge signal operating with double data rate (DDR) in semiconductor logic. Figure 8.27 shows the difference between single data rate and DDR. Figure 8.28 shows the train of input pulses and the result for the output data (Askerzade et al. 2017a).

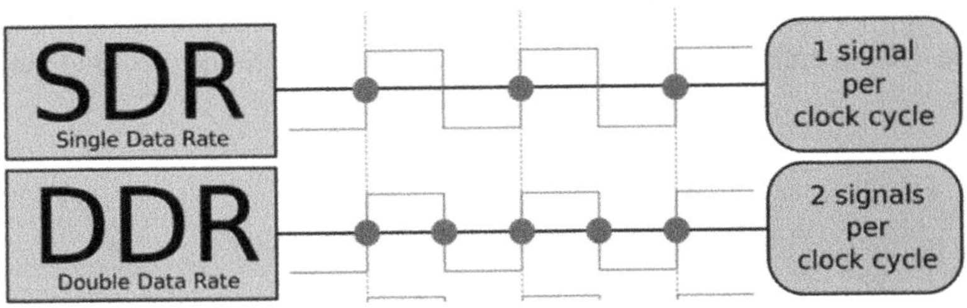

Figure 8.27. *Single data rate and double data rate principle. For a color version of this figure, see www.iste.co.uk/cresti/beyondcmos.zip*

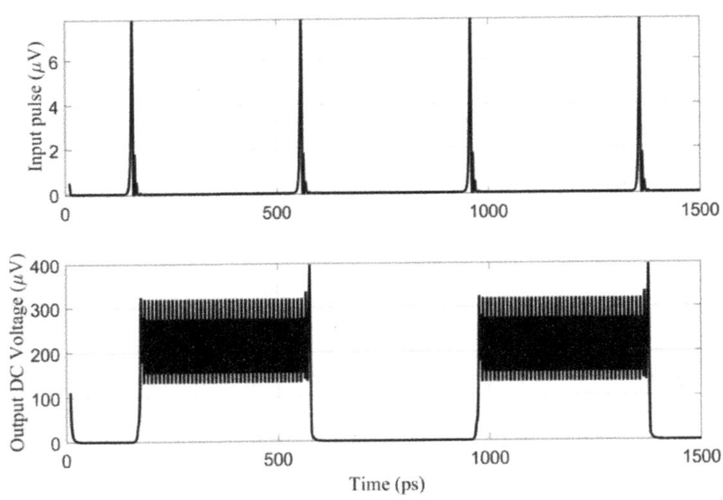

Figure 8.28. *SFQ pulse to DC conversion in an SFQDC converter cell. Note that each pulse causes a state change at the output, after a slight time delay*

8.4.1.3. *Energy-efficient SFQ architectures*

As discussed before, because of the SFQ logic fundamental principles based on the JJ physics, JJs are biased with current rather than voltage. Since there is no resistance in the circuits, their static energy consumption is zero. However, in large integrated circuits with tens of thousands of junctions, it is not practical to have a bias line for each individual junction. To maintain the appropriate bias distribution for each junction, a resistance R_{bias} is added at each bias input of a cell. This resistance is biased at a typical voltage V_b of 2.5 mV to provide a bias current of V_b/R_{bias} (the SFQ cell itself has no resistance). Consequently, the main energy consumption source of SFQ circuits mentioned earlier is the static power dissipation from these resistances that cause about 99% of the power waste.

From this observation, variants of SFQ circuits have been developed, such as ERSFQ (efficient RSFQ) (Kirichenko et al. 2011), eSFQ (Mukhanov 2011) or LRSFQ (Yuki and Yamanashi et al. 2007). In such architectures, resistances are replaced by a combination of Josephson junctions and inductances or only very large inductances, respectively. These circuits consume two orders of magnitude with less power than their SFQ counterparts, but they become bulkier or lose bias margins compared to SFQ circuits.

8.4.1.4. *Propagation and transmission of SFQ pulses*

The design of the wiring of cells and clock paths is a delicate issue for large-scale SFQ logic circuits due to their high bandwidth of about 200 GHz and limited fan-out of the SFQ cells. As a result, most of the wiring is done with active lines also known as Josephson transmission lines (JTL) (Likharev and Semenov 1991b; Brock 2001). Excess use of JTLs for wiring causes a dramatic increase in power consumption and delays of logic circuits. One of the solutions to this problem is to use passive transmission lines (PTL) (Tanaka et al. 2005). The main drawbacks of PTLs are that they consume more space especially with a four-layer process such as the AIST Standard Process (STP2) (Hidaka et al. 2006) or the RSFQ1H Niobium process from IPHT Jena (Anders et al. 2010).

8.4.1.4.1. Passive transmission lines

The mechanism of pulse propagation on microstrip lines or coplanar lines has been extensively studied. In the case of short SFQ pulses the frequency spectrum is broadband of a few hundred GHz (see section 8.2.3 and top of Figure 8.20). PTLs have a different response at different frequencies because of Cooper pair breaking at frequencies reaching the gap frequency of the superconductor, which leads to absorption and dispersion. As a result, the pulse traveling along a superconducting line is subject to a change of shape (see bottom of Figure 8.20). Note that, as the line length increases, the SFQ pulse voltage level drops more and pulses are wider. At some point, their voltage can be too low to trigger the Josephson junction placed at their output, leading to some risks of malfunction. As a result, the design of a PTL that has a good impedance match with the rest of the surrounding cells must consider its behavior up to about 165 GHz for SFQ pulse propagation. The PTL can be modeled with a ladder π-model as shown in Figure 8.29. The electrical parameters depend on the fabrication process and can be found via high-frequency simulations.

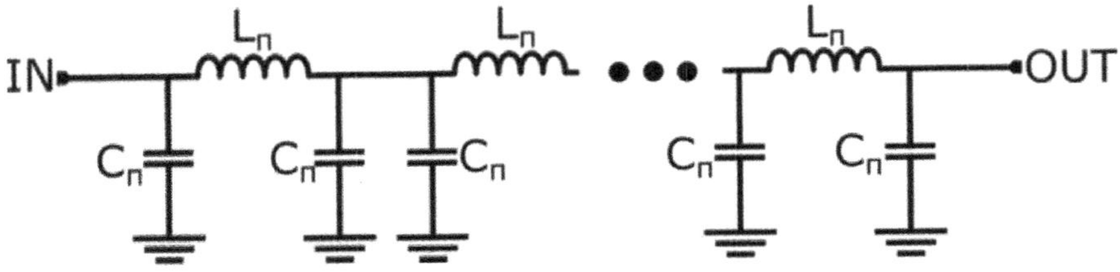

Figure 8.29. *Ladder π-model for a superconducting passive transmission line (PTL)*

8.4.1.4.2. Driver and receiver circuits

The driver and receiver circuits are also modeled using the RCSJ model. Figure 8.30 and Figure 8.31 show the driver and receiver circuits schematics matched for an unshielded 38-μm-long transmission line used in the CONNECT cell library (Yorozu et al. 2002). Each PTL needs a driver to impedance-match it with its environment. On the other end, the receiver circuit at the end of the transmission line is also necessary for impedance match and to convert the traveling pulse to an SFQ pulse again. The driver and receiver cells are designed to match the 2Ω impedance of logic cells in CONNECT library.

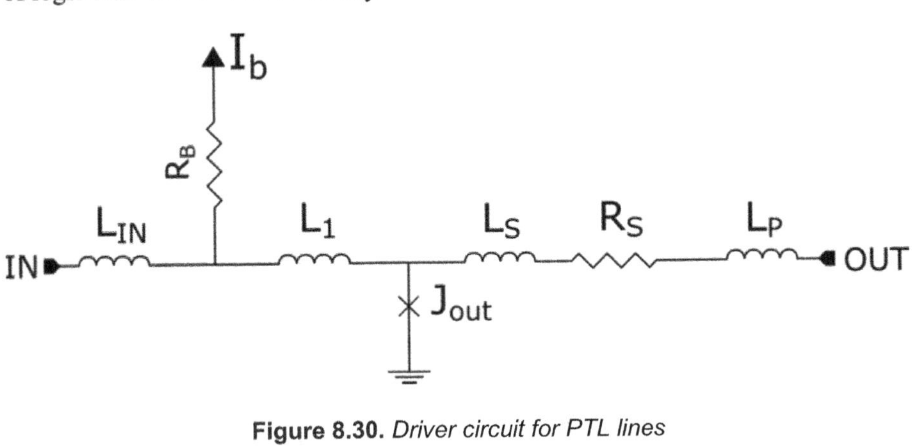

Figure 8.30. *Driver circuit for PTL lines*

In Figure 8.30, L_{IN} matches the 2Ω input impedance of the driver circuit, junction J_{out} generates the SFQ pulse to propagate on the PTL, while L_S and R_S match the input impedance of the PTL. Figure 8.31 shows the receiver circuit where L_{IN} and R_{IN} are for PTL impedance match. J_1 and J_2 constitute a buffer Josephson transmission line (JTL) stage (see the next section) to reshape the incoming pulse to an SFQ pulse.

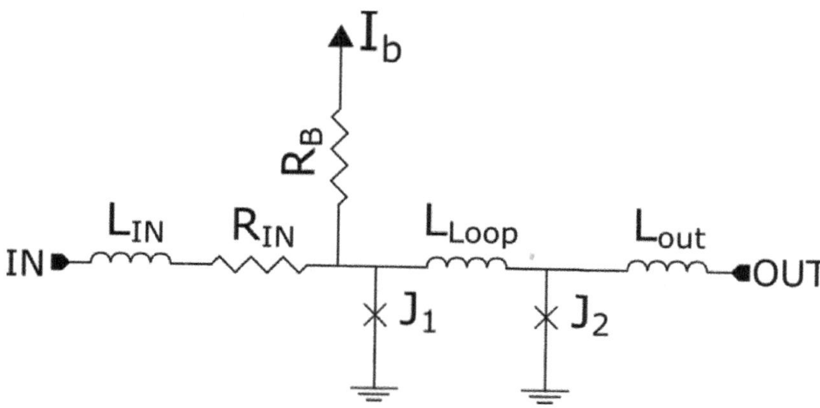

Figure 8.31. *Receiver circuit for PTL lines*

8.4.1.4.3. Josephson transmission lines (JTL)

One of the most used basic wiring cells in the SFQ technology is based on Josephson transmission lines (JTL). A JTL cell is a transmission line that also acts as a signal router and buffer. If the signal gets corrupted due to noise, the JTL cell will regenerate the shape of the SFQ pulse. Figure 8.32 shows the schematic of a JTL cell.

The JTL has two Josephson junctions, J_1 and J_2, biased at about 90% of their critical currents via the bias line, which provides a 2.5 mV voltage at the bias resistance. When a pulse comes from the input port D_{in}, it passes through the inductances until it reaches the superconductor loop involving the junctions. This loop will not store the SFQ pulse (in fact the magnetic flux Φ_0), because $LI_c < \Phi_0$. Consequently, the two junctions are triggered one by one to generate a new SFQ pulse, up to the D_{out} port. The input and output inductances are used for impedance matching at 2Ω for the circuits designed with the standard process. Over the years, many variants of JTL cells have been designed. The schematic is almost always the same, but the layout is altered in order to minimize their size for large circuits.

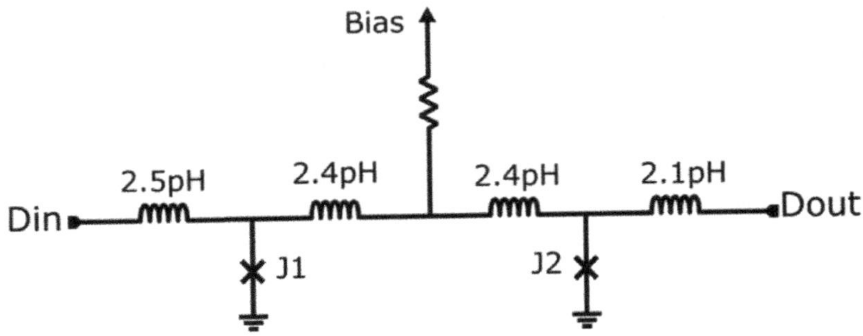

Figure 8.32. *Schematic of a JTL cell used in SFQ logic circuits*

8.4.1.5. *Splitter and merger cells*

As mentioned before, the fan-out of SFQ circuits is only one. Therefore, to distribute a signal pulse to multiple inputs, a splitter circuit is needed. The splitter can multiply an incoming signal to two or three cloned pulses.

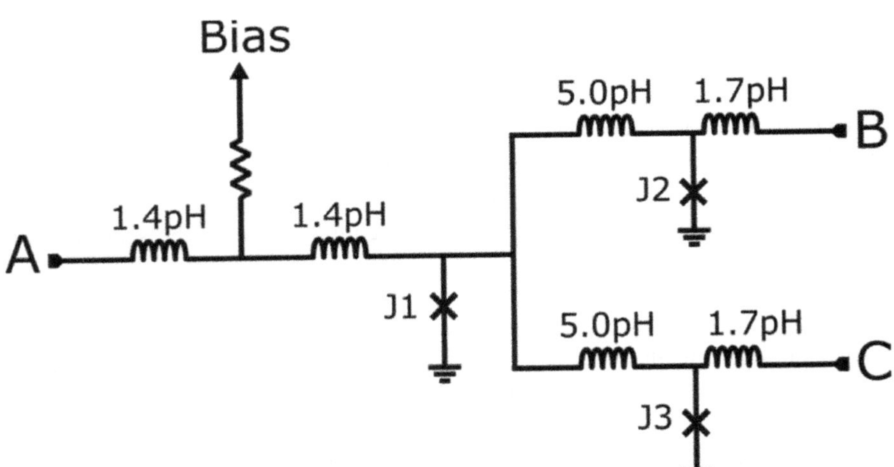

Figure 8.33. *Schematic of an SFQ splitter cell*

Figure 8.33 shows the schematic of a splitter cell used in SFQ circuits. The first part of the circuit until the bifurcation is a half-JTL buffer stage. This stage assures that the SFQ pulse has the right shape and matches the input impedance. Since there is no resistance in SFQ circuits, the voltage splits according to the two parallel inductances at the bifurcation (voltages appear at the terminals of inductances owing to Faraday's law). The inductances for both output branches are identical. Therefore, two similar signals travel through each branch and appear at ports B and C at the same time. The critical current values for J_2 and J_3 junctions are low to respond to small excitations and create SFQ pulses.

The other common wiring cell is the merger. Merger cells act as opposite of splitter cells. When two SFQ pulses arrive at the same time to the merger, it combines them into a single output pulse. However, if only one pulse comes to one of the inputs, the cell gives an output pulse as well. This cell is used when we need to combine two or more data lines to a single line.

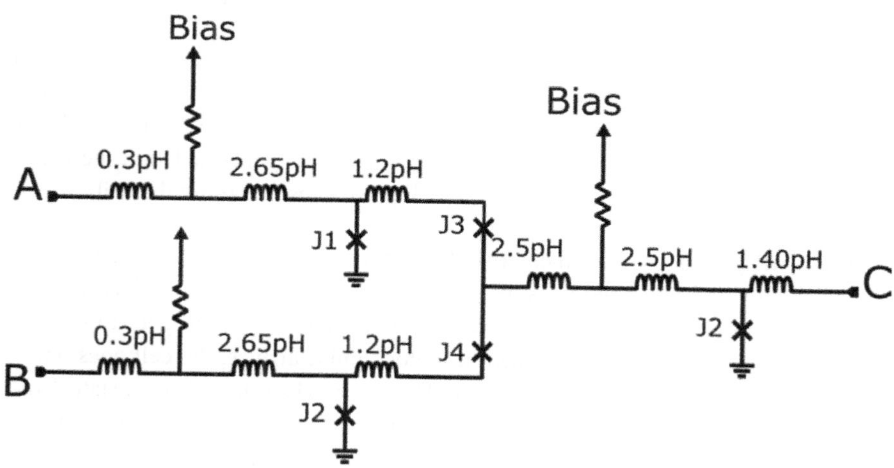

Figure 8.34. *Schematic of an SFQ merger cell*

The schematic of a merger cell with two inputs is given in Figure 8.34. Input ports A and B are, respectively, connected to the J_3 and J_4 junctions, each via a JTL buffer line. Like for the splitter cell, JTLs at inputs guarantee that the shape of SFQ pulses are appropriate and do the input impedance match. As an SFQ pulse arrives at one of the inputs, J_3 or J_4 junction transmits a portion of that pulse, which is enough to excite junction J_5 to generate a pulse. J_3 and J_4 junctions also prevent the output pulse generated by J_5 from reflecting and going back to the inputs.

When two SFQ pulses arrive simultaneously at each input, or with a very small time difference, there is not enough energy after passing through J_3 and J_4 junctions to have two SFQ pulses generated by J_5. Instead, only one pulse appears at the output.

8.4.1.6. *Logic cells*

Logic cells are the basic blocks for digital circuits and superconducting digital computing. In this section, we discuss the four main gates, AND, OR, XOR and NOT, that exist in the SFQ technology. Because of the SFQ logic's pulsed nature, all the gates in this technology are clocked.

The truth table of the AND gates tells us that its output *c* is true when inputs *a* and *b* are true. Figure 8.35 is the Moore diagram of a Josephson AND gate as designed with the standard cell library (CONNECT 2003; Nakamura et al. 2009).

The OR gate logic state is true when any of the inputs or both inputs are true. Figure 8.36 shows the Moore diagram and schematic of a clocked Josephson OR gate. When one of the inputs or both inputs become true, the state of the cell changes to 1. As the clock signal comes in state 1, an output *c* is generated, and the state of the circuit goes back to 0. At state 0, the clock pulse does not affect the cell.

The XOR gate is used to make a subtractor from adder. The XOR logic state is true when only one of the inputs is true. Figure 8.37 shows the Moore diagram of the clocked Josephson XOR gate. When one the inputs is true, the cell goes to state 1 or 2 depending on the input *a* or *b*. Then, if a clock pulse arrives, the cell goes back to state 0 with a signal generated at output *c*. However, when the cell is in state 1 or in state 2, if the other input becomes true, the cell changes to state 3. At this state, the clock pulse resets the cell to state 0 without generating any signal at output *c*.

The last logic gate is the Josephson NOT gate. The NOT logic gate is true when the input is false.

The initial state of SFQ logic cells is not defined once they are initially biased. This is because some magnetic flux can be trapped in the cells during cooling of the circuit. As a result, a clock signal must be applied first to reset all cells to their zero state.

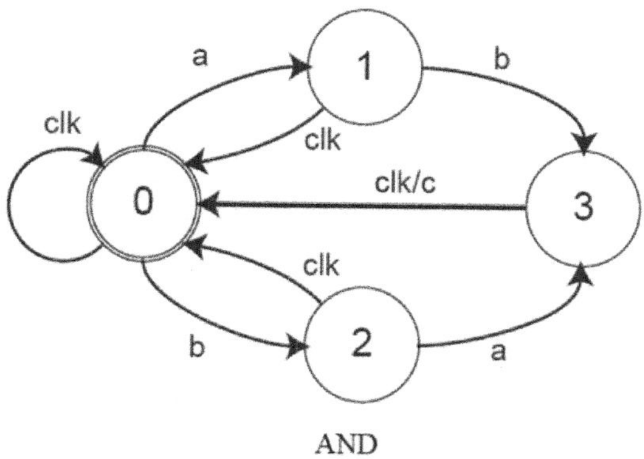

Figure 8.35. *Moore diagram of the Josephson AND gate*

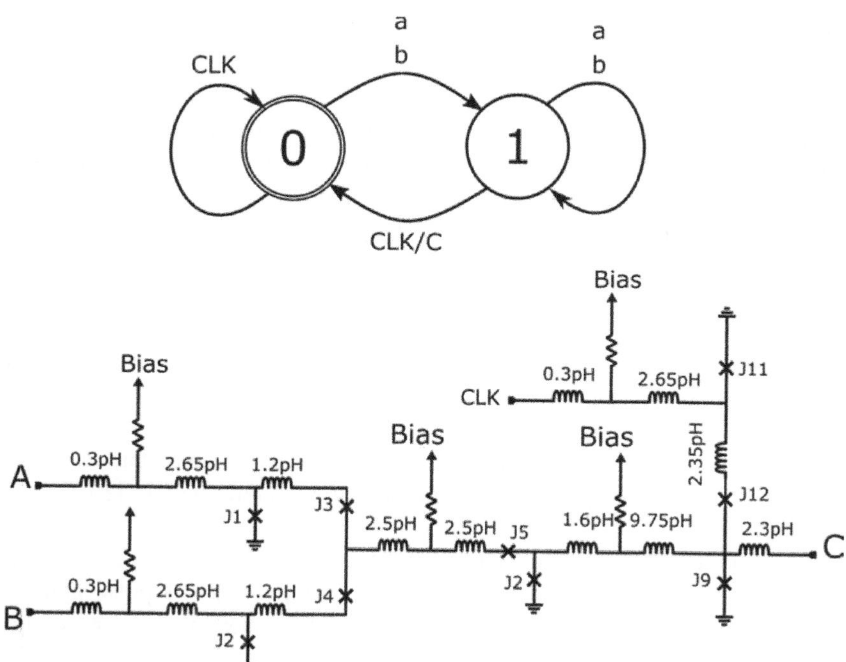

Figure 8.36. *Moore diagram (top) and schematic (bottom) of an OR gate*

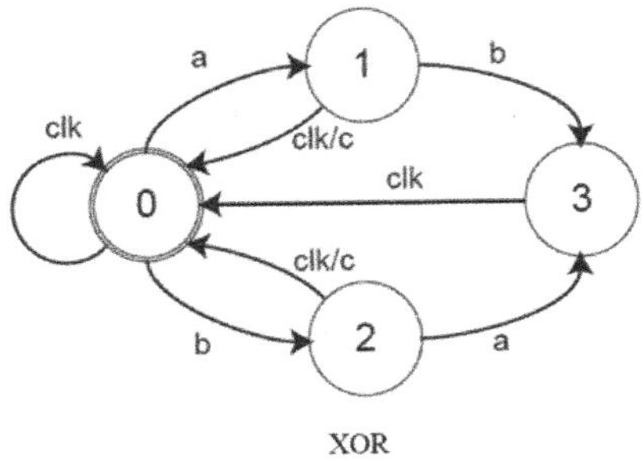

Figure 8.37. *Moore diagram of a XOR gate*

8.4.1.7. *Memory cells*

Flip-flop cells are basic memory cells in the SFQ technology. Flip-flops can be used to store data or the state of the circuit to perform various tasks such as multiplexing, demultiplexing and synchronization. There are different types of available flip-flop cells, but we only discuss the most common ones.

8.4.1.7.1. The delay flip-flop (DFF) cell

The first type of flip-flop cell that is widely used in the SFQ architecture is the D-type flip-flop also known as data or delay flip-flop (DFF). The DFF gate output is true when an input signal arrived before the clock signal. Figure 8.38 shows the Moore diagram and schematic of a clocked Josephson DFF gate. An incoming pulse at the DFF cell switches its state to 1. Then, the clock signal triggers a pulse generation at the *dout* output and resets the state to zero.

Some DFF circuits available in the SFQ technology have two outputs, the second output being complementary to the first one.

RDFF cells have a reset port to bring them back to the zero state without any pulse generation at the output, in case the RDFF is in the unwanted state 1.

Another type of DFF is the EDFF cell. Standard DFF cells can malfunction if more than one input arrives at the DFF cell input before any clock pulse signal. The EDFF cell removes this problem. Then, when the number of input pulse signals cannot be fully known during the design procedure, because of the circuit architecture, it is recommended to use EDFF cells instead of DFFs.

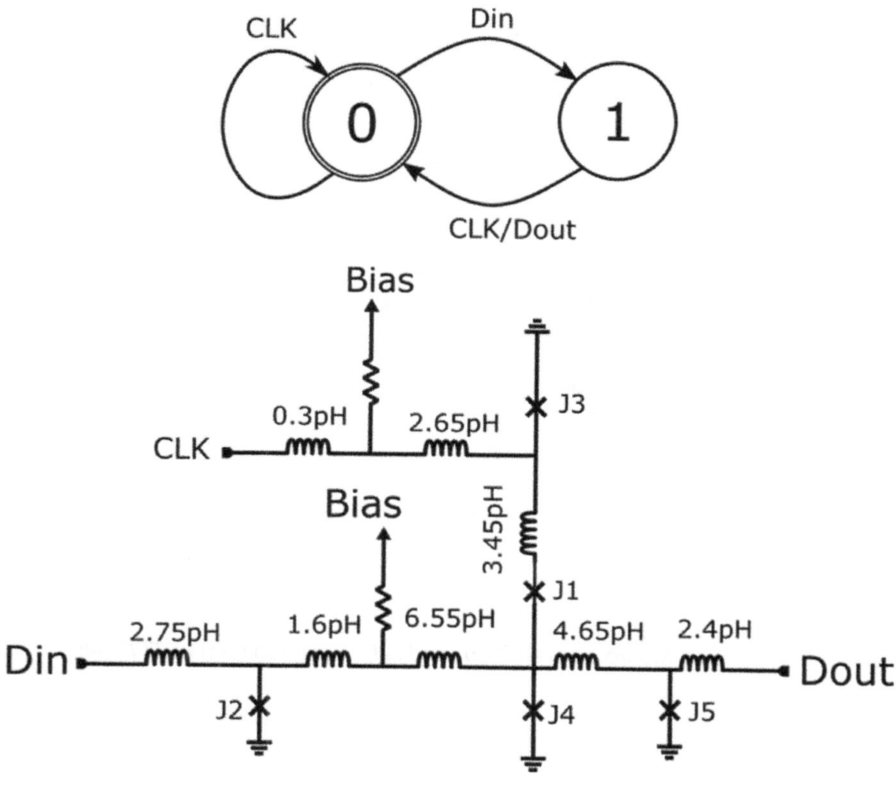

Figure 8.38. *Moore diagram (top) and schematic (bottom) of a Josephson DFF clocked gate*

8.4.1.7.2. The toggle flip-flop (TFF) cell

The toggle flip-flop cell or TFF, as its name tells, toggles for each incoming pulse. This cell is very useful to design sequential logic circuits and frequency dividers. Figure 8.39 shows the Moore diagram and schematic of a TFF gate. An input pulse at *din* port changes the state of the cell to 1, while generating an output at its first output *dout0*, while the next incoming pulse brings back the TFF to its initial state 0 and generates an output at the second, complementary, output *dout1*. As a result, each output generates a pulse for every second pulse incoming at the single TFF input. Besides, output pulses at the complementary output are in quadrature with the ones of the first output.

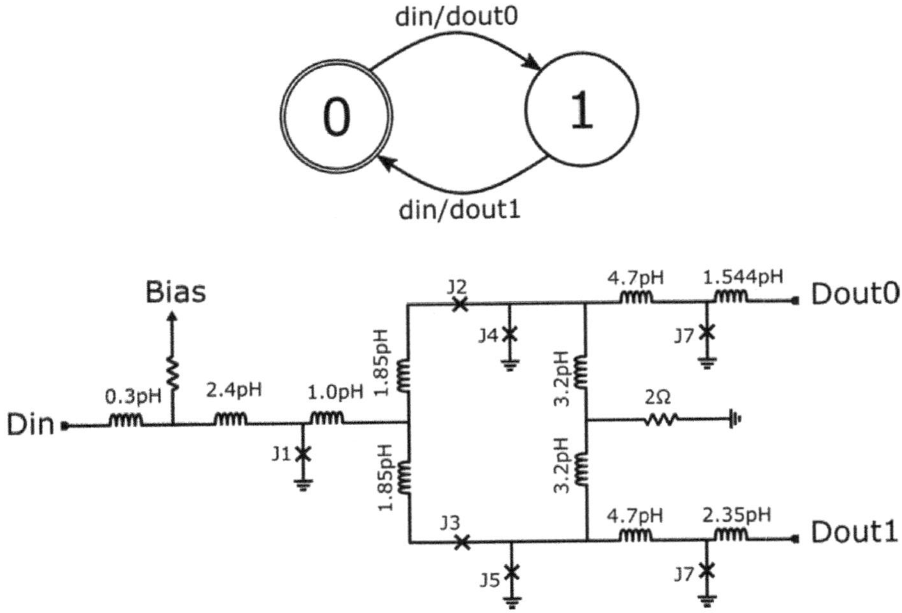

Figure 8.39. *Moore diagram (top) and schematic (bottom) of TFF cell*

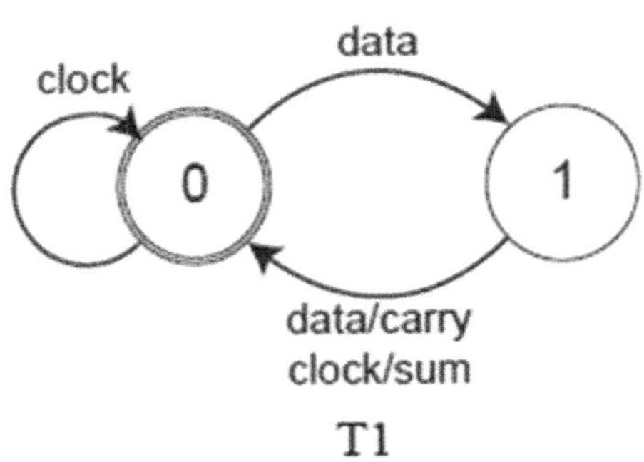

Figure 8.40. *Moore diagram of T1 cell*

The next cell is the clocked toggle flip-flop cell, or T1. It acts like an ordinary TFF, but the clock pulse can reset the state of the cell or initialize the output. This cell also acts as a one-bit full adder. Figure 8.40 shows the Moore diagram of a T1 cell. As seen in the figure, a data pulse switches the cell to state 1. Depending whether the incoming signal is from the clock line or the data line, an output pulse appears at the *sum* or *carry* output, respectively. The clock always sets the state back to zero and can be used to reset the cell. The circuit has a main superconducting storing loop, just like the TFF cell that keeps the state of the circuit. Each pulse from the *data* input changes the state of the loop. If the loop has a stored flux already, which means that the state is 1, an SFQ pulse is sent to the *carry* output port. However, in this cell, unlike for the TFF, a clock path is synchronized to the main loop so that, if it contains a stored flux, an SFQ pulse is sent to the *sum* output.

8.4.2. *Adiabatic quantum flux parametron (AQFP) logic*

The computation merits of any system are the power consumption and the speed of the system. For many years, power consumption was not the main issue, and the main focus was on increasing the speed. In recent years, due to a huge increase in computations, power consumption draws attention and is now considered the major metric of the computer design (Ball 2012; Service 2012; Takeuchi et al. 2014; Jones 2018). As mentioned before, Landauer's law states that a bit loss in the system causes energy consumption of $k_B T \ln 2$ to compensate for the entropy change of the system (Landauer 1961). This sets the theoretical limit for energy consumption of conventional logic gates such as two-input one-output gates like the OR gate. However, this limit applies only to logics that have irreversible logic operation including CMOS circuits and SFQ (Lloyd 2000; Meindl and Davis 2000).

Edward Fredkin showed that if we use reversible computers, the Landauer's theoretical limit can be surpassed, which leads to less energy consumption in the system (Fredkin and Toffoli 1982; Fredkin 1990). Since no data bit is lost during reversible computing, power consumption is smaller than the theoretical Shannon–Neuman–Landauer limit (Phillips et al. 2013; Phillips and Ulidowski 2014; Ulidowski et al. 2014). The Fredkin gate prevents the power consumption resulting from the bit loss by conserving the entropy of the system. The Fredkin gate has three inputs and three outputs and is reversible since the inputs are derived from the outputs. Many different models and physical devices have been investigated for reversible computation (Keyes and Landauer 1970; Likharev 1977; Semenov et al. 2003).

One of the most efficient computing methods is reversible computing, which allows the user to perform the needed calculations without losing any bit. Adiabatic quantum flux parametron (AQFP) logic is a form of a new superconducting logic that performs reversible computing and uses two to three orders of magnitude less power than SFQ circuits.

Figure 8.41 shows the basics of adiabatic switching. In a classical switching event, the bit passes through an energy barrier and the system needs energy to go back to the initial state. In adiabatic switching, the transition does not involve any barrier and therefore the bit is not lost. The operating principle of AQFP is based on the gate invented by Goto et al. (Hosoya et al. 1991).

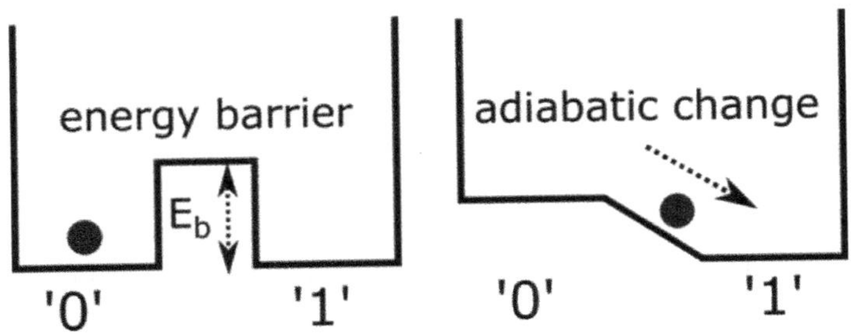

Figure 8.41. *Conventional switching (left) versus adiabatic switching (right) (Naoki Takeuchi et al. 2014). For a color version of this figure, see www.iste.co.uk/cresti/beyondcmos.zip*

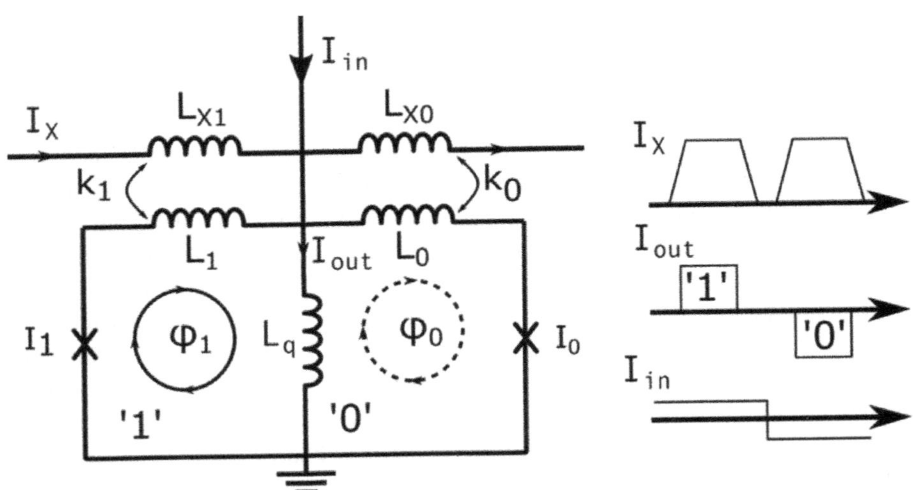

Figure 8.42. *Basic operation of an AQFP gate, designed by Goto et al.*

Figure 8.42 shows the quantum flux parametron basic gate that operates in adiabatic regime. The clock frequency of AQFP cells cannot be higher than ~5 GHz, because of their adiabatic nature. An AQFP cell is composed of two superconducting loops comprising a Josephson junction. As the bias current I_x is coupled to the loops via mutual inductances, a magnetic flux quantum will form in one of the loops depending on the polarization of the I_{in} input current. The position of the flux determines the state of the output current, which can be positive or negative. The system works adiabatically since there is no flux generation as the flux only move from one loop to another upon the change of input polarization. Besides, the bias current is AC and therefore there is no static power consumption.

8.4.3. *Towards superconducting computing*

The need for higher-speed and lower-power-consuming computers leads to searching for alternative logics to the CMOS technology. Recent advances in the field of superconductor logic technology and superconducting very large-scale integration (VLSI) circuit fabrication allow us to design complex SFQ circuits and structures with a higher number of Josephson junctions on one chip (Likharev and Semenov 1991c; Askerzade et al. 2017). This has also led to the development of logics that consume orders of less magnitude power than MOSFETs, while working at relatively higher clock frequencies (Likharev and Semenov 1991; Brock 2001; Holmes et al. 2013).

The power consumption in large-scale computers is a significant problem that limits the computing power and scalability of CMOS circuits. Data centers and large computing facilities get bigger for several reasons including growth in Internet traffic, support centers for cellular and mobile devices, cloud computing and dependency of more applications on precise simulations. Bronk et al. predicted that the energy consumption of the United States data centers will rise from 72 to 176 TWh by 2020 (Bronk et al. 2010). Reducing the energy consumption of computational data centers could save up to $15 billion each year. This estimation only accounts for energy savings and does not include the reduction in size, fewer cooling equipment and environmental benefits.

Supercomputers are one of the high-performance systems that need high power. These systems are in high demand, and their power needs increase with time. Data centers also use a lot of computing energy. DoE and DARPA are working on optimizing and reducing this power[4]. There are more than half a million data centers worldwide, and their estimated power consumption is over 40 GW (Koomey 2008; Van Heddeghem et al. 2014; Shehabi et al. 2016). There is limited information about data centers since they are mostly run by private companies and are not open to public domain.

4. See: https://top500.org.

The energy consumption in logic devices comes from various sources, like Joule heating of bias resistors, energy dissipation in interconnects, charging gate capacitors or simply by switching the circuit and bit loss. The theoretical limit for energy dissipation in logic gates comes from bit loss during gate switching events. The lost bit energy is given by the Shannon–Neuman–Landauer limit (Lloyd 2000; Meindl and Davis 2000; Markov 2014).

Many efforts have been made to develop stable superconducting processors and coprocessors (Dorojevets et al. 2001; Filippov et al. 2012a; Tang et al. 2016a). One of the very first works towards the design of arithmetic logic units (ALU) using SFQ logics was done by Semenov et al. in the 1990s (Bunyk and Semenov 1995). Their purpose was to design a bit-serial architecture with about 5,000 Josephson junctions with up to 20 GHz clock frequency. The next notable attempt at designing an SFQ microprocessor called FLUX (Dorojevets et al. 2001) was done at Stony Brook University in 2001 by Bunyk et al. They attempted to fabricate a 16-bit ALU with registers and instruction set memory for over 30 commands. The chip was designed and had over 90,000 junctions and an area of 10×15 mm^2. Bunyk et al. continued their work on the ALU architecture and designed in 2003 the FLUX-1 microprocessor (Dorojevets and Bunyk 2003). The FLUX-1 chip was designed as an 8-bit microprocessor for 20 GHz clock frequency with a parallel partitioned architecture. This design contains 63,000 Josephson junctions on a 10×10 mm^2 chip. In 2004, Kang et al. from Incheon University in Korea designed an SFQ ALU with half-adders in a serial pipeline architecture (Kim et al. 2005). In 2006, Müller et al. from Stellenbosch University in South Africa designed a superconductor ALU using gates (Gerber et al. 2007). One of the most recent and notable efforts is from Fujimaki et al. at Nagoya University, Japan, who designed in 2015 a 4-bit bit-slice ALU for a 32-bit SFQ microprocessor (Tang et al. 2016a). The fabrication of circuits was based on a 10 kA/cm^2 current density nine-layer biobium-based technological process known as the advanced process (ADP). The size of the chip was 3×1.7 mm^2 with a working clock frequency of 50 GHz. The Fujimaki laboratory group in Nagoya University designed different SFQ-based microprocessors called the CORE-e series. The CORE-e series are general-purpose microprocessors with random access memories and bit-serial architecture (Tanaka et al. 2015). Table 8.1 shows the microprocessors and ALUs designed with SFQ technology.

However, all of these circuits were tested in liquid helium cryostats. This method is not very practical for commercial use and limits the applications to research use for obvious reasons of operational cost and need for replenishment of liquid helium.

Research group	Name/architecture	Year	No. of instructions	Memory (Bit)	Performance	Power (mW)	JJ count	Area (mm^2)	Experimental demo
Stony Brook (Bunyk, Semenov 1995)	-/Serial	1994	12	-	4-bit @ 20 GHz	-	5,000	-	No
Stony Brook (Dorojevets et al. 2001)	FLUX/2-way LIW	2000	30	16 × 32	16-bit @ 20 GHz	18	90,000	10 × 15	No
Stony Brook (Dorojevets, Bunyk 2003b)	FLUX-1/2-way LIW	2002	25	16 × 32	8-bit @ 20 GHz	10	63,000	10 × 10	No
Incheon (Kim et al. 2005)	-/serial pipeline	2003	4	0	4-bit @ 5 GHz	-	931	3 × 1.5	Yes
Nagoya (Nakajima et al. 2004)	CORE 1/ Bit-Sliced	2004	7	-	8-bit @ 15.2 GHz	1.6	5,000	1.8 × 2.8	Partial
Stellenbosch (Gerber et al. 2007)	RISC/Harvard	2006	12	16 × 9	4-bit @ 20 GHz	-	8,000	5 × 5	No
Nagoya (Y. Yamanashi et al. 2007)	CORE 1β/ Bit-Sliced	2007	8	-	8-bit @ 20 GHz	3.4	-	4.7 × 4.6	Yes
Nagoya (Tang et al. 2016a)	CORE e2/ Bit-Sliced	2013	13	256 × 2	4-bit @ 50 GHz	4.63	17,220	2.85 × 2.91	Yes
Nagoya (Tanaka et al. 2015)	CORE e3/ Bit-Sliced	2014	18	-	4-bit @ 50 GHz	1.42	5,184	3.33 × 1.44	Yes
Nagoya (Ando et al. 2015)	CORE e4/ Bit-Sliced	2015	20	256 × 2	4-bit @ 50 GHz	5.6	20,532	2.82 × 3.96	Yes
TOBB ETU (Razmkhah 2018)	Bit-Parallel	2017	7 main + 5	4 × 12	4-bit @ 25 GHz	2.4	9,000	4.5 × 4.6	Partial

Table 8.1. *Superconducting arithmetic logic units (ALUs) and microprocessors with their properties*

8.4.4. *In-memory and quantum neuromorphic computing*

Artificial neural networks (ANN) have been used for various pattern recognition/classification/regression problems since the 1980s. The mathematical modeling of the network and the corresponding training process have been well established and structured through a learning algorithm called backpropagation with gradient descent.

The training of a neural network is an iterative process based on adjusting the weight connections between the fully connected neurons of consecutive layers such that, at each iteration, the cumulative error for all input–output pairs will be reduced. The training process stops when the cumulative error reaches a minimum. However, in most cases, a second independent dataset is tracked for cross-validation (CV) and the training process stops when the cumulative error of the CV data starts increasing. Almost always, the training process stops before the training dataset error is minimized. This is called "early stopping of training" and is widely adapted to avoid overfitting. The whole idea of this early exit from the training process is to maintain the generalization capabilities of the network. Otherwise, the training model will memorize the training data instead of "learning" it and that will result in very poor test or production performance of any new unseen data.

When the training stops, the weights between the neurons are stored in weight matrices. Once the weights are stored, these values will not change unless the network is retrained, either with the same or a different dataset.

Since there are no restrictions on how the weights are chosen, they can be any real number (positive or negative) based on the error optimization results of the learning process. The sole reason that the training process is iterative (through gradient descent) is the lack of a mathematical function that is associated with the underlying input–output association problem, which is almost always highly complex and non-convex (even if there may be a simplified function that may represent the association). Hence, it is not possible to define a corresponding function composed of weight values and take the partial derivatives for finding the global minimum.

So far, almost all of the general computing systems are based on von Neumann architecture and due to von-Neuman bottleneck, development of alternative devices and architectures are inevitable (Backus 1978). Von Neumann and neuromorphic architectures can outperform each other in execution of different genres of tasks.

Neuromorphic computers offer a different way of computation and a robust way of artificial intelligence (AI) emulation than conventional computers. Generally, neuromorphic processors are implemented using the Complementary metal oxide semiconductors (CMOS) with little resemblance to their biological counterparts and the challenge is to exploit a wider range of biological principles at the hardware or the model level.

On the other hand, the SFQ technology (Likharev and Semenov 1991b; Askerzade et al. 2017b) is among the best candidates for neuromorphic processor implementation technologies. Likewise, neuromorphic processor architectures are the most suitable architectures for the SFQ technology due to: a) in-memory computation (no separate memory), b) the spiking nature of neuronal signals that enables efficient computing, like in the SFQ pulsed logic and c) the very low power consumption of SFQ circuits. Nevertheless, reported SFQ-based processors (Dorojevets 2015; Tang et al. 2016b; Kirichenko et al. 2019; Ishida et al. 2020) are so far all based on von Neuman architectures even though SFQ logic is highly analogous to biological neurons.

Josephson junction-based neurons have been described in the literature from the early 1990s. Various research papers showed the implementation of brain cell characteristics, using the SFQ technology (Hidaka and Akers 1991; Mizugaki et al. 1994; Rippert and Lomatch 1997; Feldman 2000; Kondo et al. 2005; Hirose et al. 2006; Onomi et al. 2009; Crotty et al. 2010; Yamanashi et al. 2013; Segall et al. 2017; Cheng et al. 2018; Schneider et al. 2018; Shainline et al. 2018; Tzimpragos et al. 2020). However, all of them have remained at the simulation and/or single neuron level. None of them have proven to be scalable, practically implementable and trainable over the course of three decades. Besides, the SFQ technology is not self-sufficient to implement a fully functional high-performance neuromorphic processor as it still requires bias resources for individual neurons, and training parameters that must be stored during execution.

From this assessment, an SFQ–CMOS hybrid design based on leaky integrate and fire neuron (IFN) model has been demonstrated as an effective and robust way to implement a simplified biological brain cell operation. It has the fundamental features such as trainability with commodity software even after fabrication, compatibility with standard SFQ cell library, high computation power and power efficiency (Bozbey et al. 2018). Another main advantage of the designed neuron is its compatibility with the established SFQ and CMOS foundry services (Nagasawa et al. 2014; Tolpygo et al. 2019). It is experimentally shown that an SFQ-based JJ-Neuron can be implemented, reliably operated and be used to design an artificial neural network with established tools and training methods. For this purpose, a publicly available benchmarking dataset has been chosen and implemented (see Figure 8.43) using the proposed neurons with 100% match with the results of the offline training. A comparison of biological neurons (Sidiropoulou et al. 2006) with the CMOS IFN model representations (Linares-Barranco et al. 1991; Tanaka et al. 2006; Wu et al. 2015; Sourikopoulos et al. 2017), and the superconducting IFN model features shows that an SFQ neuron is six orders of magnitude faster and three orders of magnitude more energy efficient (including the cooling cost): see Figure 8.44. This is enabled by the possibility to implement Josephson junction-based soma, synapse and axon structures, which are inherently based on energy-efficient and ultra-fast spiking topologies. Furthermore, the ability to transmit data to long distances with the order of speed of the light without any power consumption by using passive transmission lines (Razmkhah and Bozbey 2016) removes the limitations on the practical chip dimensions and opens the possibilities of high-speed multi-chip module configurations (Das et al. 2017).

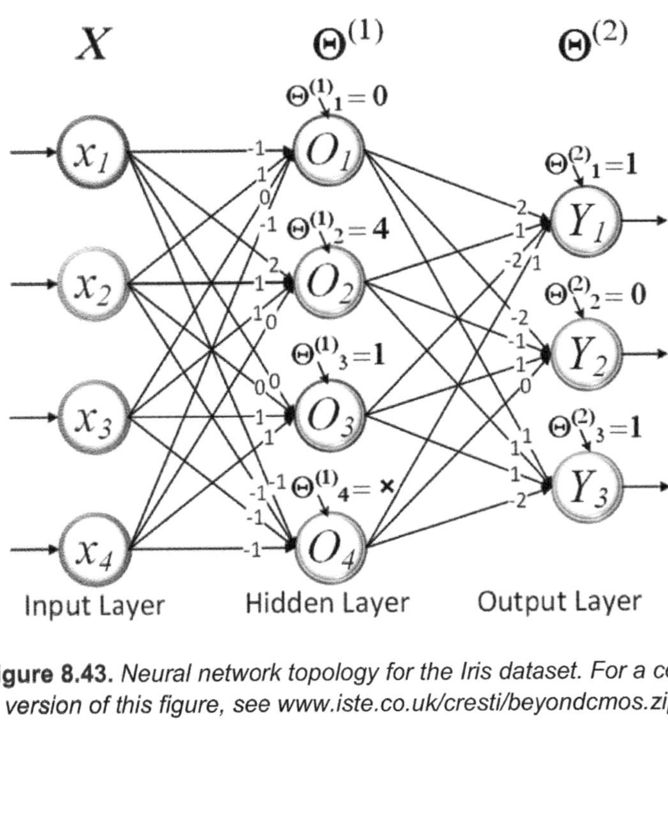

Figure 8.43. *Neural network topology for the Iris dataset. For a color version of this figure, see www.iste.co.uk/cresti/beyondcmos.zip*

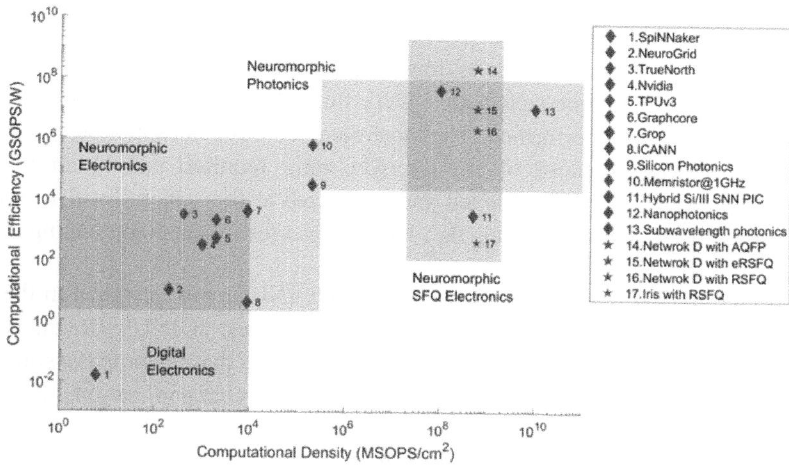

Figure 8.44. *Benchmark of energy and speed of devices for neuromorphic applications. Adapted from Zhu et al. (2020). For the power consumption of SFQ circuits, 400W/W cooling overhead is considered. For a color version of this figure, see www.iste.co.uk/cresti/beyondcmos.zip*

8.4.5. *Computer-aided design (CAD) tools*

As quantum electronics gets faster and more complex, computer-aided design (CAD) tools become a bottleneck for the design of circuits. Advances in fabrication and circuit design need more advanced analysis tools. Two approaches are possible, adapting existing CMOS tools for quantum circuits or developing new customized tools.

Open-source design and simulation programs play a crucial role in this matter since they are used by more users. More users means that limits and shortcomings of tools are found faster, more developers can contribute to advancing tools and the field is approachable for new researchers and enthusiasts.

Some projects have been launched for developing design tools. The most notable one is ColdFlux/SuperTools (Fourie et al. 2019), whose goal is to design new open-source software that encompasses all the aspects of a circuit design from basic physics to more advanced digital optimizations.

There are different levels to design a circuit. Depending on the functionalities that are needed, the user first sets the characteristics of each basic circuit element such as Josephson junctions and inductances based on the fabrication process, then designs a schematic for each sub-circuit and builds the full circuit from them. Digital simulator tools like Jsim, WRSPCICE[5], PSCAN[6] or JoSIM[7] are currently available. The next step is for the designers to draw layouts and extract their parasitic elements to back-annotate them into schematics. Once this is done, a final optimization of the whole circuit must be performed. The most complex circuits cannot be simulated in an analogue way, because of the overwhelming required simulation time and computer power. They are designed and optimized at the system level with digital tools based on customized Verilog or VHDL languages for superconducting circuits.

The JOsephson INterface Utility Software (JOINUS) was designed in the scope of ColdFlux project to perform such design sequences. JOINUS (Razmkhah and Febvre 2020) is a user-friendly, open-source interface that encompasses all of the developed software tools to help the user from the characterization of a simple Josephson junction to the simulation and optimization of a full superconducting chip. The final goal of JOINUS is to automize most of the design process so that even a relatively unexperienced user can design a full chip.

8.5. Superconducting quantum computing

8.5.1. *Epistemological approach*

Superconducting technologies previously introduced are all based on classical methods of computing, although the physics of devices used in processors is purely quantum. However, there is another way to compute that is based on a statistical approach and on the wave-particle duality, this is quantum computing. In this case, the principle of computation differs from the well-known Boolean method and is based on quantum physics. Before going into more details, it is important to remind us of how physical problems have been solved since computers were invented. A good historical example is based on an ancient, but still modern, human tradition related to war and in particular artillery (Kinard 2007). European medieval artillery has indeed greatly contributed, though not deliberately, to the development of physical models to describe the trajectory of all kinds of projectiles over the years. From the beginning, artillery is a means to achieve an objective rather than a subject of physics studies by itself. Progress to better shoot at enemies was rather empirical, based on several parameters like the mass and size of bullets, the presence of wind, the angle of the trebuchet, ballista, catapult or other cannon system. Leonardo da Vinci, Galileo, Newton, Diego Ufano or the Commandant Prosper Charbonnier

5. See: https://www.wrcad.com/wrspice.html.
6. See: https://pscan2sim.org.
7. See: https://joeydelp.github.io/JoSIM.

(Charbonnier 1908) all contributed to modeling physical phenomena associated with artillery problems, which also boosted the development of some fields in mathematics (Bézout 1793). There are two ways to solve a practical problem, either empirically from some observations though there is no rational explanation, or by modeling the observed phenomena with a more general description that applies in particular to the desired case. The first method is very practical and objective-oriented for short-term realizations. It is less general, but does the work under some assumptions. The second method, more "scientific", has led to physics theories, used to describe the world as we understand it today. It is less straightforward for everyone, as more abstraction is needed, but it is more powerful because it is more general. It is based on a description with physical variables associated with an abstraction layer relying on mathematics. In the end, the physical problem is put into equations, the equations are solved and the results are transposed back to the physical world. This is basically what a computer does: takes equations from some models and solves them. Of course, this requires the system under study to be modeled correctly and that the equations can be solved, in a way or another. Historically, modeling phenomena has been a back-and-forth process with theorists certitudes followed by unsuccessful confrontations with observed reality. This made a good experienced gunner better than a poor physicist for shooting enemies at the beginning, as the first predictions of cannon trajectories did not convince the former ones, for good reasons as we now know.

Nevertheless, all these phenomena are based on a macroscopic approach of the world and of the way to model it. When we touch the atomic or sub-atomic scale, things look different. All classical cases only result from averaged properties over time and/or space of observables associated to the quantum world. Classical theory only handles averages and not the inherent phenomena behind them. But sometimes this approach does not work anymore for some macroscopic systems. The best historical example is the diffraction and interference properties of light. It cannot be explained at all with the light considered as a photon particle. On the other hand, the pure wave formalism associated with light does not work either to justify the impact of photons on photographic plates. The wave-particle duality allows approaching such problems and, in particular, the Jaynes–Cummings model has been instrumental to progress on the understanding and resolutions of such problems. Analyzing the situation from a quantum perspective and solving, in practice and most of the time, the Schrödinger equation lead to a real complexity, with overwhelming computing times or with heavy-duty supercomputers as we know them today. If we want to solve time-dependent problems, the complexity is such that solving the system consists of predicting the past. There is a good reason for that: all systems follow in the end the same laws of physics, the system under study and also the computers that study the system. There is no way to anticipate the system evolution at the quantum scale. This is only possible with classical physics, because classical problems themselves imply that macroscopic observables follow some laws, hence postulates, that are themselves already solutions of the quantum evolution of the system in some particular, classical, cases. But when quantum situations are of interest, this trick does not work anymore. Nevertheless, solving the Schrödinger equation can be done in many situations for specific problems, including with classical computers to perform so-called quantum simulations. But there are limits to what can be done in terms of complexity.

Then, how to circumvent this fundamental issue? In fact, it is not circumvented, the experience is done, and once it is done, we know the answer, so there is no need to solve the problem. To say it differently, instead of modeling then calculating where the bullet will fall, we shoot it and then we know where it falls. It looks like this method leads us back to the middle ages or even before this, by counting on the gunner experience and leaving little place for science. Not completely in fact. What can be done is to reproduce the conditions of the artillery theater in a safe place and do the experiments to optimize things in a methodic way, then transpose learned results back to the real theater (this method is a very current method used nowadays as well, for instance, where a plane is placed into a wind tunnel to study its aerodynamics). And this is what pure quantum computing is all about in fact. This is also what is done when we perform any kind of simulation, instead of performing the corresponding experience. We mimic the system to study. We can easily mimic a classical system with simulations with a classical computer, but this is not possible to mimic a quantum system in the general case in a reasonable time unless using another quantum system. This is where the power of quantum computing resides. To understand the fundamentals, it is interesting to see how such perspectives have developed historically.

8.5.1.1. *The nuclear magnetic resonance (NMR) technique*

After quantum theory was established and validated in the early 20th century, one of the first applications of quantum physics is the nuclear magnetic resonance (NMR) technique, widely used today in magnetic resonance imaging (MRI) systems. It is based on the Larmor precession (named after Joseph Larmor) around an external magnetic field of the magnetic moment of isotope nuclei, which possess an intrinsic nuclear magnetic moment and angular momentum, like ^{1}H or ^{13}C. Polarizing the nuclei magnetically and strongly enough leads to their alignment along the external magnetic field vector, like a compass along the Earth's magnetic field. Once the external field is stopped, they return to their thermodynamic equilibrium by relaxing with a dynamic that depends on their own properties (the mass and charge of the precessing system). This is in many ways analogous to the dynamics of a spinning top or gyroscope in mechanics placed in a precessing movement due to an initial mechanical spinning moment (see Figure 8.45). The relaxation dynamics also depends on the chemical environment of the isotope nuclei through a chemical shift of the Larmor frequency due to the local magnetic field of the surrounding environment. The measurement of these shifts allows sensing the environment of magnetic isotopes in MRI systems for medical diagnostic. A single magnetic moment owns an intrinsic relaxation time until it aligns with the final direction without precessing anymore, like a spinning top. The magnetic moment represented in Figure 8.45 is analogous to the axis of the spinning top. However, in the presence of an ensemble of magnetic moments comprising a large quantity of nuclei, each of them follows its own relaxation process due to the local initial conditions and local interactions with the close environment. Measurements are global and give access only to ensemble averages from the whole system, not from each nucleus. Then, the global relaxation process of the system is associated with two relaxation times, known as *T1* and *T2*. The *T1* time results from spin–lattice interactions and corresponds to the longitudinal relaxation time of the total magnetic moment made an ensemble of magnetic moments. This is the time needed to see the total magnetic moment back to its maximum longitudinal magnetization in the case of magnetic field polarization, or away from it to end up with no longitudinal magnetization once the external magnetic field is stopped. It corresponds to the time the spinning top needs to fall down, starting from a position with vertical axis.

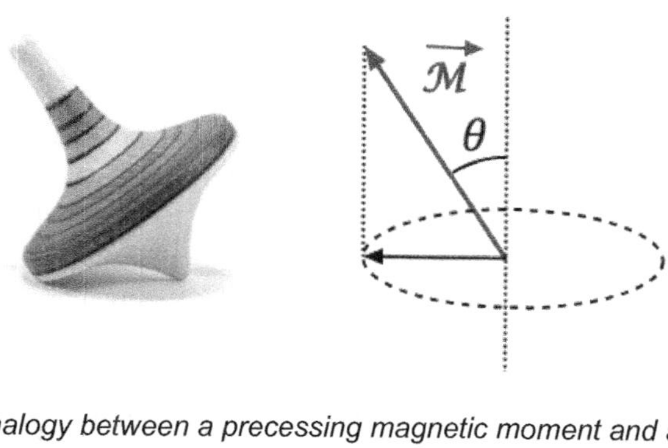

Figure 8.45. *Analogy between a precessing magnetic moment and a spinning top. For a color version of this figure, see www.iste.co.uk/cresti/beyondcmos.zip*

The *T2* time is associated with a transversal relaxation due to spin–spin interactions with other spins of the environment at the origin of magnetic field inhomogeneities affecting differently the precessing frequency of each moment. Let us point out here that we will either use the "magnetic moment" or "spin" expression to designate the same physical feature, since the spin is a mechanical moment directly dependent on the magnetic moment (their ratio is the gyromagnetic factor γ) and of inverse sign, because of the negative charge of the electron.

Starting, for instance, with four moments, all in the same initial position as shown in Figure 8.45, if precession frequencies are different for each of the four moments due to a different magnetic environment, the projections of the moments on the horizontal plane will have different directions after some time, because of their different speeds of rotation, as shown in Figure 8.46. In this case, the average moment on the horizontal plane, obtained by summing the four vectors, will have lower amplitude, reaching zero after *T2* time if we consider a large number of magnetic moments that lost their initial coherence because of individualized interactions.

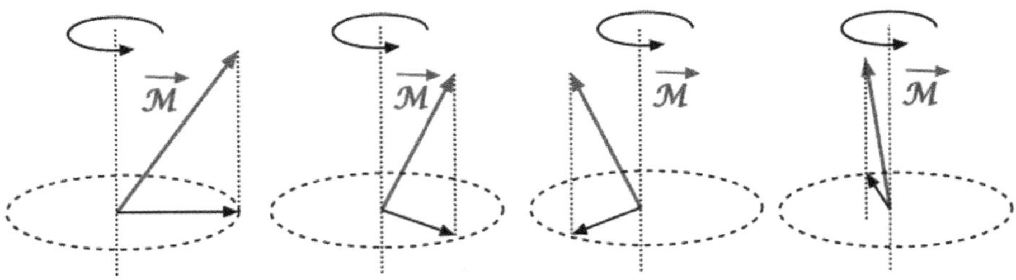

Figure 8.46. *Ensemble of four magnetic moments precessing around the vertical axis after being polarized initially by an external magnetic field. For a color version of this figure, see www.iste.co.uk/cresti/beyondcmos.zip*

T2 is usually shorter than T1. We can indeed note in Figure 8.48 that the average vertical component of the four magnetic vectors did not decrease in the shown configuration. However, if spin–lattice interactions dominate, we can imagine that some of these moments flip downward, leading to reduced ensemble-average longitudinal amplitude. Time *T1* is reached once the longitudinal average amplitude returns to zero, because there are as many downward moments as there are upward moments.

The phenomenological equations that describe the motion of the macroscopic nuclear magnetization, illustrated graphically before, are called the Bloch equations (named after Felix Bloch physicist). Note that Bloch equations are not microscopic equations and they cannot derive the motion of a single magnetic moment since this problem is of purely quantum nature, associated with the state of the atomic nucleus.

8.5.1.2. *Two-state quantum systems*

NMR is a well-mastered technique relying on quantum physics, but measurements deal with a macroscopic ensemble and are of classical nature, along with excitations brought to the system. What makes quantum computing a field of intense research and development today is the fact that technological progress currently allows us to manipulate matter at the atomic scale, for fabrication of quantum objects and also for the manipulation of their states. We can also deal with a single magnetic moment, which is a tremendous achievement since quantum physics has been established one century ago. Magnetic moments are properties of spin-1/2 particles, like electrons, protons, neutrons, neutrinos or quarks. Unlike the spinning top used before to make an analogy with magnetic moments, and unlike what NMR ensemble measurements show, the angle θ of Figure 8.45 cannot take any possible value between $-90°$ and $+90°$ when measurements are performed. This angle is instead quantized and can take only two values for a spin-1/2 particle, as was observed by the Stern and Gerlach experiment with neutral silver atoms (Gerlach and Stern 1922): the spin is up or down. In fact, the rules associated with the quantization of the magnetic moment lead to $\theta = \pm 54{,}7°$ for a spin-1/2 particle with the simple geometrical classical representation of Figure 8.45. A spin-1/2 particle is the simplest two-state, or two-level, quantum system that has been thoroughly studied since it can be solved analytically without any approximation. The two energy levels in the presence of an external magnetic field correspond to two states with spins up or spin down, observed by Stern and Gerlach.

We can make a classical analogy with the right sketch of Figure 8.45 to explain these two states: a negative magnetic moment corresponds to the spinning top turning in the other direction. Though this classical analogy is helpful to understand some of the quantum feature of two-level systems, we must point out here that there are of course limitations to it. In particular, some variables describing the quantum evolution of a spin-1/2 are fundamentally complex values, unlike in the case of the classical case.

It seems that there is a contradiction between the Stern and Gerlach experiment proving the quantization of the spin with two possible values, and the NMR method for which a range of total magnetic moment $\overrightarrow{\mathcal{M}}$ is observed over time (remember *T1* and *T2* times), even if both experiments deal fundamentally with the same quantum objects. This is because the total magnetic moment $\overrightarrow{\mathcal{M}}$ measured with the NMR technique is a statistical measurement of an ensemble of individual, quantized, magnetic moments. As a result, the direction of $\overrightarrow{\mathcal{M}}$ is either in the direction of the magnetic field $\vec{B}$ with a maximum value (the magnetization at saturation) obtained when all single magnetic moments are in the direction of the magnetic field. This corresponds to the stable state of the system (energy $\varepsilon_{mag} = -\overrightarrow{\mathcal{M}} \cdot \vec{B}$ is minimum, the compass gives the North direction of the Earth's field). Or the total magnetic moment direction is in the opposite direction of the magnetic field, with a minimum negative value obtained for the unstable macroscopic classical state (the compass gives the South direction of the Earth's field). But the magnetic moment can take any value between these two limits, depending on the proportion of spins up and spins down. Half-way in between, there is no resulting magnetic moment if the spins are equally distributed between spins up and spins down. The averaging mechanism of the measurement leads to a classical observation and to an apparent absence of quantization. Nevertheless, if we deal with a single spin-1/2 atom, quantization appears and the analogy of Figure 8.45 can be partly used to understand the evolution of a single quantized spin over time.

In the case of a single spin, we should first remark that the quantization appears only in the presence of a magnetic field (Stern and Gerlach experiment). But, in general, a single spin, like any other purely quantum system, follows a deterministic and analogue evolution over time that can be obtained by solving the time-dependent Schrödinger equation, assuming that the spin is isolated from the environment. The extremity of the magnetic moment arrow of Figure 8.45 can take any position over time as is shown in the Bloch sphere of Figure 8.47 where the state $|\psi\rangle$ of the quantum system generalizes the particular classical case of the total magnetic moment. The Bloch sphere is in fact more powerful than that, as we will see when discussing qubit properties. To stay on the spin-1/2 two-level system, which is more intuitive, the fact that only two values of the spin are observed is due to the action of measurement, which makes the spin interact with its environment. Then, depending on where the state $|\psi\rangle$ of the spin represented by two complex numbers is on the Bloch sphere when the measurement is done, we observe one state, the $|0\rangle$ state (magnetic moment up) if the magnetic moment projection on the vertical axis due to the measurement is in the northern hemisphere, or the $|1\rangle$ (magnetic moment down) state if the magnetic moment projection is in the southern hemisphere. This is a pure quantum effect as the spin can be in any direction on the sphere in the absence of measurement. Also the two states $|0\rangle$ and $|1\rangle$ are pure and constitute an orthogonal basis to project any quantum state $|\psi\rangle$. A simple way to imagine the measurement is to apply a vertical magnetic field and to sense the energy of interaction of the spin, then the sign of the spin, through the measurement of a force, which is the gradient of the energy. This is in essence the experiment of Stern & Gerlach, which separated spins in two opposite directions due to opposite forces caused by the interaction of the silver atoms with the externally applied magnetic field.

The representation of this quantum object is not just, for example, a localized electron turning around an axis, but something that can be described by the local electric and magnetic fields of a non-planar wave over a given volume (e.g. the atom's volume), which has an amplitude (proportional to the square root of the density of charge) and a phase (which has no real classical equivalent). A simple way to understand the meaning of the phase is to refer to the interference and diffraction experiments in optics: when local phases caused by different quantum objects cancel each other at a certain location there is no density of energy then no observed object or energy, while when phases add up we observe light intensity. On the Bloch sphere, it corresponds, for instance, to the superposition of two states that are represented by opposite vectors. Though the length of each vector is non-zero, their total vectorial length is zero when their phase difference is 180°.

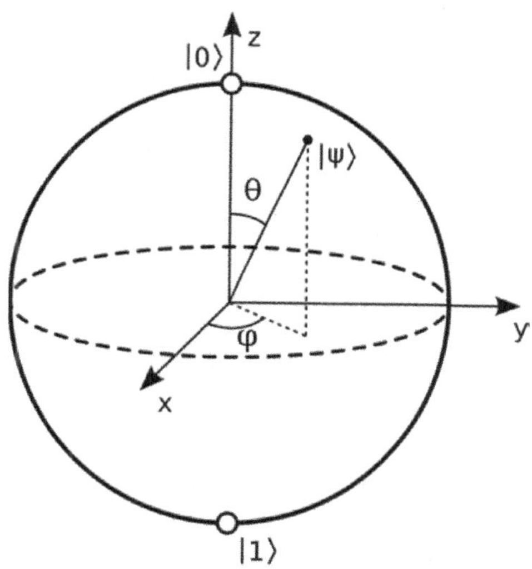

Figure 8.47. *The Block sphere is used to have a graphical representation about the vector states of a qubit which are governed by complex functions, not representable on a 2D graph*

The example based on magnetic moment is helpful to make analogy and obtain sort of an intuition of quantum phenomena. But any two-level system can be represented in the same exact way. The other usual system we are familiar with is the pendulum. We know it has an intrinsic frequency of oscillation and two classical equilibrium states, a stable state (pendulum at lowest altitude) and an instable state (pendulum pointing towards the sky). Of course, it can take any altitude in between during its dynamics. A pendulum is a resonator and also has a simple quantum equivalent with two discrete levels of energies: the low energy level (at lowest altitude) and an excited energy level (at highest altitude), that can be represented on the same Bloch sphere. In fact, there are an infinite number of levels, but we can place ourselves for simplicity in such experimental condition (a mind experiment) so that the other higher levels are not excited. We call this two-level quantum object a qu-bit or qubit. It would be a qu-trit with three levels and so on.

8.5.1.3. *The quantum bit, or qubit*

How can we use the above features to do computing? And in what way can this computing be more powerful?

In quantum computing, the information is carried by qubits (instead of bits), which are continuous complex variables as we just saw, at the difference of usual bits represented only by discrete numbers, usually "0" and "1", corresponding to two identifiable static states (for instance, voltage of 0 volt and voltage of 3 volts, or the lower and upper equilibrium positions of a pendulum). In this regard, normal bit states rely on a physical system to represent binary information but are in no way the states of a real physical system. As mentioned before, they are used to solve equations obtained from a model.

On the contrary, qubits correspond to the states of a real quantum system. For example, they can result from the interaction of a wave (light) with the states of real or artificial atoms (matter). A wave interacting with matter, if properly confined, will allow energy exchanges between atoms (at excited or at fundamental levels) and light (the number of photons of the wave changes) in a resonant way, in the same manner as a pendulum carries energy between a fundamental level (in the form of the kinetic energy of the pendulum at the lowest position) and an excited level (in the form of the potential energy of the pendulum at the highest position of the oscillation). A quantum pendulum can oscillate in all vertical planes of oscillation between the two minimum and maximum positions of the pendulum (angle θ at 0° and 180°), meaning that two variables, the azimuth and the elevation angles, are necessary to describe its position (see Figure 8.47). Since the pendulum length l is fixed, its mass can take any position on the surface of a sphere of radius l. The position of the pendulum's mass on the Bloch sphere is described by the two φ and θ continuous variables. This is an overly simplified mechanical analogy for the qubit but, again, some purely quantum quantities can be parametrized in the same way.

As a result, many more combinations can be realized simultaneously for a given number of N qubits, which can described by $2^N = e^{N \ln 2}$ complex numbers, and evolve up to a point when their state is read and destroyed in the same time, according to quantum physics rules. On the contrary, a system of N classical bits is described by N numbers (the values of each classical bit). For this to happen, the ensemble of qubits must be entangled, meaning that they should interact naturally so that their individual states depend on all the states of the other qubits of the ensemble, and do not depend of external unwanted interferences. If interference happens, which is practically always the case because of external noise sources and non-idealities of the system, there is decoherence, which at some point prevents the system to continue to evolve indefinitely in a purely deterministic way, which has fundamentally the same origin as the measurement: an interaction of the quantum system with the external world. As a result, the qubit system evolution associated with the computing task should last much less than the decoherence time.

The evolution of an ensemble of entangled qubits, which is a global quantum system, can be used for quantum computing. It does not necessarily rely on a modeling task put in equations that can be solved like for classical computing, though this is possible as well. The real power of quantum computing is to use it to solve the evolution of quantum systems, for example, in quantum chemistry to find the quantum properties of new molecules or drugs and to synthesize new ones according to some desired physical or chemical properties.

The way to perform pure quantum computation is to mimic the quantum system to study with an ensemble of artificial atoms that behave the same way, meaning which has the same or similar Hamiltonian as the system to study. Then, this artificial quantum system can be prepared into desired quantum states by experimentally turning the right buttons. Then, the system is freed to evolve naturally, i.e. quantum-mechanically, and the operator can then sense, after certain instants, in a repeatable way, the final states of the system though measurements of certain physical quantities (observables), which give probabilistic occurrences of eigenstates of the artificial quantum system. This sequence should be repeated many times since different results are obtained from the same initial solicitation due to a different environment seen each time by the qubit system. In other words, if the pendulum is launched in a certain way, it will not evolve always the same way, because it interacts with external photons that kick it and change its dynamics. Depending on the strength and dynamics of the kick, the pendulum will stay in the same hemisphere upon readout (same projection on the same static state) or will deviate strongly and make it change hemisphere (readout of the other static state). From the statistical results, we can reconstruct the full quantum dynamics of the system and use it to understand the properties of the real system that was under study.

The power of quantum computing resides in the fact that real experiments are performed when the states of qubits are prepared then the system is set free to evolve quantum-mechanically. Measuring the qubits states is a way to know how the system evolved. This technique is fundamentally different from a Boolean computation, which is a purely artificial way to solve equations by playing with integer numbers, far from the system to study.

The second advantage of quantum computing is that the physical quantum system follows the dynamics of atoms which is intrinsically very fast, much faster than the fastest available measurement techniques. This allows us to follow the evolution of the system with a limit set only by the nature of the quantum system, and not by limits due for example to the clock frequency applied to a circuit.

Nevertheless, a drawback of quantum computing, at the present state of the art, is that the quantities that are measured are analogous, then very sensitive to external noise. This was why digital computing emerged and replaced many analogue systems in the last decades. Consequently, much work is still needed to address this issue.

8.5.1.4. *Towards quantum computers*

Quantum mechanics has the oddity of being simultaneously the most successful and the most mysterious scientific theory. In 2007, the Canadian-based company, D-wave, declared that it was testing a 16-qubit quantum computer called Orion. In 2009, researcher H. Neven announced that he used the superconducting computer he founded at D-wave for pattern recognition. Although this computer is a "quantum annealer", it can be described as the first quantum computer. Later, many organizations such as IBM, Google, Microsoft and Intel announced their quantum computer development projects[8]. Organizations other than Microsoft have focused on qubits made up of superconducting Josephson junctions. Microsoft is working on Majorana-based topological qubits that also use superconductors.

Google used 12 qubits to simulate a simple chemical reaction in a quantum computer (Arute et al. 2020). Using quantum computers, we can study the physics of how electrons interact with each other in different materials in ways not possible or practical with classical computers. In a recent experiment, Google used 16 qubits on a quantum computer to show that it can simulate the dynamics of the Fermi–Hubbard model, which predicts how to form material with less energy loss as electrons move through it (Arute et al. 2020).

8. See: https://www.linkedin.com/pulse/quantum-computers-ibm-google-intel-microsoft-charles-verdon-mba.

As stated in the article "Commercialize quantum technologies in five years" published in Nature News in March 2017, superconducting qubits are the most suitable candidates for quantum computers (Mohseni et al. 2017). The creation and development of the physical foundations of quantum computers is also very important in terms of information transfer and protection. Josephson quantum bits are at the forefront of qubit candidates (Devoret et al. 2004; Wendin and Shumeiko 2007; Wendin 2017; Valiev 2008; Askerzade et al. 2017). The working principle of Josephson qubits relies on quantum effects occurring at the macroscopic scale. For instance, the quantization of magnetic flux in a superconducting ring occurs at the macroscopic level. The two energy levels required for the formation of qubit structures must be separated from other energy levels. This can be done through the use of an anharmonic oscillator thanks to the nonlinear Josephson inductance (Likharev 1986) used in Josephson qubits (Likharev 1986; Wendin and Shumeiko 2007; Wendin 2017). The Calderia–Leggett macroscopic quantum effects (Caldeira and Leggett 1983) for small-size Josephson junctions are observed at low temperatures and have been demonstrated by experimental studies (Valiev 2008).

Scaling quantum computers is crucial. With about 100 physical qubits, it is possible to work on different approaches to create logical qubits. A logical qubit allows us to store quantum data without error. This shows that the technology is ready to be scaled up and commercialized. With 1,000 qubits, it should be possible to store quantum data for about a year. Scaling up requires downsizing many components: including cables, amplifiers, filters and electronics. Recently, Google used 53 qubits to examine how a quantum entanglement wave can propagate ballistically depending on the design of the circuit and can control the propagation rate of entanglement by changing the composition of the circuit (Swingle 2018; Mi et al. 2021).

But a pure quantum computer is not a good solution because it lacks versatility, and a hybrid computer like that in Figure 8.48 is a better solution. Stages at intermediate temperatures of 4 K and 45 K are necessary before the ≈10 mK temperature stage needed for the quantum computer itself. The need for high-speed and low-power computing pushed logic architectures alternative to the complementary metal-oxide-semiconductor (CMOS) technology. Recent advances in fast single-flux quanta (SFQ) technology and superconducting large-scale integration allow us to fabricate complex SFQ circuits and structures with a high number of Josephson junctions on a single chip (Nagasawa et al. 2014; Tolpygo et al. 2015). These advances lead to the development of logics that consume power orders of magnitude less than MOSFETs, while operating at a relatively higher frequency (Likharev and Semenov 1991; Brock 2001; Holmes et al. 2013).

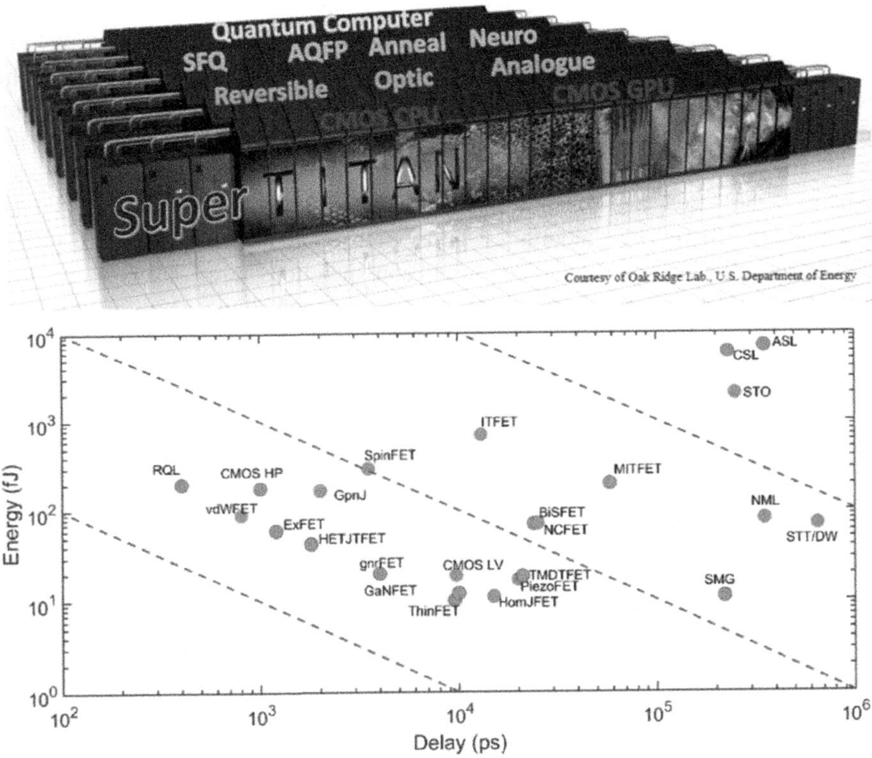

Figure 8.48. *Future of computing and switching power comparison between different technologies. (Holmes and DeBenedictis 2017). For a color version of this figure, see www.iste.co.uk/cresti/beyondcmos.zip*

8.5.2. *Superconductor quantum bits (qubits)*

To calculate the energy spectrum of superconducting qubits, we need to solve the stationary Schrödinger equation of equation [8.11] for the Josephson junction-based quantum oscillator (Nielsen and Chuang 2000; Wendin and Shumeiko 2007; Valiev 2008; Wendin 2017). *K* is the kinetic energy of the qubit, and U(φ) is the potential energy of the Josephson junction in the qubit, determined by the qubit circuit geometry.

$$\hat{H}|\psi\rangle = \mathrm{E}\,|\psi\rangle = \left(\mathrm{K} + \mathrm{U}(\varphi)\right)|\psi\rangle \qquad [8.11]$$

There are two main types of superconductor-based qubits: transmon qubits on the one hand, and flux and fluxonium qubits on the other hand. They are shown in Figure 8.49. In this figure, ϕ is phase difference across the junction, ϕ_e is the offset phase, E_c is the charging energy, E_J is the Josephson energy and E_L is the inductance energy.

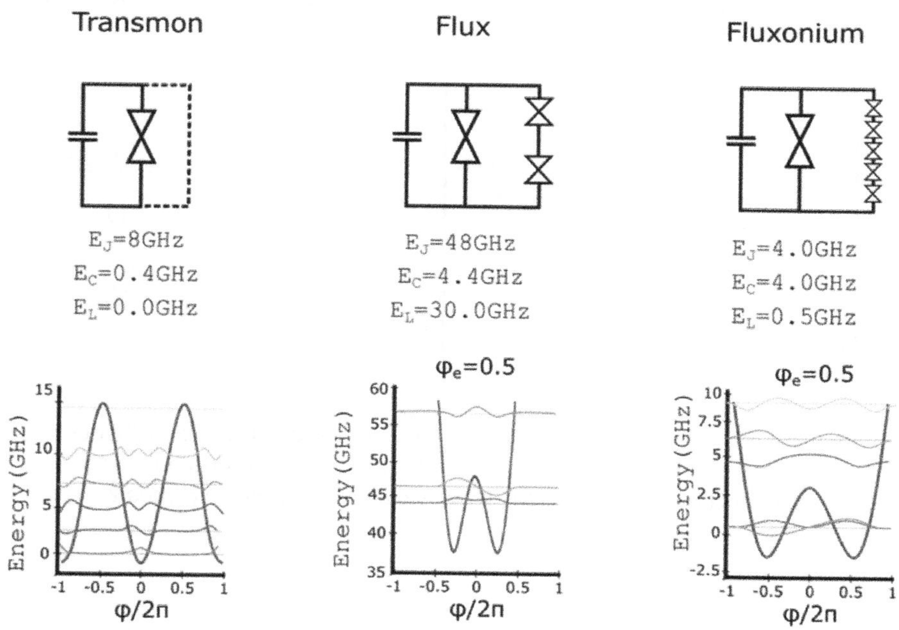

Figure 8.49. *Equivalent circuits and their associated potential energy surfaces (dark blue) with the low-lying wavefunctions (other colors) for a typical transmon, fluxonium and flux qubit. The parameter values used in the calculations are indicated and reflect values found in the literature (Siddiqi 2021). For a color version of this figure, see www.iste.co.uk/cresti/beyondcmos.zip*

8.5.2.1. *The charge qubit, transmon*

The charge qubit, also called transmon, is the simplest superconductor qubit. It is built from a single Josephson junction and is shown in Figure 8.50. The Hamiltonian of the charge qubit can be written as (Wendin and Shumeiko 2007; Wendin 2017):

$$\hat{H} = E_c\left(\hat{n} - n_g\right)^2 - E_J\left\{i_b\phi + \cos\phi\right\} \qquad [8.11]$$

Here, n̂ is the operator of the number of Cooper pairs, n_g is the amount of charges on the outer electrode (gate) and i_b is the bias current normalized to the Josephson junction critical current. Without the Josephson term, there is no forbidden band in the energy spectrum (Figure 8.50-a). Considering the Josephson term (Figure 8.50-b) leads to the shown band structure. The charge qubit was realized for the first time experimentally by Nakamura et al. (1999). The ratio of Josephson energy to Coulomb energy is E_J/E_C ~0.3. The drawback of charge qubits is that their anharmonicity (connected to the next energy level) is small. Transmon qubits have been proposed to obtain qubits with higher coherence times (Koch et al. 2007; Paik et al. 2011).

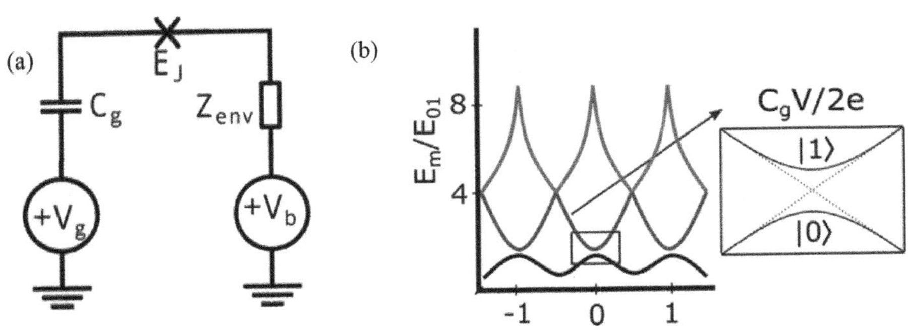

Figure 8.50. *a) Charge qubit built from a single Josephson junction (Cooper pair box). b) Energy levels and quantum states for the transmon. For a color version of this figure, see www.iste.co.uk/cresti/beyondcmos.zip*

8.5.2.2. *The flux qubit*

The schematic of a single Josephson junction flux qubit is shown in Figure 8.51. The flux qubit was first demonstrated in 2000 at Stony Brook University (Askerzade et al. 2017). The adaptation time of the flux qubit is in the orders of ~ns (Devoret et al. 2004; Canturk and Askerzade 2012). The Hamiltonian energy for flux qubit can be expressed as:

$$\widehat{H} = -E_C \frac{\partial^2}{\partial \phi^2} - E_J \left\{ \cos\phi + \frac{(\phi-\phi_e)^2}{2l} \right\} \quad [8.12]$$

where

$$l = \frac{2eLI_c}{\hbar} \text{ and } \phi_e = 2\pi \frac{\Phi_e}{\Phi_0}.$$

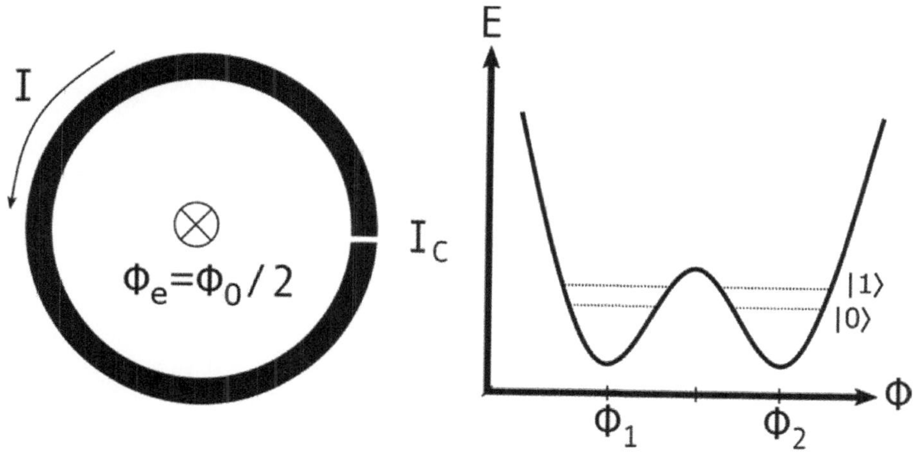

Figure 8.51. *Single-junction flux qubit (RF SQUID)*

One of the drawback properties of a single-junction flux qubit (RF SQUID) is related to the large inductance of the superconducting loop. To obtain the potential dromedary energy profile shown in Figure 8.51-b, the inductance must be large, making the system susceptible to fluctuations of the external magnetic field, which results in a smaller coherence time. The relaxation and coherence time of the RF SQUID flux qubit reaches 20 ns (Bennett et al. 2009). In order to solve this drawback, Mooij et al. (1999) used two Josephson junctions with a smaller additional critical current instead of the inductance of the loop, as shown in Figure 8.52-a.

The Hamiltonian energy for the three junction flux qubit can be written as (Mooij et al. 1999):

$$H = E_C(n_1^2 + n_2^2 + \frac{n_3^2}{\alpha + 0.5} - E_J\{2 + \alpha - \cos\phi_1 - \cos\phi_2 - \alpha\cos\phi_3\} \quad [8.13]$$

Detailed analysis shows that an external magnetic flux $\phi_e = \pi$ corresponding to one half of the flux quantum must be applied to the loop for the formation of the Bactrian potential (Figure 8.52-c). The three-junction flux qubit was experimentally demonstrated for the first time in 1999 (Orlando et al. 1999). The adaptation time of such qubits can be further improved by using a 3D QED (quantum electrodynamic) cavity (You et al. 2007; Steffen et al. 2010). Experimentally it was shown that the effect of load fluctuations decreased and the adaptation time reached ~40 ms when capacitively shunting the small Josephson junction of the three-Josephson qubit (Yan et al. 2016) (Figure 8.52-a). In this study, a half-wavelength quantum electrodynamic circuit (QED) generator circuit was used to read the qubit state.

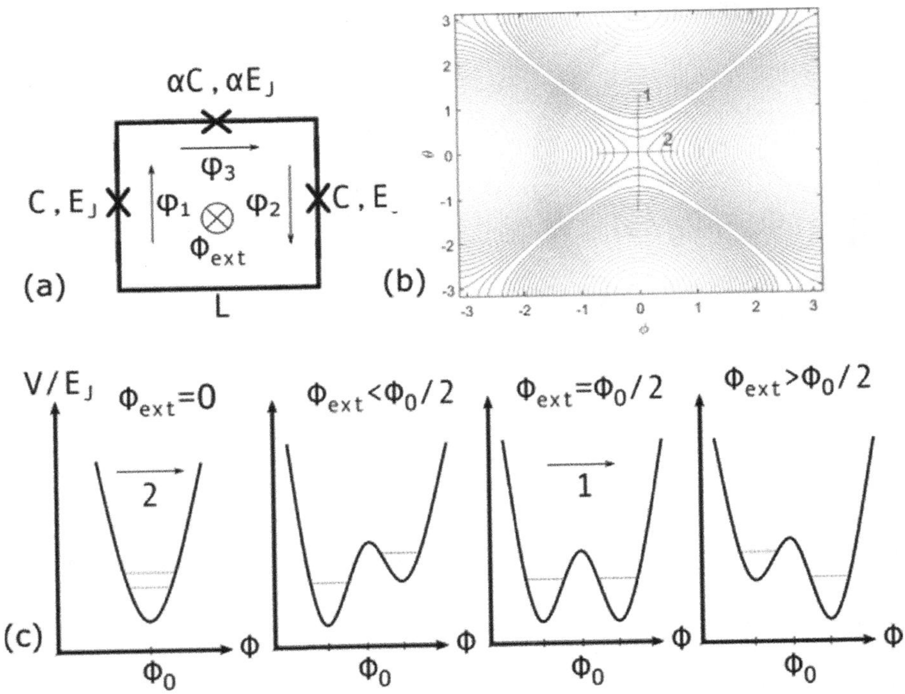

Figure 8.52. *Three-junction flux qubit*

8.5.2.3. *The fluxonium (silent flux qubit)*

Silent flux qubits are qubits in which the second harmonic takes place in the current–phase relationship. Such junctions occur in structures based on high-temperature superconductors (Askerzade 2012; Askerzade et al. 2017). Dynamics of such Josephson junctions have been shown (Canturk and Askerzade 2012; Askerzade

2015). The most important feature of silent qubits is that they do not need an external magnetic field, unlike three-junction structures. The potential energy of such a qubit (Figure 8.53) is given by equation [8.12]. The silent qubit was studied experimentally by Amin et al. (2005).

$$U(\varphi, \varphi_+) = -\frac{\Phi_0 I_{c1}}{2\pi}\left\{cos(\frac{\varphi}{2} + \varphi_+) - \frac{\alpha_1}{2}cos(\varphi + 2\varphi_+)\right\}$$

$$-\frac{\Phi_0 I_{c2}}{2\pi}\left\{cos(\frac{\varphi}{2} - \varphi_+) - \frac{\alpha_2}{2}cos(\varphi - 2\varphi_+)\right\} \quad [8.14]$$

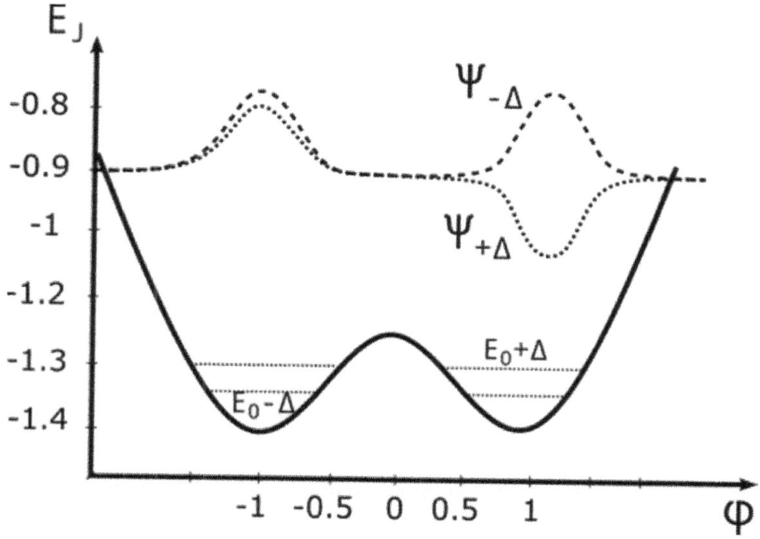

Figure 8.53. *Fluxonium potential energy profile and equation*

8.5.3. *Source of decoherence in qubits*

There are various sources of decoherence and hence error in a qubit. Figure 8.54 demonstrates some of these reasons (Siddiqi 2021). In this figure, part (a) shows the image of group of eight superconductor qubits based on Al/AlOx/Al Josephson junctions (JJs) and a Niobium (Nb) microwave embedding circuit. Every qubit is coupled to its qubit neighbor via a resonator. In part (b), surface and interfacial defect layers due to fabrication are shown. These layers could cause decoherence in qubits, and they can also be modeled. Part (c) shows non-equilibrium excitations of the superconducting films and substrate can arise from ionizing radiation and remnant magnetic fields. These noise sources generate quasiparticles, hot phonons and vortices. Part (d) demonstrates the microscopic charge and spin defects, which are believed to give rise to *1/f* fluctuations that bias the qubit circuit with a varying electric or magnetic field, or cause a jitter in the JJ inductance if defects are present in the tunnel barrier. Fluctuations in the environment can drive qubit transitions if the noise is orthogonal to the qubit axis (longitudinal relaxation) or pure dephasing if it is aligned with the qubit axis (Siddiqi 2021).

These defects and noise sources can all be modeled with a certain precision, and the effects can be seen in the experimental results (Kjaergaard et al. 2020). However, there are no tools available yet that can put all of them together and calculate the coherence times of a qubit (*T1* and *T2*) based on all the known sources.

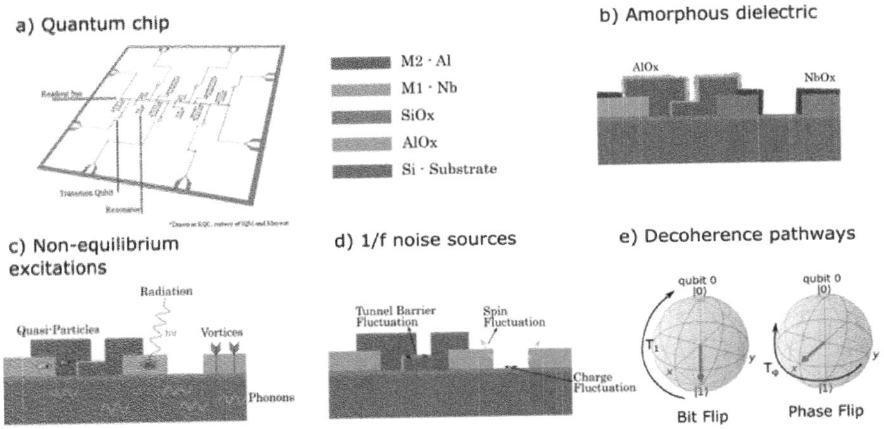

Figure 8.54. *a) A sampe qubit chip, b) surface and defect issues, c) quasi-particle excitation, d) microscopic charge defects, e) fluctuation in environment (Siddiqi 2021). For a color version of this figure, see www.iste.co.uk/cresti/beyondcmos.zip*

8.5.4. *Interface system for Josephson junction qubits*

In general, Josephson qubits can be considered as two-level artificial atoms (Nielsen and Chuang 2000; Askerzade et al. 2017). According to quantum mechanics, there can be transitions between the energy levels of a system placed in the presence of an external excitation. When investigating the spectroscopy of some Josephson qubits, a weak external source is in question (Valiev 2008; Shevchenko et al. 2012). When the source power affecting the qubit is increased, multi-photon interaction occurs (Zubayri and Sculli 2003). Until recently, dispersive QED (quantum electrodynamics) methods were used to test the state of qubits. In such a system, the qubit–resonator interaction is given by the Jaynes-Cumming Hamiltonian (Wallraff et al. 2004; Wendin 2017); see Figure 8.55.

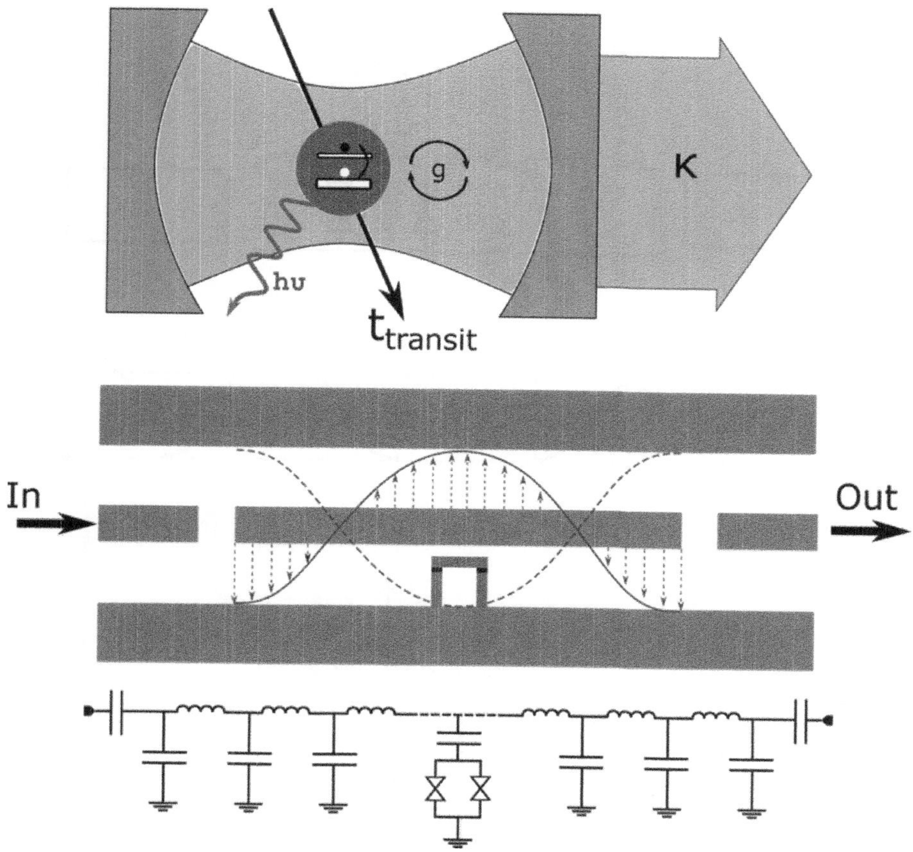

Figure 8.55. *Dispersive QED qubit control system. For a color version of this figure, see www.iste.co.uk/cresti/beyondcmos.zip*

In this case, the theory (Leek et al. 2009; Shevchenko et al. 2012) for single- and multi-photon transitions in the qubit-resonator system has been made. The most recent development in the field of control of JJ qubits consists of using the SFQ (Single-Flux Quantum) technique. This approach was proposed for the first time by Averin et al. (2006). The basic idea is based on the result of the interaction of flux quantum (fluxon) and qubit moving in the Josephson transmission line (JTL) (Figure 8.56-a,-b)

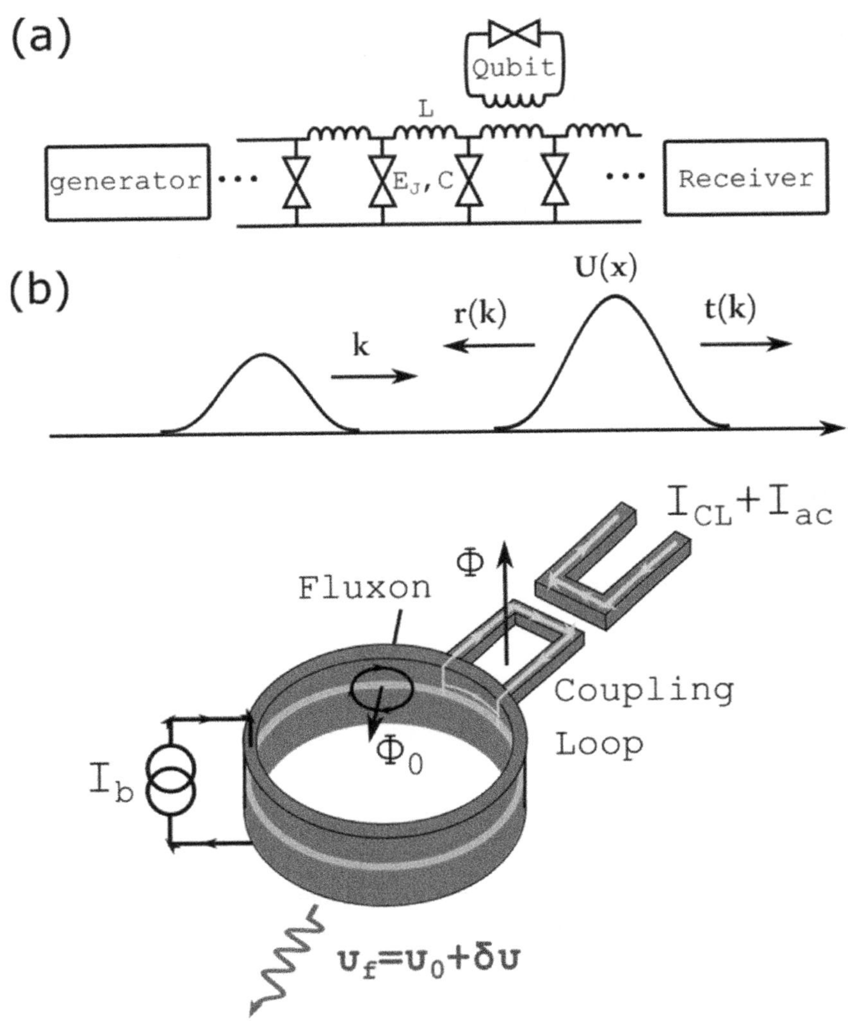

Figure 8.56. *Interaction of a flux quanta traveling on a Josephson transmission line with a qubit (Averin et al. 2006). For a color version of this figure, see www.iste.co.uk/cresti/beyondcmos.zip*

Detailed analysis of the qubit readout system was performed using the sine-Gordon equation (Soloviev et al. 2015) and also experimentally tested (Fedorov et al. 2014) (Figure 8.56-c). In such control systems, unlike in QED dispersive structures, an external microwave source is not required and a DC current source is used (Fedorov et al. 2014).

The use of flux quanta pulse technology in qubit control systems has been widely discussed (Averin et al. 2006). One of the latest developments in this field has been done in the study (Walls et al. 2007). In this approach, SFQ-based comparison circuits are proposed for qubit control. However, SFQ-based qubit readout circuits could not be implemented at that time, because the SFQ design and production technologies were not as advanced as today.

SFQ-based comparator circuits are widely used in digital and analogue integrated circuits by many groups (Askerzade and Kornev 1994a; Askerzade 2000; Brock et al. 2005; Ohki et al. 2007; Walls et al. 2007; Bozbey et al. 2009; Gordeeva and Pankratov 2009; Nakamura et al. 2009; Ebert et al. 2010; Bozbey et al. 2011; Ortlepp et al. 2012). The most important feature of SFQ-based comparator circuits is that they act as a high-frequency detector at the quantum limit (Walls et al. 2007). The dynamic properties and characterizations of the JJ balanced comparator (BC) have been extensively studied (Caldeira and Leggett 1983; Askerzade and Kornev 1994b; Bozbey et al. 2009, 2011; Nakamura et al. 2009; Canturk and Askerzade 2012). The equivalent schematic of the SFQ-based BC-qubit system is shown in Figure 8.57. Walls et al. (2007) used the Langevin–Heisenberg–Lax equations (Weiss 2012) for the RF SQUID flux qubit (Figure 8.57) to examine the resolution of the BC detector. Detailed research has highlighted that no study has ever been done with the dynamics of different flux qubits with an SFQ-based control system. In Figure 8.57, the Hamiltonian for the RF SQUID flux qubit is as follows:

$$H = \varepsilon\sigma_z + \Delta\sigma_x - \kappa\Phi_x\phi\sigma_z + H_C \qquad [8.15]$$

where σ_x and σ_z are Pauli matrices and H_C is the Hamiltonian of the comparator.

Quantum computers ideally consist of n compatible qubits, and the functions to be performed are performed as unitary transformations. However, quantum coherence is disrupted by the influence of the environment in which the qubits are placed. The environment in which the qubits are located consists of different dissipative components. These elements produce significant noise in the system. This results in incompatibility effects and a consequent loss of energy. Another factor causing incompatibility is electromagnetic radiations generated during qubit operations. To investigate the different mismatch effects, the environment of the qubit system is modeled assuming it consists of an infinite set of linear oscillators. As a result, discordance effects appear as relaxation and dephasing. The relaxation time is determined by the spectral density of the frequency fluctuations of the qubit frequency of the external medium. This spectral density is related to the properties of the medium and can be expressed by the impedance of the medium. The variation of the mismatch with time is characterized by the phase decay time. There are different approaches in the literature on minimizing incompatibility and ambient noise (Wallraff et al. 2004; Weiss 2012; Yan et al. 2016). In this project, it is foreseen that the control of the effects of incompatibility will be examined within the scope of the research. Thus, the frequency fluctuations of the relaxation time for different environments will be determined.

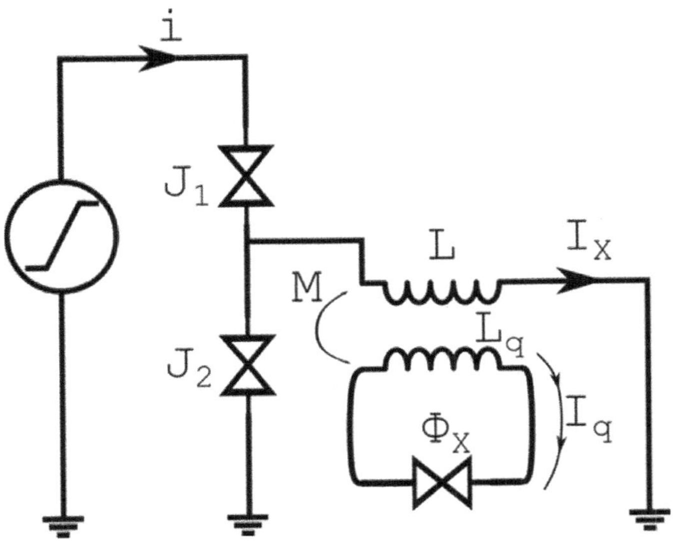

Figure 8.57. *The SFQ pulse is shown as a 2π change in phase of the generator on the left. (J1,J2) is the balanced comparator (Walls et al. 2007)*

8.5.5. *The qubit cavity*

There are many studies in the literature that make connections with optical (solid-state) qubits, as entangled real photons are easier to transport over long distances. However, optical quantum computers can become complex and large very quickly and it is not possible to integrate large numbers of qubits, so that the focus is still on superconducting qubits even if the low temperature of operation makes things more complicated. A similar method in superconductor-based qubits was first tried in 2007 (Majer et al. 2007). In this work, the qubits are connected to a coplanar waveguide (CPW), and the data are transmitted between two qubits on the same chip. Figure 8.58 shows the method used in this paper. It uses a simple CPW to combine two qubits. The problem is that the cavity is fixed, so any fabrication variation of parameters will cause errors and the entanglement has low range and fidelity.

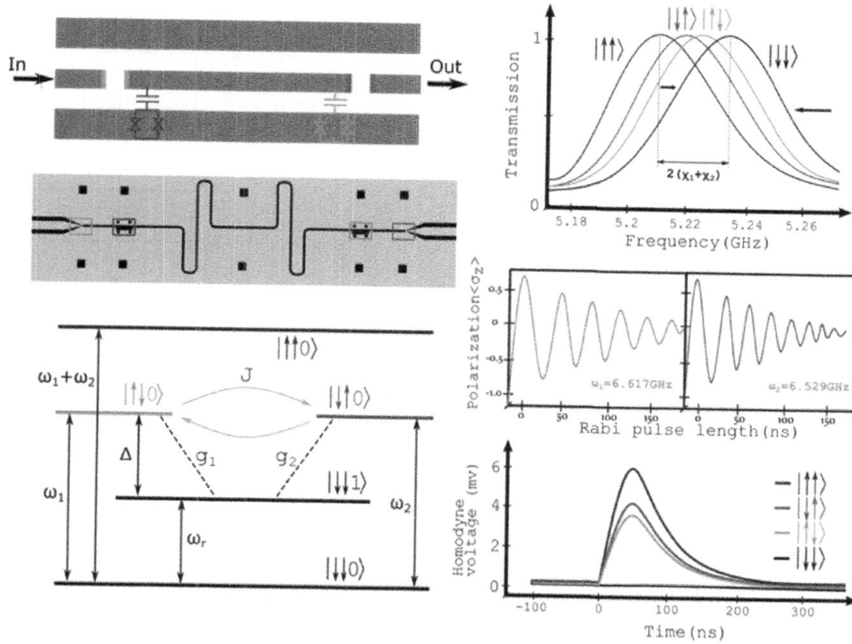

Figure 8.58. *Example and schematic for entangling two qubits on a chip (Majer et al. 2007). For a color version of this figure, see www.iste.co.uk/cresti/beyondcmos.zip*

In some recent studies, some researchers used the stability of the real photon to transmit the quantum state through the photon (Guimond et al. 2020; Mirhosseini et al. 2020). Figure 8.59 shows a work done in 2020 that converts the superconducting quantum state to a photonic quantum state and transmits the information to user via an optical channel. This method uses the stable nature of the real photon; however, when transferring data, there are problems caused by the conversion between microwave and optical photon. Part of the energy is lost in the conversion, so the system is prone to error and there is also a loss of speed in the conversion.

The 2018 study by DiCarlo et al. uses superconducting CPW to enable transmission of the quantum state (Dickel et al. 2018). They show 49% coherence and 73% Bell-state fidelity between transmon qubits on different cQED chips. However, this fidelity is not ideal at all and cannot be used to increase the number of qubits. The reason for the mentioned low fidelity is fabrication differences between chips and losses in CPWs (Dickel et al. 2018). In a 2021 study, Schoelkopf et al. proposed to reduce this problem with error correction algorithms (Burkhart et al. 2021). While this can be somewhat effective, data lost due to quantum structures is ultimately unrecoverable.

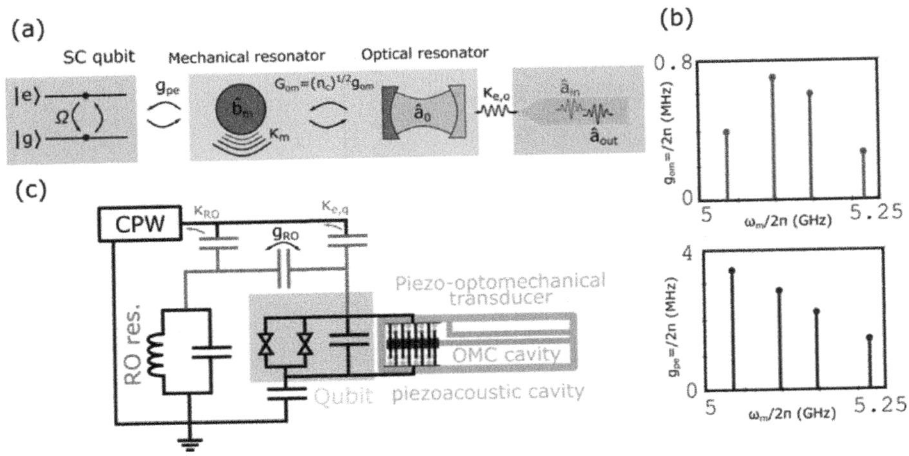

Figure 8.59. *Quantum converter setup. For a color version of this figure, see www.iste.co.uk/cresti/beyondcmos.zip*

NOTES ON FIGURE 8.59.– *a) Schematic of the microwave-optical transduction process. b) Simulated vacuum optomechanical (top) and piezoelectric (bottom) coupling rates to the hybridized mechanical modes. c) Electrical circuit representation of the integrated qubit and transducer device (Mirhosseini et al. 2020).*

A new method has used superconductor transmission lines to connect multi-qubit chips together and demonstrated a significant increase in coherence of the qubits. Gold and Rigetti et al. (Gold et al. 2021) have used superconductor transmission lines with a flip-chip method to achieve better coherence between qubits in different chips. As demonstrated in Figure 8.60, the chips are flip-chip bonded onto the carrier chip using indium bump bonds (yellow) and the qubit couplers are aligned above couplers on the carrier chip (teal) as shown in the inset as well as the cross-sectional view. b) False-colored schematic of a single QuIC including readout resonators and readout lines (magenta and green), indium bumps (yellow), flux bias lines (orange) and the qubits and paddles of the inter-chip couplers (blue). The physical qubits are labeled 0–7, while Roman numerals correspond to the design specification for the qubit; see the methodology of Gold et al. (2021) for more details. Note that component geometries are not drawn exactly as fabricated: the schematic is intended primarily to depict the circuit layout as opposed to details into individual feature geometries (Gold et al. 2021).

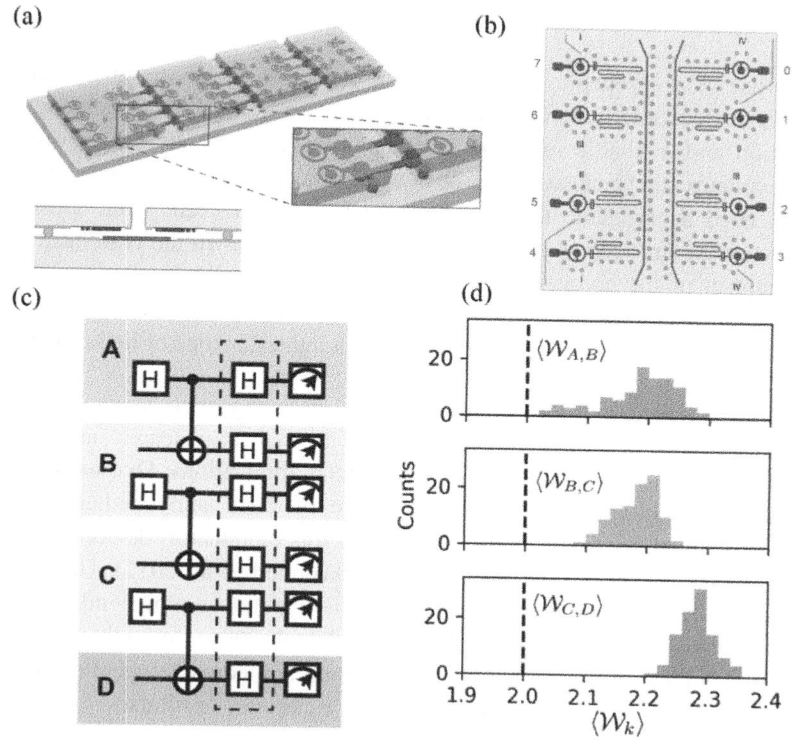

Figure 8.60. *Isometric view of the device assembly. For a color version of this figure, see www.iste.co.uk/cresti/beyondcmos.zip*

NOTES ON FIGURE 8.60.– *(a) The qubits (blue circular structures) are fabricated on the QuIC die and have one arm with a paddle-shaped coupler extending to the edge of the chip. The chips are flip-chip bonded onto the carrier chip using indium bump bonds (yellow) and the qubit couplers are aligned above couplers on the carrier chip. (b) False-colored schematic of a single QuIC including readout resonators and readout lines (magenta and green), indium bumps (yellow), flux bias lines (orange) and the qubits and paddles of the inter-chip couplers (blue). (c) Disjoint, pair-wise entanglement is generated across chip boundaries via CNOT operations on inter-chip couplers compiled to CZ gates. (d) Histograms of the average values* $\langle W_K \rangle$ *across 100 individual experimental runs using* 10_4 *shots in each measurement basis. (Courtesy of Gold et al. 2021, under HYPERLINK http://creativecommons.org/licenses/by/4.0 Creative Commons Attribution 4.0 International License)*

8.6. Cryogenic cooling

The use of a cryogenic environment for reduction of environmental noise for quantum devices and superconductor electronics is inevitable. Depending on the application and available resources, different cooling methods can be used. The cooling systems can be closed cycle, open cycle or a combination of both. Open-cycle systems need a cryogenic fluid to cool down the cold head, while closed-cycle systems use the cryogen as a compressor gas and cool down the cold head without direct contact of the cryogen. The cryogen fluid can be liquid nitrogen for 77 K, liquid helium for 4.2 K or condensed helium gas for below 10 K operations. For lower temperatures, different pressures of He^3 (a lighter isotope of helium) are used.

Due to the limited resources for helium and its rising price, as well as practical reasons, the use of the liquid Helium is favored at research and scientific applications, but it is not feasible for commercial applications. Due to advances in cryogenics technology, coolers become more efficient and smaller. Especially one-stage coolers used for high-temperature superconductor devices have been improved over years (Kusaka et al. 2013; Kirkconnell et al. 2017). Unfortunately, one-stage coolers can only reach ~35 K and low-temperature superconductor circuits need at least two stages to reach 4.2 K (Radenbaugh 2004). In addition, the use of closed-cycle cryocoolers enables the user to use intermediate temperature levels for mounting auxiliary components such as interface circuitry or filters.

There are different classes for commercial multi-stage coolers, based on Gifford-McMahon (GM) and Stirling-type pulse tube. There are many disadvantages to the moving parts inside the GM and Stirling cryocoolers. The displacer could cause vibration, reducing lifetime of cooler and may also cause heat conduction in its moving axis and heat loss due to friction. This part is removed in pulse-tube-type cryocoolers (Hashimoto et al. 2008).

Successful operation of SFQ circuits in closed-cycle refrigeration systems have already been demonstrated by groups from the USA, Japan and Europe. For instance, Hashimoto et al. used a 4 K two-stage 1W Gifford-McMahon cryocooler in order to demonstrate the function of SFQ 2×2 switch chip at 40 GHz (Webber et al. 2007). R.J. Webber et al. used a Gifford-McMahon cryocooler for a DC voltage reference from Josephson junctions at the frequency of 20 GHz for a fully automatic voltage standard system (Webber et al. 2008); they also demonstrated the function of SFQ receiver with about 11,000 junctions for 7.7 GHz satellite receiver (Hashimoto et al. 2007; Webber et al. 2009). There is also a review on four-stage 4 K Stirling-type pulse-tube coolers by Lockheed-Martin from this group at HYPRES that demonstrated the stability of cryocoolers for SFQ circuits and the advantage of multistage systems for mounting auxiliary components and HTS filters at different temperatures (Webber et al. 2008; Nakamura et al. 2009). Yoshikawa et al. also showed the functionality of GM-type cryocooler with high-throughput and high-bandwidth SFQ circuits. They demonstrated a flip-flop circuit with very good BER at 10 GHz frequency (Ortlepp and Fourie 2008). Ortlepp et al. showed the cryocooler capabilities for precise measurements (Dotsenko et al. 2009). For more details, see references Hashimoto et al. (2007), Webber et al. (2007, 2008, 2009), Dotsenko et al. (2009) and Nakamura et al. (2009).

8.7. References

Amin, M.H.S., Smirnov, A.Y., Zagoskin, A.M., Lindström, T., Charlebois, S.A., Claeson, T., Tzalenchuk, A.Y. (2005). Silent phase qubit based on d-wave Josephson junctions. *Physical Review B*, 71(6), 064516.

Anders, S., Blamire, M.G., Buchholz, F.I., Crété, D.G., Cristiano, R., Febvre, P., Fritzsch, L., Herr, A., Ili'chev, E., J. Kohlmann, J., et al. (2010). European roadmap on superconductive electronics–status and perspectives. *Physica C: Superconductivity*, 470(23–24), 2079–2126.

Ando, Y., Sato, R., Tanaka, M., Takagi, K., Takagi, N. (2015). 80-GHz operation of an 8-bit RSFQ arithmetic logic unit. In *2015 15th International Superconductive Electronics Conference (ISEC)*, 1–3.

Ansari, S., Nazifi, R., Arzefouni, M.Y., Mohajeri, R., Mirzaei, S.I., Fardmanesh, M. (2021). 3D modeling of a superconducting transition edge detector. In *2021 29th Iranian Conference on Electrical Engineering (ICEE)*, 56–60.

Arute, F., Arya, K., Babbush, R., Bacon, D., Bardin, J.C., Barends, R., Boixo, S., Broughton, M., Buckley, B.B., Buell, D.A., et al. (2020a). Hartree-fock on a superconducting qubit quantum computer. *Science*, 369(6507), 1084–1089.

Arute, F., Arya, K., Babbush, R., Bacon, D., Bardin, J.C., Barends, R., Bengtsson, A., Boixo, S., Broughton, M., Buckley, B.B. et al. (2020b). Observation of separated dynamics of charge and spin in the Fermi-Hubbard model. *arXiv:2010.07965 [quant-ph] 15 October 2020* [Online]. Available at: http://arxiv.org/abs/2010.07965 [Accessed 7 September 2021].

Askerzade, I.N. (2000). Tunnel Josephson junctions as highly sensitive comparators in stroboscopic converters. *Technical Physics*, 45(1), 66–69.

Askerzade, I.N. (2012). *Unconventional Superconductors: Anisotropy and Multiband Effects*. Springer, Berlin.

Askerzade, I.N. (2015). Effects of anharmonicity of current-phase relation in Josephson junctions (Review Article). *Low Temperature Physics*, 41(4), 241–259.

Askerzade, I.N. and Kornev, V.K. (1994a). Time resolution of balanced comparator with overdamped Josephson functions. *Radiotekhnika i Elektronika*, 39, 510–513.

Askerzade, I.N. and Kornev, V.K. (1994b). Time resolution of balanced comparator with overdamped Josephson functions. *Radiotekhnika i Elektronika*, 39, 510–513.

Askerzade, I.N., Bozbey, A., Cantürk, M. (2017a). Digital superconductivity electronics. In *Modern Aspects of Josephson Dynamics and Superconductivity Electronics*. Springer, Cham [Online]. Available at: https://link.springer.com/chapter/10.1007/978-3-319-48433-4_3 [Accessed 3 January 2018].

Askerzade, I.N., Bozbey, A., Cantürk, M. (2017b). *Modern Aspects of Josephson Dynamics and Superconductivity Electronics*, 1st edition. Springer, New York.

Askerzade, I.N., Bozbey, A., Canturk, M. (2017c). *Modern Aspects of Josephson Dynamics and Superconductivity*. Springer, Cham [Online]. Available at: http://www.springer.com/la/book/9783319484327 [Accessed 20 May 2017].

Attar, S.S., Setoodeh, S., Mansour, R.R., Gupta, D. (2014). Low-temperature superconducting DC-contact RF MEMS switch for cryogenic reconfigurable RF front-ends. *IEEE Transactions on Microwave Theory and Techniques*, 62(7), 1437–1447.

Averin, D.V., Rabenstein, K., Semenov, V.K. (2006). Rapid ballistic readout for flux qubits. *Physical Review B*, 73(9), 094504.

Awschalom, D.D., Hanson, R., Wrachtrup, J., Zhou, B.B. (2018). Quantum technologies with optically interfaced solid-state spins. *Nature Photonics*, 12(9), 516–527.

Backus, J. (1978). Can programming be liberated from the von Neumann style? A functional style and its algebra of programs. *Communications of the ACM*, 21(8), 613–641.

Baek, B., Rippard, W.H., Pufall, M.R., Benz, S.P., Russek, S.E., Rogalla, H., Dresselhaus, P.D. (2015). Spin-transfer torque switching in nanopillar superconducting-magnetic hybrid Josephson junctions. *Physical Review Applied*, 3(1), 011001.

Baghdadi, R., Allmaras, J.P., Butters, B.A., Dane, A.E., Iqbal, S., McCaughan, A.N., Toomey, E.A., Zhao, Q.-Y., Kozorezov, A.G., Berggren, K.K. (2020). Multilayered heater nanocryotron: A superconducting-nanowire-based thermal switch. *Physical Review Applied*, 14(5), 054011.

Ball, P. (2012). Computer engineering: Feeling the heat. *Nature News*, 492(7428), 174.

Barone, A. and Paternò, G. (1982). *Physics and Applications of the Josephson Effect*. Wiley, New York.

Baselmans, J. (2012). Kinetic inductance detectors. *Journal of Low Temperature Physics*, 167(3), 292–304.

Bennett, D.A., Longobardi, L., Patel, V., Chen, W., Averin, D.V., Lukens, J.E. (2009). Decoherence in rf SQUID qubits. *Quantum Information Processing*, 8(2), 217–243.

Benz, S.P. and Hamilton, C.A. (1996). A pulse-driven programmable Josephson voltage standard. *Applied Physics Letters*, 68(22), 3171–3173.

Bézout, É. (1793). *Cours de mathématiques à l'usage du corps de l'artillerie*. Pierres, Paris. DOI: 10.3931/e-rara-3687.

Birx, D. and Scalapino, D. (1979). A cryogenic microwave switch. *IEEE Transactions on Magnetics*, 15(1), 33–35.

Bogozi, A., Datye, A.V., Brzhezinskaya, M., Hijazi, Y.S., Martinez, J., Noel, J., Wu, K.H., Vlasov, Y.A., Larkins, G.L. (2005). Elastic modulus study of gold thin film for use as an actuated membrane in a Superconducting RF MEM switch. *IEEE Transactions on Applied Superconductivity*, 15(2), 980–983.

Bozbey, A., Miyajima, S., Akaike, H., Fujimaki, A. (2009). Single-flux-quantum circuit based readout system for detector arrays by using time to digital conversion. *IEEE Transactions on Applied Superconductivity*, 19(3), 509–513.

Bozbey, A., Miyajima, S., Ortlepp, T., Fujimaki, A. (2011). Design and circuit analysis of quasi-one junction SQUID comparators for low temperature detector array read-out. *Journal of Superconductivity and Novel Magnetism*, 24, 1065–1069.

Bozbey, A., Karamuftuoglu, M.A., Razmkhah, S., Ozbayoglu, M. (2018). Single flux quantum based ultrahigh speed spiking neuromorphic processor architecture. *arXiv preprint arXiv:1812.10354.*

Bray, A.J. (1974). Superconductive specific-heat transition in a magnetic field. *Physical Review B*, 9(11), 4752–4759.

Brock, D.K. (2001). RSFQ technology: Circuits and systems. *High-Speed Integrated Circuit Technology: Towards 100 GHz Logic*, 11(1), 307–362.

Brock, D.K., Martinet, S.S., Bocko, M.F., Przybysz, J.X. (2005). Design and testing of QOS comparators for an RSFQ based analog to digital converter. *IEEE Transactions on Applied Superconductivity*, 5(2), 2244–2247.

Bronk, C., Lingamneni, A., Palem, K. (2010). Innovation for sustainability in information and communication technologies (ICT). Technical Report, Inst. Public Policy, Rice Univ., Houston, Texas.

Buck, D.A. (1956). The cryotron-a superconductive computer component. *Proceedings of the IRE*, 44(4), 482–493.

Buckles, W. and Hassenzahl, W.V. (2000). Superconducting magnetic energy storage. *IEEE Power Engineering Review*, 20(5), 16–20.

Bunyk, P. and Semenov, V.K. (1995). Design of an RSFQ microprocessor. *IEEE Transactions on Applied Superconductivity*, 5(2), 3325–3328.

Bunyk, P., Leung, M., Spargo, J., Dorojevets, M. (2003). Flux-1 RSFQ microprocessor: Physical design and test results. *IEEE Transactions on Applied Superconductivity*, 13(2), 433–436.

Burkhart, L.D., Teoh, J.D., Zhang, Y., Axline, C.J., Frunzio, L., Devoret, M.H., Jiang, L., Girvin, S.M., Schoelkopf, R.J. (2021). Error-detected state transfer and entanglement in a superconducting quantum network. *PRX Quantum*, 2(3), 030321.

Buttgenbach, T.H., LeDuc, H.G., Maker, P.D., Phillips, T. (1992). A fixed tuned broadband matching structure for submillimeter SIS receivers. *IEEE Transactions on Applied Superconductivity*, 2(3), 165–175.

Cabrera, B., Clarke, R.M., Colling, P., Miller, A.J., Nam, S., Romani, R.W. (1998). Detection of single infrared, optical, and ultraviolet photons using superconducting transition edge sensors. *Applied Physics Letters*, 73(6), 735–737.

Caldeira, A.O. and Leggett, A.J. (1983). Quantum tunnelling in a dissipative system. *Annals of Physics*, 149(2), 374–456.

Canturk, M. and Askerzade, I.N. (2012). Numerical study of characteristics of externally shunted Josephson junctions with unharmonic current-phase relation. *IEEE Transactions on Applied Superconductivity*, 22(6), 1400106–1400106.

Charbonnier, P.J. (1908). *Balistique intérieure*. O. Doin, Paris.

Chen, W., Rylyakov, A.V., Patel, V., Lukens, J.E., Likharev, K.K. (1998). Superconductor digital frequency divider operating up to 750 GHz. *Applied Physics Letters*, 73(19), 2817–2819.

Chen, W., Rylyakov, A.V., Patel, V., Lukens, J.E., Likharev, K.K. (1999). Rapid single flux quantum T-flip flop operating up to 770 GHz. *IEEE Transactions on Applied Superconductivity*, 9(2), 3212–3215.

Chen, X., Li, S., Xue, X., Zheng, X. (2020). Weak microwave signal detection based on microwave photonics-enabled single-photon technology. In *Asia Communications and Photonics Conference/International Conference on Information Photonics and Optical Communications 2020 (ACP/IPOC) (2020), paper M4A.329*. Optica Publishing Group, M4A.329 [Online]. Available at: https://opg.optica.org/abstract.cfm?uri=ACPC-2020-M4A.329 [Accessed 3 June 2022].

Cheng, R., Goteti, U.S., Hamilton, M.C. (2018). Spiking neuron circuits using superconducting quantum phase-slip junctions. *Journal of Applied Physics*, 124(15), 152126.

Chesca, B., Kleiner, R., Koelle, D. (2005). SQUID theory. In *The SQUID Handbook*, Clarke, J. and Braginski, A.I. (eds). Wiley-VCH Verlag GmbH & Co, Weinheim [Online]. Available at: http://onlinelibrary.wiley.com/doi/10.1002/3527603646.ch2/summary [Accessed 17 December 2013].

Clarke, J. and Braginski, A.I. (2006). *The SQUID Handbook: Fundamentals and Technology of SQUIDs and SQUID Systems*. John Wiley & Sons, New York.

CONNECT (2003). CONNECT Cell Library Handbook. 21 June.

Crété, D., Kermorvant, J., Lemaître, Y., Marcilhac, B., Mesoraca, S., Trastoy, J., Ulysse, C. (2021). Comparison of magnetic field detectors based on SQUID and/or Josephson junction arrays with HTSc. *IEEE Transactions on Applied Superconductivity*, 31(5), 1–6.

Crotty, P., Schult, D., Segall, K. (2010). Josephson junction simulation of neurons. *Physical Review E*, 82(1) [Online]. Available at: http://arxiv.org/abs/1002.2892 [Accessed 17 December 2016].

Das, R., Bolkhovsky, V., Tolpygo, S., Gouker, P., Johnson, L., Dauler, E., Gouker, M. (2017). Large scale cryogenic integration approach for superconducting high-performance computing. In *IEEE 67th Electronic Components and Technology Conference (ECTC)*, 675–683.

Das, R.N., Yoder, J.L., Rosenberg, D., Kim, D.K., Yost, D., Mallek, J., Hover, D., Bolkhovsky, V., Kerman, A.J., Oliver, W.D. (2018). Cryogenic qubit integration for quantum computing. In *2018 IEEE 68th Electronic Components and Technology Conference (ECTC)*, 504–514.

Daunt, J.G., Mendelssohn, K.A.G., Simon, F.E. (1946). An experiment on the mechanism of superconductivity. *Proceedings of the Royal Society of London. Series A. Mathematical and Physical Sciences*, 185(1001), 225–239.

Day, P.K., LeDuc, H.G., Mazin, B.A., Vayonakis, A., Zmuidzinas, J. (2003). A broadband superconducting detector suitable for use in large arrays. *Nature*, 425(6960), 817–821.

Devoret, M.H., Wallraff, A., Martinis, J.M. (2004). Superconducting qubits: A short review. *arXiv:cond-mat/0411174*, 6 November [Online]. Available at: http://arxiv.org/abs/cond-mat/0411174 [Accessed 2 November 2018].

Dickel, C., Wesdorp, J.J., Langford, N.K., Peiter, S., Sagastizabal, R., Bruno, A., Criger, B., Motzoi, F., DiCarlo, L. (2018). Chip-to-chip entanglement of transmon qubits using engineered measurement fields. *Physical Review B*, 97(6), 064508.

Dorojevets, M. (2015). Design and demonstration of a 30 GHz 16-bit superconductor RSFQ microprocessor. Report, State University of New York at Stony Brook [Online]. Available at: http://www.dtic.mil/docs/citations/ADA624330 [Accessed 18 February 2018].

Dorojevets, M. and Bunyk, P. (2003). Architectural and implementation challenges in designing high-performance RSFQ processors: A FLUX-1 microprocessor and beyond. *IEEE Transactions on Applied Superconductivity*, 13(2), 446–449.

Dorojevets, M., Bunyk, P., Zinoviev, D. (2001). FLUX chip: Design of a 20-GHz 16-bit ultrapipelined RSFQ processor prototype based on 1.75-μm LTS technology. *IEEE Transactions on Applied Superconductivity*, 11, 326–332.

Dorojevets, M., Ayala, C.L., Yoshikawa, N., Fujimaki, A. (2013a). 8-bit asynchronous sparse-tree superconductor RSFQ arithmetic-logic unit with a rich set of operations. *IEEE Transactions on Applied Superconductivity*, 23(3), 1700104.

Dorojevets, M., Kasperek, A.K., Yoshikawa, N., Fujimaki, A. (2013b). 20-GHz 8 8-bit parallel carry-save pipelined RSFQ multiplier. *IEEE Transactions on Applied Superconductivity*, 23(3), 1300104.

Dotsenko, V.V., Delmas, J., Webber, R.J., Filippov, T.V., Kirichenko, D.E., Sarwana, S., Gupta, D., Kadin, A.M., Track, E.K. (2009b). Integration of a 4-stage 4 K pulse tube cryocooler prototype with a superconducting integrated circuit. *IEEE Transactions on Applied Superconductivity*, 19(3), 1003–1007.

Ebert, B., Mielke, O., Kunert, J., Stolz, R., Ortlepp, T. (2010). Experimentally verified design guidelines for minimizing the gray zone width of Josephson comparators. *Superconductor Science and Technology*, 23, 055005.

Engert, S., Wetzstein, O., Hofherr, M., Ilin, K., Siegel, M., Meyer, H.G., Toepfer, H. (2013). Mathematical analysis of multiplexing techniques for SNSPD arrays. *IEEE Transactions on Applied Superconductivity*, 23(3), 2501005–2501005.

Febvre, P., McGrath, W., Batelaan, P., LeDuc, H., Bumble, B., Frerking, M., Hernichel, J. (1993). A 547 GHz SIS receiver employing a submicron Nb junction with an integrated matching circuit. *1993 IEEE MTT-S International Microwave Symposium Digest*, IEEE, 771–774.

Febvre, P., McGrath, W., Batelaan, P., Bumble, B., LeDuc, H., George, S., Feautrier, P. (1994). A low-noise SIS receiver measured from 480 GHz to 650 GHz using Nb junctions with integrated RF tuning circuits. *International Journal of Infrared and Millimeter Waves*, 15(6), 943–965.

Febvre, P., Boutez, C., George, S., Beaudin, G. (1995). Models of superconducting microstrip and coplanar elements for submillimeter applications. *Millimeter and Submillimeter Waves II*, International Society for Optics and Photonics, 136–147 [Online]. Available at: https://www-spiedigitallibrary-org.gaelnomade-2.grenet.fr/conference-proceedings-of-spie/2558/0000/Models-of-superconducting-microstrip-and-coplanar-elements-for-submillimeter-applications/10.1117/12.224218.short [Accessed 3 June 2019].

Fedorov, K.G., Shcherbakova, A.V., Wolf, M.J., Beckmann, D., Ustinov, A.V. (2014). Fluxon readout of a superconducting qubit. *Physical Review Letters*, 112(16), 160502.

Feldman, M.J. (2000). A technique to demonstrate energy level quantization in a SQUID. *Physica B: Condensed Matter*, 284–288, 2127–2128.

Filippov, T., Dorojevets, M., Sahu, A., Kirichenko, A., Ayala, C., Mukhanov, O. (2011). 8-bit asynchronous wave-pipelined RSFQ arithmetic-logic unit. *IEEE Transactions on Applied Superconductivity*, 21(3), 847–851.

Filippov, T.V., Sahu, A., Kirichenko, A.F., Vernik, I.V., Dorojevets, M., Ayala, C.L., Mukhanov, O.A. (2012a). 20 GHz operation of an asynchronous wave-pipelined RSFQ arithmetic-logic unit. *Physics Procedia*, 36, 59–65.

Filippov, T.V., Sahu, A., Kirichenko, A.F., Vernik, I.V., Dorojevets, M., Ayala, C.L., Mukhanov, O.A. (2012b). 20 GHz operation of an asynchronous wave-pipelined RSFQ arithmetic-logic unit. *Physics Procedia*, 36, 59–65.

Fourie, C.J., Wetzstein, O., Ortlepp, T., Kunert, J. (2011). Three-dimensional multi-terminal superconductive integrated circuit inductance extraction. *Superconductor Science and Technology*, 24(12), 125015.

Fourie, C.J., Jackman, K., Botha, M.M., Razmkhah, S., Febvre, P., Ayala, C.L., Xu, Q., Yoshikawa, N., Patrick, E., Law, M. et al. (2019). ColdFlux Superconducting EDA and TCAD Tools Project: Overview and progress. *IEEE Transactions on Applied Superconductivity*, 29(5), 1–7.

Frasca, S. and Charbon, E. (2019). Hybrid superconductor–semiconductor electronics. *Nature Electronics*, 2(10), 433–434.

Fredkin, E. (1990). An informational process based on reversible universal cellular automata. *Physica D: Nonlinear Phenomena*, 45(1), 254–270.

Fredkin, E. and Toffoli, T. (1982). Conservative logic. *International Journal of Theoretical Physics*, 21(3–4), 219–253.

Fujimaki, A., Tanaka, M., Yamada, T., Yamanashi, Y., Park, H., Yoshikawa, N. (2008). Bit-serial single flux quantum microprocessor CORE. *IEICE Transactions on Electronics*, E91-C(3), 342–349.

Gerber, H.R., Fourie, C.J., Perold, W.J., Muller, L.C. (2007). Design of an asynchronous microprocessor using RSFQ-AT. *IEEE Transactions on Applied Superconductivity*, 17(2), 490–493.

Gerlach, W. and Stern, O. (1922). Der experimentelle nachweis der richtungsquantelung im magnetfeld. *Zeitschrift für Physik*, 9(1), 349–352.

Gerrits, T., Lita, A., Calkins, B., Nam, S.W. (2016). Superconducting transition edge sensors for quantum optics. In *Superconducting Devices in Quantum Optics*, Hadfield, R.H. and Johansson, G. (eds). Springer International Publishing, Cham. DOI: 10.1007/978-3-319-24091-6_2.

Gold, A., Paquette, J.P., Stockklauser, A., Reagor, M.J., Alam, M.S., Bestwick, A., Didier, N., Nersisyan, A., Oruc, F., Razavi, A. et al. (2021). Entanglement across separate silicon dies in a modular superconducting qubit device. *NPJ Quantum Information*, 7(1), 1–10.

Gordeeva, A.V. and Pankratov, A.L. (2009). Minimization of thermal jitter in a balanced comparator single flux quantum cell. *Journal of Applied Physics*, 103(10), 103913.

Gray, K.E. and Kampwirth, R.T. (1984). Design for a repetitive superconducting opening switch. *Cryogenics*, 24(1), 21–26.

Guimond, P.-O., Vermersch, B., Juan, M.L., Sharafiev, A., Kirchmair, G., Zoller, P. (2020). A unidirectional on-chip photonic interface for superconducting circuits. *NPJ Quantum Information*, 6(1), 1–12.

Hansen, R.C. (1990). Superconducting antennas. *IEEE Transactions on Aerospace and Electronic Systems*, 26(2), 345–355.

Hashimoto, Y., Yorozu, S., Miyazaki, T., Kameda, Y., Suzuki, H., Yoshikawa, N. (2007). Implementation and experimental evaluation of a cryocooled system prototype for high-throughput SFQ digital applications. *IEEE Transactions on Applied Superconductivity*, 17(2), 546–551.

Hashimoto, Y., Yorozu, S., Kameda, Y. (2008). Development of cryopackaging and I/O technologies for high-speed superconductive digital systems. *IEICE TRANSACTIONS on Electronics*, E91-C(3), 325–332.

Hidaka, M. and Akers, L.A. (1991). An artificial neural cell implemented with superconducting circuits. *Superconductor Science and Technology*, 4(11), 654.

Hidaka, M., Nagasawa, S., Satoh, T., Hinode, K., Kitagawa, Y. (2006). Current status and future prospect of the Nb-based fabrication process for single flux quantum circuits. *Superconductor Science and Technology*, 19(3), S138.

Hirose, T., Asai, T., Amemiya, Y. (2006). Spiking neuron devices consisting of single-flux-quantum circuits. *Physica C Superconductivity*, 445–448.

Holmes, D.S. and DeBenedictis, E.P. (2017). Superconductor electronics and the international roadmap for devices and systems. In *2017 16th International Superconductive Electronics Conference (ISEC)*, 1–3.

Holmes, D.S., Ripple, A.L., Manheimer, M.A. (2013). Energy-efficient superconducting computing #x2014: Power budgets and requirements. *IEEE Transactions on Applied Superconductivity*, 23(3), 1701610–1701610.

Hornibrook, J.M., Colless, J.I., Conway Lamb, I.D., Pauka, S.J., Lu, H., Gossard, A.C., Watson, J.D., Gardner, G.C., Fallahi, S., Manfra, M.J., Reilly, D.J. (2015). Cryogenic control architecture for large-scale quantum computing. *Physical Review Applied*, 3(2), 024010.

Hosoya, M., Hioe, W., Casas, J., Kamikawai, R., Harada, Y., Wada, Y., Nakane, H., Suda, R., Goto, E. (1991). Quantum flux parametron: A single quantum flux device for Josephson supercomputer. *IEEE Transactions on Applied Superconductivity*, 1(2), 77–89.

Irwin, K.D. and Hilton, G.C. (2005). Transition-edge sensors. In *Cryogenic Particle Detection*, Enss, C. (ed.). Springer, Berlin, Heidelberg.

Ishida, K., Tanaka, M., Nagaoka, I., Ono, T., Kawakami, S., Tanimoto, T., Fujimaki, A., Inoue, K. (2020). 32 GHz 6.5 mW Gate-Level-Pipelined 4-Bit Processor using Superconductor single-flux-quantum logic. *2020 IEEE Symposium on VLSI Circuits*, 1–2.

ITRS (2004). International technology roadmap for semiconductors 2004. Emerging research devices. Report, ITRS.

Ivanov, Z.G., Kupriyanov, M.Y., Likharev, K.K., Snigirev, O.V. (1978). Boundary conditions for the usadel equations and properties of dirty s-n-s sandwiches. *Le journal de physique colloques*, 39(C6), C6–557.

Jones, N. (2018). Data centres are chewing up vast amounts of energy. *Nature*, 561(5).

Josephson, B. (1962). Possible new effects in superconductive tunnelling. *Physics Letters*, 1(7), 251–253.

Kadin, A.M. (1999). *Introduction to Superconducting Circuits*. Wiley-Interscience, New Jersey.

Keyes, R.W. and Landauer, R. (1970). Minimal energy dissipation in logic. *IBM Journal of Research and Development*, 14(2), 152–157.

Kim, J.-Y., Kim, S., Kang, J. (2005). Construction of an RSFQ 4-bit ALU with half adder cells. *IEEE Transactions on Applied Superconductivity*, 15(2), 308–311.

Kinard, J. (2007). *Artillery: An Illustrated History of its Impact*. ABC-CLIO, Santa Barbara.

Kinev, N.V., Rudakov, K.I., Filippenko, L.V., Baryshev, A.M., Koshelets, V.P. (2019). Flux-flow Josephson oscillator as the broadband tunable terahertz source to open space. *Journal of Applied Physics*, 125(15), 151603.

Kirichenko, D.E., Sarwana, S., Kirichenko, A.F. (2011). Zero static power dissipation biasing of RSFQ circuits. *IEEE Transactions on Applied Superconductivity*, 21(3), 776–779.

Kirichenko, A.F., Vernik, I.V., Kamkar, M.Y., Walter, J., Miller, M., Albu, L.R., Mukhanov, O.A. (2019). ERSFQ 8-bit parallel arithmetic logic unit. *IEEE Transactions on Applied Superconductivity*, 29(5), 1–7.

Kirkconnell, C.S., Hon, R.C., Perella, M.D., Crittenden, T.M., Ghiaasiaan, S.M. (2017). Development of a miniature Stirling cryocooler for LWIR small satellite applications. *Tri-Technology Device Refrigeration (TTDR) II. International Society for Optics and Photonics*, 1018002 [Online]. Available at: https://www.spiedigitallibrary.org/conference-proceedings-of-spie/10180/1018002/Development-of-a-miniature-Stirling-cryocooler-for-LWIR-small-satellite/10.1117/12.2259803.short [Accessed 15 September 2017].

Kjaergaard, M., Schwartz, M.E., Braumüller, J., Krantz, P., Wang, J.I.-J., Gustavsson, S., Oliver, W.D. (2020). Superconducting Qubits: Current state of play. *Annual Review of Condensed Matter Physics*, 11(1), 369–395.

Koch, J., Yu, T.M., Gambetta, J., Houck, A.A., Schuster, D.I., Majer, J., Blais, A., Devoret, M.H., Girvin, S.M., Schoelkopf, R.J. (2007). Charge-insensitive qubit design derived from the Cooper pair box. *Physical Review A*, 76(4), 042319.

Kokabi, A., Khoshaman, A., Razmkhah, S., Hoseini, S.M., Moftakharzadeh, A., Bozbey, A., Fardmanesh, M. (2009). Response analysis of free-membrane transition-edge detectors with thin substrates. *Journal of Physics: Conference Series*, 153(1), 012017.

Kondo, T., Kobori, M., Onomi, T., Nakajima, K. (2005). Design and implementation of stochastic neurosystem using SFQ logic circuits. *IEEE Transactions on Applied Superconductivity*, 15(2), 320–323.

Koomey, J.G. (2008). Worldwide electricity used in data centers. *Environmental Research Letters*, 3(3), 034008.

Kusaka, K., Ohtake, M., Ohnishi, T., Yoshida, K., Kubo, T., Ito, H., Tsuchihashi, T. (2013). Long-term operation of the superconducting triplet quadrupoles with small cryocoolers for BigRIPS in-flight separator and RI-Beam delivery line at RIKEN. *IEEE Transactions on Applied Superconductivity*, 23(3), 4101305–4101305.

Lancaster, M., Wu, Z., Huang, Y., Maclean, T.S.M., Zhou, X., Gough, C., Alford, N.M. (1992). Superconducting antennas. *Superconductor Science and Technology*, 5(4), 277.

Landauer, R. (1961). Irreversibility and heat generation in the computing process. *IBM Journal of Research and Development*, 5(3), 183–191.

Larbalestier, D., Gurevich, A., Feldmann, D.M., Polyanskii, A. (2010). High-Tc superconducting materials for electric power applications. In *Materials for Sustainable Energy*, Dusastre, V. (ed.). World Scientific, Singapore. DOI: 10.1142/9789814317665_0046.

Larkin, T.I., Bol'ginov, V.V., Stolyarov, V.S., Ryazanov, V.V., Vernik, I.V., Tolpygo, S.K., Mukhanov, O.A. (2012). Ferromagnetic Josephson switching device with high characteristic voltage. *Applied Physics Letters*, 100(22), 222601.

Le Roux, P., Fourie, C., Razmkhah, S., Febvre, P. (2021). Accurate small signal simulation of superconductor interconnects in SPICE. *IEEE Transactions on Applied Superconductivity*, 31(5), 1–6.

Leek, P.J., Filipp, S., Maurer, P., Baur, M., Bianchetti, R., Fink, J.M., Göppl, M., Steffen, L., Wallraff, A. (2009). Using sideband transitions for two-qubit operations in superconducting circuits. *Physical Review B*, 79(18), 180511.

Leung, E.M. (2000). Superconducting fault current limiters. *IEEE Power Engineering Review*, 20(8), 15–18.

Likharev, K. (1977). Dynamics of some single flux quantum devices I: Parametric quantron. *IEEE Transactions on Magnetics*, 13(1), 242–244.

Likharev, K.K. (1986). *Dynamics of Josephson Junction and Circuits*. Gorden Breach, New York.

Likharev, K.K. and Semenov, V.K. (1991). RSFQ logic/memory family: A new Josephson-junction technology for sub-terahertz-clock-frequency digital systems. *IEEE Transactions on Applied Superconductivity*, 1(1), 3–28.

Linares-Barranco, B., Sanchez-Sinencio, E., Rodriguez-Vazquez, A., Huertas, J.L. (1991). A CMOS implementation of FitzHugh-Nagumo neuron model. *IEEE Journal of Solid-State Circuits*, 26(7), 956–965.

Lloyd, S. (2000). Ultimate physical limits to computation. *Nature*, 406(6799), 35023282.

London, F., London, H., Lindemann, F.A. (1935). The electromagnetic equations of the supraconductor. *Proc. R. Soc. Lond. A*, 149(866), 71–88.

Lvovsky, Y., Stautner, E.W., Zhang, T. (2013). Novel technologies and configurations of superconducting magnets for MRI. *Superconductor Science and Technology*, 26(9), 093001.

Ma, Q.Y. and Yang, E.S. (1990). Current controlled high Tc superconducting switch. *Cryogenics*, 30(12), 1146–1148.

Majer, J., Chow, J.M., Gambetta, J.M., Koch, J., Johnson, B.R., Schreier, J.A., Frunzio, L., Schuster, D.I., Houck, A.A., Wallraff, A. et al. (2007). Coupling superconducting qubits via a cavity bus. *Nature*, 449(7161), 443–447.

Marcilhac, B., Lemaitre, Y., Mansart, D., Mage, J. (1999). Development of superconductive microwave filters for mobile communications and filter banks. *IEEE Transactions on Applied Superconductivity*, 9(2), 4014–4017.

Marcilhac, B., Crété, D., Lemaître, Y., Mansart, D., Mage, J., Bouzehouane, K., Dolin, C., Jacquet, E., Woodall, P., Contour, J. (2001). Frequency agile microwave devices based on Y-Ba-Cu-O/Sr-Ti-O//La-Al-O structure. *IEEE Transactions on Applied Superconductivity*, 11(1), 438–441.

Markov, I.L. (2014). Limits on fundamental limits to computation. *Nature*, 512(7513), 13570.

Matthaei, G.L. (2003). Narrow-band, fixed-tuned, and tunable bandpass filters with zig-zag hairpin-comb resonators. *IEEE Transactions on Microwave Theory and Techniques*, 51(4), 1214–1219.

Mattis, D.C. and Bardeen, J. (1958). Theory of the anomalous skin effect in normal and superconducting metals. *Physical Review*, 111(2), 412–417.

Mazin, B.A. (2005). Microwave kinetic inductance detectors. PhD thesis, California Institute of Technology [Online]. Available at: http://resolver.caltech.edu/CaltechETD:etd-10042004-120707.

McCaughan, A.N., Verma, V.B., Buckley, S.M., Allmaras, J.P., Kozorezov, A.G., Tait, A.N., Nam, S.W., Shainline, J.M. (2019). A superconducting thermal switch with ultrahigh impedance for interfacing superconductors to semiconductors. *Nature Electronics*, 2(10), 451–456.

McCumber, D.E. (1968). Effect of ac impedance on dc voltage-current characteristics of superconductor weak-link junctions. *Journal of Applied Physics*, 39(7), 3113–3118.

Megrant, A., Neill, C., Barends, R., Chiaro, B., Chen, Y., Feigl, L., Kelly, J., Lucero, E., Mariantoni, M., O'Malley, P.J.J. et al. (2012). Planar superconducting resonators with internal quality factors above one million. *Applied Physics Letters*, 100(11), 113510.

Meindl, J.D. and Davis, J.A. (2000). The fundamental limit on binary switching energy for terascale integration (TSI). *IEEE Journal of Solid-State Circuits*, 35(10), 1515–1516.

Meissner, W. and Ochsenfeld, R. (1933). Ein neuer effekt bei eintritt der supraleitfähigkeit. *Naturwissenschaften*, 21(44), 787–788.

Mi, X., Roushan, P., Quintana, C., Mandra, S., Marshall, J., Neill, C., Arute, F., Arya, K., Atalaya, J., Babbush, R., et al. (2021). Information scrambling in computationally complex quantum circuits. *arXiv:2101.08870 [cond-mat, physics:hep-th, physics:quant-ph]*, 21 January [Online]. Available at: http://arxiv.org/abs/2101.08870 [Accessed 7 September 2021].

Mirhosseini, M., Sipahigil, A., Kalaee, M., Painter, O. (2020). Superconducting qubit to optical photon transduction. *Nature*, 588(7839), 599–603.

Mitchell, E.E., Hannam, K.E., Lazar, J., Leslie, K.E., Lewis, C.J., Grancea, A., Keenan, S.T., Lam, S.K.H., Foley, C.P. (2016). 2D SQIF arrays using 20 000 YBCO high R_n Josephson junctions. *Superconductor Science and Technology*, 29(6), 06LT01.

Mizugaki, Y., Nakajima, K., Sawada, Y., Yamashita, T. (1994). Implementation of new superconducting neural circuits using coupled SQUIDs. *IEEE Transactions on Applied Superconductivity*, 4(1), 1–8.

Mohseni, M., Read, P., Neven, H., Boixo, S., Denchev, V., Babbush, R., Fowler, A., Smelyanskiy, V., Martinis, J. (2017). Commercialize quantum technologies in five years. *Nature News*, 543(7644), 171.

Mooij, J.E., Orlando, T.P., Levitov, L., Tian, L., van der Wal, C.H., Lloyd, S. (1999). Josephson persistent-current qubit. *Science*, 285(5430), 1036–1039.

Mueller, O.M. and Herd, K.G. (1993). Ultra-high efficiency power conversion using cryogenic MOSFETs and HT-superconductors. In *Proceedings of IEEE Power Electronics Specialist Conference – PESC '93*, 772–778.

Mukhanov, O.A. (2011). Energy-efficient single flux quantum technology. *IEEE Transactions on Applied Superconductivity*, 21(3), 760–769.

Mukhanov, O.A., Gupta, D., Kadin, A.M., Semenov, V.K. (2004). Superconductor analog-to-digital converters. *Proceedings of the IEEE*, 92(10), 1564–1584.

Nagasawa, S., Hinode, K., Satoh, T., Hidaka, M., Akaike, H., Fujimaki, A., Yoshikawa, N., Takagi, K., Takagi, N. (2014). Nb 9-layer fabrication process for superconducting large-scale SFQ circuits and its process evaluation. *IEICE Transactions on Electronics*, E97-C(3), 132–140.

Nakajima, N., Matsuzaki, F., Yamanashi, Y., Yoshikawa, N., Tanaka, M., Kondo, T., Fujimaki, A., Terai, H., Yorozu, S. (2004). Design and implementation of circuit components of the SFQ microprocessor, CORE1. *Superconductor Science and Technology*, 17(3), 301.

Nakamura, Y., Pashkin, Y.A., Tsai, J.S. (1999). Coherent control of macroscopic quantum states in a single-Cooper-pair box. *Nature*, 398(6730), 786–788.

Nakamura, S., Numabe, H., Bozbey, A., Fujimaki, A. (2009). Current resolution of a single-flux-quantum readout circuit based on current-to-time conversion toward a flux qubit system. *IEEE Transactions on Applied Superconductivity*, 19(3), 973–976.

Nguyen, M.-H., Ribeill, G.J., Gustafsson, M.V., Shi, S., Aradhya, S.V., Wagner, A.P., Ranzani, L.M., Zhu, L., Baghdadi, R., Butters, B. et al. (2020). Cryogenic memory architecture integrating spin hall effect based magnetic memory and superconductive cryotron devices. *Scientific Reports*, 10(1), 248.

Nielsen, M.A. and Chuang, I.L. (2000). *Quantum Computation and Quantum Information*. Cambridge University Press.

Noel, J., Hijazi, Y., Martinez, J., Vlasov, Y.A., Larkins Jr, G.L. (2003). A switched high-Tc superconductor microstrip resonator using a MEM switch. *Superconductor Science and Technology*, 16(12), 1438–1441.

NSA (2005). Superconducting technology assessment [Online]. Available at: http://www.nitrd.gov/pubs/nsa/sta.pdf.

Ohki, T., Savin, A., Hassel, J., Gronberg, L., Karminskaya, T., Kidiyarova-Shevchenko, A. (2007). Balanced comparator for RSFQ qubit readout. *IEEE Transactions on Applied Superconductivity*, 17(2), 128–131.

Olofsson, G., Pagani, L., Tauber, J., Febvre, P., Deschamps, A., Encrenaz, P., Floren, H.-G., George, S., Lecomte, B., Ljung, B. et al. (1998). Low interstellar abundance of O_2 confirmed by the PIROG 8 balloon experiment. *Astronomy and Astrophysics*, 339, L81–L84.

Onnes, H.K. (1991). Further experiments with liquid helium. C. On the change of electric resistance of pure metals at very low temperatures etc. IV. The resistance of pure mercury at helium temperatures. In *Through Measurement to Knowledge*, Gavroglu, K. and Goudaroulis, Y. (eds). Springer, Dordrecht [Online]. Available at: http://link. springer.com/ chapter/10.1007/978-94-009-2079-8_15 [Accessed 11 April 2014].

Onomi, T., Kondo, T., Nakajima, K. (2009). Implementation of high-speed single flux-quantum up/down counter for the neural computationusing stochastic logic. *IEEE Transactions on Applied Superconductivity*, 19(3), 626–629.

Orlando, T.P., Mooij, J.E., Tian, L., van der Wal, C.H., Levitov, L.S., Lloyd, S., Mazo, J.J. (1999). Superconducting persistent-current qubit. *Physical Review B*, 60(22), 15398–15413.

Ortlepp, T. and Fourie, C. (2008). Cyrocooler-cooled RSFQ circuits. *Cryocooler Workshop*.

Ortlepp, T., Miyajima, S., Toepfer, H., Fujimaki, A. (2012). Josephson comparator with modified dynamic behavior for improved sensitivity. *Journal of Applied Physics*, 111(12), 123901.

Paik, H., Schuster, D.I., Bishop, L.S., Kirchmair, G., Catelani, G., Sears, A.P., Johnson, B.R., Reagor, M.J., Frunzio, L., Glazman, L.I. et al. (2011). Observation of high coherence in Josephson junction qubits measured in a three-dimensional circuit QED architecture. *Physical Review Letters*, 107(24), 240501.

Pauka, S.J., Das, K., Kalra, R., Moini, A., Yang, Y., Trainer, M., Bousquet, A., Cantaloube, C., Dick, N., Gardner, G.C. et al. (2021). A cryogenic CMOS chip for generating control signals for multiple qubits. *Nature Electronics*, 4(1), 64–70.

Pechal, M., Besse, J.-C., Mondal, M., Oppliger, M., Gasparinetti, S., Wallraff, A. (2016). Superconducting switch for fast on-chip routing of quantum microwave fields. *Physical Review Applied*, 6(2), 024009.

Phillips, I. and Ulidowski, I. (2014). Event identifier logic. *Mathematical Structures in Computer Science*, 24(2) [Online]. Available at: https://www.cambridge.org/core/ journals/mathematical-structures-in-computer-science/article/event-identifier-logic/ 190EA5E5757CF1DCFADCCF5CB83CCE88 [Accessed 2 January 2018].

Phillips, I., Ulidowski, I., Yuen, S. (2013). Modelling of bonding with processes and events. In *Reversible Computation, Lecture Notes in Computer Science. International Conference on Reversible Computation*. Springer, Berlin, Heidelberg [Online]. Available at: https://link.springer.com/chapter/10.1007/978-3-642-38986-3_12 [Accessed 2 January 2018].

Picard, E., Madrangeas, V., Bila, S., Mage, J., Marcilhac, B. (2004). Very narrow band HTS filters without tuning for UMTS communications. In *34th European Microwave Conference*, IEEE, 1113–1116.

Polonsky, S.V., Semenov, V.K., Schneider, D.F. (1993). Transmission of single-flux-quantum pulses along superconducting microstrip lines. *IEEE Transactions on Applied Superconductivity*, 3(1), 2598–2600.

Radenbaugh, R. (2004). Refrigeration for superconductors. *Proceedings of the IEEE*, 92(10), 1719–1734.

Razmkhah, S. (2018). Low power high speed computing using rapid single flux quantum circuits, PhD thesis, TOBB University of Economics and Technology, Ankara. 1 January [Online]. Available at: http://earsiv.etu.edu.tr:80/xmlui/handle/20.500.11851/3405 [Accessed 30 July 2020].

Razmkhah, S. and Bozbey, A. (2016). Design of the passive transmission lines for different stripline widths and impedances. *IEEE Transactions on Applied Superconductivity*, 26(8), 1–6.

Razmkhah, S. and Febvre, P. (2020). JOINUS: A user-friendly open-source software to simulate digital superconductor circuits. *IEEE Transactions on Applied Superconductivity*, 30(5), 1–7.

Ren, J., Sarwana, S., Sahu, A., Talalaevskii, A., Inamdar, A. (2015). Low-pass delta-delta-sigma ADC. *IEEE Transactions on Applied Superconductivity*, 25(3), 1–6.

Rippert, E.D. and Lomatch, S. (1997). A multilayered superconducting neural network implementation. *IEEE Transactions on Applied Superconductivity*, 7(2), 3442–3445.

Ritter, M.F., Fuhrer, A., Haxell, D.Z., Hart, S., Gumann, P., Riel, H., Nichele, F. (2021). A superconducting switch actuated by injection of high-energy electrons. *Nature Communications*, 12(1), 1266.

Rose-Innes, A.C. (2012). *Introduction to Superconductivity*. Elsevier, Amsterdam.

Rosenberg, D., Weber, S.J., Conway, D., Yost, D.-R.W., Mallek, J., Calusine, G., Das, R., Kim, D., Schwartz, M.E., Woods, W. et al. (2020). Solid-state qubits: 3D integration and packaging. *IEEE Microwave Magazine*, 21(8), 72–85.

Sage, J.M., Bolkhovsky, V., Oliver, W.D., Turek, B., Welander, P.B. (2011). Study of loss in superconducting coplanar waveguide resonators. *Journal of Applied Physics*, 109(6), 063915.

Schneider, M.L., Donnelly, C.A., Russek, S.E., Baek, B., Pufall, M.R., Hopkins, P.F., Dresselhaus, P.D., Benz, S.P., Rippard, W.H. (2018). Ultralow power artificial synapses using nanotextured magnetic Josephson junctions. *Science Advances*, 4(1), e1701329.

Schoelkopf, R., Phillips, T., Zmuidzinas, J. (1993). A 100 GHz Josephson mixer using resistively-shunted Nb tunnel junctions. *IEEE Transactions on Applied Superconductivity*, 3(1), 2250–2253.

Segall, K., LeGro, M., Kaplan, S., Svitelskiy, O., Khadka, S., Crotty, P., Schult, D. (2017). Synchronization dynamics on the picosecond time scale in coupled Josephson junction neurons. *Physical Review E*, 95(3), 032220.

Semenov, V.K., Danilov, G.V., Averin, D.V. (2003). Negative-inductance SQUID as the basic element of reversible Josephson-junction circuits. *IEEE Transactions on Applied Superconductivity*, 13(2), 938–943.

Service, R.F. (2012). Computer science. What it'll take to go exascale. *Science (New York)*, 335(6067), 394–396.

Shainline, J.M., Buckley, S.M., McCaughan, A.N., Chiles, J., Jafari-Salim, A., Mirin, R.P., Nam, S.W. (2018). Circuit designs for superconducting optoelectronic loop neurons. *Journal of Applied Physics*, 124(15), 152130.

Shehabi, A., Smith, S., Sartor, D., Brown, R., Herrlin, M., Koomey, J., Masanet, E., Horner, N., Azevedo, I., Lintner, W. (2016). United States data center energy usage report [Online]. Available at: https://pubarchive.lbl.gov/islandora/object/ir%3A1005775/ [Accessed 17 November 2017].

Shen, Z.Y., Wilker, C., Pang, P., Holstein, W.L., Face, D., Kountz, D.J. (1992). High-Tc-superconductor-sapphire microwave resonator with extremely high Q-values up to 90 K. *IEEE Transactions on Microwave Theory and Techniques*, 40(12), 2424–2431.

Shevchenko, S.N., Omelyanchouk, A.N., Il'ichev, E. (2012). Multiphoton transitions in Josephson-junction qubits (Review Article). *Low Temperature Physics*, 38(4), 283–300.

Shurakov, A., Lobanov, Y., Goltsman, G. (2016). Superconducting hot-electron bolometer: From the discovery of hot-electron phenomena to practical applications. *Superconductor Science and Technology*, 29(2), 023001.

Siddiqi, I. (2021). Engineering high-coherence superconducting qubits. *Nature Reviews Materials*, 6(10), 875–891.

Sidiropoulou, K., Pissadaki, E.K., Poirazi, P. (2006). Inside the brain of a neuron. *EMBO Reports*, 7(9), 886–892.

Silberhorn, C. (2007). Detecting quantum light. *Contemporary Physics*, 48(3), 143–156.

Silverstone, J.W., Santagati, R., Bonneau, D., Strain, M.J., Sorel, M., O'Brien, J.L., Thompson, M.G. (2015). Qubit entanglement between ring-resonator photon-pair sources on a silicon chip. *Nature Communications*, 6(1), 7948.

Sinclair, A.K., Schroeder, E., Zhu, D., Colangelo, M., Glasby, J., Mauskopf, P.D., Mani, H., Berggren, K.K. (2019). Demonstration of microwave multiplexed readout of DC-biased superconducting nanowire detectors. *IEEE Transactions on Applied Superconductivity*, 29(5), 1–4.

Smith, D.H., Gillett, G., de Almeida, M.P., Branciard, C., Fedrizzi, A., Weinhold, T.J., Lita, A., Calkins, B., Gerrits, T., Wiseman, H.M. et al. (2012). Conclusive quantum steering with superconducting transition-edge sensors. *Nature Communications*, 3(1), 625.

Soloviev, I.I., Klenov, N.V., Pankratov, A.L., Revin, L.S., Il'ichev, E., Kuzmin, L.S. (2015). Soliton scattering as a measurement tool for weak signals. *Physical Review B*, 92(1), 014516.

Sonnet Software (n/d). Sonnet [Online]. Available at: http://www.sonnetsoftware.com/ [Accessed 4 June 2019].

Sourikopoulos, I., Hedayat, S., Loyez, C., Danneville, F., Hoel, V., Mercier, E., Cappy, A. (2017). A 4-fJ/Spike artificial neuron in 65 nm CMOS technology. *Frontiers in Neuroscience*, 11, 123.

Steffen, M., Kumar, S., DiVincenzo, D.P., Rozen, J.R., Keefe, G.A., Rothwell, M.B., Ketchen, M.B. (2010). High-coherence hybrid superconducting qubit. *Physical Review Letters*, 105(10), 100502.

Swingle, B. (2018). Unscrambling the physics of out-of-time-order correlators. *Nature Physics*, 14(10), 988–990.

Tagirov, L.R. (1999). Low-field superconducting spin switch based on a superconductor/ferromagnet multilayer. *Physical Review Letters*, 83(10), 2058–2061.

Takeuchi, N., Yamanashi, Y., Yoshikawa, N. (2014a). Reversible computing using adiabatic superconductor logic. In *Reversible Computation, Lecture Notes in Computer Science. International Conference on Reversible Computation.* Springer, Cham [Online]. Available at: https://link.springer.com/chapter/10.1007/978-3-319-08494-7_2 [Accessed 26 December 2017].

Takeuchi, N., Yamanashi, Y., Yoshikawa, N. (2014b). Reversible logic gate using adiabatic superconducting devices. *Scientific Reports*, 4, 6354.

Tanaka, M., Kondo, T., Nakajima, N., Kawamoto, T., Yamanashi, Y., Kamiya, Y., Akimoto, A., Fujimaki, A., Hayakawa, H., Yoshikawa, N. et al. (2005). Demonstration of a single-flux-quantum microprocessor using passive transmission lines. *IEEE Transactions on Applied Superconductivity*, 15(2), 400–404.

Tanaka, H., Morie, T., Aihara, K. (2006). An analog CMOS circuit for spiking neuron models. *International Congress Series*, 1291, 217–220.

Tanaka, M., Takata, K., Kawaguchi, T., Ando, Y., Yoshikawa, N., Sato, R., Fujimaki, A., Takagi, K., Takagi, N. (2015). Development of bit-serial RSFQ microprocessors integrated with shift-register-based random access memories. In *2015 15th International Superconductive Electronics Conference (ISEC)*, 1–3.

Tanaka, M., Kozaka, M., Kamiya, K., Üşenmez, K., Aydoğan, E.C., Razmkhah, S., Bozbey, A., Fujimaki, A. (2021). SFQ parallel encoders promising for video imaging with superconductor stripline detectors. *IEEE Transactions on Applied Superconductivity*, 31(1), 1–6.

Tang, G.M., Takata, K., Tanaka, M., Fujimaki, A., Takagi, K., Takagi, N. (2016). 4-bit bit-slice arithmetic logic unit for 32-bit RSFQ microprocessors. *IEEE Transactions on Applied Superconductivity*, 26(1), 1–6.

Tinkham, M. (1996). *Introduction to Superconductivity*. McGraw-Hill, New York.

Tolpygo, S.K., Bolkhovsky, V., Weir, T.J., Galbraith, C.J., Johnson, L.M., Gouker, M.A., Semenov, V.K. (2015). Inductance of circuit structures for MIT LL superconductor electronics fabrication process with 8 niobium layers. *IEEE Transactions on Applied Superconductivity*, 25(3), 1–5.

Tolpygo, S.K., Bolkhovsky, V., Rastogi, R., Zarr, S., Day, A.L., Golden, E., Weir, T.J., Wynn, A., Johnson, L.M. (2019). Advanced fabrication processes for superconductor electronics: Current status and new developments. *IEEE Transactions on Applied Superconductivity*, 29(5), 1–13.

Tolpygo, S.K., Golden, E.B., Weir, T.J., Bolkhovsky, V. (2021). Inductance of superconductor integrated circuit features with sizes down to 120 nm. *Superconductor Science and Technology*, 34(8), 085005.

Tzimpragos, G., Vasudevan, D., Tsiskaridze, N., Michelogiannakis, G., Madhavan, A., Volk, J., Shalf, J., Sherwood, T. (2020). A computational temporal logic for superconducting accelerators. In *Proceedings of the Twenty-Fifth International Conference on Architectural Support for Programming Languages and Operating Systems*, ASPLOS '20. Association for Computing Machinery, New York, 435–448. DOI: 10.1145/3373376.3378517.

Ulidowski, I., Phillips, I., Yuen, S. (2014). Concurrency and reversibility. In *Reversible Computation, Lecture Notes in Computer Science. International Conference on Reversible Computation*. Springer, Cham [Online]. Available at: https://link.springer.com/chapter/10.1007/978-3-319-08494-7_1 [Accessed 2 January 2018].

Ullom, J.N. and Bennett, D.A. (2015). Review of superconducting transition-edge sensors for x-ray and gamma-ray spectroscopy. *Superconductor Science and Technology*, 28(8), 084003.

Usadel, K.D. (1970). Generalized diffusion equation for superconducting alloys. *Physical Review Letters*, 25(8), 507–509.

Valiev, K.A. (2005). Quantum computers and quantum computations. *Physics-Uspekhi*, 48(1), 1.

Van Heddeghem, W., Lambert, S., Lannoo, B., Colle, D., Pickavet, M., Demeester, P. (2014). Trends in worldwide ICT electricity consumption from 2007 to 2012. *Computer Communications*, 50(Supplement C), 64–76.

Wallraff, A., Schuster, D.I., Blais, A., Frunzio, L., Huang, R.-S., Majer, J., Kumar, S., Girvin, S.M., Schoelkopf, R.J. (2004). Strong coupling of a single photon to a superconducting qubit using circuit quantum electrodynamics. *Nature*, 431(7005), 162–167.

Walls, T.J., Averin, D.V., Likharev, K.K. (2007). Josephson junction comparator as a quantum-limited detector for flux qubit readout. *IEEE Transactions on Applied Superconductivity*, 17(2), 136–141.

Webber, R.J., Burroughs, C.J., Radparvar, M. (2007). Performance of a cryocooled Nb DC programmable voltage standard at 4 K. *IEEE Transactions on Applied Superconductivity*, 17(4), 3857–3861.

Webber, R.J., Dotsenko, V., Talalaevskii, A., Miller, R., Tang, J.C. (2008). Operation of superconducting digital receiver circuits on 2-Stage gifford-mcmahon cryocooler. *Advances in Cryogenic Engineering*, 42, 927–932.

Webber, R.J., Dotsenko, V.V., Delmas, J., Kadin, A.M., Track, E.K. (2009). Evaluation of a 4 K 4-stage pulse tube cryocooler for superconducting electronics. In *Cryocoolers 15th International Cryocooler Conference, CO, USA*, 657–664.

Weiss, U. (2012). *Quantum Dissipative Systems*. World Scientific, Singapore.

Wendin, G. (2017). Quantum information processing with superconducting circuits: A review. *Reports on Progress in Physics*, 80(10), 106001.

Wendin, G. and Shumeiko, V.S. (2007). Quantum bits with Josephson junctions (Review Article). *Low Temperature Physics*, 33(9), 724–744.

Wu, X., Saxena, V., Zhu, K., Balagopal, S. (2015). A CMOS spiking neuron for brain-inspired neural networks with resistive synapses and in situ learning. *IEEE Transactions on Circuits and Systems II: Express Briefs*, 62(11), 1088–1092.

Yamanashi, Y., Tanaka, M., Akimoto, A., Park, H., Kamiya, Y., Irie, N., Yoshikawa, N., Fujimaki, A., Terai, H., Hashimoto, Y. (2007a). Design and implementation of a pipelined bit-serial SFQ microprocessor, CORE1 #946. *IEEE Transactions on Applied Superconductivity*, 17(2), 474–477.

Yamanashi, Yuki, Nishigai, T., Yoshikawa, N. (2007b). Study of LR-loading technique for low-power single flux quantum circuits. *IEEE Transactions on applied superconductivity*, 17(2), 150–153.

Yamanashi, Y., Umeda, K., Yoshikawa, N. (2013). Pseudo sigmoid function generator for a superconductive neural network. *IEEE Transactions on Applied Superconductivity*, 23(3), 1701004–1701004.

Yan, F., Gustavsson, S., Kamal, A., Birenbaum, J., Sears, A.P., Hover, D., Gudmundsen, T.J., Rosenberg, D., Samach, G., Weber, S. et al. (2016). The flux qubit revisited to enhance coherence and reproducibility. *Nature Communications*, 7, 12964.

Ye, L., Gopman, D.B., Rehm, L., Backes, D., Wolf, G., Ohki, T., Kirichenko, A.F., Vernik, I.V., Mukhanov, O.A., Kent, A.D. (2014). Spin-transfer switching of orthogonal spin-valve devices at cryogenic temperatures. *Journal of Applied Physics*, 115(17), 17C725.

Yeh, J.-H. and Anlage, S.M. (2013). In situ broadband cryogenic calibration for two-port superconducting microwave resonators. *Review of Scientific Instruments*, 84(3), 034706.

Yorozu, S., Kameda, Y., Terai, H., Fujimaki, A., Yamada, T., Tahara, S. (2002). A single flux quantum standard logic cell library. *Physica C: Superconductivity*, 378, 1471–1474.

You, J.Q., Hu, X., Ashhab, S., Nori, F. (2007). Low-decoherence flux qubit. *Physical Review B*, 75(14), 140515.

Zhang, D., Liang, G.-C., Shih, C.F., Lu, Z.H., Johansson, M.E. (1995). A 19-pole cellular bandpass filter using 75-mm-diameter high-temperature superconducting thin films. *IEEE Microwave and Guided Wave Letters*, 5(11), 405–407.

Zhu, J., Zhang, T., Yang, Y., Huang, R. (2020). A comprehensive review on emerging artificial neuromorphic devices. *Applied Physics Reviews*, 7(1), 011312.

Zmuidzinas, J. and LeDuc, H. (1992). Quasi-optical slot antenna SIS mixers. *IEEE Transactions on Microwave Theory and Techniques*, 40(9), 1797–1804.

Zubayri, M.S. and Sculli, M.A. (2003). *Quantum Optics*. Cambridge University Press.

9

All-Optical Chips

Frank BRÜCKERHOFF-PLÜCKELMANN[1], Johannes FELDMANN[2] and Wolfram PERNICE[1]

[1] *Institute of Physics, University of Muenster, Germany*

[2] *Department of Materials, University of Oxford, UK*

9.1. Introduction

Artificial intelligence has become a part of everyday life in recent years and is considered a key technology of the 21st century. One famous example is the program Alpha Go developed by the Google Deepmind team, which was able to beat one of the world's best Go players in 2016. Since the game was too complex to be solved deterministically with existing computers, artificial intelligence was used, which was previously trained using games played by humans (Silver et al. 2016). Its successor, AlphaGo Zero, does not even require a training set but just the rules of the game to become better at Go and other games than human players (Silver et al. 2017). Consequently, AlphaGo Zero beats AlphaGo in 100 games with a final score of 100:0. This achievement indicates that artificial intelligence algorithms are capable of finding solutions for complex problems on their own. Nowadays, the underlying technology is crucial for image and speech recognition, autonomous driving and medical diagnostics among others (Amato et al. 2013; Lane et al. 2017; Babiker et al. 2019).

The drawback of this development is that the amount of data generated and processed is increasing steadily and therefore hardware requirements are becoming increasingly demanding (Batra et al. 2018). Especially high data throughput is

---

For a color version of all figures in this chapter, see www.iste.co.uk/cresti/beyondcmos.zip.

Beyond-CMOS, coordinated by Alessandro CRESTI.

challenging processors based on the conventional von Neumann architecture, in which the processing unit is separated from the memory, to their limits. In this architecture, data needs to be shuffled back and forth between the two units, resulting in lower computational speeds and higher energy consumption, also known as the von Neumann bottleneck (Backus 1978). For example, the simulation of a mouse-scale artificial neural network with 2.5 million neurons is 9,000 times slower than the actual biological system, while simultaneously requiring 40,000 times more energy (Benjamin et al. 2014).

To address these challenges, new computational hardware is being developed to meet the steadily growing demands of artificial intelligence technology, most noteworthy Google's tensor processing unit and Microsoft's Brainwave project. The Brainwave project uses field programmable gate arrays (FPGAs) to implement the algorithms directly at the hardware level and also to reconfigure the hardware during the operation (Linn 2016). The tensor processing unit is instead optimized for multiply and accumulate (MAC) operations and is based on systolic arrays to minimize data shuffling (Jouppi 2017).

However, both approaches are based on electrical data processing and thus eventually the bandwidth is limited. By switching to the optical domain and nanophotonic circuits, the available bandwidth increases drastically. Moreover, parallelization is inherently possible by exploiting wavelength division multiplexing and in-memory computing can be achieved with phase change materials. In the following, a brief introduction to nanophotonic circuits is given and the fundamental devices needed for photonic computing are explained. Then, the framework of phase change photonics is explained, which combines integrated photonic circuits with phase change materials. Based on this, two prototypes for optical data processing are presented. First, we illustrate an optimized photonic tensor core for matrix vector multiplication, and second an optical artificial neural network, which mimics the complete artificial neural network at the hardware level.

9.2. Nanophotonic circuits

In 1841, J. Colladon discovered that light can be guided along a falling water jet due to the refractive index difference between water and air. Nowadays, the same effect is used for large-scale fiber-optic networks, which build the backbone of high-speed telecommunication. With modern lithography methods, optical waveguides can also be fabricated on a nanometer scale, thus enabling the creation of integrated photonic circuits. Similar to electrons in copper wires in integrated electronic circuits, photons are guided through the waveguides in nanophotonic circuits. In combination with active devices to modify the circuitry on demand, they pave the way to high-speed optical data processing. In the following, the working principle of dielectric waveguides is explained. Based on this, two major passive components in photonic circuits, namely directional couplers and ring resonators, are introduced.

9.2.1. *Dielectric waveguides*

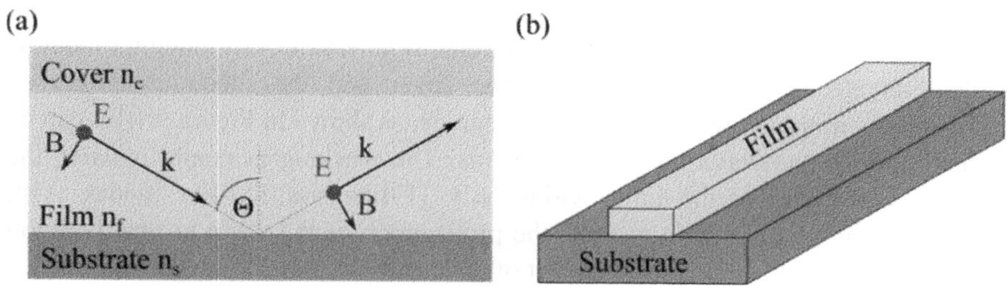

Figure 9.1 *Dielectric optical waveguides*

NOTES ON FIGURE 9.1.– *(a) The working principle of a two-dimensional slab waveguide based on total internal reflection is shown. The slab waveguide consists of three different regions, where the refractive index of the film layer is larger than the refractive index of the substrate and the cover. Therefore, light can be confined in the film region if the angle of incident Θ is larger than the critical angle of total internal reflection. (b) A sketch of a ridge waveguide is shown. The working principle of a three-dimensional ridge waveguide is similar to that of the slab waveguide. However, the wave is now confined in two directions, and thus the ridge waveguide serves as the optical counterpart of an electric wire.*

Similar to light guided in optical fibers based on total internal reflection, dielectric waveguides exploit the refractive index difference between at least two different dielectric media to confine light and are used to route light in nanophotonic circuits. Figure 9.1(a) explains the working principle of a two-dimensional slab waveguide in the framework of ray optics. A film layer with refractive index n_f is placed in between the substrate and the cover with refractive indices n_s, $n_c < n_f$. Therefore, total internal reflection occurs at the film–substrate and film–cover interface for angles of incidence larger than the critical angle of Θ_c=arcsin($n_f/n_{s,c}$). In this way, an electromagnetic wave can propagate in a zig-zag course along the propagation direction. Whether such a wavevector k exists for a wavelength λ depends on the refractive indices, the waveguide geometry and λ itself. For a slab waveguide, two different types of modes, namely the transverse electric (TE) and transverse magnetic (TM), exist, which can be distinguished by the orientation of the fields. For a TE mode, the electric field and for a TM mode, the magnetic field is orientated perpendicular to the propagation direction. The effective refractive index defines the phase propagation speed of the wave, which is given by the ratio between the propagation constant β and the vacuum wavevector k_0. Only modes with an effective index larger than the refractive index of the substrate and the cover are guided inside the waveguide. It is important to note that even though a mode is guided, its electric field expands into the cover and the substrate and drops exponentially in these regions. A ridge waveguide, as shown in Figure 9.1(b), can be used to confine the wave in two dimensions. The working principle remains the same; however, due to the finite width, only TE-like and TM-like modes exist. Overall, the ridge waveguide restricts the propagation of the wave to one direction and thus serves as the optical counterpart of an electric wire in an integrated circuit.

9.2.2. *Basic photonic devices*

At the basic level, there are two categories of photonic devices: passive components, which impact the wave propagation by changing the waveguide shape, as, for example, waveguide couplers, and active components, which actively control the beam propagation such as electro-optical modulators. Figure 9.2 shows two important passive photonic devices, which are required for optical data processing and routing. First, there are waveguide couplers, such as the directional coupler shown in Figure 9.2(a). Two waveguides, A and B, are brought in close vicinity to each other, such that the mode profiles of the individual waveguides overlap. Therefore, they form a coupled superstructure and power can be transferred from one waveguide to the other. The beam propagation inside a directional coupler can be calculated in the framework of coupled mode theory (Little and Huang 1995). For the special case of two identical waveguides and the starting condition $A_{in} = 1$ and $B_{in} = 0$, the power splitting ratio between the two waveguides is shown in Figure 9.2(b). In this condition, complete transfer of the light from one waveguide to the other is possible. The required length of the coupling region L_C depends on the gap between the waveguides and the modal overlap between the individual waveguide in general. For other starting conditions, the superposition principle can be applied to solve the problem.

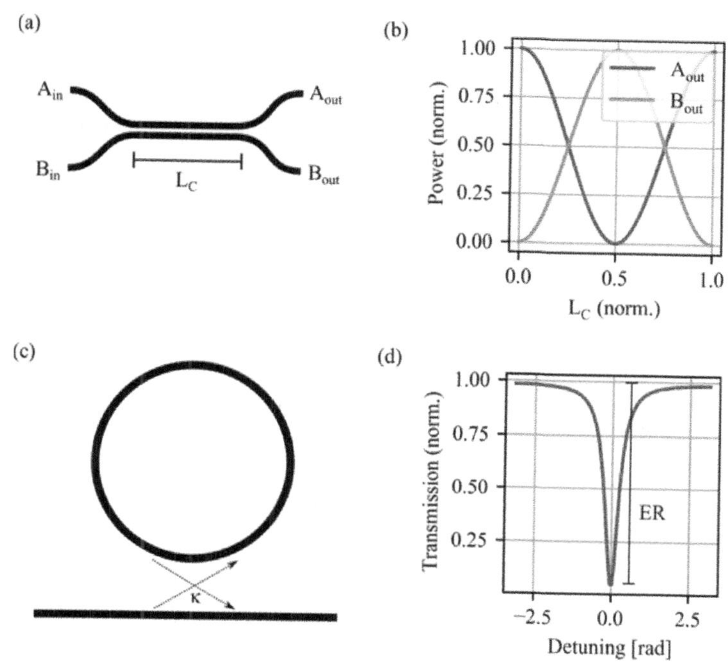

Figure 9.2. *Basic passive photonic structures*

NOTES ON FIGURE 9.2.– *(a) A sketch of a directional coupler is shown. Two waveguides are placed close to each other such that light can couple from one waveguide to the other via the evanescent field. By choosing the length of the coupling region L_C and the gap between the waveguides, a specific amount of power is transferred from one waveguide to the other. (b) The power splitting ratio between the waveguide outputs A_{out} and B_{out} is shown, assuming identical waveguides and the starting condition $A_{in}=1$, $B_{in}=0$. In this scenario, an arbitrary splitting ratio can be achieved. (c) A sketch of a waveguide-coupled ring resonator is shown. A fraction of the power is coupled from the waveguide to the ring resonator. Depending on the detuning from the resonance position, the coupling strength k and the round-trip losses α, the wave stays inside the ring resonator for several round trips. An exemplary transmission spectrum of the waveguide-coupled ring resonator is shown in (d). The extinction ratio (ER) becomes maximal if the coupling strength equals the round-trip losses.*

Another important class of passive devices is wavelength sensitive devices, which are used for wavelength division multiplexing (WDM) among other applications. For WDM, several electromagnetic waves of different wavelengths are first combined with a multiplexer, then processed together and again demultiplexed (split up according to their wavelength). This enables a degree of parallelization when processing data in the optical domain, which is fundamentally not allowed in electronics. One example of a wavelength sensitive device is the waveguide-coupled ring resonator, as shown in Figure 9.2(c). If a wavelength is resonant with the resonator, the phase accumulated in one resonator round trip is a multiple of 2π. Therefore, an optical signal propagates several round trips inside the resonator depending on the detuning from the resonance wavelength, the coupling strength k between the waveguide and the ring and the losses per round trip in the ring resonator α. Figure 9.2(d) shows the transmission through the waveguide as a function of the detuning from the resonance position. If the wavelength fulfills the resonance condition and the medium is not lossless, the transmission is reduced since the effective propagation length is increased. The extinction ratio (ER) becomes maximal if the coupling strength equals propagation losses, also known as critical coupling. A second waveguide can be added to the ring resonator to use it as an add–drop filter for multiplexing (Bogaerts et al. 2012).

9.3. Phase change photonics

Phase change photonics is the combination of nanophotonic circuits with phase change materials (PCMs), enabling new compact and energy-efficient ways of data processing in photonic devices. PCMs have been known for decades and are, for example, used commercially in rewritable optical data storage media as DVD-RW (Wuttig and Yamada 2007) and Blu-Ray RE. This class of materials can be reversibly switched between two states of matter that are stable at room temperature and come along with a large change in its optical and electronic properties. In photonic circuits, PCMs are used as a switchable element that alters the beam propagation inside a waveguide by adding an active component to the passive photonic devices described previously and therefore enabling new functionalities for reconfigurable circuits and data processing.

9.3.1. *Switching dynamics of phase change materials*

PCMs can be switched between two states, an amorphous state with no long-range order between the atoms and an ordered crystalline state. In the amorphous state, the atoms are mainly bonded by covalent bonds, strongly localizing the electrons and decreasing the conductivity (Lucovsky and White 1973). In contrast, resonant bonds, in which electrons are shared between several atoms, are present in the crystalline state, delocalizing the electrons and thus increasing the conductivity (Lencer et al. 2011; Wuttig et al. 2017). Depending on the exact PCM composition, the states are stable over several months (Rios et al. 2015) and the phase change also strongly alters the optical properties. Figure 9.3(b) shows how the PCM can be switched between the states. In order to switch it to the crystalline state, the material needs to be heated above the glass transition temperature such that the atoms have sufficient energy to arrange in the energetically preferred ordered way but without melting it. Afterwards, the material is cooled down slowly again. In contrast, the material is switched to the amorphous state by heating it above the melting temperature to destroy the crystalline order. If the material is cooled down fast enough, the atoms are frozen into the unordered amorphous state. The cooling rate to switch a material to the amorphous state must be of the order of 1–100 K/ns (Peng et al. 1997; Redaelli et al. 2008). Switching of the material can be induced via electrically heating the PCM (Saxena et al. 2019), heating the material with external heater structures (Zheng et al. 2020) and also optically if the material is absorptive (Cheng et al. 2017; Li et al. 2019).

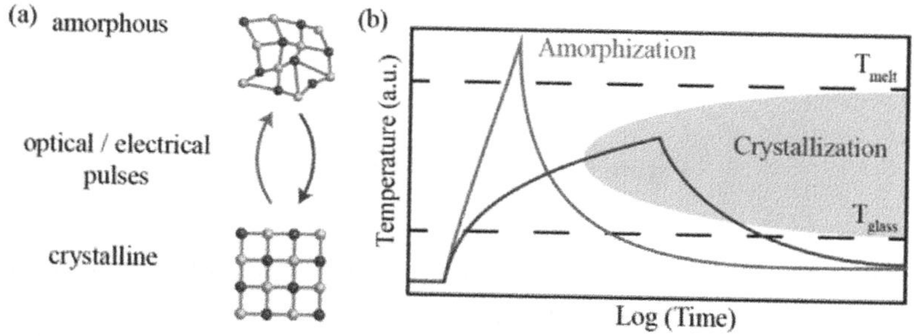

Figure 9.3. *Switching dynamics of phase change materials*

NOTES ON FIGURE 9.3.– *(a) A sketch of the two phases of a phase change material is shown. No long-range order is present in the amorphous state, whereas the atoms are ordered in a lattice in the crystalline state. Depending on the exact material composition, the refractive index can vary greatly between the two states. (b) The switching between the amorphous and crystalline state is shown. In order to switch a phase change material to the amorphous state, the material needs to be heated above the melting temperature and cooled down rapidly afterwards. In contrast, the PCM only needs to be heated above the glass transition temperature and cooled down slowly to switch it to the crystalline state.*

9.3.2. Waveguide-coupled phase change materials

For photonic applications, PCMs, which show a stark contrast in the refractive index between the amorphous and crystalline state, are of great interest. A well-studied example is GST, which belongs to the popular group of chalcogenide solids based on germanium (Ge)–antimony (Sb)–tellurium (Te) alloys (Abdollahramezani et al. 2020). For telecom wavelengths, the refractive index of the amorphous and crystalline state is (Stegmaier et al. 2016):

$$n_{\text{aGST}} = 3.94 + \text{i}0.045 \qquad n_{\text{cGST}} = 6.11 + \text{i}0.83 \qquad [9.1]$$

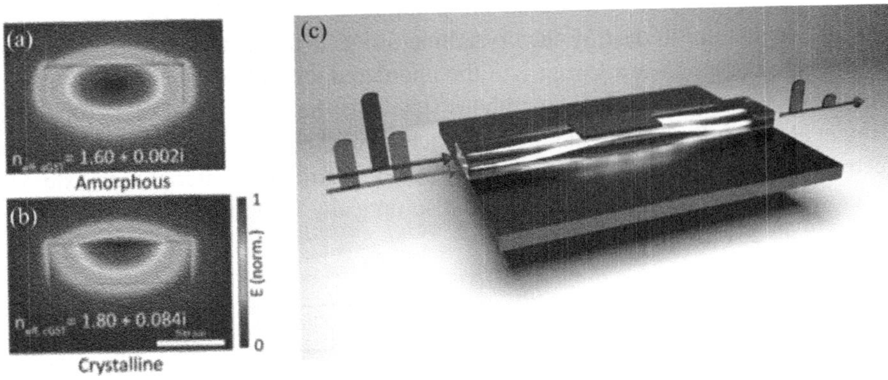

Figure 9.4. *Phase change photonics*

NOTES ON FIGURE 9.4.– *(a), (b) The fundamental mode of a 330 nm high and 1.2 µm broad silicon nitride waveguide on silicon oxide coupled to a 10 nm layer of GST covered with 10 nm ITO. In the amorphous state, GST barley absorbs, resulting in a small imaginary part of the mode's effective index. In contrast, the mode strongly absorbs in the crystalline state. (c) A model of a waveguide-coupled phase change material cell is shown. The cell can be switched with high-power write pulses and the transmission level measured with low-power read pulses.*

In order to use a PCM in a photonic integrated circuit, a small patch of the material is deposited on top of the waveguide and covered with passivation layer to avoid oxidation. Figures 9.4(a) and (b) show the electric field distribution of the fundamental mode in the cross-section of a silicon nitride waveguide with a height of 330 nm and a width of 1.2 µm on silicon oxide, coupled to a 10 nm layer of GST and covered by 10 nm indium tin oxide (ITO). In its amorphous state, the PCM only slightly impacts the mode profile and the mode's effective index. In contrast, the mode is pulled upwards to the PCM in the crystalline state, since the real part of the crystalline GST is much larger than that of silicon nitride. The change in the real part of the effective index leads to a difference in the accumulated phase when propagating through the PCM cell, whereas the difference in the imaginary part leads to a contrast in absorption:

$$\alpha_{\text{SiN-aGST}} = -0.07\frac{dB}{\mu m} \qquad \alpha_{\text{SiN-cGST}} = -2.96\frac{dB}{\mu m} \qquad [9.2]$$

Since both GST states are absorbing for telecommunication wavelengths, they can also be heated and therefore switched with optical pulses.

Figure 9.4(b) schematically shows the interaction between the waveguide and the PCM. First, the PCM is in the crystalline state, leading to a low transmission of the low-power read pulse. Then, a high-power write pulse is applied to melt the PCM and switch it to the amorphous state. Afterwards, a read pulse is sent again, now showing high transmission. An advantage of optical switching and reading of PCM cells is the reduced impact of the inherent randomness and drift present in PCMs (Gong et al. 2018). In the optical domain, the wave evanescently couples to the PCM, and the measurement does not depend on conductive filaments in the PCM. Due to the optical readout, any changes in the PCM configuration smaller than the wavelength of the pulse are averaged and therefore their impact is greatly reduced. For optical switching, energies below 20 pJ have been demonstrated (Feldmann et al. 2017). In combination with the non-volatility of the GST, it offers an energy-efficient memory and computing unit for optical chips. Besides the fully amorphous or crystalline transmission levels in the PCM cell, many intermediate states can be reached by tuning the fractions of crystalline and amorphous parts within the PCM. Up to 64 different transmission levels have already been demonstrated for a PCM cell (Wu et al. 2020).

Although GST is a well-studied PCM, many other examples can be found, especially in the ternary phase diagram of Ge:Sb:Te. Monoatomic PCMs are of special interest, which could lead to high switching speeds and increase the endurance of the PCM cell due to less degradation of the material (Salinga et al. 2018; Cheng et al. 2021). Another class of PCMs is not absorptive in both phase states in the telecom range but has a stark contrast in the real part of the refractive index (Delaney et al. 2020). These materials enable compact and low-loss phase shifters.

9.4. Photonic tensor core

Building on the basic phase change photonic cell described in the previous section, nanophotonic circuits for data processing in the form of a tensor core can be developed. The photonic tensor core is a nanophotonic circuit designed to carry out matrix vector multiplications, which are the backbone of many image recognition and signal analysis algorithms, among others (He et al. 2016). In the framework of phase change photonics, the required multiply and accumulate (MAC) operations are directly integrated into the nanophotonic circuit. Therefore, the computation is performed in-memory and completely avoids the data shuffling needed in the conventional von Neumann architecture. In combination with the large bandwidth available in the optical domain and the inherent capability for parallelization by WDM, unprecedented performance in the PMAC/s range for matrix vector multiplications can be potentially achieved (Feldmann et al. 2021).

9.4.1. *Optical multiply and accumulate operations*

The accumulation (or addition) of two positive numbers can be carried out in the optical domain by encoding the two summands in the power of two laser pulses, as shown in Figure 9.5(a). The two pulses with power S_1 and S_2 overlap temporally and spatially and are due to the superposition principle combined, such that the summed power S can be measured, for example, with a photodetector. For the electric field of two overlapping pulses with frequencies w_1 and w_2, the following equation holds:

$$\mathrm{E}\,sin(w_1 t) + \mathrm{E}\,sin(w_2 t) = 2\mathrm{E}\,sin\left(\frac{w_1+w_2}{2}t\right)cos\left(\frac{w_1-w_2}{2}t\right) \qquad [9.3]$$

The total electric field consists of two terms: a fast-oscillating with the average frequency of the individual pulses and a slow-oscillating with half of the detuning between the two frequencies. Since infrared wavelengths correspond to frequencies of the order of 193 THz, the fast-oscillating term will be averaged out by a photodetector. However, the beat term is resolved by the detector if the detuning is on the same order as the detector bandwidth and thus can impact the matrix vector multiplication. Therefore, the frequency detuning between the laser pulses gives an upper limit for the applicable modulation and detection speed.

Using pulses of different frequencies has the advantage that the accumulation is independent of the relative phase, hence easing the requirements on the fabrication of the integrated photonic circuit and its operation. Therefore, this concept is called “incoherent” accumulation. Another approach is to use pulses of the same frequency and perform the accumulation by fixing the phase between them. In this method, the phase is used as another parameter besides the pulse power and hence also enables negative-valued accumulations. This approach is called “coherent” accumulation.

The easiest way to perform incoherent accumulation with an optical device is to use a waveguide coupler as the directional coupler shown in Figure 9.5(b). The two summands S_1 and S_2 are sent into the upper and lower arm of the directional coupler from the left side. Of the two output arms on the right, the upper one is tapered away and the lower one is used to detect the summed power S. For an n:m splitter, the summed power is:

$$S = \frac{1}{n+m}(n \cdot S_1 + m \cdot S_2) \quad [9.4]$$

Therefore, the losses caused by the accumulation unit with a 50:50 splitter are at least 3 dB due to the design of the device. In order to avoid these losses, WDM can be deployed where both pulses have a different frequency. In theory, this approach is lossless; however, photonic devices for multiplexing, e.g. ring resonators or Bragg gratings (Mu et al. 2019), are more complex, since they depend on the phase propagation of the individual pulses. Consequently, they are also more sensitive to fabrication imperfections and temperature changes.

Similar to the accumulation, the multiplication is carried out in a transmission measurement through the basic phase change cell described in section 9.3.2. The first factor is given by the power of the laser pulse, and the second is given by the absorption of the PCM as shown in Figure 9.5(c). Depending on the state of the PCM, it absorbs a fraction of the input pulse. Thus, the power of the output pulse corresponds to the product of the initial pulse power and the PCM's transmission state, and is detected by a photodiode. This operation is directly carried out in the nanophotonic circuit by using a waveguide-coupled PCM, as sketched in Figure 9.6(c). The multiplication is passive in the sense that it is only based on evanescent coupling between the waveguide and the PCM cell. Therefore, it does not impose any limit on the modulation speed.

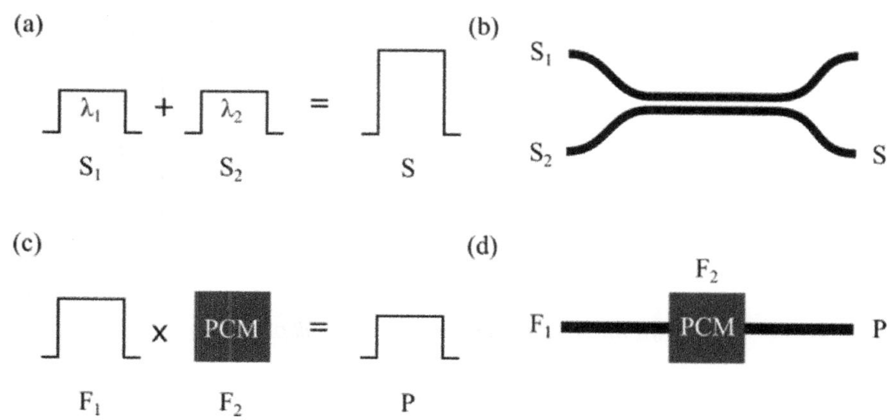

Figure 9.5. *Photonic accumulation and multiplication*

NOTES ON FIGURE 9.5.– *(a) The concept of an optical accumulate operation is shown. Two summands, S_1 and S_2, are encoded in the power of two pulses with different wavelengths and overlapped. A photodetector then detects the summed power S of both pulses. The accumulate operation is directly implemented in the nanophotonic circuit with a directional coupler as in (b). (c) The concept of an optical multiplication operation is shown. The first factor F_1 is again encoded in the power of an optical pulse, and the second F_2 in the phase configuration of a phase change material cell. The pulse power is reduced by the phase change material cell and the product pulse power is detected with a photodetector. The multiplication is carried out using a waveguide-coupled phase change material cell, as shown in (d).*

9.4.2. Design of the photonic tensor core

The full matrix vector multiplication is explained in Figure 9.6(a) and consists of several MAC operations carried out within one time step with the photonic tensor core. Since the accumulation uses incoherent superposition, each component of the input vector x has a different wavelength. Moreover, only positive input vectors can be used, and all matrix elements must be positive. A reference matrix and vector can also be used to perform negative-valued operations.

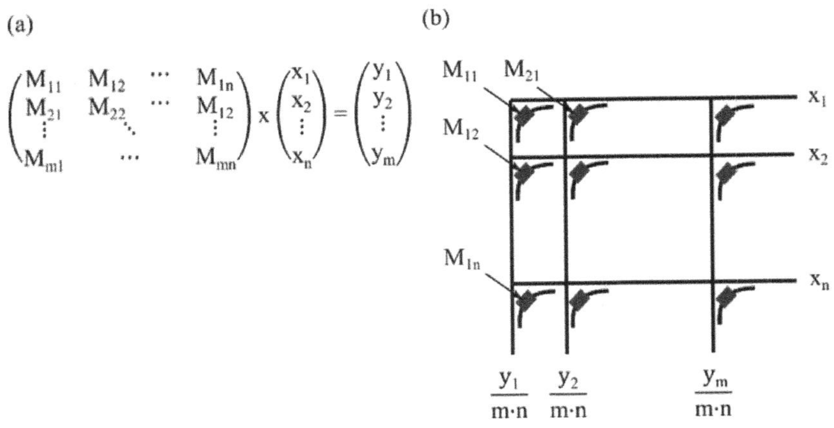

Figure 9.6. *Photonic matrix vector multiplication*

NOTES ON FIGURE 9.6.– *(a) The mathematical operation carried out by the photonic tensor core is illustrated. In this design of the photonic tensor core, all matrix elements and the input vector must be positive. (b) The schematic circuit of the photonic tensor core is shown. The input vector is sent to the horizontal input waveguides as laser pulses of different power. Afterwards, they are partially transferred to the vertical output waveguides corresponding to the output vector. During the coupling between the waveguides, the pulses are multiplied with the corresponding matrix element encoded in PCM cells.*

Figure 9.6(b) shows a sketch of the full photonic tensor core. The input vector elements are sent into the matrix from the right side in the form of laser pulses with powers corresponding to the individual vector component. Thus, the number of horizontal input waveguides is equal to the number of input vector components. By analogy, for every component of the output vector, one vertical output waveguide exists. At the end of each output waveguide, a photodetector can be placed to convert the optical signal back to an electrical one. The matrix itself connects the horizontal input waveguides with the vertical output waveguides with directional couplers. For the M_{ij} matrix cell, 1/i of the power in the horizontal input waveguide is coupled to the bend waveguide with a directional coupler. Afterwards, the pulse is weighted with a PCM cell, whose transmission corresponds to the M_{ij} matrix element. Then, 1/j of the power is coupled to the vertical output waveguide through a directional coupler. This fractional coupling between the horizontal and vertical waveguides ensures that all matrix cells equally contribute to the output vector.

Since the photonic tensor core uses incoherent accumulation with directional couplers, the output vector is 1/mn of the real output vector for an (m×n) matrix, where the factor 1/m is due to coupling-induced losses and can be avoided theoretically using multiplexing techniques. Since this factor is constant for a given matrix size and the same for each vector component, it can be easily removed via calibration.

It is important to note that a full matrix vector multiplication is carried out in a single time step. The speed at which the multiplications can be carried out is only limited by the bandwidth of the optical modulators and photodetectors, which can be up to 50–100 GHz.

9.4.3. *Parallel computing by wavelength division multiplexing*

One major advantage of the photonic tensor core in comparison to electrical circuits is the available bandwidth. The bandwidth is limited only by the transparency window of the material platform used, as well as the wavelength dependence of the devices and propagating modes. However, only a small fraction of the bandwidth is required for a single matrix vector multiplication due to the speed limitations of the electronic pulse generation and detection. In order to make use of the full bandwidth, several matrix vector multiplications can be carried out in parallel.

In the optical domain, parallelization is inherently possible via WDM. Since the MAC operations performed in the photonic tensor core are linear, vectors can be encoded across different wavelength ranges and processed by the same photonic tensor core in parallel. Without multiplexing and by taking an n-component input vector, the components of the first vector use the wavelengths $\lambda_1,\ldots, \lambda_n$ and are sent into the corresponding rows of the matrix. A second vector $x^{(2)}$ is multiplied in parallel by using the wavelengths $\lambda_{n+1},\ldots, \lambda_{n2}$. Since the first component of both $x^{(1)}$ and $x^{(2)}$ must be sent to the first row, they are combined with a multiplexer. The number of vectors that can be processed at the same time is determined by the spectral bandwidth of the system. At the output of the matrix, the combined pulses of all wavelengths must be reassigned to the correct matrix vector multiplication, which is achieved by demultiplexers. In the two-vector example, the demultiplexer has two outputs, where all pulses in the wavelength range λ_1–λ_n are routed to the first output and all pulses in the wavelength range λ_{n+1}–λ_{2n} are routed to the second output. The summed pulse power can then be remeasured individually with a photodetector to obtain all output vectors.

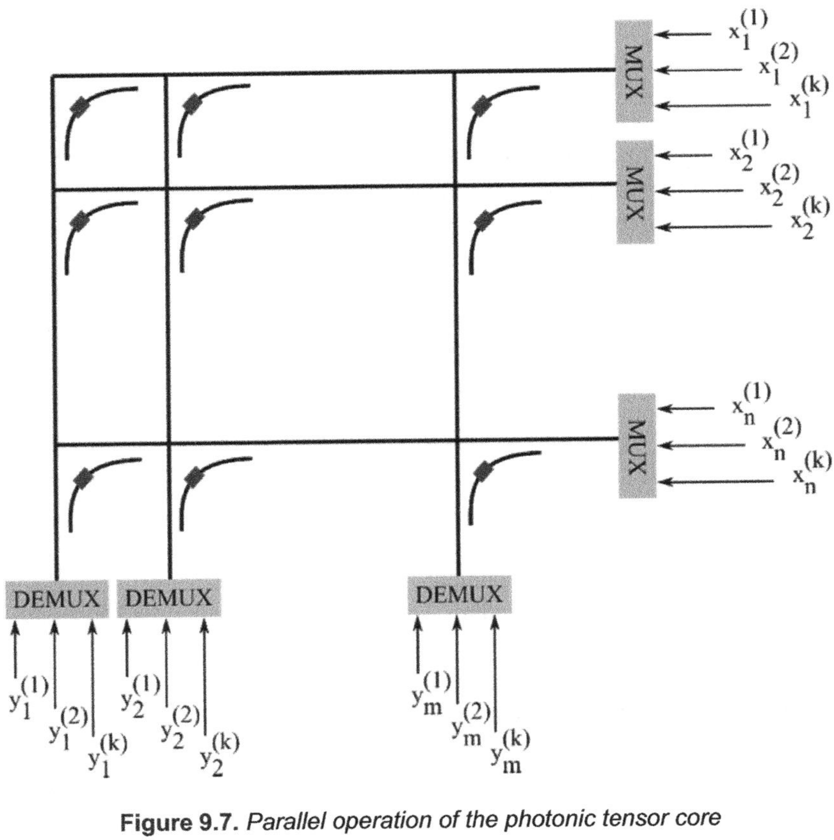

Figure 9.7. *Parallel operation of the photonic tensor core*

NOTES ON FIGURE 9.7.– *Since MAC operations are linear operations, parallelization by wavelength division multiplexing is possible. In order to calculate the matrix vector product for a fixed matrix and input vectors x^1 to x^k, the first components of all vectors are multiplexed together and sent to the first row. Equivalent operations are carried out for the other input components. At the output, the signal is demultiplexed to obtain the output vectors y^1 to y^k.*

9.4.4. *Photonic tensor core prototype*

The large bandwidth available in the optical domain combined with PCM-based in-memory computing and parallelization with WDM enables computational speed in the PMAC/s (10^{15} MAC operations per second) range with energy requirements as low as 17 fJ/MAC, potentially outperforming conventional architectures as well as application-specific integrated circuits such as Google's TPU by orders of magnitude (Feldmann et al. 2021). However, photonic processing units are still in the developmental phase, with promising prototypes already demonstrated (Shen et al. 2017; Lin et al. 2018; Xu et al. 2021). Figure 9.8 shows the experimental setup used in the work of Feldmann et al. (2021), which achieved 2 TMAC/s (2×10^{12} MAC operations per second). Figure 9.8(d) shows a sketch of the entire setup. In order to avoid using several lasers and improve scalability, a frequency comb, as shown in Figure 9.8(b), built in silicon nitride is used (Raja et al. 2019), which provides many sharp wavelength lines with fixed channel spacing shown in Figure 9.8(c). The frequency comb output is demultiplexed and the individual wavelengths are modulated corresponding to the vector components. Subsequently, the signals are again combined using a multiplexer and sent into the on-chip MAC unit, where the pulses are multiplied with PCM cells and accumulated incoherently as described in the previous section. A picture of an exemplary 16x16 matrix is shown in Figure 9.8(d). Light is coupled to the on-chip MAC unit using total internal reflection couplers (Gehring et al. 2019), allowing access to a wide spectral range for encoding the input vectors. The waveguide-coupled PCM is placed on a crossing, in order to enable a switching individual matrix element with optical pulses. Finally, the output of the on-chip MAC unit is demultiplexed and converted to electrical signals with photodetectors.

A well-suited application of the photonic tensor core is convolution processing for image classification. Here, multiple matrix vector multiplications are carried out to highlight specific features in an image, for example edges as shown in Figure 9.8(e). For a given filter, the values encoded in the matrix elements remain constant. Thus, the matrix PCM cells do not need to be reprogrammed, enabling energy-efficient and high-speed operation.

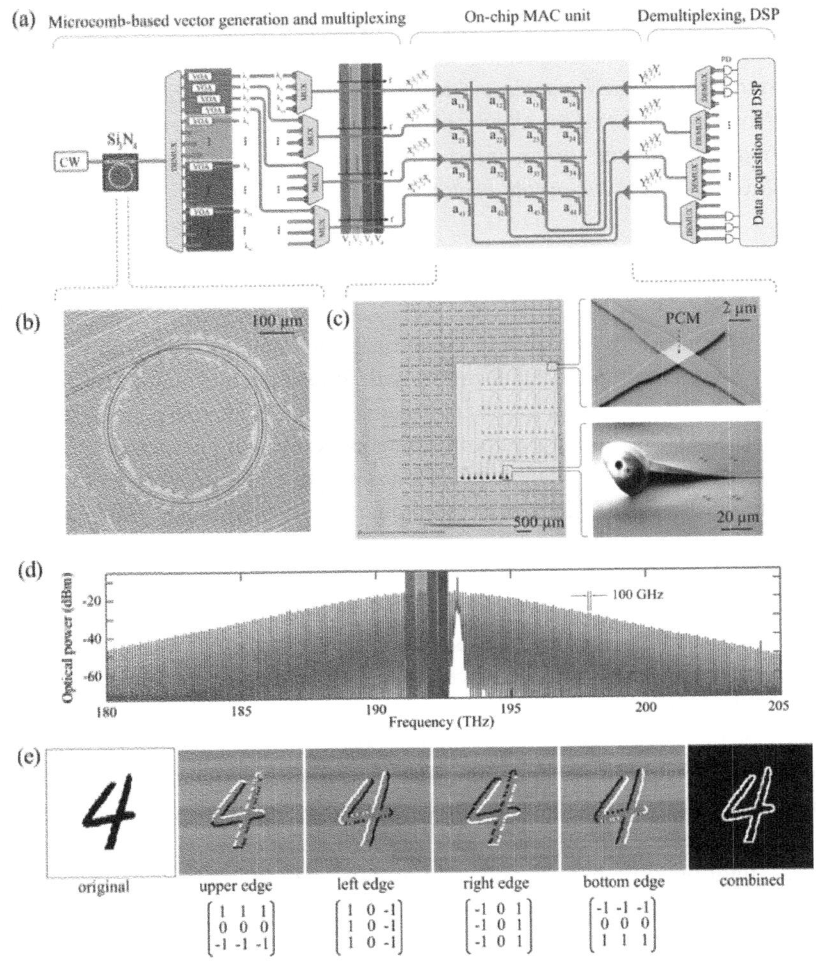

Figure 9.8. *Photonic tensor core architecture*

NOTES ON FIGURE 9.8.– *(a) A sketch of the complete photonic tensor core is shown. The frequency comb in (b) is used to generate the different wavelengths for all input vector components (d). The wavelengths are demultiplexed and individually modulated and again multiplexed together to be sent into the on-chip MAC unit that carries out the MAC operations. The optical circuit of a 16x16 matrix is shown in (c). Three-dimensional total internal reflection couplers are used to couple the light into the matrix. The PCM patches corresponding to the matrix elements are placed on waveguide crossings such that they can be switched optically. Finally, the output signals of the MAC unit are again demultiplexed to the corresponding output vector entries and the intensities measured. One possible application of the photonic tensor core is edge detection for image classification (e). Picture taken from Feldmann et al. (2021).*

9.5. Optical artificial neural network

In contrast to the photonic tensor core, which is designed to carry out a specific operation, an optical artificial neural network aims to mimic the complete artificial neural network at the hardware level and operates solely in the optical domain. In this way, only one optical input should be provided and the optical output must be evaluated, reducing any electrical–optical conversion to a minimum. In the following, a brief overview of artificial neural networks is given and a method for optical spike generation is presented, which is crucial for the operation of the artificial neural network. In section 9.5.3, a prototype for an optical spiking neuron is presented, which can be extended to large-scale optical neural networks.

9.5.1. *Artificial neural networks*

From a mathematical viewpoint, an artificial neural network is a nonlinear function F_P:R^n->R^m, which depends on a set of free parameters P. The free parameters are determined during a training process such that the artificial neural network solves the given task properly. Solving in this context means that the artificial neural network assigns the right output activation A_{out}=$F_P(A_{in})$ to an input activation A_{in}. For example, the pixel values of an image could serve as an input activation and the output activation is large if a dog is in the picture, whereas it is small if there is none.

The concept of the artificial neural networks is vaguely inspired by the working principle of a biological brain, as shown in Figure 9.9(a). It consists of several layers, where each layer contains multiple neurons, which are the building blocks of a neural network, represented by the red dots. Each neuron has an activation assigned to it. The first layer of the artificial neural network is the interface to the real world. In the example of processing an image, the activations of the neurons in the first layer are therefore the pixel values of the picture. The input layer is followed by a number of hidden layers. The activations of the neurons in a hidden layer depend on the neurons' activations in the previous layer and the strength of the connections between the neurons. If all neurons of a layer are connected to all neurons of the subsequent layer, the neuron is called fully connected. Finally, there is the output layer, which presents the result of the neural network and can consist of multiple neurons. In the example presented above, the first neuron of the output layer would only have a high activation if a dog was in the picture. In contrast, a second neuron in the output layer would only have a high activation if there was a cat in the picture. Therefore, the whole neural network classifies the animals (or more generally objects) in the input image.

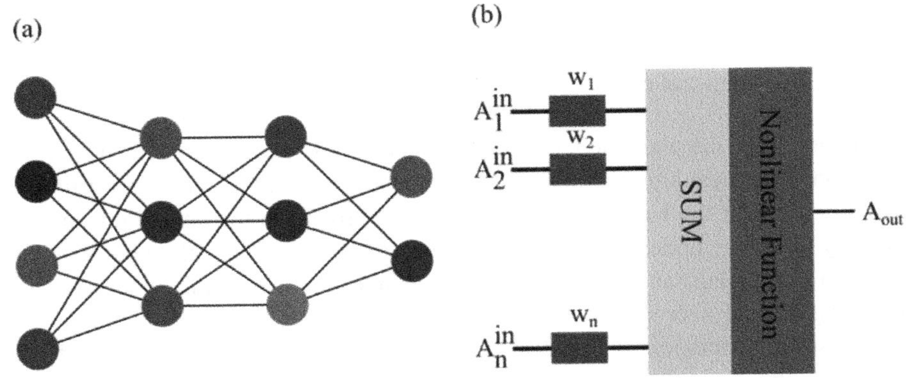

Figure 9.9. *Artificial neural networks*

NOTES ON FIGURE 9.9.– *(a) A sketch of a fully connected artificial neural network is shown. Each red dot corresponds to an artificial neuron. The neural network consists of an input layer, which is the interface to the real world, potentially several hidden layers and an output layer, which shows the result of the artificial neural network. The neural network is fully connected if all neurons of one layer are connected to all neurons of the subsequent layer. (b) The working principle of a single neuron is shown. The input activations A_1 to A_n are individually weighted by the weights w_1 to w_n and added together. Afterwards, the neuron activation A_{out} is calculated via a nonlinear function.*

Figure 9.9(a) explains how the activation of the neuron is calculated depending on the activations of all neurons in the previous layer. First, the input activations are multiplied individually with the corresponding weights, which represent the strength of the connection between the neurons. The weights are also the free parameters P of the artificial neural network. Next, the weighted input activations are summed up, and the neurons output activation is calculated with a nonlinear activation function. Without applying a nonlinear function in each neuron, all the layers of the neurons could be condensed to a single layer and the nonlinearity therefore adds computational complexity. Typical examples of the activation function are the sigmoid function and the rectified linear unit function. Both generate a significant output activation only if the weighted input signals are above a certain threshold value.

Before a specific task, such as pattern recognition, can be carried out, the artificial neural network must be trained. In this process, the connectivity between the neurons and the threshold of their activation functions are set such that the artificial neural network fulfills the given task properly. At the basic level, there are two types of training: supervised and unsupervised learning. For supervised learning, a training dataset with correct pairs of input and output activations is needed and the artificial neural network function is fitted to this training dataset by finding the set of free parameters P that minimizes the deviation between the output activations of the neural network and the correct output activations from the training dataset. There are several ways to train the neural network via supervised learning. Most prominent for this task is the backpropagation algorithm (LeCun et al. 1998). Typically training the neural network is a very time consuming task. If no training dataset is available, unsupervised learning techniques are applied. The artificial neural network uses a learning rule to adapt the connectivity between the neurons during its operation, which, for example, can be used to extract an unknown pattern from a data stream. This can be achieved by using a proper learning rule such as the Hebbian learning rule inspired by biological neurons according to "What fires together wires together" (Bi and Poo 2001).

9.5.2. *Nonlinear activation unit*

The main building block of an artificial neural network is a single neuron, as shown in Figure 9.9(b). The weighting and summation of the input activations are multiply and accumulate operations, for which the same principles as for the photonic tensor core described in section 9.4 hold. However, instead of detecting the summed input power with a photodetector, a nonlinear activation must be developed according to the input power. Moreover, the signal must be optical and not electrical in order to use it as an input activation for subsequent layers.

This functionality is achieved with the nonlinear activation unit shown in Figure 9.10(a). A ring resonator is coupled to a probe waveguide and the transmission through the probe waveguide is measured with pulses of fixed frequency. As previously described, the transmission through a waveguide coupled to a ring resonator depends on the detuning from the resonance position of the resonator, the coupling strength to the ring and the losses inside the ring resonator. The power of the output pulse thus depends on the properties of the ring resonator and represents the activation of the neuron. The waveguide-coupled ring resonator becomes a nonlinear activation unit by including a crossing with a PCM patch to the resonator. In this way, the PCM simultaneously impacts the ring resonator and can be switched via the external input waveguide. The PCM controls the ring properties in two ways. First, the phase accumulated in the PCM region depends on the PCM state, and thus the resonance positions also depend on the PCM state. Second, the PCM affects the losses inside the ring resonator and therefore the maximal ER since the coupling strength between the probe waveguide and the ring resonator remains fixed. By sending the accumulated and weighted input activation pulses into the input waveguide arriving at the PCM, the output pulse of the probe waveguide depends on the total input power. The transmission through the probe waveguide for different input powers is shown in Figure 9.10(b). Due to the switching threshold of the PCM, a minimal input power is required before the probe waveguide transmission changes. At a certain input power threshold, the PCM becomes partially amorphous. The activation unit is designed in such a way that the detuning from the resonance position increases and the maximal ER decreases, resulting in a stark change of transmission of the order of 10 dB. The full measurements can be found in the publication of Feldmann et al. (2019).

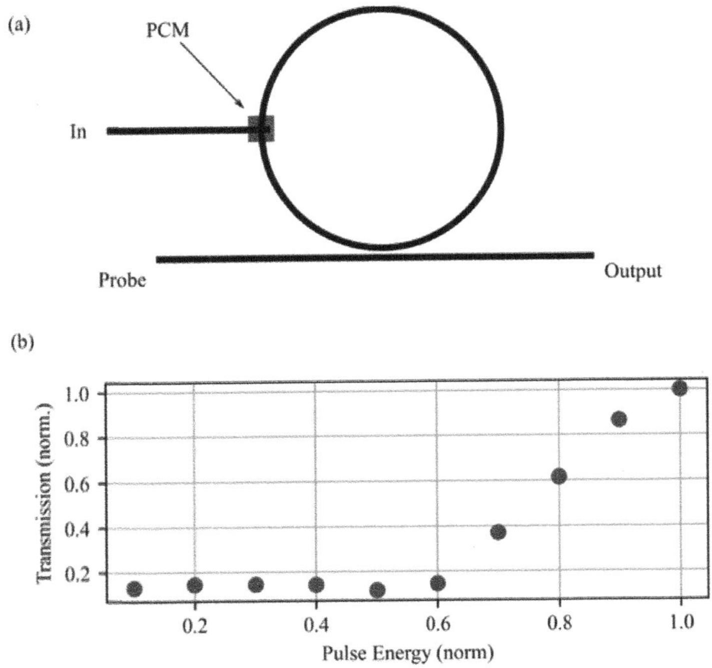

Figure 9.10. *Photonic Activation Function*

NOTES ON FIGURE 9.10.– *(a) The photonic device used for spike generation is shown. If the input pulse power exceeds the switching threshold of the PCM, the PCM becomes amorphous and thus strongly changes the properties of the ring resonator. Afterwards, the transmission of the ring is probed with a low-power optical pulse, which can then serve as the input pulse for subsequent layers. (b) The input energy-dependent transmission of the probe waveguide is shown. The shape is similar to that of a rectified linear unit with a maximum contrast of about 10 dB. The full measurement data are published in Feldmann et al. (2019).*

9.5.3. *Optical neuron prototype*

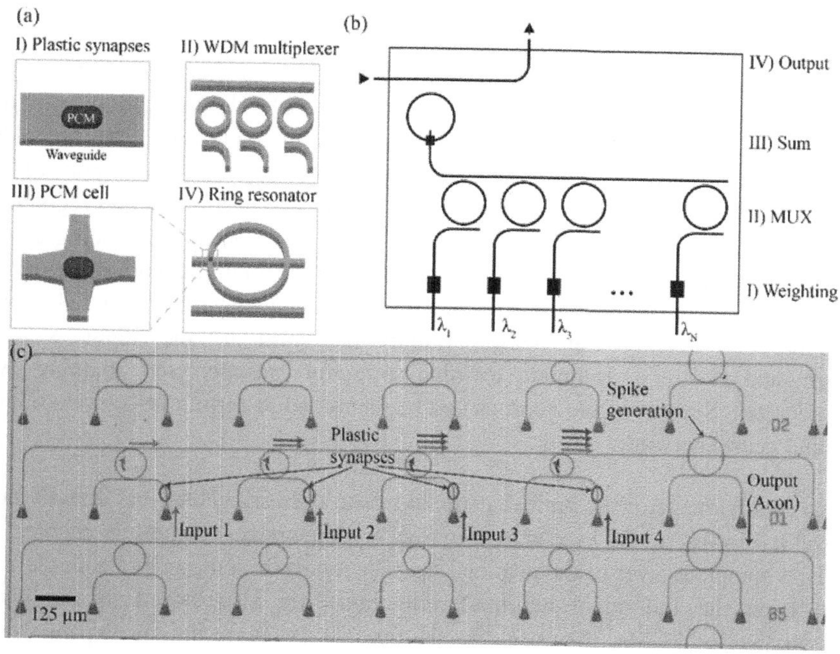

Figure 9.11. *All-optical neural network*

NOTES ON FIGURE 9.11.– *The all-optical neuron consists of the photonic elements depicted in (a). Waveguide-coupled PCMs are employed to weight the inputs and rings are used to perform the incoherent accumulation via wavelength multiplexing. The activation unit described above with a crossing containing the PCM cell determines the activation. (b) A sketch of the nanophotonic integrated circuit is shown. The input activations A_1 to A_n are encoded in the wavelengths λ_1 to λ_n, weighted individually with phase change material cells and accumulated incoherently with a ring multiplexer. The summed power is sent to the nonlinear activation unit and the neuron activation is probed. (c) Optical micrograph of several neurons with four inputs is shown. Picture taken from Feldmann et al. (2019).*

The nonlinear activation unit can be used to implement a complete optical neuron, which can then serve as a building block for large-scale optical neural networks. Figure 9.11(b) shows the different optical devices required to build the optical neuron. First, waveguide-coupled PCM cells are used to emulate the weights. Second, ring resonators (with different radii) in add–drop filter configuration perform incoherent accumulation of the weighted signals based on WDM and the activation unit itself described in the previous section consists of the PCM on the crossing inside a ring resonator. A sketch of the photonic integrated circuit of a single neuron with N inputs is shown in Figure 9.11(c). As for the photonic tensor core, the inputs are encoded in different wavelengths λ_1 to λ_n and weighted individually by waveguide-coupled PCM cells. Then, the input activations are added together incoherently with ring resonators and sent into the input port of the activation unit. Afterwards, a probe pulse is sent into the output waveguide to determine the neuron transmission. Figure 9.11(c) shows an optical micrograph of the real photonic circuit built on the silicon nitride platform in the work of Feldmann et al. (2019). Here, the weights were set manually to achieve the intended neuron behavior. By feeding the neuron output backward to the input weights, implementation of a Hebbian learning rule for unsupervised learning was demonstrated. Several of the neurons can be combined to form a large-scale optical artificial neural network.

A key advantage of the optical artificial neural network is that it is directly built on the hardware level. In combination with in-memory computing enabled by PCMs, a complete layer of the artificial neural network is computed in a single time step. Since the individual neurons' activations are also stored in the phase configuration of PCMs, they are directly used to calculate the subsequent layer, avoiding any data shuffling. Therefore, the operation speed of the optical artificial network is only dependent on the number of layers and the switching times of PCMs, but not on the number of neurons in a single layer, making it an ideal candidate for computing large-scale artificial neural networks. Moreover, the PCM weights no longer change once the artificial neural network is trained. Since the weighting of the input activations is also performed in-memory, this reduces the energy consumption to a minimum.

9.6. Challenges and outlook

Optical data processing has the potential to handle unprecedented data throughput due to the large bandwidth available in the optical domain. In combination with in-memory computing implemented with PCMs, data shuffling-induced bottlenecks are avoided and energy-efficient computing becomes possible.

First prototypes for a photonic tensor core for matrix vector multiplication and an optical artificial neural network have already been demonstrated and show the possibilities of optical data processing (Shen et al. 2017; Lin et al. 2018; Feldmann et al. 2019, 2021; Xu et al. 2021). Nevertheless, there are several challenges to overcome before optical systems can compete with commercial application-specific integrated circuits.

First, the footprint of photonic devices is generally larger than the footprint of electrical devices. Since photonic chips typically operate at telecom wavelengths, even on high refractive index platforms such as silicon-on-insulator (SOI), the waveguide width measures several 100 nm and minimal bend radii for ring resonators of the order of several micrometers are needed (Vlasov and McNab 2004). In the presented prototype systems made from SiN, even waveguide widths of 1.2 μm and ring radii of the order of 50 μm are required. In contrast, modern electric integrated circuits can be fabricated in a 5 nm process, thus reducing the size of a single transistor to a few atoms (Hiramoto 2019). Moreover, the size reduction of electronic integrated circuits described by Moore's law is not inherently possible in the optical domain, simply because there is a minimum waveguide width required for a given platform and wavelength to be able to confine light in a waveguide. Since working in the infrared is desirable for many reasons, new approaches are required to decrease the footprint of photonic devices. One is the development of three-dimensional architectures and plasmonic structures (Koonath and Jalali 2007; Rudé et al. 2015). Another one is the use of WDM. In this way, one photonic integrated circuit is used to carry out many operations in parallel, effectively increasing the computational density.

Second, all components required to run the optical system need to be built on-chip. For the presented prototypes, only the photonic integrated circuit itself was implemented on-chip, while other components, especially lasers, detectors and modulators, were off-chip components. Even though this approach is feasible in a laboratory setting, the required space and the costs are too high for commercial applications. This problem can be solved by switching from the passive silicon nitride platform to active platforms such as SOI or indium phosphide. On these platforms available through commercial foundry services, active components such as lasers, modulators and detectors can also be built on-chip and controlled and readout via electrical signals, thus paving the way to compact and affordable photonic chips (Crosnier et al. 2017).

Finally, the interface between existing electrical circuits and all optical circuits has to be optimized. In particular, conventional architectures work with a binary digital system, whereas the control and output signals of the photonic devices are analog. Digital-to-analog converters can be used for this task; nevertheless, due to the electric conversion inside them, they build an electrical bottleneck and also have high power consumption.

Altogether, all optical chips are well suited to fulfill the rapidly increasing hardware demands of artificial intelligence systems due to their large bandwidth, inherent possibility of parallelization and in-memory computing. First prototypes have already been demonstrated and give a promising outlook to future ways of computation in the optical domain. The main challenge now is the step from laboratory-based prototypes to commercial chips that can challenge their electronic counterparts.

9.7. References

Abdollahramezani, S., Hemmatyar, O., Taghinejad, H., Krasnok, A., Kiarashinejad, Y., Zandehshahvar, M., Alù, A., Adibi, A. (2020). Tunable nanophotonics enabled by chalcogenide phase-change materials. *Nanophotonics*, 9(5). doi: 10.1515/nanoph-2020-0039.

Amato, F., López, A., Peña-Méndez, E.M., Vaňhara, P., Hampl, A., Havel, J. (2013). Artificial neural networks in medical diagnosis. *Journal of Applied Biomedicine*, 11(2), 47–58. doi: 10.2478/v10136-012-0031-x.

Babiker, M.A.A., Elawad, M.A.O., Ahmed, A.H.M. (2019). Convolutional neural network for a self-driving car in a virtual environment. *Proceedings of the International Conference on Computer, Control, Electrical, and Electronics Engineering 2019, ICCCEEE 2019*. doi: 10.1109/ICCCEEE46830.2019.9070826.

Backus, J. (1978). Can programming be liberated from the von Neumann style? A functional style and its algebra of programs. *Communications of the ACM*, 21(8), 613–641. doi: 10.1145/359576.359579.

Batra, G., Jacobson, Z., Madhav, S., Queirolo, A. (2018). Artificial-intelligence hardware: New opportunities for semiconductor companies [Online]. Available at: https://www.mckinsey.com/industries/semiconductors/our-insights/artificial-intelligence-hardware-new-opportunities-for-semiconductor-companies.

Benjamin, B.V., Gao, P., McQuinn, E., Choudhary, S., Chandrasekaran, A.R., Bussat, J.M., Alvarez-Icaza, R., Arthur, J.V., Merolla, P.A., Boahen, K. (2014). Neurogrid: A mixed-analog-digital multichip system for large-scale neural simulations. *Proceedings of the IEEE*, 102(5), 699–716. doi: 10.1109/JPROC.2014.2313565.

Bi, G.Q. and Poo, M.M. (2001). Synaptic modification by correlated activity: Hebb's postulate revisited. *Annual Review of Neuroscience*, 24(February), 139–166. doi: 10.1146/annurev.neuro.24.1.139.

Bogaerts, W., de Heyn, P., van Vaerenbergh, T., de Vos, K., Kumar Selvaraja, S., Claes, T., Dumon, P., Bienstman, P., van Thourhout, D., Baets, R. (2012). Silicon microring resonators. *Laser and Photonics Reviews*, 6(1), 47–73. doi: 10.1002/lpor.201100017.

Cheng, Z., Ríos, C., Pernice, W.H.P., David Wright, C., Bhaskaran, H. (2017). On-chip photonic synapse. *Science Advances*, 3(9), 1–7. doi: 10.1126/sciadv.1700160.

Cheng, Z., Milne, T., Salter, P., Kim, J.S., Humphrey, S., Booth, M., Bhaskaran, H. (2021). Antimony thin films demonstrate programmable optical nonlinearity. *Science Advances*, 7(1), 1–27. doi: 10.1126/sciadv.abd7097.

Crosnier, G., Sanchez, D., Bouchoule, S., Monnier, P., Beaudoin, G., Sagnes, I., Raj, R., Raineri, F. (2017). Hybrid indium phosphide-on-silicon nanolaser diode. *Nature Photonics*, 11(5), 297–300. doi: 10.1038/nphoton.2017.56.

Delaney, M., Zeimpekis, I., Lawson, D., Hewak, D.W., Muskens, O.L. (2020). A new family of ultralow loss reversible phase-change materials for photonic integrated circuits: Sb 2 S 3 and Sb 2 Se 3. *Advanced Functional Materials*, 30, 1–10. doi: 10.1002/adfm.202002447.

Feldmann, J., Stegmaier, M., Gruhler, N., Riós, C., Bhaskaran, H., Wright, C.D., Pernice, W.H.P. (2017). Calculating with light using a chip-scale all-optical abacus. *Nature Communications*. 8(1), 1–8. doi: 10.1038/s41467-017-01506-3.

Feldmann, J., Youngblood, N., Wright, C.D., Bhaskaran, H., Pernice, W.H.P. (2019). All-optical spiking neurosynaptic networks with self-learning capabilities. *Nature*, 569(7755), 208–214. doi: 10.1038/s41586-019-1157-8.

Feldmann, J., Youngblood, N., Karpov, M., Gehring, H., Li, X., Stappers, M., Le Gallo, M., Fu, X., Lukashchuk, A., Raja, A.S. et al. (2021). Parallel convolutional processing using an integrated photonic tensor core. *Nature*, 589(7840), 52–58. doi: 10.1038/s41586-020-03070-1.

Gehring, H., Blaicher, M., Hartmann, W., Varytis, P., Busch, K., Wegener, M., Pernice, W.H.P. (2019). Low-loss fiber-to-chip couplers with ultrawide optical bandwidth. *APL Photonics*, 4(1). doi: 10.1063/1.5064401.

Gong, N., Idé, T., Kim, S., Boybat, I., Sebastian, A., Narayanan, V., Ando, T. (2018). Signal and noise extraction from analog memory elements for neuromorphic computing. *Nature Communications*, 9(1), 1–8. doi: 10.1038/s41467-018-04485-1.

He, K., Zhang, X., Ren, S., Sun, J. (2016). Deep residual learning for image recognition. *Proceedings of the IEEE Computer Society Conference on Computer Vision and Pattern Recognition*, December, 770–778. doi: 10.1109/CVPR.2016.90.

Hiramoto, T. (2019). Five nanometre CMOS technology. *Nature Electronics*, 2(12), 557–558. doi: 10.1038/s41928-019-0343-x.

Jouppi, N. (2017). What makes TPUs fine-tuned for deep learning? [Online]. Available at: https://cloud.google.com/blog/products/ai-machine-learning/what-makes-tpus-fine-tuned-for-deep-learning?hl=es.

Koonath, P. and Jalali, B. (2007). Multilayer 3-D photonics in silicon. *Optics InfoBase Conference Papers*, 15(20), 12686–12691. doi: 10.1364/oe.15.012686.

Lane, N.D., Bhattacharya, S., Mathur, A., Georgiev, P., Forlivesi, C., Kawsar, F. (2017). Squeezing deep learning into mobile and embedded devices. *IEEE Pervasive Computing*, 16(3), 82–88. doi: 10.1109/MPRV.2017.2940968.

LeCun, Y., Bottou, L., Bengio, Y., Haffner, P. (1998). Gradient-based learning applied to document recognition. *Proceedings of the IEEE*, 86(11), 2278–2323. doi: 10.1109/5.726791.

Lencer, D., Salinga, M., Wuttig, M. (2011). Design rules for phase-change materials in data storage applications. *Advanced Materials*, 23(18), 2030–2058. doi: 10.1002/adma.201004255.

Li, X., Youngblood, N., Ríos, C., Cheng, Z., Wright, C.D., Pernice, W.H., Bhaskaran, H. (2019). Fast and reliable storage using a 5 bit, nonvolatile photonic memory cell. *Optica*, 6(1), 1. doi: 10.1364/optica.6.000001.

Lin, X., Rivenson, Y., Yardimci, N.T., Veli, M., Jarrahi, M., Ozcan, A. (2018). All-optical machine learning using diffractive deep neural networks. *Science*, 361, 1004–1008.

Linn, A. (2016). Project brainwave: How Bing & Azure use AI in the cloud [Online]. Available at: https://blogs.microsoft.com/ai/project_brainwave_catapult_moonshot/.

Little, B.E. and Huang, W.-P. (1995). Coupled-mode theory for optical waveguides. *Progress in Electromagnetic Research*, 217–270 [Online]. Available at: http://www.jpier.org/PIER/pier.php?volume=10.

Lucovsky, G. and White, R.M. (1973). Effects of resonance bonding on the properties of crystalline and amorphous semiconductors. *Physical Review B*, 8(2), 660–667. doi: 10.1103/PhysRevB.8.660.

Mu, D., Qiu, H., Jiang, J., Wang, X., Fu, Z., Wang, Y., Jiang, X., Yu, H., Yang, J. (2019). A four-channel DWDM tunable Add/Drop demultiplexer based on silicon waveguide Bragg gratings. *IEEE Photonics Journal*, 11(1), 1–8. doi: 10.1109/JPHOT.2019.2897359.

Peng, C., Cheng, L., Mansuripur, M. (1997). Experimental and theoretical investigations of laser-induced crystallization and amorphization in phase-change optical recording media. *Journal of Applied Physics*, 82(9), 4183–4191. doi: 10.1063/1.366220.

Raja, A.S., Voloshin, A.S., Guo, H., Agafonova, S.E., Liu, J., Gorodnitskiy, A.S., Karpov, M., Pavlov, N.G., Lucas, E., Galiev, R.R. et al. (2019). Electrically pumped photonic integrated soliton microcomb. *Nature Communications*. 10(1), 1–8. doi: 10.1038/s41467-019-08498-2.

Redaelli, A., Pirovano, A., Benvenuti, A., Lacaita, A.L. (2008). Threshold switching and phase transition numerical models for phase change memory simulations. *Journal of Applied Physics*, 103(11). doi: 10.1063/1.2931951.

Rios, C., Stegmaier, M., Hosseini, P., Wang, D., Scherer, T., Wright, C.D., Bhaskaran, H., Pernice, W.H.P. (2015). Integrated all-photonic non-volatile multi-level memory. *Nature Photonics*. 9(11), 725–732. doi: 10.1038/nphoton.2015.182.

Rudé, M., Simpson, R.E., Quidant, R., Pruneri, V., Renger, J. (2015). Active control of surface plasmon waveguides with a phase change material. *ACS Photonics*, 2(6), 669–674. doi: 10.1021/acsphotonics.5b00050.

Salinga, M., Kersting, B., Ronneberger, I., Jonnalagadda, V.P., Vu, X.T., Le Gallo, M., Giannopoulos, I., Cojocaru-mirédin, O., Mazzarello, R., Sebastian, A. (2018). Monatomic phase change memory. *Nature Materials*, 17, 681–685. doi: 10.1038/s41563-018-0110-9.

Saxena, N., Persch, C., Wuttig, M., Manivannan, A. (2019). Exploring ultrafast threshold switching in In 3 SbTe 2 phase change memory devices. *Scientific Reports*, 9(1)1–6. doi: 10.1038/s41598-019-55874-5.

Shen, Y., Harris, N.C., Skirlo, S., Prabhu, M., Baehr-Jones, T., Hochberg, M., Sun, X., Zhao, S., Larochelle, H., Englund, D. et al. (2017). Deep learning with coherent nanophotonic circuits. *Nature Photonics*, 11(7), 441–446. doi: 10.1038/nphoton.2017.93.

Silver, D., Huang, A., Maddison, C.J., Guez, A., Sifre, L., Van Den Driessche, G., Schrittwieser, J., Antonoglou, I., Panneershelvam, V., Lanctot, M. et al. (2016). Mastering the game of Go with deep neural networks and tree search. *Nature*, 529(7587), 484–489. doi: 10.1038/nature16961.

Silver, D., Schrittwieser, J., Simonyan, K., Antonoglou, I., Huang, A., Guez, A., Hubert, T., Baker, L., Lai, M., Bolton, A. et al. (2017). Mastering the game of Go without human knowledge. *Nature*, 550(7676), 354–359. doi: 10.1038/nature24270.

Stegmaier, M., Ríos, C., Bhaskaran, H., Pernice, W.H.P. (2016). Thermo-optical effect in phase-change nanophotonics. *ACS Photonics*, 3(5), 828–835. doi: 10.1021/acsphotonics.6b00032.

Vlasov, Y.A. and McNab, S.J. (2004). Losses in single-mode silicon-on-insulator strip waveguides and bends. *Optics Express*, 12(8), 1622. doi: 10.1364/opex.12.001622.

Wu, C., Yu, H., Lee, S., Peng, R., Takeuchi, I., Li, M. (2020). Programmable phase-change metasurface for multimode photonic convolutional neural network. *Nat. Commun.*, 12(96), 1–18. doi: 10.1109/ipc47351.2020.9252309.

Wuttig, M. and Yamada, N. (2007). Phase-change materials for rewriteable data storage. *Nature Materials*, 6(11), 824–832. doi: 10.1038/nmat2009.

Wuttig, M., Bhaskaran, H., Taubner, T. (2017). Phase-change materials for non-volatile photonic applications. *Nature Photonics*, 11(8), 465–476. doi: 10.1038/nphoton.2017.126.

Xu, X., Tan, M., Corcoran, B., Wu, J., Boes, A., Nguyen, T.G., Chu, S.T., Little, B.E., Hicks, D.G., Morandotti, R. et al. (2021). 11 TOPS photonic convolutional accelerator for optical neural networks. *Nature*, 589(7840), 44–51. doi: 10.1038/s41586-020-03063-0.

Zheng, J., Fang, Z., Wu, C., Zhu, S., Xu, P., Doylend, J.K., Deshmukh, S., Pop, E., Dunham, S., Li, M. et al. (2020). Nonvolatile electrically reconfigurable integrated photonic switch enabled by a silicon PIN diode heater. *Advanced Materials*, 32(31), 1–8. doi: 10.1002/adma.202001218.

List of Authors

Jonathan BACKMAN
Integrated Systems Laboratory
ETH Zurich
Switzerland

Kaustav BANERJEE
Department of Electrical and
Computer Engineering
University of California
Santa Barbara
USA

Frédéric BONELL
CNRS
CEA
Grenoble INP
IRIG-Spintec
Université Grenoble Alpes
France

Frank BRÜCKERHOFF-PLÜCKELMANN
Institute of Physics
University of Muenster
Germany

Wei CAO
Department of Electrical and
Computer Engineering
University of California
Santa Barbara
USA

Alessandro CRESTI
CNRS
Grenoble INP
IMEP-LAHC
Université Grenoble Alpes
Université Savoie Mont Blanc
Grenoble
France

Bruno DLUBAK
Unité Mixte de Physique
CNRS
Thales
Université Paris-Sud
Université Paris-Saclay
Palaiseau
France

Pascal FEBVRE
IMEP-LAHC
Université Savoie Mont Blanc
Le Bourget du Lac
France

Johannes FELDMANN
Department of Materials
University of Oxford
UK

Sara FIORE
Integrated Systems Laboratory
ETH Zurich
Switzerland

Thomas GUILLET
Catalan Institute of Nanoscience and Nanotechnology (ICN2)
CSIC and The Barcelona Institute of Science and Technology (BIST)
Spain

Matthieu JAMET
CNRS
CEA
Grenoble INP
IRIG-Spintec
Université Grenoble Alpes
France

Cedric KLINKERT
Integrated Systems Laboratory
ETH Zurich
Switzerland

Joris LACORD
CEA
LETI
Université Grenoble Alpes
France

Youseung LEE
Integrated Systems Laboratory
ETH Zurich
Switzerland

Mathieu LUISIER
Integrated Systems Laboratory
ETH Zurich
Switzerland

Marco PALA
Université Paris-Saclay
CNRS
Centre de Nanosciences et de Nanotechnologies
Palaiseau
France

Wolfram PERNICE
Institute of Physics
University of Muenster
Germany

Sasan RAZMKHAH
Department of Electrical and Electronics Engineering
TOBB University of Economics and Technology
Ankara
Turkey

Pierre SENEOR
Unité Mixte de Physique
CNRS
Thales
Université Paris-Sud
Université Paris-Saclay
Palaiseau
France

Juan F. SIERRA
Catalan Institute of Nanoscience and Nanotechnology (ICN2)
CSIC and The Barcelona Institute of Science and Technology (BIST)
Spain

Christian STIEGER
Integrated Systems Laboratory
ETH Zurich
Switzerland

Josef SVĚTLÍK
Catalan Institute of Nanoscience and Nanotechnology (ICN2)
CSIC and The Barcelona Institute of Science and Technology (BIST)
Spain

Áron SZABÓ
Integrated Systems Laboratory
ETH Zurich
Switzerland

Sergio O. VALENZUELA
Catalan Institute of Nanoscience and Nanotechnology (ICN2)
CSIC and The Barcelona Institute of Science and Technology (BIST)
and
Institució Catalana de Recerca i Estudis Avançats (ICREA)
Barcelona
Spain

Diogo C. VAZ
CIC nanoGUNE BRTA
Donostia-San Sebastián
Spain

Steven A. VITALE
MIT Lincoln Laboratory
Lexington
Massachusetts
USA

Dominique VUILLAUME
Institute of Electronic Microelectronics and Nanotechnology (IEMN)
CNRS
Villeneuve d'Ascq
France

Index

O, P

Q, R, S

T, V